T0258870

Fishes of the Sea of Japan
and the Adjacent Areas of the Sea of Okhotsk
and the Yellow Sea

(continued on p. iv)

Fishes of the Sea of Japan and the Adjacent Areas of the Sea of Okhotsk and the Yellow Sea

PART 4

Teleostomi

XXIX.   Perciformes

2. Blennioidei–13. Gobioidei

(CXLV. Fam. Anarhichadidae–CLXXV. Fam. Periophthalmidae)

G.U. Lindberg and Z.V. Krasyukova

Bruce B. Collette
Scientific Editor

RUSSIAN TRANSLATIONS SERIES
71

1989
A.A. BALKEMA/ROTTERDAM

Translation of:

Ryby Yaponskogo morya i Sopredel'nykh Chastei Okhotskogo i Zheltogo Morei.
Nauka Publishers, Leningrad, 1975

© 1989 Copyright reserved

Translator: Dr. B.R. Sharma
General Editor: Dr. V.S. Kothekar

ISBN 90 6191 415 9

UDC 597.583.1 (265.3/.5) (083.71)

Part Four of the *Keys* . . . completes the review of the largest and economically richest order of fishes, Perciformes, begun in Part Three of this series. It includes 12 suborders with keys and brief descriptions of 230 species, and information on their ecology and distribution. This work should prove invaluable to ichthyologists and, like the other Parts, serve as a basis for utilizing fish stocks of the seas of the Far East. The present volume also provides a basis for zoogeographic generalizations.

Foreword to the English Edition

The Smithsonian Institution Libraries, in cooperation with the National Science Foundation, has sponsored the translation into English of this and hundreds of other scientific and scholarly studies since 1960. The program, funded with Special Foreign Currency, represents an investment in the dissemination of knowledge to which the Smithsonian Institution is dedicated.

This volume concludes the four-part treatise on the fishes of the Sea of Japan and adjacent areas begun by G.U. Lindberg in 1959. Included are accounts of the remainder of the order Perciformes that was begun in Part Three with the suborder Percoidei. Twelve suborders are treated here: Blennioidei (nine families, forty-six genera, ninty-six species), Ophidioidei (three families, six genera and species), Ammodytoidei (two families, genera, and species), Callionymoidei (two families, five genera, fifteen species), Siganoidei (*Siganus fuscescens*), Acanthuroidei (two families, four genera and species), Trichiuroidei (two families, seven genera and species), Scombroidei (three families, twelve genera, twenty-three species), Luvaroidei (*Luvarus imperialis*), Tetragonuroidei (*Tetragonurus cuvieri*), Stromateoidei (four families, eight genera, ten species), and Gobioidei (five families, thirty-five genera, sixty-three species). Not covered are the twelve most advanced orders of the fishes in Lindberg's classification including the Scorpaeniformes, Pleuronectiformes, Tetraodontiformes, and Lophiiformes. It is unfortunate that Professor Lindberg did not complete the study of these families before his death in 1976 (see obituary in *Copeia*, 1977, 612–13).

The three preceding parts were translated for the Smithsonian Institution Libraries and the National Science Foundation by the Israel Program for Scientific Publications. Part One was originally published in 1959 and the English translation appeared in 1967. Part Two was published in 1965 and its English version in 1969. Part Three was published in 1969 and the English translation appeared in 1971. Although the present volume may seem dated, because of the lapse of twelve years since the publication of the original Russian version, it is still valuable as an entry into the extensive Russian literature on this subject which is not adequately known by Western ichthyologists. The author thoroughly covers the literature through about 1969. References to later works include 1970 (16), 1971 (7), 1972 (3), and one 1974 paper added in the addendum.

Ichthyologists specializing in various groups have generously read through the relevant parts of the translations and indicated where the original Russian needed to be checked. The specialists include M. Eric Anderson of the California Academy of Sciences (on Zoarcidae), Daniel M. Cohen of the Los Angeles County Museum of Natural History (on Ophidioidei), National Science Foundation Postdoctoral Fellow Edward O. Murdy (on Gobioidei), Victor G. Springer of the National Museum of Natural History (on Blennioidei), and Betsy B. Washington of the National Marine Fisheries Service Systematics Laboratory (on Ammodytoidei). I have reviewed the entire text to correct minor errors and omissions in the translation and I have also made some minor changes in the interest of readability.

Bruce B. Collette, Research Associate
Department of Vertebrate Zoology and
Director, Systematics Laboratory
National Marine Fisheries Service
National Museum of Natural History
Washington, D.C. 20560

Foreword

3 The purpose of the present volume and also the boundaries of the water bodies covered by it have been outlined in the Introduction to Part One, published in 1959 (and subsequently, with Parts Two and Three, translated into English). In Part One a map of the Sea of Japan and adjacent waters is presented together with an index of geographic names taken from the *Sea Atlas* (vol. I, 1950).

Part Four of the *Keys*... includes (except for the suborder Percoidei, published in Part Three) members of the following suborders of the order Perciformes: Blennioidei, Ophidioidei, Ammodytoidei, Callionymoidei, Siganoidei, Acanthuroidei, Trichiuroidei, Scombroidei, Luvaroidei, Tetragonuroidei, Stromateoidei, and Gobioidei. Representatives of these groups have been found in the Sea of Japan, the Sea of Okhotsk, and the Yellow Sea. A key to all the suborders of the order Perciformes is given in Part Three.

The 12 suborders are represented in the Sea of Japan by 30 families with 106 genera and 190 species. In addition, 14 genera and 31 species have been described from adjacent waters. Some genera and species have also been considered which are found in fresh waters; they represent 1 suborder, 5 families, 8 genera, and 11 species, which occur in the water bodies under consideration.

As in Part Three, in designating genera we did not verify the correctness of the "type species", as implied in the decisions of the International Zoological Congress. Whenever the term "type" is used, it implies the type specimen based on data available in published literature.

Like the three earlier parts, this part too is primarily a review of literature. However, for the comparison of descriptions of families, genera, and species presented here, we examined fishes in the collections of the Institute of Zoology, Academy of Sciences of the USSR, Museum of the Pacific Fisheries and Oceanography Research Institute (TINRO), and the Kuril-Sakhalin Expedition (KSE) of 1947–1949. As a result of this examination, additions and emendments were made to the descriptions and in the keys, or new keys compiled in accordance with the additional data obtained.

The literature reviewed by us, in addition to that included in the previous parts of the *Keys*..., has been cited under the diagnosis of each suborder.

Unfortunately, the data presented in the list of fish names from the northeastern part of the Atlantic Ocean (Clofnam, 1973) could not be used, since by the time this list was received the manuscript was already in press.

The format in Part Four is the same as that followed earlier: brackets and breviers indicate that the particular species, genus, or family has not yet been found in the Sea of Japan; numbered forms indicate that they are known for the Yellow Sea or the southern part of the Sea of Okhotsk (up to 50°00′ N); and unnumbered forms unknown even in this region, but considered by us possible inhabitants of adjacent waters.

The body length, given at the end of the description of a species, if not specifically designated otherwise, is always the absolute maximum known to us.

Information on the biology is usually omitted but references in which this information is available are listed.

The suborders Blennioidei and Scombroidei have been compiled by Z.V. Krasyukova and the others by G.U. Lindberg. Finalization of the manuscript for the press was done jointly.

Names of the expeditions from which material was examined during the writing of Part Four of this *Key* ... have been abbreviated as follows:

GEVO — Hydrographic Expedition of the Pacific Ocean
DVE — Far East Expedition of the Department of Agriculture
KSE — Kuril-Sakhalin Marine Multi-Disciplinary Expedition of the Institute of Zoology, Academy of Sciences of the USSR and TINRO
GGI — Expedition of the State Hydrological Institute and TINRO
ZIN — Hydrobiological Expedition of the Zoological Institute of the Academy of Sciences of the USSR in the Sea of Japan

Original sketches and diagrams were prepared by the artist M.M. Zharenkov.

The authors are grateful to A.P. Andriyashev for his invaluable advice and assistance in the editing of the manuscript.

Contents

Systematic List of Species and Subspecies of Fishes of the Sea of Japan and Adjacent Waters

2. Suborder Blennioidei

15 Body never deep; either oblong or moderately or highly elongate. Maxillae not firmly attached to premaxillae and hence protractile. Number of rays in dorsal and anal fins equal to number of corresponding neural and hemal spines of vertebrae. If more than one dorsal fin, then first dorsal usually with spines, although thin and flexible; if without spiny rays and body oblong, caudal fin any shape but not bifurcate. Pelvic fins usually absent. If pelvics present, either jugular or mental (always slightly anterior to bases of pectoral fins), and usually poorly developed—either without spiny ray or with small spine difficult to distinguish, and with segmented, usually unbranched, soft rays numbering 2-4, rarely 1. If pelvic fins relatively well developed, with 1 spine and 3-4 bi- or trifurcate soft rays, their length much less than half length of pectorals, except in the genus *Leptoclinus* in which length more than half (Lindberg, 1971).

Key to Families of Suborder Blennioidei[1]

1 (2). Teeth on vomer, palatines, and along sides of lower jaw large and massive; anterior part of lower jaw and premaxillae with strong conical canines. Dorsal fin with flexible spiny rays at least in anterior part; posterior rays thickened and hard. Scales very small, embedded in skin. Branchiostegal membranes broadly attached to isthmus. Pelvics not present.[2] Body moderately elongated. Rays in dorsal and anal fins less than 100. Exceptionally, body may be elongated or eel-like, and number of rays in fins notably increased to around 240 (*Anarhichthys*). . . . . .
. . . . CXLV. **Anarhichardidae.**

2 (1). Such teeth not present.

3 (18). Spiny rays of dorsal fin usually more numerous than soft rays; latter often not present at all. If soft rays present, caudal fin more or less distinguishable and not entirely confluent with dorsal and anal fins.

4 (5). Mouth superior; oral slit almost vertical (Figure 3). Dorsal fin with long base and only spiny rays of moderate height. Pelvic fins absent. Scales usually absent. Gill openings small.

---

[1]Lindberg, 1971.

[2]In *Andamia* (Salariinae, Blenniidae) pelvics short, less than half length of pectorals, but adapted for climbing along rocks. Sucker near lower lip possibly also serves the same purpose.

Branchiostegal membranes fused with isthmus. Teeth fairly strong................................ CXLVI. **Cryptacanthodidae.**

5 (4). Mouth not superior; oral slit horizontal or oblique, but not vertical.

6 (7). Caudal peduncle short but well defined. Dorsal and anal fins not confluent with caudal fin. Base of anal fin short, less than half length of base of dorsal fin. Lower jaw thickened and protrudes slightly beyond upper jaw (Figure 4). Pelvic fins absent. Minute scales not only present on body but also continue onto fins .... **[Zaproridae].**

7 (6). Caudal peduncle poorly developed or absent; if present, base of anal fin long, much more than half length of base of dorsal fin; dorsal and anal fins usually connected with caudal fin or almost confluent with it. Lower jaw not thickened.

8 (15). Tropical, heat-loving marine blennies. Soft (segmented) rays in pelvic, dorsal, and anal fins usually simple, unbranched. Pelvic fins long,[3] more than half length of pectoral fins, with barely developed spiny ray and 1–3 unbranched, well-defined, rarely rudimentary, soft rays. Lateral line, if present, always single, passing above pectoral fin. Number of principal rays of caudal fin 13 or less. 2 nostrils on each side.

9 (10). Dorsal fins usually well separated, 3 (Figure 6). Cirri not present on occiput. Scales usually ctenoid. Rays of pectoral fins usually branched, rays of caudal fin always branched....
.... CXLVII. **Tripterygiidae.**

10 (9). Dorsal fins 1 or 2. Usually cirri present not only above eyes, but also on occiput. Scales, if present, not ctenoid. Rays of pectoral fins never branched. Rays of caudal fin mostly unbranched.

11 (12). First dorsal fin short except in members of the tribe Clinini, in which first dorsal fin continuous and long. Lachrymal (preorbital) bone anteriorly broad, posteriorly not reaching vertical from middle of eye. Scales present, regular, or some other shape.
.... **[Clinidae].**

12 (11). First dorsal fin long. Lachrymal (preorbital) bone posteriorly reaches vertical from middle of eye or even farther. Scales absent, except in *Neoclinus* (Figure 8) (Chaenopsidae), in which small random scales occur.

13 (14). Anterior teeth on jaws often resemble incisors, never enlarged or comb-shaped. Teeth on jaws usually conical but some may be enlarged and curved. Mouth large. Upper jaw usually extends

[3]Relatively long pelvic fins known in *Leptoclinus* (Stichaeidae).

distinctly beyond eyes. Body distinctly elongate . . . .
. . . . CXLVIII. **Chaenopsidae.**

14 (13). Anterior teeth on jaws large, comb-shaped. One or more canines usually located behind former. Mouth comparatively small; upper jaw generally does not extend beyond eyes. Body relatively short . . . . CXLIX. **Blenniidae.**

15 (8). Northern, cold-loving marine blennies. Soft (segmented) rays in pelvic (if present), dorsal, and anal fins usually unbranched; if rays in pelvic fins branched, then so highly reduced that average number of segments in ray does not exceed 5 except in *Leptoclinus* (Stiochaeidae). Pelvic fins short (except in *Leptoclinus*), less than half length of pectoral fins, with one poorly developed spine and 2-4 branched or rudimentary unbranched rays covered with skin and difficult to discern. Lateral line, if present, not simple; often branched or with additional branches; dorsal branch passes above pectoral fin and does not curve down to extend along middle of body. Number of principal rays of caudal fin, if not reduced, 13 or more. One nostril on each side of snout.

16 (17). Caudal fin with branched principal as well as marginal rays. Pelvic fins, if present, rudimentary, with 1 short spine and 1 small unbranched ray . . . . CL. **Pholididae.**

17 (16). Caudal fin with only principal rays branched (Figure 26). Pelvic fins, if present, more or less developed, with 1 spine and 2-4 soft rays . . . . CLI. **Stichaeidae.**

18 (3). Spiny rays usually absent in dorsal fin; if present, fewer in number than soft rays and, moreover, caudal fin in this case distinctly confluent with dorsal and anal fins (*Zoarces, Krusensterniella, Zoarchias, Neozoarces*).

19 (20). Gill openings large, extending far forward. Branchiostegal membranes well separated and only rarely narrowly connected anteriorly and slightly attached to isthmus or not attached at all.[4] Scales absent. Vent distinct and located farther behind base of pectoral fins than in Carapidae [Ophidioidei]. Pelvic fins present or absent . . . . **[Lycodapodidae].**

20 (19). Gill openings small or moderate in size, not extending far forward.[5] Branchiostegal membranes broadly connected, sometimes forming fold across isthmus, and attached to latter. Pelvic fins, if present, thoracic. Pectoral fins large, at least half head

[4]Gill openings and branchiostegal membranes similar in *Lycogramma* (Zoarcidae), but this genus has scales. Shmidt (1950: 108) considers the family Lycodapodidae a subfamily of Zoarcidae.

[5]Except for *Lycogramma* (Zoarcidae).

length; if slightly smaller, then dorsal fin with spiny rays, or gill openings very small but equal to size of pupil................
.... CLII. **Zoarcidae.**

CXLV. Family ANARHICHADIDAE[6]—Wolffishes

Body moderately or highly elongate, compressed laterally, covered with very small thin scales embedded in skin, not overlapping. Head not covered with scales. Barbels not present. Snout blunt. Mouth large. Upper jaw slightly protractile. Teeth large, differentiated into three major types (caninelike, conical, and molariform); strongest teeth located on vomer and its opposite parts, the dentaries (see Figure 28). Suprapharyngeal teeth located on three pads. Teeth of jaws and palatines adapted for rupturing and crushing organisms with hard shells (mollusks, crustaceans, echinoderms). Gill openings widely separated by isthmus. Gill slits broad. Opercular siphon present. Rays of branchiostegal membranes 6-8 (often 7). Dorsal fin single, continues from occiput to caudal fin or confluent with it (*Anarhichthys[7]*). All rays of dorsal fin spiny, or soft throughout fin, or become thicker and harder toward posterior end. Pectoral fins large, round, inserted low. Pelvic fins absent. Rays of pectoral and caudal fins bi- or trifurcate, as are some of the posteriormost rays of anal fin. Remaining rays of anal fin unbranched. Anal fin with one undeveloped spine at orgin. Number of pores in canals on lateral line of head variable. Canal of lateral line on body absent; seismosensory papillae free located in form of upper and middle branches.

Two genera. One genus known from the Sea of Japan.

1. Genus *Anarhichas* Linné, 1758—Wolffishes

Anarhichas Linné, Syst. Nat., ed. X, 1758: 247 (type: *A. lupus* L.).

Body moderately elongate, maximum body depth (even in larvae) not more than 9 times in body length. Caudal fin completely separate from anal fin but connected with dorsal through highly reduced rays. Vertebrae and rays in dorsal and anal fins less than 100 (precaudal vertebrae not more than 31). Posterior reduced rays of dorsal fin hard.[8] Branchiostegal rays (6) 7. Lateral line bifurcate; upper branch located only on trunk, lower branch continues almost to base of caudal fin and distinctly visible in live fish in water (Barsukov, 1959).

[6]Description of family taken in part from the monographs of Barsukov (1959) and Makushok (1958).

[7]Distributed along the Pacific coast of North America from Kodiak Island (Alaska) to San Diego (Barsukov, 1959).

[8]Soft rays of anal fin consist of very large number (several dozen) of short segments; posteriormost rays of anal fin branched, others unbranched.

4 species, 1 known from the Sea of Japan.

1. *Anarhichas orientalis* Pallas, 1811—Bering Wolffish (Figure 1)

Anarhichas orientalis Pallas, Zoogr. Rosso-Asiat., 3, 1811: 77, pl. 3. (eastern Kamchatka). Popov, Dokl. Akad. Nauk SSSR, 14, 1931: 380–385, pl. I. Taranetz, Kratkii Opredelitel'..., 1937: 158. Shmidt, Ryby Okhotskogo Morya, 1950: 66. Andriyashev, Ryby Severnykh Morei SSSR, 1954: 227, figs. 115, 116. Makushok, Stichaeoidea..., 1958: 62, 122. Barsukov, Sem. Zubatok, 1959: 110, fig. 29, pls. I-XX.

Anarhichas lepturus, Shmidt, Ryby Vostochnykh Morei..., 1904; 206. Soldatov and Lindberg, Obzor..., 1930: 486.

37100. Tatar Strait, Datta Bay. August 28 [*sic*]. Fisherman Sosyukin. 1 specimen.

D LXXXVI-LXXXVIII; A 53-55; P 20-22; C 23-25; vertebrae 88-89 (precaudal 29-31, caudal 57-59); branchiostegal rays 7. Maximum body depth of adults about 5 times in its absolute length, in juvenile fish about 7 times in this length. Length of head from tip of snout to posterior angle of opercle about 22% of standard length (in young fish about 18%). Width of pectoral fin base in adults more than 60% of body depth (at origin of anal fin). Caudal fin rounded. Color dark chocolate-brown, usually without
20 spots and stripes. Sides of young fish 15 to 17 cm (Figure 2) with 3-4 dark longitudinal stripes consisting of fused irregular spots spreading from head almost to base of caudal fin.[9] Bases of dorsal and anal fins with one dark longitudinal stripe each, interrupted at some places. Above this stripe 8-9 very large but dull spots located on dorsal fin, not reaching upper margin of fin. Caudal fin without spots and stripes. Top of head and cheeks with dark spots, but spots on cheeks smaller.

Biology almost unstudied.

Length to 112 cm, but possibly even longer (Barsukov, 1959: 112).

Distribution: In the Sea of Japan known from Tatar Strait, Datta Bay (Taranetz, 1935: 98); near the southwestern coast of Sakhalin, along the coast of Hokkaido, and along the southwest coast of Sakhalin (Ueno, 1971: 83); Gulf of Ushoro (Kobayashi, 1962: 258); Sea of Okhotsk, Bering Sea (Shmidt, 1950: 66).

CXLVI. Family CRYPTACANTHODIDAE—Wrymouths

Body elongate and slender, laterally compressed, naked or covered with minute cycloid scales. Lateral line absent or consists only of open pores without tubules. Head oblong, almost rectangular. Mucous canals well delineated on lower jaw and preopercle. Head on upper side flat, with deep round pits (fossae) in space between eyes and behind them. Mouth

---

[9]Color of young fish described by Taranets (1935).

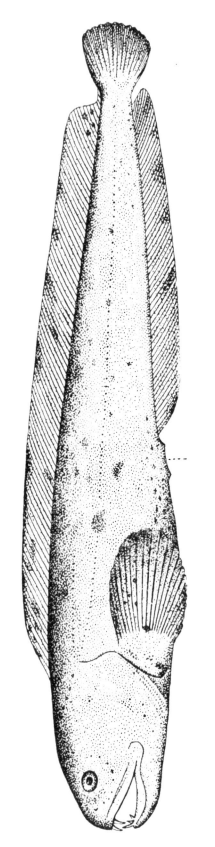

Figure 1. *Anarhichas orientalis*—Bering wolffish. Length about 630 mm. Gulf of Norton, Bering Sea (from Andriyashev, 1954).

19

large, very oblique. Lower jaw massive, protrudes forward. Premaxillae not movable. Jaws with fairly sharp conical teeth, large teeth on vomer, and sometimes on palatines. Branchiostegal membranes attached to isthmus. Gill openings sometimes continue downward toward front. Dorsal fin long, covered with skin, composed entirely of fairly long, spiny rays. Dorsal and anal fins connected with caudal fin through membrane. Pelvic fins absent. Pectorals short (Soldatov and Lindberg, 1930).

4 genera. 1 genus known from the Sea of Japan.

Key to Genera of Family Cryptacanthodidae

1 (2). Body covered with scales. Lateral line present. Isthmus narrow. Palatines with teeth. Pacific coast of North America . . . .
. . . . [**Delolepis** Bean, 1882].[10]

2 (1). Body naked.

3 (4). Lateral line present. Isthmus broad. Palatines with teeth. . . . . . .
. . . . 1. **Cryptacanthoides** Lindberg.

4 (3). Lateral line reduced.

5 (6). Isthmus narrow. Palatines with teeth . . . .
. . . . [**Cryptacanthodes** Storer, 1839].[11]

6 (5). Isthmus fairly broad. Palatines without teeth . . . .
. . . . [**Lyconectes** Gilbert, 1895].[12]

1. Genus *Cryptacanthoides* Lindberg, 1930—Wrymouths

Cryptacanthoides Lindberg. In: Soldatov and Lindberg, Obzor..., 1930: 482 (type: *C. bergi* Lindberg).

Body highly elongate; anterior part moderately and posterior part highly compressed laterally. Maximum body depth 14.4 times in standard length. Body naked. Lateral line consists of about 80 minute open pores arranged along median line of body. Poorly defined anal papilla present. Head elongate; rectangular on upper side, with well-developed pits. Snout short, blunt; nostrils tubular, located immediately behind premaxillaries. Eyes small, deeply hidden in ocular pit; interorbital space broad. Mouth broad, almost vertical, lower jaw protrudes far forward. Premaxillae with two rows of blunt conical teeth; teeth in outer row large, in inner row smaller. Dentaries with one row of teeth along side and two rows near symphysis. Vomer and palatines with teeth. Five teeth, similar to jaw teeth, located on head of vomer; palatines with one similar tooth each. Gill openings do not continue downward anteriorly, their lower end at level of lower margin of pectoral base. Branchiostegal membranes

[10]Pacific coast of North America (Norman, 1957: 470).
[11]Distributed along the Atlantic coast of North America (Norman, 1957: 470).
[12]Known from near the Aleutian Islands (Norman, 1957: 470).

connected and broadly attached to triangular isthmus; branchiostegal rays 6; gill rakers short. Pectoral fins short; pelvic fins absent. Dorsal fin origin above base of pectoral fin and consists of only spiny rays. Anal fin with two spiny rays, others soft and branched. Dorsal and anal fins connected with caudal; latter well developed, long, and pointed (Soldatov and Lindberg, 1970).

1 species. Also found in the Sea of Japan.

1. *Cryptacanthoides bergi* Lindberg, 1930—Berg's Wrymouth (Figure 3)

Cryptacanthoides bergi Lindberg. In: Soldatov and Lindberg. Obzor..., 1930: 484, figs. 66, 67 (Peter the Great Bay). Matsubara, Fish Morphol. and Hierar., 1955: 754, fig. 272.

Lyconectes ezoensis Hikita and Hikita, Japan. J. Ichthyol., 1, 2, 1950: 140, 1 fig.

25131. Peter the Great Bay. June 7, 1932. Generozova. 1 specimen.

31831. Sea of Okhotsk, Aniva Bay. September 23, 1947. G.U. Lindberg. 1 specimen.

31832. Sea of Okhotsk, Aniva Bay. September 23, 1947. G.U. Lindberg. 2 specimens.

D LXIX; A II 47; P 12; C 18; *l. l.* 80.

Width of head much more than maximum depth of body. Interorbital space less than snout length. Maxillary does not reach anterior margin of eye. Eye 2.5 times in the interorbital space and 13.6 times in head length. Body naked. Lateral line with about 80 open pores. Basic body color light yellow. Chocolate-brown stripe extends under dorsal fin along body; similar stripe above anal fin, but lighter in color. Series of spots along sides around each pore of lateral line formed by concentration of pigment of same color; second series of smaller and lighter spots located below first series. Dorsal fin with chocolate-brown spots. Pectoral, anal, and caudal fins without spots (Soldatov and Lindberg, 1930).

Length 218 mm (Hikita and Hikita, 1950).

Distribution: In the Sea of Japan known from Peter the Great Bay (Soldatov and Lindberg, 1930: 484); Tartar Strait (Taranets, 1937b: 158); near Sado Island (Honma and Sugihara, 1963: 6); Toyama Bay (Katoh et al., 1956: 322); Wakasa Bay (Takegawa and Morino, 1970: 382); San'in region (Katoh et al., 1956: 322). Reported from the eastern coast of Sakhalin (Ueno, 1971: 84); Aniva Bay (our specimens); Volcano Bay (Sato and Kobayashi, 1956: 15); Kushiro (Hikita and Hikita, 1950: 140).

[Family ZAPRORIDAE—Prowfishes][13]

Body slightly stout, compressed laterally. Caudal peduncle short but

[13]Position of this family in the classification still remains somewhat uncertain. According to American authors (McAllister and Krejsa, 1961), who have examined this aspect carefully, Zaproridae is close to Stichaeidae and should be located between it and Anarhichadidae.

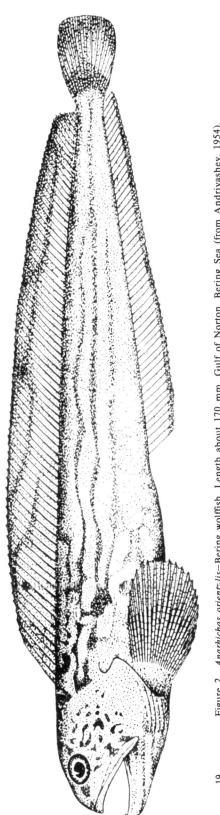

19

Figure 2. *Anarhichas orientalis*—Bering wolffish. Length about 170 mm. Gulf of Norton, Bering Sea (from Andriyashev, 1954).

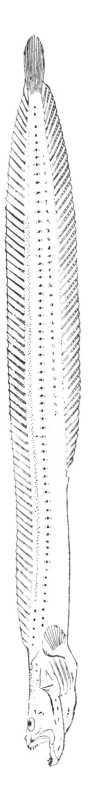

21

Figure 3. *Cryptacanthoides bergi*—Berg's wrymouth. Length 206 mm. Sea of Japan, Peter the Great Bay (Soldatov and Lindberg, 1930).

well defined; dorsal and anal fins not connected with caudal. Base of anal fin short, distinctly less than half basal length of dorsal fin. Head short; lower jaw stout and protrudes slightly forward beyond upper jaw. Lateral line absent. Pelvics absent. Scales small, cover body and membranes of all fins over 2/3 their height.

1 genus and 1 species. Northern part of the Pacific Ocean.

[Genus *Zaprora* Jordan, 1896—Prowfish]

Zaprora Jordan, Proc. Calif. Acad. Sci., 1896: 202 (type: *Z. silenus* Jordan).

Characters of genus same as in description of family.

1 species. Northern part of the Pacific Ocean, from California to Alaska, and southern coast of Hokkaido (Kushiro).

[*Zaprora silenus* Jordan, 1896—Prowfish] (Figure 4)

Zaprora silenus Jordan, Proc. Calif. Acad. Sci., 1896: 203, pl. 20 (Vancouver Island). Jordan and Evermann, Fish. N. and M. Amer., 3, 1898: 2850. Chapman and Townsend, Ann. Mag. Nat. Hist., 11, 2, 1938: 89-117, figs. 1-10 (osteology). Ueno, Japan. J. Ichthyol., 3, 2, 1954: 79, fig. 2. Ueno and Abe, Bull. Hokk. Reg. Fish. Res. Lab., 28, 1964: 20, figs. 13, 14.

25499. Sea of Okhotsk, Kamchatka. 1932. M.N. Krivobok. 1 specimen.

39003. Bering Sea, 57°50′ N, 179°14′ E. September 9, 1967. L.S. Kodolov. 1 specimen.

39004. Bering Sea, 54°38′ N, 170°50′ E. September 21, 1967. V.V. Fedorov. 1 specimen.

D 53-56; A 24-28; P 22-24; branchiostegal rays 5; gill rakers on first gill arch 8 + 17 (Ueno and Abe, 1964).

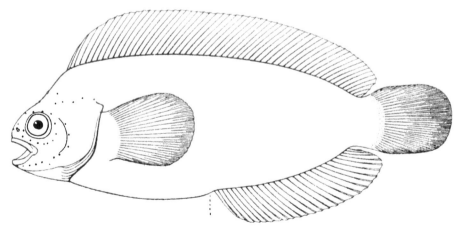

Figure 4. *Zaprora silenus*—prowfish. Length 131 mm. No. 39004. Bering Sea.

This lone member of the family is characterized by the presence of a short but well-developed caudal peduncle and short anal fin. Length of base of anal fin more than 2 times in length of dorsal fin base. Small mouth, with stout lower jaw. Numerous pores on head. It may be mentioned that the caudal peduncle is not shown in the sketch given by Ueno (1954a, Fig. 2); this may have been an oversight on the part of the artist, since the caudal peduncle is distinct in the photograph published in the paper by Ueno and Abe (1964, Fig. 13).

Length to 725 mm (Jordan and Evermann, 1898: 2850).

Distribution: Not found in the Sea of Japan. 3 specimens reported from Kushiro, Pacific coast of Hokkaido (Ueno and Abe, 1964: 20). Described from the waters of British Columbia (Jordan); recorded from California up to the Gulf of Alaska (Clemens and Wilby, 1961: 234). Found in the Bering Sea (Nos. 39003, 39004).

CXLVII. Family TRIPTERYGIIDAE—Triplefins

Body short and covered with relatively large ctenoid scales.

Distinguished from related families, especially Clinidae, by presence of dorsal fin divided into three distinct parts: First part immediately behind head with 3-4 fine flexible, spiny rays; second part consists of unsegmented, flexible, hard rays; and third part consists of segmented soft rays. Anal fin with fairly long base and soft rays. Pelvics always present. Pectorals large. Caudal fin rounded, relatively large, and separate from dorsal and anal fins; principal rays branched. Head generally with prenasal and supraorbital cirri. Inhabit tropical and temperate waters of the world oceans, but not recorded from the eastern coast of the Pacific Ocean. Mostly littoral fishes found in the surf region of coral reefs and rock crevices. Generally elude catches.

4-5 genera, 1 genus known from the Sea of Japan.

1. Genus *Tripterygion* Risso, 1826—Triplefins

Tripterygion Risso, Hist. Nat. Europe, Méridionale, 3, 1826: 241 (type: *T. nasus* Risso). Jordan and Snyder, Proc. U.S. Nat. Mus., 25, 1902: 444 (synonymy). Matsubara, Fish Morphol. and Hierar., 1955: 730.

Body covered with ctenoid scales. Lateral line complete or incomplete. Mouth moderate in size, jaws almost equal in length. Cirri absent on occiput. Eyes large. Dorsal fin in 3 parts: First part with 3-4 flexible spiny rays; second part with 10-24 rays; and third soft part with 7-15 rays. Caudal fin rounded. Anal fin long. Pectorals also long, their lower rays simple and stout (Jordan and Snyder, 1902a). Male and female differ in shape and size of anus and genital papilla (Tomiyama, 1951) (Figure 5, A, B, a, b).

Many species. 2 found in the Sea of Japan.

12

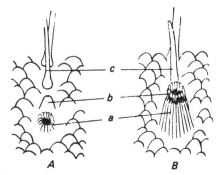

Figure 5. Anus and genital papilla in *Tripterygion* (Matsubara, 1955).
A—male; B—female; a—anus; b—genital papilla; c—first spiny ray of anal
fin.

Key to Species of Genus Tripterygion from the Sea of Japan[14]

1 (2). D III + XIV 10; A 21. Body with dark vertical stripes..........
.... 1. **T. etheostoma** Jordan and Snyder.
2 (1). D III + XVII 12; A 27. Body light-colored, caudal fin usually dark.
.... 2. **T. bapturum** Jordan and Snyder.

24

1. *Tripterygion etheostoma* Jordan and Snyder, 1902—Striped Triplefin
   (Figure 6)

Tripterygion etheostoma Jordan and Snyder, Proc. U.S. Nat. Mus., 25,
1902: 444, fig. 1 (Misaki). Matsubara, Fish Morphol. and Hierar., 1955:
730, fig. 263. Fowler, Synopsis..., 1958: 155. Abe, Enc. Zool., 2. Fishes,
1958: 118, fig. 348 (color figure).

Enneapterygius etheostoma Snyder, Proc. U.S. Nat. Mus., 42, 1912: 518.
Jordan, Tanaka and Snyder, J. Coll. Sci. Tokyo Univ., 33, 1, 1913: 378, fig.
339.

D III, XIV–XV, 9–10; A I 19–21; P 15–16; transverse rows of scales
along side of body 34–37. Gill rakers on first gill arch 6, very short. Head
naked, 4 times in standard length, depth 4.5 times. Depth of caudal
peduncle almost 3 times in head length. Diameter of eyes 3.5 times in
head length, interorbital space 8 times, and snout 3 times. Teeth small,
arranged in bands on jaws and vomer. Branchiostegal membranes form
broad fold across isthmus. Coloration of female: body yellowish-white,
with 6 dark chocolate-brown vertical bars across it, irregular chocolate-
brown spots on head, all fins except pectorals with numerous chocolate-
brown spots forming striped pattern; pectorals with barely discernible
spots. Coloration of male: body very dark except for narrow white spaces
behind second and third dorsal fins; vertical bars contrast sharply with

[14]From Jordan and Snyder, 1902a.

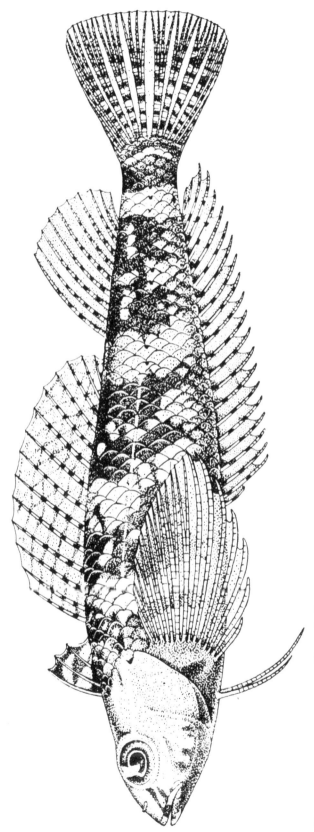

Figure 6. *Tripterygion etheostoma*—Striped triplefin. Length 65 mm. Misaki (Jordan and Snyder, 1902a).

dull background color of remaining part of body; fins, excluding caudal, almost black; second dorsal fin with narrow white stripe along margin; soft dorsal and anal fins with large white spots at posterior end. Caudal peduncle of female similar in color to that of male (Jordan and Snyder, 1902a).

Length, up to 65 mm (Jordan and Snyder, 1902a).

Distribution: In the Sea of Japan known from near Sado Island (Honma and Sugihara, 1963: 5); Toyama Bay (Katayama, 1940: 24); Wakasa Bay (Takegawa and Morino, 1970: 382); San'in region (Mori, 1956a: 21); Cheju-do Island (Mori, 1952; 127). Along the Pacific coast of Japan reported from Tiba Prefecture to Okinawa (Matsubara, 1955: 730).

2. *Tripterygion bapturum* Jordan and Snyder, 1902—Black-Tailed Triplefin (Figure 7)

Tripterygion bapturum Jordan and Snyder, Proc. U.S. Nat. Mus., 25, 1902: 447, fig. 2 (Misaki). Matsubara, Fish Morphol. and Hierar., 1955: 731. Abe, Enc. Zool., 2, Fishes, 1958: 118, fig. 347 (color figure).

D III, XVII, 12; A I 26; P 17; transverse series of scales 43; *l. l.* 28. Head naked, 4.2 (and depth 6) times in standard length. Depth of caudal peduncle, diameter of eye, and snout length 3.25 times in head length. Body more elongate than in *T. etheostoma*. Color of specimens preserved in alcohol fairly distinct. Body without dark stripes,[15] light-colored, yellowish, each scale with dark edging, large light chocolate-brown blotch located on operculum, snout and lips dark, first dorsal fin with blotches, second fin dark along base, third fin with a few dark blotches, and anal fin light-colored with series of dark minute blotches near base. Caudal fin blackish with white border along posterior margin and at base (Jordan and Snyder, 1902a).

27 Length, to 70 mm (Abe, 1958).

Distribution: In the Sea of Japan known from near Sado Island (Honma and Sugihara, 1963: 5); Toyama Bay and San'in region (Katoh et al., 1956: 322). Found along the Pacific coast of Japan (Matsubara, 1955: 731).

CXLVIII. Family CHAENOPSIDAE

This family (according to Clark Hubbs—a subfamily of Blenniidae) includes several genera of combtooth blennies which were scattered in various families of this suborder before its revision (Hubbs, 1953: 12). The character that distinguishes this family from the other families of the suborder is the nature of the first suborbital (lachrymal) bone, which extends far forward and forms the entire lower margin of the orbit. Unlike all the genera of the family Blenniidae, all the genera of Chaenopsidae

[15]Body and median fins with stripes in sketch by Abe (1958).

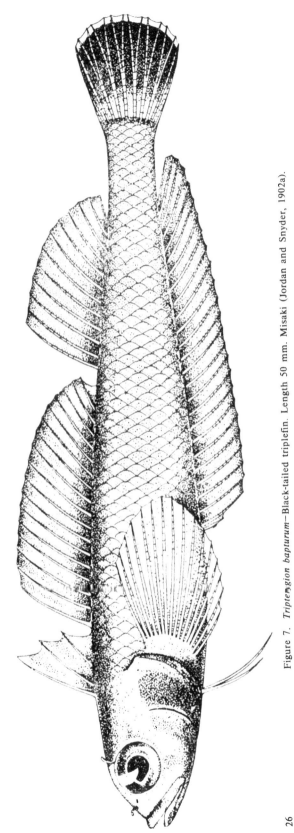

Figure 7. *Tripterygion bapturum*—Black-tailed triplefin. Length 50 mm. Misaki (Jordan and Snyder, 1902a).

have a few conical teeth on the jaws, and some may be slightly enlarged and curved. The teeth in the anterior part of the jaws are often similar to incisors but not arranged in the form of a comb. The upper jaw generally continues far beyond the vertical from the posterior margin of the eye. Body usually elongate, slightly eel-like.

Hubbs included 4 tribes in Chaenopsinae (which we have raised to a family), of which one, Neoclinidi, is found along the American and Asian coasts of the Pacific Ocean.

One genus known from the waters of Japan and the Sea of Japan.

1. Genus *Neoclinus* Girard, 1858

Neoclinus Girard, Fishes: In Pacific Railroad Expl. and Surv., 10, 4, 1858: 114 (type: *N. blanchardi* Girard). Cl. Hubbs, Copeia, 1, 1953: 12.

Zacalles Jordan and Snyder (nec *Zacalles* Foerster, 1868, Insecta, Proc. U.S. Nat. Mus., 25, 1902: 448 (type: *Z. bryope* Jordan and Snyder).

Calliblennius Barbour, Proc. Biol. Soc. Washington, 25, 1912: 187 (type: *Zacalles bryope* Jordan and Snyder).

The characteristic feature of this genus, the only member of the tribe Neoclinidi, is the presence of scales and lateral line on the body.

Four species found in the Pacific Ocean, of which one recorded from the waters of Japan and the Sea of Japan.

1. *Neoclinus bryope* (Jordan and Snyder, 1902a) (Figure 8)

Zacalles bryope Jordan and Snyder, Proc. U.S. Nat. Mus., 25, 1902: 448, fig. 3 (Misaki).

Calliblennius bryope Barbour, Proc. Biol. Soc. Washington, 25, 1912: 187.

Neoclinus bryope Cl. Hubbs, Stanford Ichthyol. Bull., 4, 2, 1952: 52; Copeia, 1, 1953: 17.

22862. Sagami Bay. April 9, 1901. P.Yu. Shmidt. 6 specimens.

23416. Sagami Bay. April 11, 1901. P.Yu. Shmidt. 1 specimen.

D XXIV-XXVI (XXV-XXVI) 14-18 (16-17); A II 28-31 (29-30); P 12-15 (14); *l. l.* 19-23 (20-23); gill rakers $10 + 10 = 20$ (Hubbs, 1953: 17).

Head naked. Body covered with very thin cycloid scales, often fused with skin, about 21 scales in oblique row. Space above lateral line on thorax and abdomen naked. Lateral line short, continues to apex of pectoral fin.

Differs from the closely related species, *N. stephensae* Cl. Hubbs from California, in height of spiny rays in dorsal fin almost equal to body depth, and distinctly greater than height of anal fin; absence of cirri on occiput; 28 and in coloration. Membrane between first and second spiny rays of dorsal fin with ocellar spot (absent in Californian species).

Standard length less than 100 mm (Hubbs, 1953).

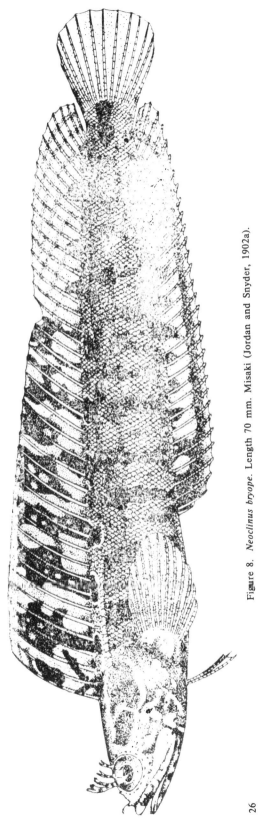

Figure 8. *Neoclinus bryope*. Length 70 mm. Misaki (Jordan and Snyder, 1902a).

Distribution: In the Sea of Japan reported from Sado Island (Honma, 1952: 226); Toyama Bay (Katayama, 1940: 24); San'in region (Mori, 1956a: 21). Pacific coast of Japan from Tokyo to Tosa Bay of Shikoku Island (Matsubara, 1955: 736).

CXLIX. Family BLENNIIDAE—Combtooth Blennies

Body naked, not covered with scales. Spiny and soft rays of dorsal fin almost equal; 1 or 2 spines in anal fin; caudal fin free, with 10-15 principal rays; pelvics jugular, each with small spiny and 2-4 soft simple rays. Mouth not protractile. Jaws with single row of thin, close-set teeth; curved (or straight—Springer, 1968) canine teeth present on posterior end of lower jaw. Palatines usually without teeth (Regan, 1912).

Distinguished from the related family Clinidae by bones of infraorbital ring strong; preorbital (lachrymal) continues posteriorly to vertical from middle of eye or further; and other 3 bones also strong, well developed, with height less than width (Hubbs, 1952).

Cirri or tentacles usually present on head. Segmented rays in pelvic and unpaired fins usually simple and unbranched. Lateral line, if present, always single, passing above pectoral fins.

Springer (1968), on the basis of a detailed study of the skeleton of members of Blenniidae, divided the family into 2 subfamilies—Blenniinae and Nemophidinae. The former included 3 tribes: Blenniini, Omobranchini, and Salariini:

Many genera, 4 known from the Sea of Japan.

Key to Genera of Family Blenniidae[16]

1 (4). Dermal cirri or tentacles present on head. Gill openings large.
2 (3). Canine teeth present on posterior part of both jaws. Solitary strong tooth present on vomer. Teeth on jaws curved, incisor-shaped, and immovable.....................1. **Blennius** Linné.
3 (2). Canine teeth absent on posterior part of jaws. Teeth absent on vomer and palatines. Teeth on jaws minute, movable, and set in fleshy gums.........................2. **Istiblennius** Whitley.
4 (1). Dermal cirri or tentacles absent on head. Gill opening small, located above pectoral fins.
5 (6). Posterior part of both jaws with one curved canine tooth each (Figure 12, A) ....3. **Omobranchus** Valenciennes.
6 (5). Posterior end of lower jaw with well-developed, curved, canine tooth; upper jaw without such tooth or tooth very poorly developed (Figure 12, B)....................4. **Dasson** Jordan and Hubbs.

---

[16]Key includes only genera known from the Sea of Japan and adjacent parts of the Yellow Sea and the Sea of Okhotsk.

1. Genus *Blennius* Linné, 1758

Blennius Linné, Syst. Nat., ed. X, 1758: 256 (type: *B. ocellaris* L.).

Body elongate, compressed laterally, naked, without scales. Head short; profile usually bluntly rounded. Mouth small, horizontal. Jaws with single row of long, fine, curved, close-set teeth; posterior part of jaws with canine teeth. Premaxillae immovable. Gill opening broad, continues forward. Branchiostegal membranes either not attached to isthmus or form broad fold across it. Dorsal fin continuous, with small notch between spiny and soft parts. Pectoral fins moderate in size. Pelvic fins well developed, inserted anterior to vertical from base of pectorals (Jordan and Snyder, 1902a).

Several species. One reported from the Sea of Japan.

1. **Blennius yatabei** Jordan and Snyder, 1900 (Figure 9)

Blennius yatabei Jordan and Snyder, Proc. U.S. Nat. Mus., 1900: 374, pl. XIX (Misaki). Jordan and Snyder, Proc. U.S. Nat. Mus., 1902: 451, fig. 4. Wang and Wang, Fish of Shangtung, III, 1935: 209, fig. 35. Abe, Enc. Zool., 2, Fishes, 1958: 117, fig. 345 (color figure).

22863. Sagami Bay. April, 1901. P.Yu. Shmidt. 2 specimens.

D XII, 16; A I, 19; P 14; V I, 2. Body depth 3.5 to 4.0 times in its total length. Supraorbital dermal tentacle long, its length equal to diameter of eye,[17] and with 4–6 branches.

Color of fish preserved in alcohol olive-green to chocolate-brown. Body with minute blackish spots. In live fish spots on lower part of body yellowish, tips of supraorbital tentacle brick-red.

Morphological changes in structure of rays of anal fin occur in males during spawning (Tomiyama, 1952a).

Biology of this species very poorly studied; development of lateral line in larvae would be of particular interest (Iwai, 1963).

Length, to 90 mm (Abe, 1958).

Distribution: In the Sea of Japan known from Pusan (Mori, 1952: 126); Sado Island (Honma and Sugihara, 1963: 6); Toyama Bay (Katayama, 1940: 24); Wakasa Bay (Takegawa and Morino, 1970: 382), San'in region (Katoh et al., 1956: 322). Southern Japan (Jordan and Snyder, 1902a: 450). Cheju-do Island (Uchida and Yabe, 1939: 13); Gulf of Chihli (Bohai) in the Yellow Sea (Zhang et al., 1955: 164). Pacific coast of Japan (Shmidt, 1931: 147). Ryukyu Island (Aoyagi, 1955: 78).

2. Genus *Istiblennius* Whitley, 1943

Istiblennius Whitley, Austral. Zool., 10, 2, 1943: 185 (type: *Salarias mülleri* Klunzinger, 1880).

---

[17]This ratio changes with age (Tomiyama, 1950b).

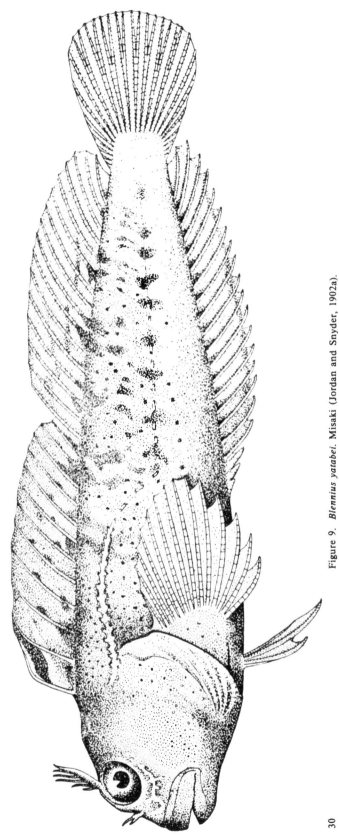

Figure 9. *Blennius yatabei*. Misaki (Jordan and Snyder, 1902a).

Distinguished from the closely related genera *Salarias* and *Entomacrodus*, which are not found in the waters under study, by high dorsal fin with deep notch connected with caudal fin (anal fin not connected with caudal fin), absence of large recurved tooth at posterior end of lower jaw, and simple unbranched supraorbital dermal tentacle (Whitley, 1943). Differs from the genera *Blennius* and *Omobranchus* in absence of canine

31 teeth in posterior part of jaws, absence of teeth on vomer and palatines, and movability of teeth on jaws.

Several species. Two known from the Sea of Japan.

Key to Species of Genus Istiblennius[18]

1 (2). Head with high dermal crest on occiput. D XIII 21; A 23. Color dark chocolate-brown, with spots and bands with greenish sheen.
.... 1. **I. enosimae** (Jordan and Snyder).

2 (1). Head without dermal crest on occiput. D XII 16; A 20. Color chocolate-brown with transverse dark stripes and numerous white dots and bands................... 2. **I. stellifer** (Jordan and Snyder).

1. *Istiblennius enosimae* (Jordan and Snyder, 1902) (Figure 10)

Scartichthys enosimae Jordan and Snyder, Proc. U.S. Nat. Mus., 25, 1902: 460, fig. 9 (Misaki).

Salarias enosimae, Fowler, Synopsis..., 1958: 207.

Istiblennius enosimae, Matsubara, Fish Morphol. and Hierar., 1955: 750, fig. 270. Abe, Enc. Zool., 2. Fishes, 1958: 114, fig. 335 (color figure).

D XIII 21; A I 22; P 14. Head 5.1 times and body depth almost 5 times in standard length. Depth of caudal peduncle 2 times and diameter of eyes 4.5 times in head length (Jordan and Snyder, 1902a).

Secondary sexual characters well expressed in adult male and female (Tomiyama, 1959).

Length, to 150 mm (Abe, 1958).

Distribution: Found in the Sea of Japan; known from the central part of Honshu south up to Ryukyu Islands (Matsubara, 1955: 750). Known from Cheju-do Island (Mori, 1952: 126).

2. *Istiblennius stellifer* (Jordan and Snyder, 1902) (Figure 11)

Scartichthys stellifer Jordan and Snyder, Proc. U.S. Nat. Mus., 25, 1902: 461, fig. 10 (Japan).

Entomacrodus stellifer, Springer, Proc. U.S. Nat. Mus., 122, 1967: 49, pl. 4.

Salarias stellifer, Fowler, Synopsis..., 1958: 217.

Istiblennius stellifer, Matsubara, Fish Morphol. and Hierar., 1955: 749. Abe, Enc. Zool., 2. Fishes, 1958: 115, fig. 337 (color figure).

---

[18]From Jordan and Snyder, 1902a.

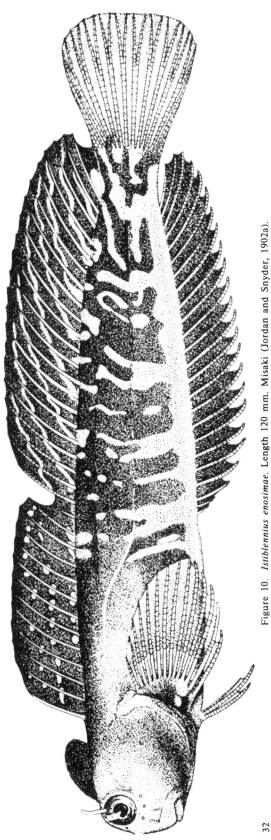

Figure 10. *Istiblennius enosimae*. Length 120 mm. Misaki (Jordan and Snyder, 1902a).

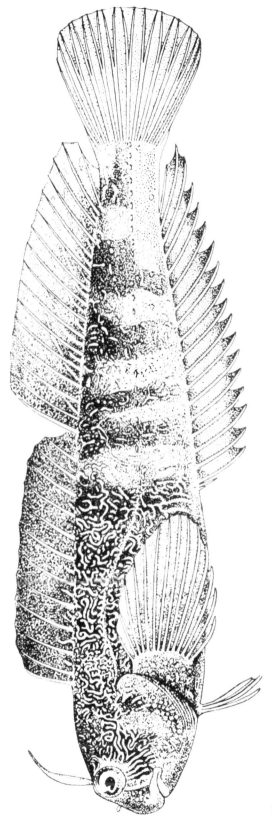

Figure 11. *Istiblennius stellifer.* Wakano-ura (Jordan and Snyder, 1902a).

D XII 16; A I 19[19]; P 14. Head and body depth 4.5 times in standard length. Depth of caudal peduncle 9.5 times and diameter of eyes 4.5 times in head length (Jordan and Snyder, 1902a). Vertebrae 34–36, in specimens from Japan more often 35 (Springer, 1967).

Length, to 90 mm (Abe, 1958).

Distribution: In the Sea of Japan known from near Sado Island (Honma, 1952: 226); Toyama Bay (Katayama, 1940: 24); near Cheju-do Island (Mori, 1952: 126). Along the Pacific coast of Japan reported for Tiba Prefecture, Tsuruga Bay, and Wakano-ura Bay (Matsubara, 1955: 749).

3. Genus *Omobranchus* Valenciennes, 1836

Omobranchus Ehrenberg. In: Valenciennes, Hist. Nat. Poiss., 11, 1836: 287 (type: *O. fasciolatus* Ehrenberg).

Aspidontus Quoy and Gaimard, voy. "Astrolabe," Zool., 3, 1835: 719 (type: *A. taeniatus* Quoy and Gaimard).

Head blunt, mouth small, as in *Blennius*, gill opening above base of pectoral fin, dorsal fin entire, canines at posterior end of both jaws very long (Figure 12, A), and dermal crest on head very poorly developed[20] (Swainson, 1839).

Many species. Three known from the Sea of Japan.

Key to Species of Genus Omobranchus[21]

1 (4). Dermal crest on head between eyes very poorly developed, some-times not discernible. Sides of body either with longitudinal or transverse dark stripes.

2 (3). Body sides with longitudinal dark stripes. Body grayish to choco-late-brown. D XII 22; A I 22........ 1. **O. japonicus** (Bleeker).

3 (2). Body sides with transverse dark bands in anterior part. Body yellowish or orange. D XII 22; A I 24 ....
.... 2. **O. elegans** (Steindachner).

4 (1). Dermal crest on head well defined between eyes in both male and female. Sides of body with about 10 indistinct dark transverse bands. D XII 21; A II 23 .... 3. **O. uekii** (Katayama).

1. *Omobranchus japonicus* (Bleeker, 1869) (Figure 13)

Petroscirtes japonicus Bleeker, Versl. Kon. Akad. Wet. Amst., 3, 1869: 246 (Jedo). Fowler, Synopsis..., 1958: 245.

Omobranchus japonicus, Matsubara, Fish Morphol. and Hierar., 1955: 739, fig. 267. Abe, Enc. Zool., 2, Fishes, 1958: 117, fig. 344 (color figure).

[19]Often 18 for specimens from Japan (Springer, 1967).

[20]In *O. uekii*, described by Katayama (1941), dermal crest on head well defined.

[21]Matsubara, 1955: 739.

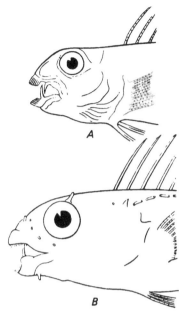

Figure 12. Teeth in *Omobranchus* (A) and *Dasson* (B)
(from Matsubara, 1955).

Aspidontus dasson Jordan and Snyder, Proc. U.S. Nat. Mus., 25, 1902:
456, fig. 8.

Aspidontus japonicus, Jordan and Snyder, Proc. U.S. Nat. Mus., 25,
1902: 458.

D XII 21–22; A I 22; P 13; V 2.

Color dark chocolate-brown in anterior part and light chocolate-brown
in posterior part of body. Body sides with 4 dark parallel stripes (Fowler,
1958). Young differ in color; future dark stripes along sides of body visible
as usual spots or dots (Tomiyama, 1952b).

Length, to 90 mm (Abe, 1958).

Distribution: In the Sea of Japan known from Pohang (Mori, 1952:
126); San'in region (Katoh et al., 1956: 322); Yamaguchi Prefecture
(Yoshida and Ito, 1957: 268); Korean Peninsula (Matsubara, 1955: 739).

2. *Omobranchus elegans* (Steindachner, 1876) (Figure 14)

Petroscirtes elegans Steindachner, Ichthyol. Beitr., 5, 1876: 169
(Nagasaki). Fowler, Synopsis..., 1958: 243.

Aspidontus elegans, Jordan and Snyder, Proc. U.S. Nat. Mus., 24, 1902:
453, fig. 6.

35   *Omobranchus elegans*, Matsubara, Fish Morphol. and Hierar., 1955:
740. Abe, Enc. Zool., 2. Fishes, 1958: 116, fig. 342 (color figure).

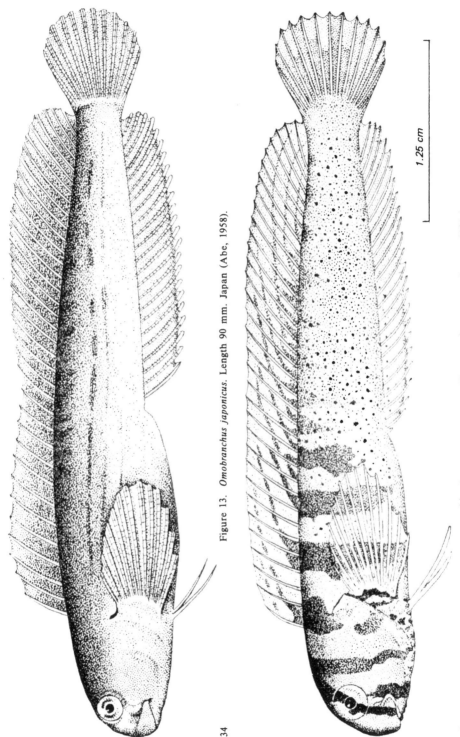

Figure 13. *Omobranchus japonicus*. Length 90 mm. Japan (Abe, 1958).

Figure 14. *Omobranchus elegans*. Hakodate (Jordan and Snyder, 1902a).

1.25 cm

34

34

22864. Sagami Bay. April 11, 1901. P.Yu. Shmidt. 6 specimens.

D XII 22; A I 24; P 13. Head 5 times and depth 5.5 times in standard length; gill rakers on first gill arch from 8 to 10, small (Jordan and Snyder, 1902a).

Ovoviviparity was proposed earlier for these fishes. Studies by Tomiyama (1950a) confirmed that fertilized eggs are laid.

Length, to 80 mm (Abe, 1958).

Distribution: In the Sea of Japan known from Pusan (Mori, 1952: 126); coast of Japan from Hakodate (Matsubara, 1955: 740) to San'in region (Mori, 1956a: 21). Southern coast of Korean Peninsula (Mori and Uchida, 1934: 21); Cheju-do Island (Uchida and Yabe, 1939: 13). Yellow Sea, Gulf of Chihli (Bohai) (Zhang, 1955: 163). Along the Pacific coast of Japan reported from Misaki, Wakano-ura, and Enoshima (Jordan and Snyder, 1902a: 455) and Nagasaki (Okada and Matsubara, 1938: 397).

3. *Omobranchus uekii* (Katayama, 1941) (Figure 15)

Petroscirtes uekii Katayama, Zool. Mag., 53, 12, 1941: 591–592, fig. 1 (Toyama Bay).

Omobranchus uekii, Matsubara, Fish Morphol. and Hierar., 1955: 740.

D XII 21; A 23; P 13; V 2. Head 4.7 times, depth 4.3 times, caudal fin 5.0 times, pectoral fins 5.0 times, and pelvic fins 6.9 times in standard length. Eyes 4.8, snout 7.0 and interorbital space 4.5 times in head length. Body elongate, highly compressed laterally (Katayama, 1941). Dermal crest present on head with 3 dark transverse stripes. Body sides, except for caudal peduncle, with about 10 indistinct dark transverse stripes (Matsubara, 1955).

Length 69 mm (Katayama, 1941).

Distribution: In the Sea of Japan known from Toyama Bay (Katayama, 1941: 591) and Wakasa Bay (Takegawa and Morino, 1970: 382). Reported from the Pacific coast of Japan (Matsubara, 1955: 740).

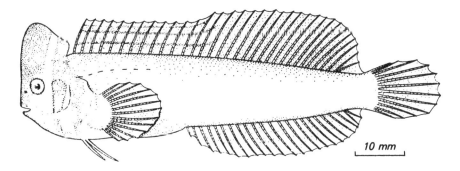

10 mm

Figure 15. *Omobranchus uekii*. Toyama Bay (Katayama, 1941).

4. Genus *Dasson* Jordan and Hubbs, 1925

Dasson Jordan and Hubbs, Mem. Carnegie Mus., 10, 2, 1925: 318 (type: *Aspidontus trossulus* Jordan and Snyder).

Distinguished from related genera by movability of close-set teeth on 36 jaws, slightly recurved canine in posterior part of lower jaw, and small canine at posterior end of upper jaw (see Figure 12, B); gill openings reduced to small pore on upper side of base of pectoral fin; and height of dorsal fin equal throughout length (Jordan and Hubbs, 1925).

Several species. One known from the Sea of Japan.

1. ***Dasson trossulus*** (Jordan and Snyder, 1902) (Figure 16)

Aspidontus trossulus Jordan and Snyder. Proc. U.S. Nat. Mus., 25, 1902: 455, fig. 7 (Japan).

Dasson trossulus, Matsubara, Fish Morphol. and Hierar., 1955: 742. Fowler, Synopsis..., 1958: 238.

Omobranchus trossulus, Abe, Enc. Zool., 2, Fishes, 1958: 116, fig. 341 (color figure).

D X 21; A I 19; P 13. Length of head slightly more than 4 times and body depth 4.75 times in standard length. Height of caudal peduncle 2.4 times, eyes 3.6 times, and interorbital space and snout slightly more than 3 times in head length.

Body color grayish, with 2 longitudinal violet-black stripes, and broad blackish stripe extending from tip of snout to base of caudal fin; width of this stripe in anterior part of body equal to diameter of eye. Dorsal fin with broad dark stripe along base, its upper part with dark spots and stripes; anal fin with 5 large dark blotches; vertical dark stripe located on base of caudal fin; pectoral and pelvic fins light-colored (Jordan and Snyder, 1902a).

Sexual dimorphism and change in color with age described by Tomiyama (1951).

Length, to 200 mm (Jordan and Snyder, 1902a).

Distribution: In the Sea of Japan known from Pusan (Mori, 1952: 126); 37 Sado Island (Honma, 1955: 225); Toyama Bay (Katayama, 1940: 248); Wakasa Bay (Takegawa and Morino, 1970: 382); San'in region (Mori, 1956a: 22). Along the Pacific coast of Japan reported from Misaki (Jordan and Snyder, 1902a), Tiba Prefecture, and everywhere southward (Matsubara, 1955: 742).

CL. Family PHOLIDIDAE—Gunnels

Body elongate, highly compressed laterally, covered with overlapping scales. Head small, without dermal tentacles and crests,[22] naked (only the

---

[22]Sometimes a skin fold forms in the interorbital space during fixation, which may be mistaken for a crest (Makushok, 1958).

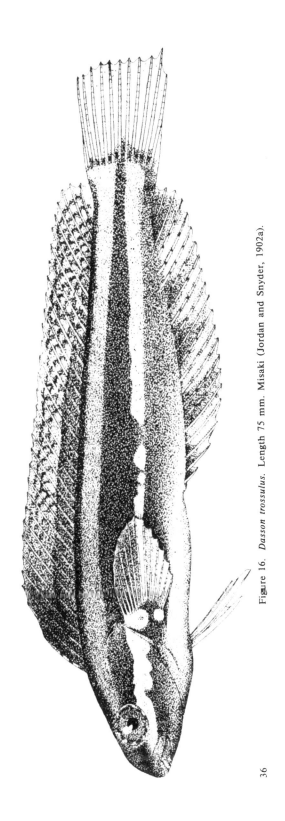

Figure 16. *Dasson trossulus*. Length 75 mm. Misaki (Jordan and Snyder, 1902a).

body of *Pholis nebulosus* covered with scales).[23] Mouth small and oblique. Teeth present on vomer, absent on palatines. Stomach straight, not distinctly demarcated from intestine, which forms small loops. Pyloric ceca absent. Branchiostegal membranes broadly fused, not attached to isthmus. Rays of branchiostegal membranes 5.[24] Opercular siphon present.

Seismosensory canals normally developed, opening on outer side through constant number of pores for entire family: nasal pores 2, interorbital pore 1 (sometimes closed), postorbital pores 6, occipital pores 3, suborbital pores 6, preopercular pores 5, mandibular pores usually 4 (in *Ph. gunellus* and *Ph. dolichogaster* only 3) (Figure 17, A, B).[25] Lateral line of body present, represented by middle branch of open seismosensory papillae, readily distinguished in fresh specimens but not discernible in preserved ones. Vertebrae 84–107; precaudal vertebrae 36–60; vertebrae short and asymmetrical. Dorsal fin with short nonflexible spines. One or two spines present at origin of anal fin. Membranes of dorsal and anal fins fused to a great extent with caudal fin. Pectorals with 10–15 rays, vertical base, small (*Pholis, Apodichthys*), rudimentary (*Xererpes*), or absent (*Ulvicola*). Pelvic fins rudimentary (I 1 in *Pholis*)[26] or absent.

Small fishes inhabiting the littoral zone where they remain in clumps of algae, or under rocks in pools formed at low tide. Parents care for laid eggs. Feed on minute benthic invertebrates (Makushok, 1958).

Two subfamilies,[27] 4 genera.

38   *Key to Subfamilies and Genera of Family Pholididae*[28]

1 (2). Two spines at origin of anal fin. Pelvic fins present. Precaudal vertebrae not more than 47 (subfamily Pholidinae) . . . .
. . . . 1. **Pholis** Scopoli.

2 (1). One spine at origin of anal fin. Pelvic fins absent. Precaudal vertebrae not less than 50 (subfamily Apodichthyinae).[29]

---

[23]The single differentiating feature of *Enedrias nebulosus* from species of *Pholis* gave Makushok (1958) a basis for including this species in *Pholis*, since the other characters of this genus totally incorporate those of *Enedrias*.

[24]Erroneously stated as 4 by Boulenger (1904: 711).

[25]McAllister (1968: 145) indicated 5 to 6 (7). Sometimes an increase in number of preopercular pores up to 6–8 is observed (Makushok, 1958: 108).

[26]Completely disappear in some specimens of *Pholis fasciatus* (Jenson, 1942: 44).

[27]Hubbs (1927) was the first to separate subfamilies.

[28]From Makushok (1958: 110). The genus *Gunnellops* Bleeker, included by some authors (Jordan and Snyder, 1901; Soldatov and Lindberg, 1930; Norman, 1958; Fowler, 1958) in the families Pholididae or Blenniidae, was included in the family Cepolidae by A.Ya. Taranetz (unpublished manuscript). This was probably taken into account by Makushok (1958), who included this genus in the family Pholididae. In the addendum to Part Three of our *Fishes of the Sea of Japan . . . .* which includes the family Cepolidae, this question has been analyzed.

[29]All 3 genera of this subfamily are monotypic.

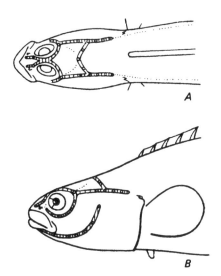

A

B

37   Figure 17. Seisomosensory system of head in Pholidae (Makushok,
                                    1958).

A–dorsal view; B–lateral view.

3 (4). Spine of anal fin very large, with deep notch toward front and con-
       vex on posterior side, and set in a well-developed dermal recess.
       ....[**Apodichthys** Girard, 1854].[30]

4 (3). Spine of anal fin less developed, usual in shape, dermal recess not
       expressed.

5 (6). Pectoral fins present..... [**Xererpes** Jordan and Gilbert, 1895].[31]

6 (5). Pectoral fins absent........[**Ulvicola** Gilbert and Starks, 1897].[32]

1. Genus *Pholis* Scopoli, 1777–Gunnels

Pholis Gronow, Zoophylaceum, 1765: 78 (not binomial).

Pholis Scopoli, Introd. Hist. Nat., 1777: 456 (type: *Blennius gunnellus*
L.). Jordan and Snyder, Proc. U.S. Nat. Mus., 25, 1902: 470. Soldatov and
Lindberg, Obzor..., 1930: 450. Taranetz, Kratkii Opredelitel'..., 1937:
154. Andriyashev, Ryby Severnykh Morei SSSR, 1954: 252. Matsubara,
Fish Morphol. and Hierar., 1955: 761. Fowler, Synopsis..., 1958: 262.
Makushok, Stichaeoidea..., 1958: 62.

Enedrias Jordan and Gilbert. In: Jordan and Evermann, Fish. N. and M.
Amer., 3, 1898: 2414 (type: *Gunnellus nebulosus* Temminck and Schlegel).

Distinguished from closely related genera by presence of 2 (not 1)
spines at origin of anal fin, fewer precaudal vertebrae (not more than 47),

[30]Pacific coast of North America.
[31]Reported from California.
[32]Reported from California.

and presence of well-developed pectoral fins (rudimentary in *Xererpes* and absent in *Ulvicola*).

Most species of *Pholis* are distributed along the northern coast of the Pacific Ocean. *Ph. gunnellus* is endemic to the northern part of the Atlantic Ocean where *Ph. fasciatus* is also distributed (Greenland).

About 10 species, 5 known from the Sea of Japan.

Key to Species of Genus Pholis[33]

1 (10). Head naked, not covered with scales.

2 (3). Pectoral fins small, 3-4 times in head length. Dorsal fin with about 93 spiny rays, anal fin with 2 spines and 46-48[34] soft rays. Body sides with 2 longitudinal rows of dark markings which form single stripe, with light-colored longitudinal stripe between them (Figure 18). All fins light-colored ....
.... 1. **Ph. pictus** (Kner).

39   3 (2). Pectoral fins moderate in size, 1.25 to 3 times in head length.

4 (7). Dorsal and anal fins completely confluent with caudal fin, do not form notch at connections with it. Body distinctly compressed laterally.

5 (6). Dorsal fin with about 93 spiny rays. Pectoral fins 2.5-2.7 times in head length. Body red. Narrow dark stripe continues from eye to base of pectoral fin ....
.... 2. **Ph. dolichogaster dolichogaster** (Pallas).

6 (5). Dorsal fin with about 82 spiny rays. Pectoral fins 3 times in head length. Body gray, with stripes. Yellow stripe with dark margins continues from eye through base of pectoral fin ....
.... 2a. **Ph. dolichogaster taczanowskii** (Steindachner).

7 (4). Dorsal and anal fins form notch at connections with caudal fin. Body relatively less compressed laterally.

8 (9). Dorsal fin with 84-89 spiny rays, anal fin with 2 spiny and 40-44 soft rays. Dorsal fin with several rectangular dark spots. Body sides of adults dark red..
.... 3. **Ph. fasciatus** (Bloch and Schneider).

9 (8). Dorsal fin with about 77 spiny rays, anal fin with 2 spiny and about 35 soft rays. Dorsal fin without rectangular dark spots. Body usually yellowish-green on back and yellow or orange on belly, sometimes red or chocolate-brown. Body sides with about 20 poorly defined transverse dark stripes. Dark stripe continues downward from eye. All fins with red stripes. Gray V-shaped pattern with black edge located between eye and occiput. About

[33]Matsubara, 1955, with additions.
[34]Range of A: 43-51; D 86-95.

14 red spots located along base of dorsal fin, which are partially or completely edged with black . . . . . . 4. **Ph. ornatus** (Girard).

10 (1). Head covered with scales. Dorsal fin with 76–84 spiny rays, anal fin with 2 spiny and 37–44 soft rays.[35]

11 (12). Pectoral fins 2.6 times in head length. Between 13–17 dark spots located along back and base of dorsal fin . . . .
. . . . 5. **Ph. nebulosus** (Temminck and Schlegel).

12 (11). Pectoral fins 1.25 times in head length. 15 or more, paired vertical dark spots along back. Each pair separated by narrow light-colored band . . . . 6. [**Ph. fangi** Wang and Wang].

1. *Pholis pictus* (Kner, 1868)—Decorated Gunnel (Figure 18)

Urocentrus pictus Kner, Sitzungsb. Denkshr. Acad. Wiss., 58, 1868: 51, pl. 7, fig. 21 (Singapore—erroneous; possibly De-Kastri Bay).

Pholis pictus, Jordan and Snyder, Proc. U.S. Nat. Mus. 25, 1902: 471, fig. 15. Shmidt, Ryby Vostochnykh Morei . . . , 1904: 172. Soldatov and Lindberg, Obzor . . . , 1930: 451. Shmidt, Ryby Okhotskogo Morya, 1950: 73. Matsubara, Fish Morphol. and Hierar., 1955: 761, fig. 281. Fowler, Synopsis . . . , 1958: 263, fig. 21.

12405. Sea of Okhotsk, Aniva Bay. 1901. P.Yu. Shmidt. 1 specimen.

12406. Sea of Japan, Peter the Great Bay. April 5, 1900. P.Yu. Shmidt. 1 specimen.

12407. Sea of Japan, Peter the Great Bay. May 6, 1900. P.Yu. Shmidt. 1 specimen.

12408. Sea of Okhotsk, Aniva Bay. August 24, 1901. P.Yu. Shmidt. 2 specimens.

13098. Sea of Okhotsk, Aniva Bay. 1899. V. Brazhnikov. 2 specimens.

13099. Liman of Amur River. 1902. V . Brazhnikov. 1 specimen.

17806. Tatar Strait, De-Kastri Bay. May 18, 1911. Lyaskovskii. 1 specimen.

17807. Tatar Strait, De-Kastri Bay. November 9, 1911. Lyaskovskii. 1 specimen.

17808. Tatar Strait, De-Kastri Bay. May 12, 1912. Derbek. 1 specimen.

18932. Primor'e, Vladimir Bay. June 17, 1913. DVE. 1 specimen.

18933. Tatar Strait. May 1, 1913. DVE. 1 specimen.

40 18934. Primor'e, Olga Bay. May 8 and June 14, 1913. DVE. 2 specimens.

18935. Tatar Strait, Vanino Bay. May 2, 1913. DVE. 1 specimen.

18936. Sea of Japan, Peter the Great Bay. June 12, 1913. DVE. 10 specimens.

18937. Tatar Strait. May 18, 1912. DVE. 2 specimens.

18938. Primor'e, Vladimir Bay. June 23, 1913. DVE. 1 specimen.

[35]According to Pavlenko (1910: 47) A III 34 for *P. nebulosus* requires confirmation.

18939. Ussuriisk Bay. October 1, 1912. DVE. 1 specimen.

26145-26150. Peter the Great Bay (Preobrazhenie Bay, Kvandagou, Petrov Island).

30487. Western Sakhalin, Shirokaya Pad'. August 10-22, 1933. A. Kuznetsov. 13 specimens.

31607-31611. Tatar Strait, Antonovo. August, 1946. KSE. 16 specimens.

34768. Western Sakhalin, Shirokaya Pad'. August 22, 1933. 8 specimens.

35127. Kuril Islands, Iturup Island. August 22, 1953. V. Makushok. 4 specimens.

D LXXXVI-XCV; A II 43-51; P 10-12; V I 1; vertebrae 98-101, of which 41-45 precaudal. Species highly variable in number of rays as well as color. Most characteristic coloration given in key to species. Sometimes spots of upper and lower rows fused, forming pattern all over body. Vertical dark stripe continues through midpoint of eye, broadens slightly in interorbital space, and fuses with analogous stripe on other side. Broad light-colored stripe located behind eye also merges on occiput. Dorsal fin confluent with caudal fin and forms small notch (Soldatov and Lindberg, 1930). Buccal cavity with broad palatine and narrow mandibular membranes. Upper part of gill opening with siphon formed of dermal process of operculum and broad thick fold of skin on trunk, which continues to margin of pectoral fin. Upper lip undivided, lower lip divided medially (Shmidt, 1950). Body color of live fish greenish-gray with blackish spots, head gray, almost black on upper side; yellow strip continues behind eyes from lower margin of preopercle upward and through occiput, which is bordered by black stripes. Predominant color reddish-yellow, dorsal fin reddish, bluish dots located along base. Caudal, anal, pectoral, and pelvic fins reddish. Elements of pattern in younger specimens arranged somewhat differently: brown spots on back fuse into single dark stripe of stretched narrow spots separated by narrow intervals (Shmidt, 1904).

Specimens of KSE (44, Nos. 41663-41686) 24 to 222 mm long conform
41 to description given above. These fish were caught in Tatar Strait, near Moneron Island, in Aniva Bay and near Kunashir and Šiaškotan Islands in July and August, 1947-1949 at depths ranging from 3 to 110 m, with water temperature near bottom usually 3.0 to 8.7°C, and bottom comprising rocks, stones, sand, silted sand, and in some places black silt. At places of catches the biocenosis comprised *Zostera marina*, *Z. nana*, and *Laminaria*. *Ph. pictus* was often present among red algae, *Ahnfeltia*. Flounders and soles (*Lepidopsetta bilineata*, *Limanda punctatissima*, *Pleuronectes herzensteini*), *Hypsagonus quadricornis*, and *Opistocentrus ocellatus* were incidental in the catches.

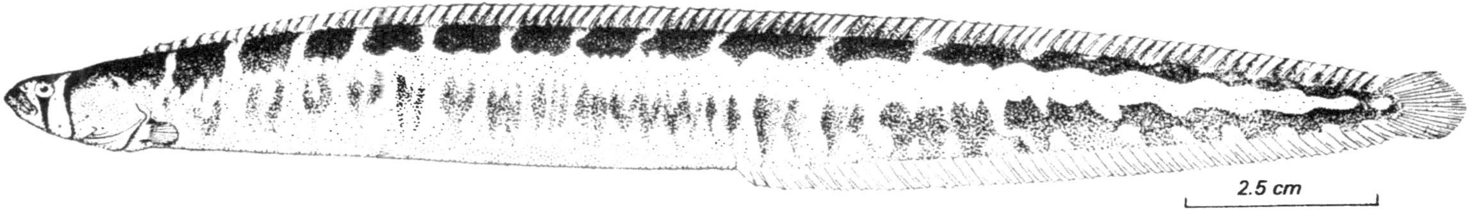

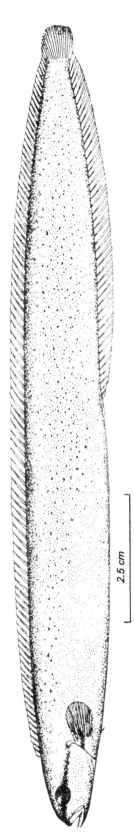

Figure 18. *Pholis pictus*—ornamented gunnel. De-Kastri Bay (Jordan and Snyder, 1902a).

2.5 cm

40

Figure 19. *Pholis dolichogaster*—stippled gunnel. Komandor Islands (Jordan and Evermann, 1900).

2.5 cm

42

Length, to 308 mm (Shmidt, 1904).

Distribution : In the Sea of Japan known from Peter the Great Bay to the Gulf of De-Kastri (Soldatov and Lindberg, 1930: 451) (in Primor'e Nos. 18932, 18934, 18938); Tatar Strait, coast of Sakhalin (Ueno, 1971: 84); Toyama Bay (Katayama, 1940: 25). Sea of Okhotsk, liman of Amur River, Aniva Bay, off southern Kuril Islands, Iturup Island (Shmidt, 1950: 73). Possibly also the Bering Sea (Jordan, Tanaka and Snyder, 1913: 388).

2. ***Pholis dolichogaster dolichogaster*** (Pallas, 1811)—Stippled Gunnel (Figure 19)

Blennius dolichogaster Pallas, Zoogr. Rosso-Asiat., 3, 1811: 175, pl. 2, fig. 2 (Kamchatka).

Pholis dolichogaster, Jordan and Snyder, Proc. U.S. Nat. Mus., 25, 1902: 471. Soldatov and Lindberg, Review..., 1930: 452. Taranetz, Kratkii Opredelitel'..., 1937: 154; Izv. TINRO, 12, 1937: 39 (Shirokaya Pad'). Andriyashev, Ryby Severnykh Morei SSSR, 1954: 254, fig. 132. Matsubara, Fish Morphol. and Hierar., 1955: 761.

D XCI-XCIII; A II 44-47; P 13-14 (Shmidt, 1904). Distinguished from related species by color and relatively developed pectoral fins (about 2 times in head length). Color of some fish uniformly red (color often preserved in alcohol). Sometimes color chocolate-brown to olive-green (yellowish in alcohol). Small dark spots located along body sides, and sometimes hazy transverse stripes also. Several whitish round spots arranged in middle of body, with three to four dark minute spots along sides. Narrow dark stripe (not discernible or barely so in alcohol) extends from eye along slit toward base of pectoral fin. Dorsal and anal fins darkish, with light-colored stripes (Soldatov and Lindberg, 1930).

Specimens of KSE (40, Nos. 41641-41649) 39 to 173 mm long conform to description given here. X-rays of 16 specimens showed: D XC-XCIII; A II 45-49. Specimens caught by KSE near eastern coast of Sakhalin and off Šiaškotan Island in July, August, and September, 1947-1949 at depths of 2.5 to 148 m, with water temperature near bottom −0.3 to +5.3°C, and bottom comprising rocks, stones, sand, and black silt. At places of catches biocenosis comprised *Zostera marina* and *Suberites domuncula*. Incidental fishes in the catches: *Icelus uncinalis crassus, Malacocottus zonurus zonurus, Eumesogrammus praecisus, Artediellus dydymovi*, and *Psychrolutes paradoxus*.

Length, 250 m (Andriyashev, 1954).

Distribution: In the Sea of Japan known from northern part, De-Kastri Bay (Shmidt, 1950: 74); Shirokaya Pad' (Taranetz, 1937a: 39). In the Sea of Okhotsk found near the eastern coast of Sakhalin and off Šiaškotan Island (specimens of KSE). Bering Sea (Taranetz, 1937b: 154).

2a. **Pholis dolichogaster taczanowskii** (Steindachner, 1880)—Gray Gunnel (Figure 20)

Centronotus taczanowskii Steindachner, Ichthyol. Beitr., 9, 1880: 24, pl. 3, fig. 1 (Strelok Strait, Peter the Great Bay).

Pholis taczanowskii, Jordan and Evermann, Fish N. and M. Amer., 3, 1898: 2416. Jordan and Snyder, Proc. U.S. Nat. Mus., 25, 1902: 473. Soldatov and Lindberg, Obzor..., 1930: 452. Matsubara, Fish Morphol. and Hierar., 1955: 761.

Pholis dolichogaster taczanowskii, Taranetz, Kratkii Opredelitel'..., 1937: 154.

18947. Peter the Great Bay. 1907-1908. V. Brazhnikov. 1 specimen.

41640. Kuril Islands, Kunashir Islands. July 19, 1951. O.G. Kusakan. 1 specimen.

D LXXXII-LXXXIV; A II 45.

Distinguished from *Ph. dolichogaster dolichogaster* by smaller number of rays in dorsal fin and gray color (Soldatov and Lindberg, 1930: 452).

Head 9 times and depth 10 times in standard length. Teeth conical, with blunt cusps. Dorsal fin very low. Snout slightly longer than longitudinal diameter of eye.

Specimen No. 41640 was caught in the littoral zone from boulder dumps.

Length, to 120 mm.

Distribution: In the Sea of Japan known from Peter the Great Bay, near Pusan, Wonsan, and Hakodate (Taranets, 1937b: 154). In the Sea of Okhotsk reported from the southwestern and eastern coasts of Sakhalin (Ueno, 1971: 84) and known near Kunashir Island (KSE).

3. **Pholis fasciatus** (Bloch and Schneider, 1801)—Banded Gunnel (Figure 21)

Centronotus fasciatus Bloch and Schneider, Syst. Ichthyol., 1801: 165, pl. XXXVII, fig. 1 (Tranquebar).

Pholis fasciatus, Jordan and Snyder, Proc. U.S. Nat. Mus. 25, 1902: 473 (Aomori). Pavlenko, Ryby Zaliva Petr Velikii, 1910: 48. Soldatov and Lindberg, Obzor..., 1930: 453. Taranetz, Kratkii Opredilitel'..., 1937: 154. Matsubara, Fish Morphol. and Hierar., 1955: 671.

41662. Sea of Okhotsk, Aniva Bay. July 20, 1947. Z.I. Petrova. 5 specimens.

D LXXXIV; A II 40; V I 1; P 12; C 18; Br. 5 (Bloch and Schneider, 1801). Distinguished from other Far East species by rather larger pectoral fins, equal to half head length, and peculiarities of coloration of dorsal fin. In *Ph. fasciatus* 9 to 12 dark longitudinal stripes extend along back and often on body sides, which continue onto dorsal fin. Posterior transverse stripes reach anal fin without continuing or, as shown

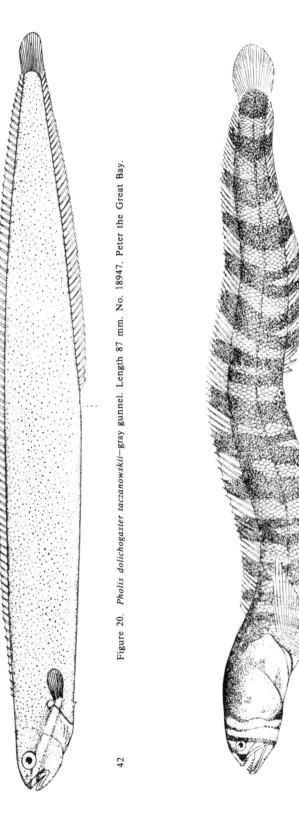

Figure 20. *Pholis dolichogaster taczanowskii*—gray gunnel. Length 87 mm. No. 18947. Peter the Great Bay.

42

Figure 21. *Pholis fasciatus*—banded gunnel. Length 38 mm. Tranquebar (Bloch and Schneider, 1801).

42

in the drawing of the type specimen, continuing onto fin. Pectorals and caudal fin without spots or stripes. Two dark stripes originate behind eyes from top of head down to its lower surface, which are separated by yellow space. Basic body color in live fish reddish, in alcohol yellowish, with dark sinuous stripes reaching belly; transverse connections between these stripes sometimes impart reticulate pattern to body (Soldatov and Lindberg, 1930). Ventrally, scales do not continue beyond pelvics, in front of which surface is naked (Shmidt, 1950: 75). Our specimens (length 83 to 123 mm) were caught at a depth of 23 m from among clumps of red algae. Shrimps, *Strongylocentrotus pulchellus, Asterias amurensis*, and *Spirontocaris* were incidental in the catches.

Length to 150 mm (Soldatov and Lindberg, 1930).

Distribution: In the Sea of Japan known from Peter the Great Bay (Pavlenko, 1910: 48); Oshoro Bay (Kobayashi, 1962: 258); along the Sea of Japan, coast of Hokkaido (Ueno, 1971: 84); and near Aomori (Jordan and Snyder, 1902a: 473). Sea of Okhotsk, Bering Sea, Arctic Ocean (Taranets, 1937b: 154). Greenland (Shmidt, 1950: 75).

44 4. ***Pholis ornatus*** (Girard, 1854)—Saddleback Gunnel (Figure 22).

Gunnellus ornatus Girard, Proc. Acad. Nat. Sci., Philad., 1854: 49 (California).

Pholis ornatus, Jordan and Evermann, Fish. N. and M. Amer., 3, 1898: 2419; 4, 1900, fig. 833. Shmidt, Ryby Vostochnykh Morei..., 1904: 174. Soldatov and Lindberg, Obzor..., 1930: 454. Taranetz, Kratkiĭ Opredelitel'..., 1937: 155. Matsubara, Fish Morphol. and Hierar., 1955: 761.

18950. Peter the Great Bay. October 10, 1912. DVE, 2 specimens.

34341. Western coast of Sakhalin, Shirokaya Pad'. August 18, 1933. A. Kuznetsov. 1 specimen.

37295. Sea of Japan, Ussuriisk Gulf. 1962. ZIN. 1 specimen

41687. Sea of Japan, Tatar Strait. September 16, 1933. A. Kuznetsov. 1 specimen.

41688. Sea of Japan, Tatar Strait. 1929. GGI. 1 specimen.

41689. Sea of Japan, Tatar Strait. September 20, 1933. A. Kuznetsov. 1 specimen.

D LXXIV-LXXIX; A II 35-38; V I 1 (Makushok, 1958).

Distinguished from related species by presence on dorsal fin of 12 to 14 red spots with dark characteristic margin (in form of parentheses).

Body highly compressed laterally, maintaining almost equal height up to origin of anal fin, and notably reducing toward caudal fin. Head small, somewhat rounded. Maxilla extending slightly beyond anterior margin of orbit. Dorsal and anal fins low. Origin of anal fin almost midway between base of pectoral fin and end of caudal fin (Soldatov and Lindberg, 1930).

Length to 300 mm (Soldatov and Lindberg, 1930).

Distribution: In the Sea of Japan known from Peter the Great Bay (Pavlenko, 1910: 48); Olga Bay, near Cape Lazarev, Nevel'sk Strait (Soldatov and Lindberg, 1930: 454); Tatar Strait (specimens of KSE); along the Sea of Japan coast of Hokkaido (Ueno, 1971: 84) and Wakasa Bay (Takegawa and Marino, 1970: 383). Described from the coast of California. Known from San Francisco to the Bering Sea and Kamchatka.

5. ***Pholis nebulosus*** (Temminck and Schlegel, 1845)—Scalyhead Blenny (Figure 23)

Gunnellus nebulosus, Temminck and Schlegel, Fauna Japonica, Poiss., 1845: 138, pl. 73, fig. 2 (Japan).

Enedrias nebulosus, Jordan and Snyder, Proc. U.S. Nat. Mus., 25, 1902: 468, fig. 14. Shmidt, Ryby Vostochnykh Morei..., 1904: 175. Pavlenko, Ryby Zaliva Petr Velikii, 1910: 47. Soldatov and Lindberg, Obzor..., 1930: 450. Taranetz, Kratkii Opredelitel'..., 1937: 154. Matsubara, Fish Morphol. and Hierar., 1955: 761. Fowler, Synopsis..., 1958: 260, fig. 20.

Pholis nebulosus, Makushok, Stichaeoidea..., 1958: 62.

6602. Sea of Japan, Vladivostok. 1883. I.S. Polyakov. 3 specimens.

11602. Korean Peninsula, Choson-man Gulf. 1897. F.F. Busse. 3 specimens.

12427. Korean Peninsula, Choson-man Gulf. October 18, 1896. A.A. Bunge. 1 specimen.

12429–12430. Sea of Japan, Peter the Great Bay, Zolotoi Roz. April 20, 1900. P.Yu. Shmidt.

13088. Korean Peninsula, Pusan. March, 1901. P.Yu. Shmidt. 4 specimens.

18839. Sea of Japan, Primor'e, Olga Bay. June 14, 1913. DVE. 11 specimens.

18840–18843. Sea of Japan, Peter the Great Bay. 1912–1913. DVE. 1 specimen.

18844. Peter the Great Bay. 1907–1908. V.K. Brazhnikov. 3 specimens.

20473–20476. Peter the Great Bay. June–August, 1896. M.Ya. Yankovskii. 7 specimens.

22176. Peter the Great Bay. July 25, 1927. E.P. Rutenberg. 1 specimen.

25987–26012. Peter the Great Bay. August–September, 1934. ZIN. 11 specimens.

30709. Peter the Great Bay. May 18, 1914. A.I. Cherskii. 4 specimens.

41650. Sea of Okhotsk, Kunashir Island. July 28, 1951. O.G. Kusakin. 1 specimen.

41651. Sea of Okhotsk, Kunashir Island. July 7, 1951. O.G. Kusakin. 3 specimens.

41652. Sea of Okhotsk, Kunashir Island. July 8, 1954. Shchegolev. 1 specimen.

D LXXVI-LXXXIV; A II 37-44; P 11-12 (Soldatov and Lindberg, 1930: 450).

Distinguished from closely related species (as well as *Ph. fangi*) by presence of scales on head. Pelvic fins small, with one spine and one soft ray. Pectoral fins relatively small, rounded, with 12 to 15 rays. Numerous (more than 12) dark spots along back and base of dorsal fin characteristic.

Length to 220 mm (Soldatov and Lindberg, 1930).

Distribution: In the Sea of Japan known off the Korean Peninsula, Pusan, Wonsan (Mori, 1952: 129); Choson-man Gulf (Nos. 11602, 12427, 12428); Chongjin (Mori, 1956a: 21); Peter the Great Bay to Olga Bay (Soldatov and Lindberg, 1930: 450); western Sakhalin, Sea of Japan, coast of Hokkaido (Ueno, 1971: 84); Toyama Bay (Katayama, 1940: 25); Sado Island (Honma, 1952: 226); Wakasa Bay (Takegawa and Morino, 1970: 382); and farther south up to Fukuoka region, northern Kyushu (Tsukahara, 1967: 299); and Nagasaki (Shmidt, 1931b: 147). Yellow Sea (Wang and Wang, 1935: 215); Gulf of Chihli (Bohai) (Zhang et al., 1955: 167). Sea of Okhotsk, coast of Hokkaido and southern Kuril Islands (Ueno, 1971: 84). Pacific coast of Japan (Jordan and Hubbs, 1925: 320).

6. [*Pholis fangi* Wang and Wang, 1935]–Fang's Gunnel (Figure 24)

Enedrias fangi Wang and Wang, Contr. Biol. Lab. Sci. China, 11, 6, 1935: 215, fig. 39 (Chefoo). Fowler, Synopsis..., 1958: 262.

Pholis fangi, Makushok, Stichaeoidea..., 1958: 62.

Enedrias nebulosus, Zhu et al., Ryby Vostochno-Kitaiskogo Morya, 1963: 379, fig. 285.

D LXXVIII; A II 39; P 15; V I 1; gill rakers on lower part of arch 11. Close to *Ph. nebulosus* but differs in long pectorals with length less than 2 times in head length (in *Ph. nebulosus* more than 2 times); longer head length, 7.4 times (versus 8.5) in standard length; convex profile of head (versus flat); larger diameter of eye; location of origin of anal fin at vertical from 36th ray of dorsal fin (versus 38th); and characteristic body coloration (depicted in figure and described in key)

It should be noted that the fish described by Chinese ichthyologists (Zhu et al., 1963, fig. 285), fully corresponds to the morphological characters and coloration of *Ph. fangi*, although the drawing is labeled *Ph. nebulosus* (probably by mistake).

Length of type specimen 121 mm. Length of specimen described by Chinese ichthyologists, 160 mm.

Distribution: Not found in the Sea of Japan. Described from Chefoo (Wang and Wang, 1935: 215); indicated for the Gulf of Chihli (Bohai) (Zhang, 1955: 168). Commonly found also in the East China Sea (Zhu et al., 1963: 379).

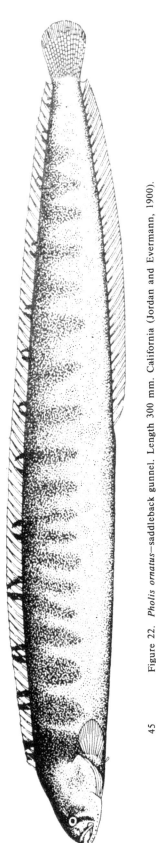

Figure 22. *Pholis ornatus*—saddleback gunnel. Length 300 mm. California (Jordan and Evermann, 1900).

45

Figure 23. *Pholis nebulosus*—Scalyhead blenny. Length 168 mm. Japan (Temminck and Schlegel, 1845).

45

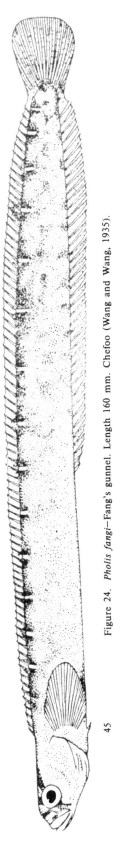

Figure 24. *Pholis fangi*—Fang's gunnel. Length 160 mm. Chefoo (Wang and Wang, 1935).

45

CLI. Family STICHAEIDAE–Pricklebacks

Makushok, Stichaeoidea..., 1958: 67.

Body moderately or highly elongate, usually covered with overlapping scales. Head without dermal tentacles or with several dermal processes (Chirolophinae), or with dermal crest (Alectriinae, *Cebidichthys*[36]). Head naked (in most forms), scales present only on cheeks. Mouth usually small, conical, oblique or horizontal. Upper as well as lower lip well developed. Mental processes absent. Teeth on jaws arranged in one or several rows (especially on upper jaw), conical (except for most members of Chirolophinae). In most forms pyloric ceca well developed (usually 4–6). Branchiostegal membranes usually widely united, not attached to isthmus (situation different in Lumpeninae, *Anoplarchus*, and some members of Stichaeinae). Branchiostegal rays 6 or 5 (Alectriinae, some members of Opisthocentrinae). Gill openings continue far beyond upper 47 margin of base of pectorals. Well-developed siphon or upper notch present. Seismosensory canals of head (Figure 25, A-I) generally well developed; pores usually arranged in single row.[37]

Preopercular pores usually 6, 4 on lower jaw. Vertebrae from 43 (*Stichaeopsis hopkinsi*[38]) to 110–113 (*Azygopterus*) and more (*Eulophias*). Precaudal vertebrae 13–45. All rays of dorsal fin spiny or its posterior part with soft rays (*Dictyosoma, Cebidichthys, Eulophias*). Origin of anal fin with 1–5 small spines (degree of development highly variable from species to species, from poorly developed spinules to fully formed spines). Bases of pectoral fins vertical (except in some members of Opisthocentrinae); fins usually large, sometimes highly reduced or even absent; number of rays varies from 8 to 21. Pelvic fins present or absent.

Makushok (1958: 54) reported that the "family Stichaeidae has no specific structural character (that is found) in all its members" and that "a comparative analysis of different systems of organs indicates a mosaic pattern of characters in the family due to sharply directed morphological adaptations and not to phylogenetic differences."

Eight subfamilies, 30 genera, and 54 species. In the Sea of Japan all 8 subfamilies, 19 genera, and 34 species known; in adjacent waters 1 genus and 4 species.

Makushok (1958: 63) considered the family Stichaeidae close to the families Ptilichthyidae, Pholididae, and Anarhichadidae, and placed these in the superfamily Stichaeoidea, giving the characters detailed below.

Body moderately long or highly oblong, covered with minute cycloid

---

[36]Genus *Cebidichthys* from the subfamily Xiphisterinae is not found in our waters; it is known only from the coast of California.

[37]Arrangement differs in Stichaeinae and *Dictyosoma*.

[38]Absent in our waters; known from the coast of California.

44

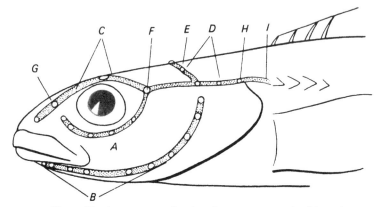

Figure 25. Arrangement of main seismosensory canals of head in
Stichaeidae.

A—suborbital canal; B—mandibular-opercular canal; C—supraorbital
canal; D—postorbital canal; E—occipital canal, or supratemporal
commissure; F—anterior postorbital pore; G—anterior and posterior nasal
pores; H—posterior suprahumeral pore; I—commencement of trunk canal
of lateral line.

scales, which continue on head and fins or are reduced to varying degrees
(if scales do not continue on head, then restricted on upper side to
occipital canal). Each side with one small unpaired nostril at the
end of a long olfactory tubule. Gill arches 4, slit behind last arch
not closed. Oral valves (palatine and mandibular) usually present
(completely reduced only in the genera *Anisarchus* and *Lumpenella*).
48 Swim bladder absent. Seismosensory canals of head open exteriorly
through pores. Most forms without seismosensory canals in trunk (present
only in some groups of Stichaeidae). Number of vertebrae varies from 43
(*Stichaeopsis hopkinsi*[39]) to 240 (Ptilichthyidae[40]) and more (*Anarhichthys*).
Dorsal and anal fins long, continuous, their membranes reaching base of
caudal fin and in several cases fused to a lesser or greater extent with it.
Dorsal fin usually with only spiny rays, but some members of Stichaeidae
and *Ptilichthys* with soft rays in posterior part of dorsal fin. Anal fin
originates immediately behind vent, its first ray usually nonsegmented, in
the form of a spinule or spine. Preanal distance usually less than half total
body length. Caudal fin never bifurcate (usually rounded-oval), principal
rays 12–15 (usually 13–14). Pectoral fins, if present, with 8–21 rays.
Pelvic fins thoracic, with 1 spine and maximum 4 soft rays; completely
reduced in most groups. Soft rays of fins usually bi- or trifurcate
(Makushok, 1958: 63).

[39]Known from waters off eastern coast of Kamchatka and Vancouver Island.
[40]Distributed in northern part of the Pacific Ocean.

Benthic, predominantly coastal fishes; some members of various groups encountered at considerable depths (*Anarhichas latifrons* and *Lumpenella*). Many species display parental care by protecting laid eggs. Eggs laid on bottom, usually large. Juvenile fish littoral-pelagic. Predominantly small fishes (except for wolffishes). Usually feed on minute benthic invertebrates. Exceptionally predaceous (for example, *Stichaeus grigorjewi*). Typical phytophagous forms probably do not exist (Makushok, 1958: 65).

Distribution: Known from northern part of the Pacific Ocean (most forms) and the Atlantic Ocean, as well as the Arctic Ocean. Southern boundary of area of distribution passes through southern part of the Sea of Japan, central California, Cape Cod, and Cape La Manche, coinciding with the northern boundary of the area of distribution of tropical blennies.

Key to Families of Superfamily Stichaeoidea[41]

1 (6). Teeth not differentiated in shape. Head of vomer not elongate; vomerine teeth, if present, conical.

2 (5). In caudal fin only principal rays branched (Figure 26, A).

3 (4). Body not greatly elongate (no more than 134 vertebrae). Gill openings on lower side continue much above upper margin of base of pectoral fins; siphon or upper notch of operculum present (Figure 27, A, B). Branchiostegal rays 5 or 6. Spiny rays of dorsal fin connected through membrane. Dermal mental process absent.
.... 1. **Stichaeidae.**

4 (3). Body greatly elongate (vertebrae 238—240). Gill openings on lower side barely reach lower margin of base of pectoral fins; siphon or upper notch absent in operculum. Branchiostegal rays 3.[42] Spiny rays of dorsal fin not connected by membrane. Dermal mental process present .... **[Ptilichthyidae].**[43]

5 (2). In caudal fin, not only principal rays but adjoining procurrent rays also branched (Figure 26, B)..................... 2. **Pholididae.**

6 (1). Teeth sharply differentiated in shape. Head of vomer highly elongate, with large molars (Figure 28, A-C) .... 3. **Anarhichadidae.**

Key to Genera of Family Stichaeidae[44]

1 (44). Pectoral fins present.

---

[41]From Makushok (1958: 66), with modifications.

[42]McAllister (1968: 145) indicated 6.

[43]One genus and 1 species (*P. goodei* Bean, 1882). Few captures. Eastern coast of Kamchatka, environs of Unalaska Island, and Vancouver Island (Puget Sound) (Makushok, 1958: 117).

[44]From Makushok (1958: 68), with some modifications.

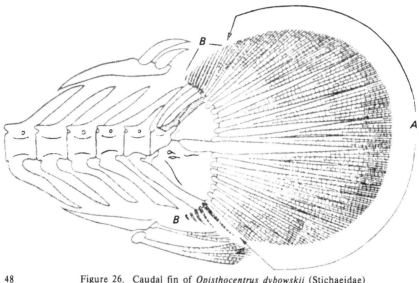

Figure 26. Caudal fin of *Opisthocentrus dybowskii* (Stichaeidae) (Makushok, 1958).

A—principal rays; B—procurrent rays.

2 (43). Vertebrae not more than 94. Caudal fin more or less separate and with at least 12 (12–15) principal rays.

3 (10). Trunk with well-developed seismosensory canals, usually taken as "lateral line." Pelvic fins present (subfamily Stichaeinae).

4 (5). Trunk with one seismosensory canal on each side of body (upper) .... 1. **Stichaeus** Reinhardt.

5 (4). Trunk with two or more seismosensory canals.

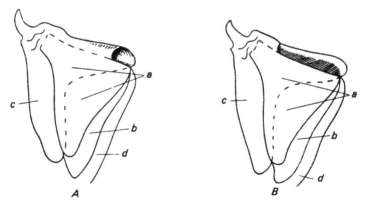

Figure 27. Postero-upper part of operculum of Stichaeidae (Makushok, 1958).

A—siphon; B—upper notch; a—opercle; b—subopercle; c—cleithrum; d—branchiostegal membrane.

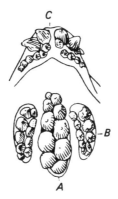

Figure 28. Teeth of *Anarhichas orientalis* (Anarhichadidae) (Barsukov, 1959).

A—vomerine teeth; B—palatine teeth; C—teeth on interpremaxillae

6 (7). Posterior part of anal fin with 2-3 spines (Figure 29, A). Pectoral fins with 18 rays.................. 2. [**Eumesogrammus** Gill].

7 (6). Posterior part of anal fin without spines. Pectoral fins with no more than 16 (15-16) rays.

8 (9). Branches or anastomoses with blind ends proceed randomly from longitudinal seismosensory canals of trunk (Figure 30, A-C). Mandibular pores 4. Branchiostegal membranes not attached to isthmus............3. **Stichaeopsis** Kner and Steindachner.

50 9 (8). Vertical branches with blind ends proceed regularly from longitudinal seismosensory canals of trunk (Figure 31). Mandibular pores 3. Branchiostegal membranes fused for most part with isthmus.............. 4. **Ernogrammus** Jordan and Evermann.

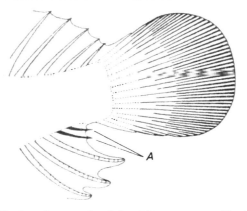

Figure 29. Posterior part of anal fin of *Eumesogrammus* (Shmidt and Andriyashev, 1935).

A—spines.

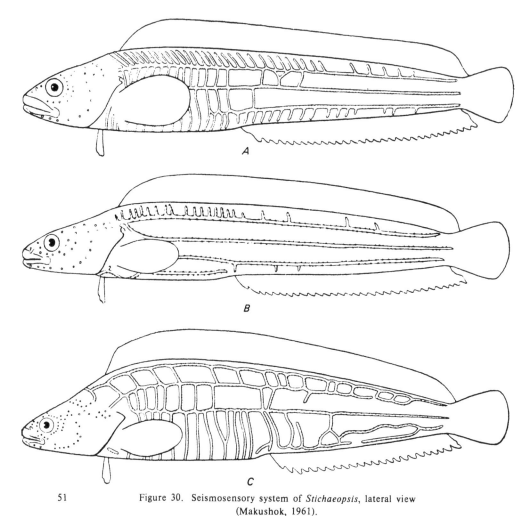

51

Figure 30. Seismosensory system of *Stichaeopsis*, lateral view
(Makushok, 1961).

A—*S. nevelskoi*; B—*S. epallax*; C—*S. nana*.

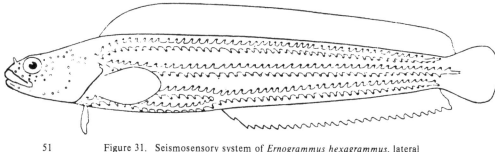

51

Figure 31. Seismosensory system of *Ernogrammus hexagrammus*, lateral
view (Makushok, 1961).

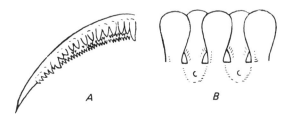

Figure 32. Jaw teeth of Chirolophinae (Makushok, 1958).
A—*Bryozoichthys lysimus*; B—*Chirolophis japonicus*.

10 (3). Combination of different characters: a) trunk with well-developed seismosensory canals but pelvic fins absent; b) seismosensory canals absent (sometimes one short canal present) but pelvic fins usually present.

11 (18). Head with dermal processes in form of barbels and cirri on upper side (Subfamily Chirolophinae).

12 (17). Pelvic fins well developed (I 4 or I 3), their soft rays branched, similar to soft rays of anal fin. Supraorbital cirri 2 pairs.

13 (14). Teeth on jaws conical, arranged in several rows (Figure 32, A); teeth also present on vomer and palatines .... 5. **Bryozoichthys** Whitley.

14 (13). Teeth on jaws incisor-shaped, arranged in 2 irregular rows (Figure 32, B); teeth absent on vomer and palatines.

15 (16). Opercular siphon present (Figure 27, A). Postorbital pores 7, mandibular pores 4 .... 6. **Chirolophis** Swainson.

16 (15). Opercular siphon absent (replaced by upper notch; see Figure 27, B). Postorbital pores 5, mandibular pores 3 .... 7. **Soldatovia** Taranetz.*

17 (12). Pelvic fins poorly developed (I 2) or absent, their soft rays, like those of anal fin, simple, not branched. Supraorbital cirri 3 pairs. .... [**Gymnoclinus** Gillbert and Burke, 1912].[45]

18 (11). Head without dermal processes in form of barbels and cirri on upper side (if such present, in form of snout-occipital crest; see *Alectrias cirratus*).

19 (36). Pectoral fins large (less than 2 times in head length) with at least 12 (12–21) rays. Pelvic fins usually present (absent in *Kasatkia* and *Opisthocentrus*).

20 (29). Gill openings continue forward ventrally; branchiostegal membranes narrowly attached to isthmus. Postorbital, occipital, and

*Spelling of author names in taxonomic divisions is sometimes at variance with spelling in text and bibliography since the Israeli orthography has been followed in this translation— General Editor.

[45]Reported off Bering Island.

suborbital canals of head completely reduced (Figure 33). Mandibular pores not more than two (subfamily Lumpeninae).

21 (26). Outer ray of pelvic fins and first ray of anal fin in form of poorly developed spinules.

22 (25). Lower rays of pectoral fins shorter than middle rays. Teeth absent on vomer.

23 (24). Gill openings do not reach forward to vertical from posterior margin of eye. Oral valves (palatine and mandibular) present. Mandibular pores 2. Principal rays of caudal fin 13 (6 + 7). Scales cover entire cheek.................... 8. **Lumpenus** Reinhardt.

52

24 (23). Gill openings continue beyond vertical from middle of eye. Oral valves absent. Mandibular pore single. Principal rays of caudal fin 12 (6 + 6). Scales on cheeks greatly reduced.............
....9. **Anisarchus** Gill.

25 (22). Lower rays of pectoral fins much longer than middle rays. Teeth present on vomer........................ 10. **Leptoclinus** Gill.

26 (21). Outer ray of pelvic fins and at least first two rays of anal fin in form of spines.

27 (28). Scales on head present only on cheeks. Palatines with teeth. Oral valves present .... 11. **Acantholumpenus** Makushok.

28 (27). Scales cover entire head. Oral valves absent ....
.... 12. **Lumpenella** Hubbs.

29 (30). Gill openings do not continue forward ventrally; branchiostegal membranes broadly fused and not attached to isthmus. Postorbital and occipital canals of head developed and suborbital canal, if reduced, only partially so. Mandibular pores at least 3 (subfamily Opisthocentrinae).

30 (31). Branchiostegal rays 6. Interorbital pore 1, preopercular pores 5. Pectoral fins with not more than 13 rays....................
.... 13. **Lumpenopsis** Soldatov.

31 (30). Branchiostegal rays 5. Interorbital pores at least 3 (3–5), preopercular pores 6. Pectoral fins with at least 17 (17–21) rays.

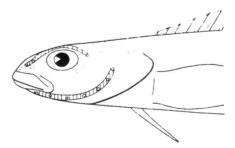

Figure 33. Seismosensory system of head in *Lumpenus fabricii*, lateral view (Makushok, 1958).

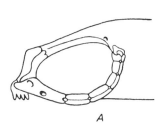
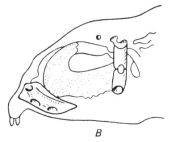

A

B

Figure 34. Suborbital ossicles of Opisthocentrinae (Makushok, 1958).
A–*Kasatkia memorabilis*; B–*Opisthocentrus dybowskii*.

32 (33). Suborbital canal continuous (Figure 34, A). All spiny rays of
dorsal fin equal in thickness and rigid.....................
........................14. **Kasatkia** Soldatov and Pavlenko.

33 (32). Suborbital canal interrupted in middle (Figure 34, B). Spiny rays
of dorsal fin gradually thicken posteriorly, anterior rays flexible.

34 (35). Pelvic fins present.....................15. **Ascoldia** Pavlenko.

35 (34). Pelvic fins absent...................16. **Opisthocentrus** Kner.

36 (19). Pectoral fins small (more than 2 times in head length) or highly
reduced, with not more than 12 (8-12) rays. Pelvic fins absent.

37 (42). Trunk without seismosensory canals; seismosensory papillae
open in form of middle and upper branches. Branchiostegal rays
5. Scales cover only posterior half of body or completely absent.
Spiny rays of dorsal fin anteriorly slender and flexible, gradually
thickening and becoming rigid posteriorly (subfamily Alectriinae).

38 (41). Scales present on posterior part of body. Palatines with teeth.
Dermal crest on head reaches occiput. Suborbital canal well
developed.

39 (40). Branchiostegal membranes not attached to isthmus.........
.........................17. **Alectrias** Jordan and Evermann.

40 (39). Branchiostegal membranes broadly attached to isthmus (gill
openings wide)...................[**Anoplarchus** Gill, 1861].[46]

41 (38). Body completely devoid of scales. Palatines without teeth.
Dermal crest does not reach occiput. Suborbital canal highly
reduced ....18. **Pseudalectrias** Lindberg.

42 (37). Trunk with seismosensory canals. Branchiostegal rays 6. Scales
cover entire body except head (subfamily Xiphisterinae). Entire
body with dense network of seismosensory canals. Snout-
occipital crest absent or rudimentary.......................
...................................19. **Dictyosoma** Schlegel.

[46]Reported from Alaska to San Francisco. Absent in western part of the Pacific Ocean
(Lindberg, 1938).

43 (2). Vertebrae at least 130. Caudal fin completely confluent with dorsal and anal fins, with not more than 10 rays (most probably including procurrent rays also) (Figure 99) (subfamily Eulophiinae) . . . . 20. **Eulophias** H.M. Smith.

44 (1). Pectoral fins absent (subfamily Azygopterinae) . . . .
. . . . [**Azygopterus** Andriashev and Makushok, 1955].[47]

1. Subfamily Stichaeinae

Makushok, Stichaeoidea . . . , 1958: 77.

Body moderately elongate, compressed laterally (anteriorly rounded in several forms), covered with overlapping scales. Head large, naked, without dermal processes. Mouth moderate or large (*Stichaeus grigorjewi*). Vomer and palatines with teeth. Branchiostegal membranes broadly fused in most forms and not attached to isthmus; however, in *Ernogrammus* membranes attached to isthmus and in *Stichaeus punctatus* and *S.* 54 *ochriamkini* narrowly fused with each other, but not attached to isthmus. Nasal pores 2 or 3 (4-5 in *Stichaeopsis ncna*). Pores on posterior part of supraorbital, postorbital, occipital, and posterodorsal part of suborbital canals arranged in 2 rows in most forms (Figure 35, A-E). Preopercular pores 6 (only in *Stichaeopsis nana* 10-12 and in *Ernogrammus* usually 7), mandibular pores 4 or 3 (*Ernogrammus*). The most characteristic peculiarity of this subfamily, manifested in a highly variable form (supraspecific variability), is the structure of the canals in the lateral line of the body and the extent of their anastomoses. Vertebrae 46-60, precaudal vertebrae 13-20. Dorsal fin with stiff spines. Anal fin with 1, often 2 spines at origin. Principal rays of caudal fin 12-14. Membrane of anal fin only touches caudal fin; membrane of dorsal fin continues slightly onto base. Pectoral fins large, with 14-18 rays. Pelvic fins well developed (I 4; I 3), their soft rays, like soft rays of other fins, divide 2-3 times.

Most species distributed along the Asian coast of northern part of the Pacific Ocean (mainly in the Sea of Japan and the Sea of Okhotsk).[48] 55 Usually found on stony-pebbled bottom in the littoral zone with clumps of fucoids and laminaria, up to depths of 100 to 200 m. Feed on minute benthic invertebrates, mostly crustaceans.[49]

5 genera. 3 in the Sea of Japan and 1 in adjacent waters.

1. Genus *Stichaeus* Reinhardt, 1837

Stichaeus Reinhardt, Dansk. Vidensk. Selsk. Nat.-Math. Abhandl..

---

[47]Reported from Alaska to San Francisco. Absent in western part of the Pacific Ocean.

[48]Ranges of *Stichaeus punctatus* and *Eumesogrammus praecisus* extend even to the western part of the Atlantic Ocean.

[49]Except for *Stichaeus grigorjewi*; no doubt this species became predaceous.

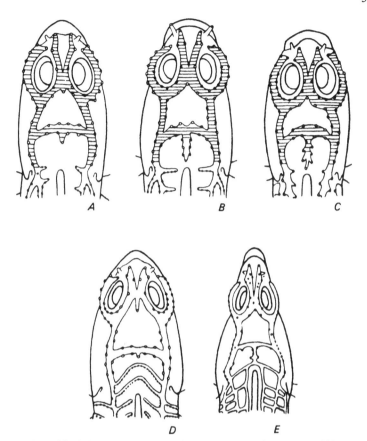

Figure 35. Seismosensory system of head, dorsal view (Makushok, 1958).
A—*Eumesogrammus praecisus*; B—*Stichaeopsis epallax*; C—*Ernogrammus hexagrammus*; D—*Stichaeopsis nevelskoi*; E—*Stichaeopsis nana*.

1837: 109 (type: *Blennius punctatus* Fabricius). Jordan and Evermann, Fish. N. and M. Amer., 3, 1898: 2439. Jordan and Snyder, Proc. U.S. Nat. Mus., 25, 1902: 495. Soldatov and Lindberg, Obzor..., 1930: 467. Andriyashev, Ryby Severnykh Morei SSSR, 1954: 229. Matsubara, Fish Morphol. and Hierar., 1955: 766. Fowler, Synopsis..., 1958: 281. Makushok, Stichaeoidea..., 1958: 78 (comparative notes).

Dinogunellus Herzenstein, Mélanges Biologiques..., 13, 1, 1890: 119. Jordan and Snyder, Proc. U.S. Nat. Mus., 25, 1902: 497. Soldatov and Lindberg, Obzor..., 1930: 469.

Body moderately elongate, compressed laterally, covered with minute cycloid scales, head naked. Teeth on jaws, vomer, and palatines.[50] Mouth

[50]Teeth of upper jaw arranged in two to four rows, of lower jaw in single row, which bifurcates near symphysis (Makushok, 1958: 15).

conical,[51] jaws equal, or lower jaw protrudes slightly forward. Head with numerous regularly arranged pores in canals of seismosensory system. "Lateral line" represented by upper seismosensory canal originating from postorbital canal and extending to caudal part of body, not reaching caudal base. Posterior part of anal fin without spiny rays. Opercular siphon present (Figure 27, A). Branchiostegal rays 6.[52]

4 species. All known from the Sea of Japan.

Key to Species of Genus Stichaeus[53]

1 (4). Seismosensory canal of body with one row of pores; each scale of "lateral line" with single pore.

2 (3). Number of pores in seismosensory canal 72–86. Dorsal fin with 24–30 rays anterior to vertical from last pore of canal. Lower part of head usually with pattern... [**S. punctatus punctatus** (Fabricius)].

3 (2). Number of pores in seismosensory canal 60–73. Dorsal fin with 21–23 rays anterior to vertical from last pore of canal. Lower part of head usually without pattern.... 1. **S. p. pulcherrimus** Taranetz.

4 (1). Seismosensory canal of body with two rows of pores; each scale of "lateral line" with two pores.

5 (6). Dorsal fin with 52–56 rays; anal fin with 42–45 rays. Upper lip very fleshy. Suborbital seismosensory canal with 6 pores .... 2. **S. grigorjewi** Herzenstein.

6 (5). Dorsal fin with less than 52 rays; anal fin with less than 42 rays. Upper lip not fleshy. Suborbital seismosensory canal with 2–3 pores.

7 (8). At least 15 rays in dorsal fin posterior to vertical from last pore of seismosensory canal of trunk. Branchiostegal membranes not attached to isthmus, narrowly fused with each other. Dorsal fin with 3 or 4 roundish dark blotches .... 3. **S. ochriamkini** Taranetz.

8 (7). Less than 15 rays in dorsal fin posterior to vertical from last pore of seismosensory canal of trunk; branchiostegal membranes not attached to isthmus, broadly fused with each other. Dark blotches not present on dorsal fin .... 4. **S. nozawae** Jordan and Snyder.

[*Stichaeus punctatus punctatus* (Fabricius, 1780)—Arctic Shanny]
(Figure 36)

Blennius punctatus Fabricius, Fauna Greenland, 1780: 153 (western Greenland).

Stichaeus punctatus, Shmidt, Ryby Vostochnykh Morei..., 1904: 188

[51]Except for *S. grigorjewi*, in which mouth semisuperior.

[52]Genus *Stichaeus* is the least specialized in the family Stichaeidae (Andriyashev, 1954; Makushok, 1958, 1961b).

[53]From Taranetz (1937: 156), with some modifications.

(comparative characters). Taranetz, Kratkii Opredelitel'..., 1937: 156. Shmidt, Ryby Okhotsokogo Morya, 1950: 68 (characters and confirmation of characters). Andriyashev, Ryby Severnykh Morei SSSR, 1954: 230, fig. 117.

D XLVI-XLIX; A I-II 32-35; principal rays of caudal fin 7 + 6-7; P 15-16; V I 4; vertebrae 51-55; precaudal 14-16, caudal 36-40 (Makushok, 1958: 120).

Specimens from the Bering Sea with 72-86 pores in seismosensory canal of trunk. Number of rays in dorsal fin anterior to last pore of canal 24-30; distance from this pore to caudal fin 23-36% of standard length. Interorbital space 31-38% of eye diameter. Lower part of head usually with pattern (Taranetz, 1935: 96).

D XLVI-L; A I-II, 35-36; P 14-16 (Shmidt, 1904; 189). Shmidt compared Pacific specimens with those collected from the Arctic Ocean and northern part of the Atlantic Ocean and showed that there were no significant differences between them.

Color of five fish bright red, sides darkened with diffuse brownish spots; head with series of sinuous brown lines on opercle and subopercle, and 3-4 sinuous stripes from eye downward to chin. Arrangement and number of sinuous lines on head not similar. Dorsal fin reddish, darkened, with gray diffuse dots, and with 5 ocellar black or bluish tinged spots in which yellow spot occurs in posterior part. Spots almost equidistant. Anal fin reddish, with about 10-14 diffuse black spots. Pectoral fins with minute brown dots forming indistinct bands (Shmidt, 1904: 189).

Makushok (1958, 1961b), like Andriyashev (1954), considers *S. p. punctatus* the most primitive member of the genus, characterized by: position of narrowly fused branchiostegal membranes which are not attached to isthmus; presence of siphon on operculum (noted for this species by Shmidt in 1950); distinct thickening of spiny rays from anterior toward posterior end of dorsal fin (all rays of this fin rigid); presence of one or two poorly developed spiny rays at origin of anal fin; pelvic fin with single spine and 4 soft rays; caudal fin with 13-14 principal rays; branchiostegal rays six; relatively large eyes; terminal mouth; and single seismosensory canal which continues slightly beyond midpoint of standard length.

Distribution: Unknown in the Sea of Japan. Described from the west coast of Greenland. Known in the Chukchi Sea (Cape Lisfurne). In the Bering Sea found along both coasts. Along the Pacific coast of North America continues to the Prince of Wales Island (56°N) (Andriyashev, 1954: 231).

1. *Stichaeus punctatus pulcherrimus* Taranetz, 1935 (Figure 37)
   Stichaeus punctatus pulcherrimus Taranetz, Vestn. Dal'nevost. Fil.

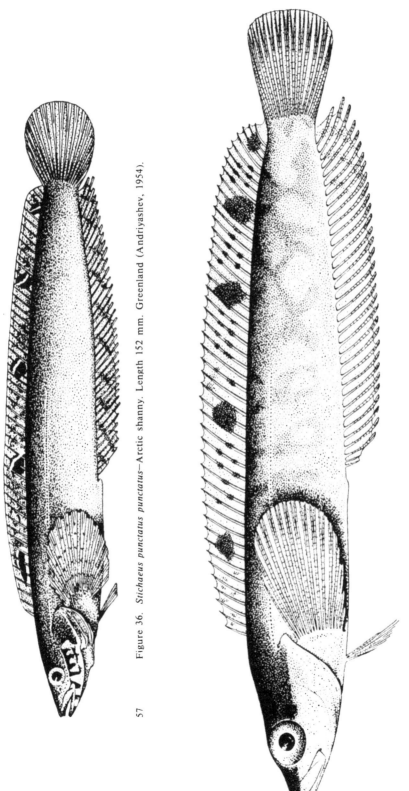

Figure 36. *Stichaeus punctatus punctatus*—Arctic shanny. Length 152 mm. Greenland (Andriyashev, 1954).

57

Figure 37. *Stichaeus punctatus pulcherrimus*. Length 162 mm. No. 35166. Sea of Okhotsk.

57

Akad. Nauk SSSR, 13, 1935: 96 (Sea of Okhotsk); Kratkii Opredelitel'...,
1937: 156. Shmidt, Ryby Okhotskogo Morya, 1950: 68 (comparative
characters).

Stichaeus punctatus, Shmidt, 1904: 188-190 (in part). Pavlenko, Ryby
Zaliva Petr Velikii, 1910: 52. Soldatov and Lindberg, Obzor..., 1930: 468.
Matsubara, Fish Morphol. and Hierar., 1955: 766. Fowler, Synopsis...,
1958: 282.

29480. Sea of Japan, Peter the Great Bay. 1911. V.K. Soldatov. 2
specimens.

35166. Sea of Okhotsk, Ayan Bay. August 3, 1916. GEVO. 11
specimens.

58 39865. Sea of Okhotsk, Terpenia Gulf. September 8, 1947. KSE. G.U.
Lindberg. 1 specimen.

39866. Šiaškotan Island. September 13, 1949. KSE. B.E. Bykhovskii. 2
specimens.

40282. Sea of Okhotsk, Penzha Bay, Yama Bay. September 1, 1908.
F.Ya. Derbek. 1 specimen.

40283. Sea of Okhotsk, coast of western Kamchatka. August 14, 1914.
GEVO. 2 specimens.

D XLV-XLIX; A 32-35 (Soldatov and Lindberg, 1930: 468).

Seismosensory canal of trunk opens through 60-73 pores. Dorsal fin
with 21-23 rays anterior to vertical from last pore of canal. Distance from
this pore to base of caudal fin 36-42% of standard length. Interorbital
space 21-24% of diameter of eye. Lower part of head usually without
pattern. Length 117 mm (Taranetz, 1935: 96).

Shmidt (1950: 68) reported that in addition to the characters given by
A.Ya. Taranetz, he noted the following: "Number of pores in the lateral
line 60-76 (in *S. p. punctatus*, 77-88); head length 24.2-26.0% of standard
length (in *S. p. punctatus* 20-22.6%); body depth near anus 13.5-16.3% of
standard length (in *S. p. punctatus* 12.7-13.3%). Body deeper and less
elongate than in *S. p. punctatus*. Lower lip with fairly broad gap in middle;
in anterior half broader but highly attenuate toward corner of mouth.
Upper lip with small notch in middle." Shmidt (1904: 189) also reported
that the coloration of live specimens of *S. p. pulcherrimus* from Peter the
Great Bay distinctly differs from that in specimens of *S. p. punctatus*
caught in northern regions of the Pacific and Atlantic oceans: "Sides dark
with diffuse brownish spots. Dorsal fin with 4 bluish spots, second spot
among 5 absent.[54] Head without sinuous lines; entire head sharply
divided, however, by horizontal line through lower margin of pupil into
two parts—brownish above this line and whitish below it, without a single

---

[54]Soldatov and Lindberg (1930: 468) reported: "Young fish with five dark oval spots on the
dorsal fin; in adults these are indistinct or disappear."

58

dark spot." The considerable data obtained by DVE and TONS (Soldatov and Lindberg, 1930: 468), as well as our data (KSE) reveal that these fish dwell on a highly variable bottom: stone, sand, silt, stones with shells, and silt with shells. An analysis of our data concerning the major characters of the subspecies is presented in Table 1. Morphological parameters which 59 are not characteristic of *S. p. pulcherrimus* and bring these specimens close to the type subspecies (*S. p. punctatus*) have been italicized in the Table. In all likelihood, representatives of *S. p. pulcherrimus* found in the southern part of the Sea of Okhotsk (Nos. 39865, 39866) and in Peter the Great Bay are intermediate between the sharply distinct subspecies taken from the extreme northern regions of their area of distribution. The ecological conditions of Terpenia Gulf and Šiaškotan Island are probably such that intermediate forms could appear there as described by Lindberg and Andriyashev (1938) for the subspecies *Icelus spiniger intermedius* under the conditions of the western part of the Bering Sea and in the Sea of Okhotsk near the southwestern coast of Kamchatka. A study of this phenomenon would be of considerable zoogeographic interest.

TABLE 1

No.	Number of spec.	Locality	Number of pores in seismosensory canal of trunk	Number of rays in dorsal fin anterior to vertical from last pore of seismosensory canal of trunk
35166	11	Sea of Okhotsk; Ayan Gulf	58–70	24-27 in 5 specimens 20-30 in 6 specimens
29480	2	Sea of Japan, Peter the Great Bay	69–69 [*sic*]	22-*30*
39865	1	Sea of Okhotsk, Terpenia Gulf	*86*	20
39866	2	Šiaškotan Island	*81–86*	21–22

Distribution: In the Sea of Japan known from Peter the Great Bay and Tatar Strait. Sea of Okhotsk, Shantar Islands, from Nikolai Bay to Ayan Bay (collections of DVE and TONS), Terpenia Gulf (No. 39865), and Šiaškotan Island (No. 39866).

2. *Stichaeus grigorjewi* Herzenstein, 1890—Grigorev's Prickleback (Figure 38)

Stichaeus grigorjewi Herzenstein, Mélanges Biologiques..., , 13, 1, 1890: 119 (Volcano Bay, Hokkaido). Taranetz, Dokl. Akad. Nauk SSSR, 1, 3, 1936: 144; Kratkii Opredelitel'..., 1937: 157. Makushok, Stichaeoidea..., 1958: 60 (subgenus *Dinogunellus* Herzenstein). Matsubara, Fish Morphol. and Hierar., 1955:·766.

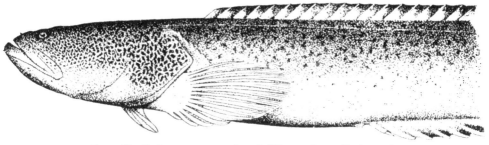

Figure 38. *Stichaeus grigorjewi*. Length 500 mm. Japan (Jordan and Snyder, 1902a).

Stichaeus elongatus Sakamoto, J. Imp. Fish. Inst., 26, 1, 1930: 15, fig. 1.

Dinogunellus grigorjewi, Jordan and Snyder, Proc. U.S. Nat. Mus., 25, 1902: 497, fig. 27. Shmidt, Ryby Vostochnykh Morei..., 1904: 191. Pavlenko, Ryby Zaliva Petr Velikii, 1910: 52. Soldatov and Lindberg, Obzor..., 1930: 469.

22187. Sea of Japan, Ussuriisk Bay. September 15, 1927. E.P. Rutenberg. 1 specimen.

37007. Kuril Islands, Iturup Island. August 15, 1953. V.M. Makushok. 1 specimen.

39867. Kuril Islands. Iturup Island. September 4, 1948. KSE. Semenov. 3 specimens.

39868. Kuril Islands. September 24, 1948. KSE. G.U. Lindberg. 1 specimen.

39869. South Kuril Strait. September 3, 1948. KSE. G.U. Lindberg. 1 specimen.

39870. Sea of Japan, Tatar Strait. August 12, 1949. KSE. G.U. Lindberg. 1 specimen.

40050. Sea of Japan, Peter the Great Bay. May 15, 1970. V.V. Fedorov. 1 specimen.

D LIII–LVI; A I 41–43; principal rays of caudal fin 6 + 6–7; P 14; V I 3; vertebrae 58–61: precaudal 15–16, caudal 42–45 (Makushok, 1958: 12).

Body depth $9\frac{1}{2}$–$7\frac{5}{8}$ times in total length. Head highly depressed dorsally, length* $2\frac{1}{4}$ and width $1\frac{7}{8}$–$1\frac{4}{7}$ times in head length and $4\frac{1}{2}$–$4\frac{1}{3}$ times in body length. Eyes situated on top of head and their diameter 15 times in the head length, 11–$11\frac{1}{2}$ times in the postorbital space and $1\frac{2}{3}$–$1\frac{4}{7}$ in the interorbital space. Mouth very large when open; upper jaw extends distinctly beyond vertical from posterior margin of eye. Vomer and palatines with well-developed teeth. Lateral line consists of paired pores and continues from upper end of gill slit along dorsum, disappearing toward posterior end of body (Herzenstein, 1890: 119).

*Error in Russian text; should read "depth"—General Editor.

Body covered with minute scales. Head naked. Lower jaw protrudes. All rays of dorsal fin spiny; length of middle rays equal to snout length (Soldatov and Lindberg, 1930: 469). Anal fin with one small spine.

According to the description given by Shmidt (1904: 191), specimens from Peter and Great Bay and the eastern coast of Sakhalin are distinguished by a smaller number of spiny rays in the dorsal fin (LII versus LIV—LVI) and arrangement and larger size of teeth on lower jaw.

Lower jaw with 2 canines anterior to group of symphysial teeth (Figure 39, A, a). These canines are notably shifted forward and protrude when mouth is closed. Body color in live fish chocolate-brown to gray, dorsal side densely covered with brown to almost black spots (Shmidt, 1904: 191).

In our 10 specimens (113-400 mm in length) D LII-LIII; A I 42; diameter of eye 9-10.8% of head length. These fish were caught in the South Kuril and Tatar Strait at depths of 51-83 m on a silty bottom. The large collections of DVE and TONS (Soldatov and Lindberg, 1930: 469) indicate that this species dwells on other types of bottoms as well (pure silt, sand, stones). Catches of *S. grigorjewi* also contained many flatfishes

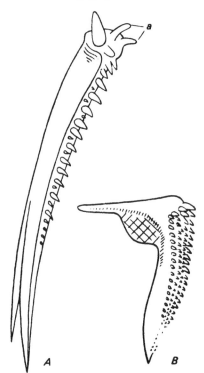

Figure 39. Jaw teeth of *Stichaeus grigorjewi*. (Makushok, 1958).
A—lower jaw; a—canines; B—upper jaw.

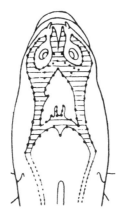

Figure 40. Seismosensory system of head of *Stichaeus grigorjewi*, dorsal view (Makushok, 1958).

(*Limanda aspera, Hippoglossoides elassodon robustus, Pleuronectes quadrituberculatus, Acanthopsetta nadeshnyi*), cod, eelpout (*Lycodes tanakae*), great sculpin (*Myoxocephalus polyacanthocephalus*) and antlered sculpin (*Enophrys diceraus*).

This species is the only predaceous representative of the family and exhibits, as pointed out by Makushok (1958), several structural features: upper position of the eyes, large superior mouth, branchiostegal membranes broadly fused, and occurrence of 2 rows of pores in the preorbital canal (Figure 40).

Found to a depth of 150 m (Sakamoto, 1930). Consumed by man (Abe, 1958: III).

Length, to 510 mm (Soldatov and Lindberg, 1930).

Distribution: In the Sea of Japan known from Pusan and Chongjin (Mori, 1952: 128); Peter the Great Bay and farther north to De-Kastri Bay (Shmidt, 1904: 192); reported from western Sakhalin coast near Shirokaya Pad' (Taranetz, 1937b, 157); known off Hokkaido (Matsubara, 1955: 766); Sado Island (Honma, 1952: 226); Toyama Bay (Sakamoto, 1930: 15); and San'in region (Mori, 1956a: 21). In the Sea of Okhotsk found near Iturup Island and in the South Kuril Strait (collections of KSE). Along the Pacific coast of Japan found in Volcano Bay (Sato and Kobayashi, 1956: 14) and farther south to Tokyo (Jordan and Snyder, 1902a: 497). In the Yellow Sea recorded from Namp'o (Mori, 1952: 128). China (Fowler, 1958: 283).

3. ***Stichaeus ochriamkini*** Taranetz, 1935—Okhryamkin's Prickleback (Figure 41)

Stichaeus ochriamkini Taranetz, Vestn. Dal'nevost. Fil. Akad. Nauk SSSR, 13, 1935: 96 (Sea of Japan); Dokl. Akad. Nauk SSSR, 1, 3, 1936: 144; Kratkii Opredelitel'..., 1937: 157. Shmidt, Ryby Okhotskogo Morya, 1950: 68.

18730. Sea of Japan, Peter the Great Bay. October 16, 1912. 1 specimen.

31675. Western coast of Sakhalin, Antonovo. September, 1946. B.E. Bykhovskii. 1 specimen.

36847. Sea of Japan, Peter the Great Bay. November 20, 1925. 6 specimens.

39824. Sea of Okhotsk, Aniva Bay. September 23, 1947. KSE. G.U. Lindberg. 1 specimen.

39825. Sea of Okhotsk, Aniva Bay. August 3, 1947. KSE. O.A. Skarlato. 1 specimen.

39826. Sea of Okhotsk, Aniva Bay. 1947. KSE. M.N. Politskii. 3 specimens.

39827. Sea of Japan, Moneron Island. August 18, 1948. KSE. G.B. Semenova. 1 specimen.

39828. Tatar Strait. August 24, 1949. KSE. G.U. Lindberg. 4 specimens.

39829. Tatar Strait. August 24, 1949. KSE. G.U. Lindberg. 1 specimen.

39830. Tatar Strait. August 28, 1949. KSE. G.U. Lindberg. 1 specimen.

39831. Tatar Strait. August 23, 1949. KSE. G.U. Lindberg. 2 specimens.

39832. Tatar Strait. August 19, 1949. KSE. G.U. Lindberg. 1 specimen.

39833. Šiaškotan Island. September 10, 1949. KSE. G.U. Lindberg. 1 specimen.

39874. West coast of Sakhalin. September 27, 1948. KSE. G.B. Semenova. 1 specimen.

40285. West coast of Sakhalin, Aleksandrovsk. September 20, 1933. A. Kuznetsov. 1 specimen.

40307. West coast of Sakhalin. September 27, 1948. KSE. 1 specimen.

D XLVI-XLVIII; A I-II 32-34; principal rays of caudal fin 7 + 6-7; P 14-15; V I 4; vertebrae 50-53: precaudal 14-15, caudal 37-39 (Makushok, 1958: 120). D XLV-XLVIII; A 33-37; P 14-15. Dorsal fin with 15-21 rays posterior to vertical from end of lateral line. Head length 20.2-24.5% and body depth above origin of base of anal fin 11.5-14.1% of standard length. Diameter of eye 21.3-29.3% of head length. Interorbital space 20-33% (41%) of diameter of eye. Lateral line consists of two rows of pores. Lower part of head without pattern. Dorsal fin with one spot in anterior part and three spots in posterior part (Taranetz, 1935).

Very similar to *S. punctatus* but differs in structure of lateral line (2 rows of pores versus one), and from *S. nozawae* in number of rays of dorsal fin posterior to vertical from end of lateral line (15-21 versus 9-13), narrow interorbital space, shape of head, and presence of spot on dorsal fin.

Makushok (1958) reported that, like *S. punctatus*, this species is also primitive compared to other members of the genus. Opercular siphon present. Branchiostegal rays 6. Anal fin origin with 1-2 poorly developed spines. Branchiostegal membranes narrowly fused and not attached to isthmus.

63 Our specimens (30-130 mm in length): D XLVI-XLVII; A I 32-34; dorsal fin with 16-22 rays posterior to vertical from end of lateral line. Head length 20-28% of standard length. In most specimens arrangement of spots on dorsal fin similar to description by Taranetz (1953), but some with different arrangement, of these spots: two rounded dark spots in anterior part of fin and three in posterior part, or one spot in anterior part of fin and four spots in posterior part, or only three spots on dorsal fin. Often a row of minute dark spots extends along base of dorsal fin.

Fish in the collections of the Kuril-Sakhalin Expedition were more often found on a sandy bottom, sometimes slightly silted or with pebbles, gravel, and stones; often in thickets of red algae; and occasionally in clumps of *Zostera*. Catches of *S. ochriamkini* also contained a large number of flounders and walleye pollock, as well as cod and Jordan's sculpin (*Triglops jordani*). Isolated specimens of herring, *Lycodes palearis fasciatus, Enophrys diceraus,* and *Podothecus* were also found. Among invertebrates, catches included a large number of shrimps (*Pandalus* and *Spirontocaris*), often sea urchins (*Strongylocentrotus pulchellus* and *S. droebachiensis*), sea cucumbers (*Cucumaria japonica*), and starfishes (*Asterias amurensis*). Sometimes crab (*Paralithodes camtschatica*) was also present. Animals recovered from clumps of red algae were likewise red in color. Temperature of water near bottom at sites of catches ranged from 2.0 to 3.5°C, rarely 13°C.

Length, 133 mm.

Distribution: In the Sea of Japan known from Peter the Great Bay and up to the northern part of Tatar Strait, near the west coast of Sakhalin and the west coast of Moneron Island. In the Sea of Okhotsk found in Aniva Bay and the South Kuril Strait (our specimens).

4. *Stichaeus nozawae* Jordan and Snyder, 1902—Nozawa's Prickleback (Figure 42)

Stichaeus nozawae Jordan and Snyder, Proc. U.S. Nat. Mus., 25, 1902: 496, fig. 26 (Otaru, Hokkaido). Soldatov and Lindberg, Obzor..., 1930: 468. Taranetz, Dokl. Akad. Nauk SSSR, 1, 3 (80), 1936: 144; Kratkii Opredelitel'..., 1937: 157. Matsubara, Fish Morphol. and Hierar., 1955: 766. Fowler, Synopsis..., 1958: 284 (synonymy).

39806. South Kuril Strait. September 3, 1948, KSE. G.B. Semenova. 2 specimens.

39807. South Kuril Strait. September 2, 1948. KSE G.B. Semenova. 1 specimen.

39808. Tatar Strait. September 24, 1948. KSE. G.U. Lindberg. 4 specimens.

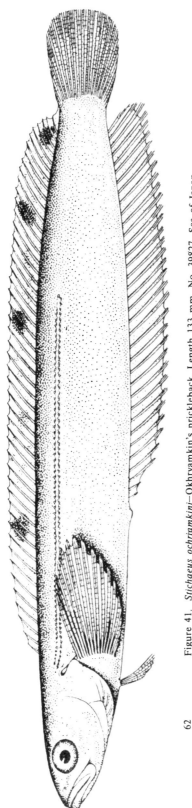

Figure 41. *Stichaeus ochriamkini*—Okhryamkin's prickleback. Length 133 mm. No. 39827. Sea of Japan.

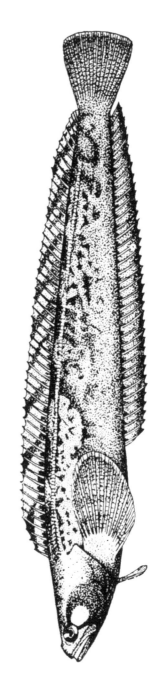

Figure 42. *Stichaeus nozawae*—Nozawa's prickleback. Length 255 mm. Japan (Jordan and Snyder, 1902).

39809. Sea of Okhotsk, Aniva Bay. August 30, 1947. KSE. G.U. Lindberg. 1 specimen.

39810. Sea of Okhotsk, Aniva Bay. August 17, 1947. KSE. G.U. Lindberg. 1 specimen.

39811. Tatar Strait. August 23, 1949. KSE. G.U. Lindberg and M I Legeza. 1 specimen.

39812. Tatar Strait. August 28, 1949. KSE. G.U. Lindberg and M.I. Legeza. 1 specimen.

39813. Tatar Strait. August 26, 1949. KSE. G.U. Lindberg and M.I. Legeza. 2 specimens.

39814. Tatar Strait. August 26, 1949. KSE. G.U. Lindberg. 1 specimen.

39815. Tatar Strait. August 23, 1949. KSE. G.U. Lindberg and M.I. Legeza. 3 specimens.

39816. Tatar Strait. August 26, 1949. KSE. G.U. Lindberg and M.I. Legeza. 1 specimen.

39817. Tatar Strait. August 18, 1949. KSE. G.U. Lindberg and M.I. Legeza. 2 specimens.

39818. South Kuril Strait. September 19, 1949. KSE. G.U. Lindberg and M.I. Legeza. 2 specimens.

39819. Sea of Okhotsk, east coast of Sakhalin. September 30, 1949. KSE. G.U. Lindberg and M.I. Legeza. 1 specimen.

39820. Šiaškotan Island. September 18, 1949. KSE. G.U. Lindberg and M.I. Legeza. 2 specimens.

39821. Kunashir Island. September 18, 1949. KSE. G.U. Lindberg and M.I. Legeza. 2 specimens.

39822. South Kuril Strait. September 18, 1949. KSE. G.U. Lindberg and M.I. Legeza. 3 specimens.

39823. Šiaškotan Island. September 14, 1949. KSE. E.F. Gur'yanova. 1 specimen.

39790. Sea of Okhotsk, Aniva Bay. September 21, 1947. KSE. Z.I. Petrova. 1 specimen.

39791. Tatar Strait. August 24, 1949. KSE. G.U. Lindberg. 1 specimen.

39792. Tatar Strait. August 14, 1949. KSE. M.I. Legeza. 1 specimen.

39793. Sea of Okhotsk, Aniva Bay. August 15, 1947. KSE. Z.I. Petrova. 1 specimen.

39794. West coast of Sakhalin, Kholmsk. August 4, 1947. KSE. G.U. Lindberg. 2 specimens.

64 39795. Sea of Okhotsk, Aniva Bay. July 28, 1947. KSE. G.U. Lindberg. 17 specimens.

39796. Tatar Strait, Antonovo. August 6, 1947. KSE. G.U. Lindberg. 1 specimen.

39797. Sea of Okhotsk, Aniva Bay. September 14, 1948. KSE. G.U. Lindberg. 1 specimen.

39798. Sea of Okhotsk, Aniva Bay. September 13, 1948. KSE. G.U. Lindberg. 1 specimen.

39799. La Perouse Strait. September 13, 1948. KSE. G.U. Lindberg. 1 specimen.

39800. South Kuril Strait. September 4, 1948. KSE. G.U. Lindberg. 1 specimen.

39801. Tatar Strait. September 30, 1948. KSE. G.U. Lindberg. 1 specimen.

39802. La Perouse Strait. September 13, 1948. KSE. G.B. Semenova. 2 specimens.

39803. South Kuril Strait. September 3, 1948. KSE. G.B. Semenova. 1 specimen.

39804. La Perouse Strait. August 24, 1948. KSE. G.B. Semenova. 5 specimens.

39805. Sea of Okhotsk, Aniva Bay. September 13, 1948. KSE. G.B. Semenova. 3 specimens.

39806-7. South Kuril Strait. September 2-3, 1948. KSE. 3 specimens.

39808. Tatar Strait. September 24, 1948. KSE. 4 specimens.

39809-10. Sea of Okhotsk, Aniva Bay. August, 1947. KSE. 2 specimens.

40046. Sea of Japan. November 20, 1925. Expedition ship *Gidrolog.* 1 specimen.

40048. Sea of Japan, Peter the Great Bay. May 24, 1960. V.V. Barsukov. 2 specimens.

40049. Sea of Japan, Pos'et Bay. 1967. 1 specimen.

40286. Sea of Japan, Peter the Great Bay. November 16, 1925. Hydrographic ship *Gidrograf.* 1 specimen.

40294. Sea of Japan. June 17, 1970. Fishing trawler *Milogradovo.* V.V. Fedorov. 2 specimens.

40295. South Kuril Strait. September 3, 1948. KSE. G.B. Semenova. 1 specimen.

40296. Sea of Okhotsk, Aniva Bay. August 6, 1947. KSE. G.U. Lindberg. 1 specimen.

40297. Sea of Japan. June 17, 1970. Fishing trawler *Milogradovo.* V.V. Fedrov. 4 specimens.

20498. Sea of Okhotsk, Aniva Bay. July, 1947. Z.I. Petrova. 1 specimen.

D XLIII-LI; A I-II 33-37; principal rays of caudal fin 6 + 6-7; P 14-15; V I 3; vertebrae 48-55: precaudal 14-16, caudal 34-41 (Makushok, 1958: 120).

D LI; A I 37. Head length 5 times and maximum body depth $6\frac{2}{3}$ times in standard length. Depth of caudal peduncle $3\frac{2}{3}$ times, snout $5\frac{1}{2}$ times, longitudinal diameter of eye $5\frac{1}{2}$ times, and interorbital space 10 times in head length.

Body compressed laterally, head usually small and pointed. Eyes large, set in upper anterior part on sides of head, interorbital space convex, suborbital region narrow. Lower jaw protrudes slightly forward beyond upper jaw, lips thin, maxilla extends beyond vertical from posterior margin of eye. Teeth on jaws form narrow bands, outer teeth slightly enlarged; tip of each jaw with 2 canines which are stronger on lower jaw; vomer and palatines with narrow band of minute teeth. Gill openings V-shaped; branchiostegal membranes form fold across isthmus. Pseudo-branch large. Gill rakers short, 3 + 9. Nostrils with small tubules. Head without dermal processes. Body covered with minute smooth scales, membrane of dorsal and base of caudal fins covered with minute scales; head naked. Lateral line simple, extends from upper corner of gill opening along upper part of body and terminates near caudal fin; pores of lateral line arranged in 2 rows. Dorsal fin originates above gill opening and terminates without fusing with caudal fin. Anal fin originates at vertical from 14th ray of dorsal fin. Caudal fin slightly rounded, its length $1\frac{1}{2}$ times in head length. Pectoral fins rounded lower rays shorter than upper ones. Pelvic fins small, pointed, 3 times in head length (Jordan and Snyder, 1902a).

Seismosensory system of head (Figure 43, A) differs slightly from that of *S. punctatus* (Figure 43, B) and *S. grigorjewi* (Figure 40).

In the 83 specimens examined by us (length 46 to 596 mm) from the collections of the Kuril-Sakhalin Expedition: D XLI–LI; A I 31–39; dorsal fin with 5 to 12 rays posterior to vertical from end of lateral line; ratio of diameter of eye to head length 17–23%. Our material was collected from various depths (6.5 to 118 m). Bottom at sites of catches varied—sandy with gravel, pebbles and gravel, rarely silted sand, silt and argillaceous silt.

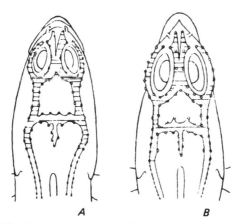

A   B

65   Figure 43. Seismosensory system of head, dorsal view (Makushok, 1958).
A—*Stichaeus nozawae*; B—*Stichaeus punctatus*.

68

65 Some specimens were collected from clumps of red algae. All catches of
S. nozawae contained an abundance of invertebrates as well as various fish
species. Predominant fishes were flounders (*Lepidopsetta, Acanthopsetta,
Limanda, Glyptocephalus,* etc.), *Podothecus gilberti, Triglops jordani,
Enophrys diceraus,* and *Hemilepidotus gilberti.* Frequent among the
invertebrates were starfishes (*Asterias amurensis*), mollusks (*Leda*), rarely
brittle stars, sea urchins, hydroids, and bryozoans. Water temperature at
bottom in places of catches ranged from 2 to 12°C.

Length, to 600 mm.

Distribution: In the Sea of Japan known from Peter the Great Bay,
Pos'et Bay, Tatar Strait, west coast of Sakhalin, and La Perouse Strait.
Reported from Otaru, Hokkaido (Jordan and Snyder, 1902a: 496). In the
Sea of Okhotsk found near east coast of Sakhalin, Aniva Bay, South Kuril
Strait, and off Kunashir and Šiaškotan islands (our specimens).

2. [Genus *Eumesogrammus* Gill, 1864]

Eumesogrammus Gill, Proc. Acad. Nat. Sci., Philad., 16, 1864: 209
(type: *Clinus praecisus* Kröyer). Andriyashev, Ryby Severnykh Morei
SSSR, 1954: 231 (synonymy, description). Matsubara, Fish Morphol. and
Hierar., 1955: 765. Fowler, Synopsis..., 1958: 284 (description).

Close to the genus *Stichaeus*, but differs in presence of 4 branches of
lateral line (Figure 44, A, B) and structure of anal fin, in which last 2-3

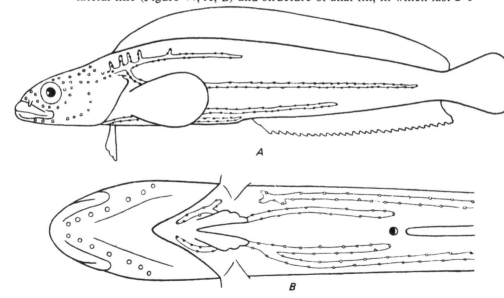

Figure 44. Seismosensory system of *Eumesogrammus praecisus*
(Makushok, 1961).

A—lateral view; B—ventral view.

rays convert into short, spiny rays. One amphiboreal species (Andri-yashev, 1954).

66 1. [*Eumesogrammus praecisus* (Kröyer, 1836)] (Figure 45)

Clinus praecisus Kröyer, Naturhist. Tidsskr., 1, 1836: 25 (Greenland).

Ernogrammus storoshi, Shmidt, Ryby Vostochnykh Morei..., 1904: 193 (eastern Sakhalin). Soldatov and Lindberg, Obzor..., 1930: 464.

Eumesogrammus praecisus, Shmidt, Ryby Okhotskogo Morya, 1950: 67, fig. 2. Andriyashev, Ryby Morei SSSR, 1954: 231, fig. 119. Matsubara, Fish Morphol. and Hierar., 1955: 765. Fowler, Synopsis..., 1958: 285.

39875. Sea of Okhotsk, Terpenia Peninsula, Nizmennaya Inlet. September 9, 1947. KSE. G.U. Lindberg. 2 specimens.

39876. Sea of Okhotsk, Terpenia Peninsula, Nizmennaya Inlet. September 9, 1947. KSE. G.U. Lindberg. 2 specimens.

39877. Sea of Okhotsk, east of Mordvinov Bay. September 4, 1947. KSE. G.U. Lindberg. 2 specimens.

D XLVII-XLIX; A II 29-30 II-III; P 18; V I 3; vertebrae 50-52 (15-16 + 34-36) (Makushok, 1958: 120).

Seismosensory canals of body 4 (Figure 44, A, B). In addition to upper and middle canals, lower and ventral canals also present which usually anastomose at base of pectoral fins; ventral canals on either side of body not connected with each other and terminate blindly before base of pelvic fins near anal opening; few vertical branches originate from upper canal and extend toward base of dorsal fin (Makushok, 1961b). Upper part of gill opening with siphon formed from dermal process of operculum and short fold of trunk. Process of operculum closes siphon on lower side and closely situated above pectoral fin (Shmidt, 1950).

Body color in live fish chocolate-brown to gray with darker and often indistinct bands or spots. Head dark; dark stripe extends from eye obliquely downward and backward and is bordered by two light-colored stripes. Dorsal fin with black ocellar spot. Other fins dark, with light-colored border.

Found at depths of 40-142 m, on stony-pebbled bottom, at positive temperatures close to zero (often 1-2°C), and salinity 32-33°/₀₀ (Andri-yashev, 1954).

In the 6 specimens examined by us (length 53 to 158 mm) from the collections of the Kuril-Sakhalin Expedition: D XLVII-XLVIII; A II 30 II-III; P 18. These fish were caught from a sandy bottom, sometimes with an admixture of pebbles, gravel, and stones, sometimes slightly silted. In catches of *E. praecisus* stones were often scooped up with hydroids, bryozoans, sponges, and acorn barnacles, as well as many specimens of *Chionecoetes opilio* and *Gorgonocephalus*. Fishes included: *Gymnelis haemifasciatus, Melletes papillio, Stelgistrum steinegeri, Aspidophoroides bartoni, Taurocottus bergi, Crystallias matsushimae, Leptoclinus maculatus diaphoanocarus*, and many eelpouts.

Length, to 230 m (Andriyashev, 1954).

Distribution: Unknown from the Sea of Japan. In the Sea of Okhotsk near Terpenia Peninsula and Cape Svobodnyi in Mordvinov Bay (material of KSE). Found along the Asian coast of the Bering Sea north up to Bering Strait (Andriyashev, 1954: 233). Greenland: Hudson Bay (Taranetz, 1937b: 156).

67   **3. Genus *Stichaeopsis* Kner and Steindachner, 1870**

Stichaeopsis Kner and Steindachner, Sitzb. Acad. Wiss., 61, 1870: 441 (type: *S. nana* Kner and Steindachner). Soldatov and Lindberg, Obzor..., 1930: 466. Taranetz, Dokl. Akad. Nauk SSSR, 1, 3, 1936: 141; Kratkii Opredelitel'..., 1937: 155. Shmidt, Ryby Okhotskogo Morya, 1950: 66. Fowler, Synopsis..., 1958: 291.

Plagiogrammus Bean, Proc. U.S. Nat. Mus., 16, 1893: 699.

Ozorthe Jordan and Evermann, Fish N. and M. Amer., 3, 1898: 2441. Soldatov and Lindberg, Obzor..., 1930: 465.

Body moderately elongate, highly compressed laterally, without scales. Head short and pointed. Jaws equal, with pointed teeth; teeth absent on vomer and palatines. All rays of dorsal fin spiny, only one anterior ray with flexible apex, others rigid. Pelvics inserted anterior to base of pectoral fins. Lateral lines three (Soldatov and Lindberg, 1930).

4 species, 3 known from the Sea of Japan.[55]

Key to Species of Genus Stichaeopsis[56]

1 (2). Head and body highly compressed laterally. Interorbital space highly raised, almost crestate. Anal fin with not more than 26 rays.
.... 1. **S. nana** Kner and Steindachner.

2 (1). Head and anterior part of body almost round in cross section. Interorbital space almost flat or concave. Anal fin with at least 28 rays.

3 (4). Vertical branches of lateral lines, if present, extend only from first (upper) line toward base of dorsal fin, and from third (lower) line to base of anal fin (Figure 30, B)..................................
.... 2. **S. epallax** (Jordan and Snyder).

4 (3). All lateral lines with branches (Figure 30, A). Branches originating from right and left of second (middle) lateral line anterior to anal fin often meet on abdomen.......... 3. **S. nevelskoi** (Schmidt).

---

[55]Fourth species of this genus, *Stichaeopsis hopkinsi* (Bean, 1894) found along the coast of California.

[56]Taranetz (1937b), with modifications.

1. *Stichaeopsis nana* Kner and Steindachner, 1870–Dwarf Prickleback (Figure 46)

Stichaeopsis nana Kner and Steindachner, Sitzb. Acad. Wiss., 1870: 21 (De-Kastri Bay). Jordan and Snyder, Proc. U. S. Nat. Mus., 25, 1902: 495 (description). Soldatov and Lindberg, Obzor..., 1930: 467, fig. 64. Taranetz, Dokl. Akad. Nauk, SSSR, 1, 3, 1936: 142; Kratkii Opredelitel'..., 1937: 156, fig. 95. Matsubara, Fish Morphol. and Hierar..., 1955: 766. Fowler, Synopsis..., 1958: 295 (description and synonymy).

Ozorthe dictyogrammus, Jordan and Snyder, Proc. U. S. Nat. Mus., 25, 1902: 493, fig. 25. Soldatov and Lindberg, Obzor..., 1930, 465, fig. 64. Matsubara, Fish Morphol. and Hierar., 1955: 766.

31676. West coast of Sakhalin, Antonovo. August 28, 1946. KSE. 1 specimen.

31731. West coast of Sakhalin, Antonovo. August 10, 1946. KSE. 4 specimens.

39878. Little Kuril range, Zelenyi Island. September 11, 1949. KSE. 1 specimen.

39879. Little Kuril range, Šiaškotan Island, Krabovaya Inlet. 1949. KSE. 5 specimens.

39880. Little Kuril range, Šiaškotan Island, Dimitrova Inlet. September 5, 1949. KSE. 2 specimens.

39881. Little Kuril range, Šiaškotan Island, Krabovaya Inlet. August 6, 1949. KSE. 2 specimens.

39882. Little Kuril range, Šiaškotan Island, Krabovaya Inlet. July 11, 1949. KSE. [no. of specimens not given].

68 39883. Kunashir Island, south coast of Kuril. August 20, 1951. V. Shchegolev. 1 specimen.

39884. Kunashir Island, south coast of Kuril. August 19, 1951. O.K. Kusakin. 5 specimens.

40306. Kunashir Strait, Kunashir Island. June 23, 1969. 1 specimen.

D XLIV; A I 24–25; P 16; V I 4; vertebrae 48–50 (19–20 + 29–30); three seismosensory canals of trunk with numerous anastomoses (Figure 30, C); distal ends of vertical branches of upper and lower canals connected through horizontal anastomoses, forming additional longitudinal canals right at base of dorsal and anal fins. Unpaired canal formed as secondary structure on ventral surface of body (Figure 47) (Makushok, 1961b).

Makushok (1961b) has reported several anatomical features distinguishing *S. nana* from other closely related species. These are: replacement of opercular siphon by postero-upper notch (Figure 27, B), reduction in first spiny ray of anal fin, and bifurcation of fourth (inner) ray of pelvic fin.

Body color light chocolate-brown with reddish or chocolate-brown spots. Anal and dorsal fins with dark border along margin. Dark stripes

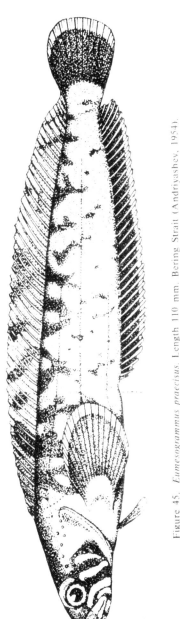

Figure 45. *Eumesogrammus praecisus*. Length 110 mm. Bering Strait (Andriyashev, 1954).

66

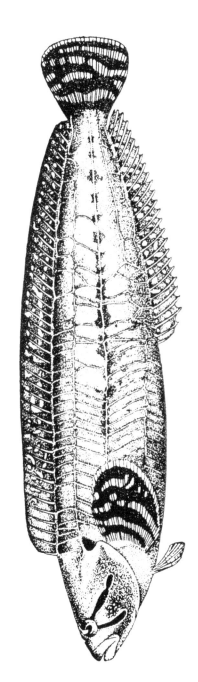

Figure 46. *Stichaeopsis nana*—dwarf prickleback. Length 100 mm. Japan (Jordan and Snyder, 1902a).

68

with light-colored margins originate from eye in direction toward pelvic fin. Pectoral and caudal fins with light-colored stripes. Rounded black spot above upper end of gill opening (Soldatov and Lindberg, 1930).

In 23 specimens collected during the Kuril-Sakhalin Expedition (length 32 to 276 mm), morphological characters corresponded to the description of this species. However, some fish had 46 spiny rays (versus 44) in the dorsal fin and 23–25 soft rays (versus 24–25) in the anal fin. The fish examined by us were caught on different types of bottoms: stones with a large admixture of broken shells, rocks with sand, empty shells and gravel, silted sand and sandy silt.

Length, to 200 mm (Soldatov and Lindberg, 1930).

Distribution: In the Sea of Japan known from Peter the Great Bay, De-Kastri Bay (Soldatov and Lindberg, 1930: 466); west coast of Sakhalin (KSE); and reported from Hakodate (Matsubara, 1955: 766). In the Sea of Okhotsk found near Kunashir Island and the Little Kuril ridge (KSE). Along the Pacific coast of Japan indicated for Nemuro (Matsubara, 1955: 766).

69 2. **Stichaeopsis epallax** Jordan and Snyder, 1902 (Figure 48)

Ernogrammus epallax Jordan and Snyder, Proc. U.S. Nat. Mus., 25, 1902: 491, fig. 24 (Otaru, Sea of Japan). Soldatov and Lindberg, Obzor..., 1930: 464. Matsubara, Fish Morphol. and Hierar., 1955: 766.

Stichaeopsis epallax, Taranetz, Kratkii Opredelitel'..., 1937: 156. Shmidt, Ryby Okhotskogo Morya, 1950: 67. Fowler, Synopsis..., 1958: 292.

39834. Little Kuril ridge, Yurii Island. August 19, 1947. KSE. 4 specimens.

39835. Kunashir Island. August 11, 1951. O.G. Kusakin and V. Shchegolev. 1 specimen.

39863. West coast of Sakhalin, Antonovo. August 7, 1952. M.I. Legeza. 1 specimen.

40051. Kunashir Island. June 30, 1969. A.N. Golikov. 1 specimen.

D XLVI–XLVIII; A II 31–33; P 15; V I 3; vertebrae 50–51 (17–18 + 33–34) (Makushok, 1958: 120).

Head long, pointed, 5 times in standard length. Interorbital space narrow, concave; lower jaw protrudes slightly forward. Body covered with minute oblong cycloid scales. Head and body without distinct stripes or spots. Pectorals with 4–5 indictinct darkish stripes (Soldatov and Lindberg, 1930). More vertical branches originate from upper canal of lateral line and extend toward base of dorsal fin than found in *Eumesogrammus praecisus*; a large number originate from lower canal and extend toward base of anal fin; ventral canals meet in front of base of

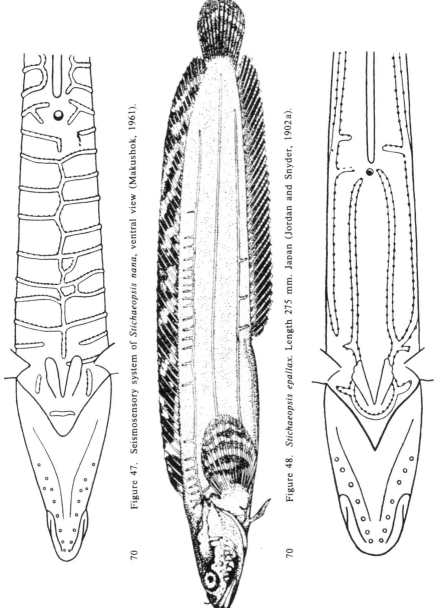

Figure 47. Seismosensory system of *Stichaeopsis nana*, ventral view (Makushok, 1961).

70

Figure 48. *Stichaeopsis epallax*. Length 275 mm. Japan (Jordan and Snyder, 1902a).

70

Figure 49. Seismosensory system of *Stichaeopsis epallax*, ventral view (Makushok, 1958).

70

pelvic fins and terminate blindly near anal fin[57] (Figure 49). Anal fin with 2 thin spines (Makushok, 1961b).

Dwells at depths of 100 m in the Sea of Okhotsk and remains in littoral zone in the Sea of Japan (Shmidt, 1950).

In the specimens collected during the Kuril-Sakhalin Expedition: D XLVII–XLVIII; A II 31–33. These were collected from shallow waters (10–20 m), where the bottom consisted of sand and stones.

Length, to 275 mm (Soldatov and Lindberg, 1930).

Distribution: In the Sea of Japan known from Peter the Great Bay (Soldatov and Lindberg, 1930: 464); in Tatar Strait, near the Soviet Gavan (DVE); and off the west coast of Sakhalin (KSE). Described from Otaru (Jordan and Snyder, 1902a: 492); known from Hakodate (Snyder, 1912: 449); Toyama Bay (Katayama, 1940: 25); and off Sado Island (Honma, 1963: 21). South Kuril Strait (KSE).

3. *Stichaeopsis nevelskoi* (Schmidt, 1904) (Figure 50)

Ozorthe nevelskoi, Shmidt, Ryby Vostochnykh Morei..., 1904: 194. Soldatov and Lindberg, Obzor..., 1930: 466. Matsubara, Fish Morphol. and Hierar., 1955: 766.

Stichaeopsis nevelskoi, Taranetz, Dokl. Akad. Nauk SSSR, 1, 3, 1936: 143; Kratkii Opredelitel'..., 1937: 156. Shmidt, Ryby Okhotskogo Morya, 1950: 66, pl. III, fig. 2. Fowler, Synopsis..., 1958: 293.

39864. Sakhalin, west coast. August 23, 1952. M.I. Legeza. 1 specimen.

D XLVI–XLVIII; A II 30–32; P 16; V I 3; vertebrae 52–54 (17–18 + 34–37) (Makushok, 1958: 120). Head short, broad at occiput. Eyes large, their maximum diameter equal to snout length. Jaws, vomer, and palatines with minute sharp teeth. Dorsal fin consists of spiny rays only, almost equal in height throughout length, commencing slightly behind origin of pectoral fin. Anal fin slightly notched along margin. Caudal fin rounded at end. Pectoral fins long, rounded at end, their length almost equal to length of head. Length of pelvic fins equal to distance from snout tip to anterior margin of pupil (Shmidt, 1904).

Vertical branches of upper and lower canals of lateral lines respectively originate near bases of dorsal and anal fins, and are more numerous than in *S. epallax*. Vertical canals attain greater development, especially in anterior part of body. Upper, middle, and posterior part of lower canal discernible among longitudinal canals; of abdominal paired canals only an insignificant rudiment remains near pelvic fins, together with rudiment of unpaired abdominal canal (Figure 51) (Makushok, 1961b).

---

[57]Pinchuk (1974: 951) pointed out deviations in structure of seismosensory system of trunk of this species.

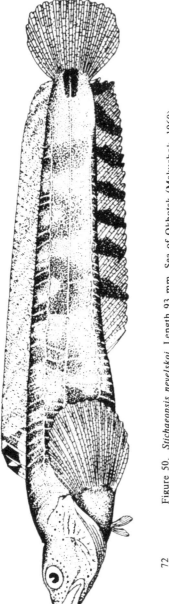

Figure 50. *Stichaeopsis nevelskoi*. Length 93 mm. Sea of Okhotsk (Makushok, 1960).

72

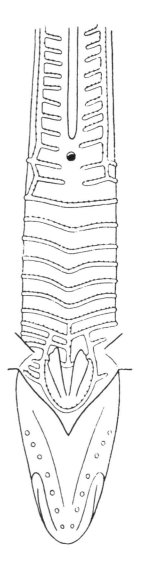

Figure 51. Seismosensory system of *Stichaeopsis nevelskoi*, ventral view (Makushok, 1961).

72

Specimens preserved in alcohol chocolate-brown, with minute black spots on head; stripes on head not evident. Sides of trunk with 5 broad dark bands. Membrane of dorsal fin marked with irregular dark stripes between which large light-colored spots occur, and large black roundish spot located on anterior end of dorsal fin. Light-colored anal fin with 7 dark brown oblique stripes. Rays of pectoral fins and marginal rays of caudal fin with dark spots (Shmidt, 1904).

Dwells at depths of 35–80 m (in the Sea of Okhotsk) and 100 m in the Sea of Japan (Shmidt, 1950).

Length, to 235 mm.

Distribution: In the Sea of Japan known from Tatar Strait north to De-Kastri Bay, and off the west coast of Sakhalin. In the Sea of Okhotsk found in Tauisk Bay and off the east coast of Sakhalin near Cape Bellingshausen; known off west coast of Kamchatka (Shmidt, 1950: 67).

4. Genus *Ernogrammus* Jordan and Evermann, 1898

Ernogrammus Jordan and Evermann, Bull. U. S. Nat. Mus., 47, 1898: 2441 (type: *Stichaeus enneagrammus* Kner). Jordan and Snyder, Proc. U.S. Nat. Mus., 25, 1902: 489. Shmidt, Ryby Vostochnykh Morei..., 1904: 192. Soldatov and Lindberg, Obzor..., 1930: 463. Matsubara, Fish Morphol. and Hierar., 1955: 765.

This genus is similar to *Stichaeopsis* and hence Taranetz (1936) included the solitary species of *Ernogrammus* as a synonym of *Stichaeopsis*. But Makushok (1958, 1961b) established that the single member of the genus *Ernogrammus* (*E. hexagrammus*) from the viewpoint of structure of the lateral line is not related through transitional froms with species of *Stichaeopsis*. Moreover, *E. hexagrammus* differs from them in a large branchiostegal membrane fused with the isthmus, presence of 3 mandibular pores (versus four), and 7 preopercular pores (versus 6).

1 species. Also known from the Sea of Japan.

1. *Ernogrammus hexagrammus* (Schlegel, 1845) (Figure 52)

Stichaeus hexagrammus Schlegel. In: Temminck and Schlegel, Fauna Japonica, 1842–1850: 136, pl. 73, fig. 1.

Stichaeopsis hexagrammus, Taranetz, Dokl. Akad. Nauk SSSR, 1, 3, 1936: 143; Kratkii Opredelitel'..., 1937; 156. Fowler, Synopsis..., 1958: 293.

Ernogrammus enneagrammus, Jordan and Evermann, Bull. U. S. Nat. Mus., 47, 1898: 2441.

Ernogrammus hexagammus, Jordan and Snyder, Proc. U. S. Nat. Mus., 25, 1902: 490, fig. 23. Shmidt, Ryby Vostochnykh Morei..., 1904: 192. Soldatov and Lindberg, Obzor..., 1930: 463. Matsubara, Fish Morphol. and Hierar., 1955: 765.

39871. Kunashir Island. July 28, 1951. O. Kusakin. 2 specimens.

39872. Kunashir Island. August 8, 1951. O. Kusakin and B. Shchegolev. 18 specimens.

39873. Kunashir Island. August 11, 1951. O. Kusakin and B. Shchegolev. 1 specimen.

39874. West coast of Sakhalin. 1948. KSE. 1 specimen.

73   D XLII-XLIII; A I 28-30; C 6-7 + 6-7; P 14; V I 4; vertebrae 46-47 (15 + 31-32) (Makushok, 1958: 120).

In addition to characters given in the description of the genus, let us mention that in *E. hexagrammus* the first spine of the anal fin is completely reduced (Makushok, 1958: 34) and the seismosensory system of the trunk represented by 4 longitudinal canals on each side of the body; short vertical branches with blind ends (not anastomosed) originate from these canals on both sides regularly (one on each myomere), which open at their distal ends through large pores. The upper and middle canals are not connected, while the ventral canals are connected with each other anterior to the base of the pelvic fins and, furthermore, anastomose twice with the lower canals (Figure 53) (Makushok, 1961b: 243).

Length, to 130 mm (Abe, 1958: 111).

Distribution: In the Sea of Japan known from Pusan (Mori, 1952: 128); Peter the Great Bay (Soldatov and Lindberg, 1930: 463); north to De-Kastri Bay and Cape Tyk at Sakhalin; off Sado Island (Honma, 1963: 21); Toyama Bay (Katayama, 1940: 25); San'in region (Mori, 1956a: 21); along Pacific coast of Japan from Hokkaido to Nagasaki (Okada and Matsubara, 1938: 402). Gulf of Chihli (Bohai) (Zhang et al., 1955: 170). Sea of Okhotsk (Jordan and Snyder, 1902a: 491). Our specimens, Kunashir Island and west coast of Sakhalin.

2. Subfamily Chirolophinae

Makushok, Stichaeoidea..., 1958: 81.

Body moderately elongate, compressed laterally, covered with imbricate scales. Head naked (*Chirolophis japonicus* has scales on cheeks), short, blunt, and upper side attenuate from sides; underside broad, covered, similar to adjacent part of body, with numerous cirrose processes and tentacles, among which 2 supraorbital pairs very characteristic.[58] Mouth small, not very wide. Teeth on jaws conical, arranged in several rows (Figure 32, A) or incisor-shaped and arranged in two alternate rows, their flat cusps forming a continuous cutting edge (Figure 32, B). Teeth on vomer and palatines absent.[59] Seismosensory canals of head well

[58]This is probably a camouflaging adaptation in forms living in clumps of coastal algae (Makushok, 1958: 15).

[59]Present only in *Bryozoichthys*.

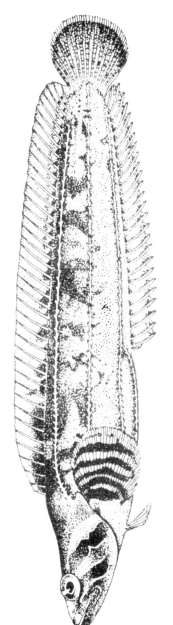

72   Figure 52.   *Ernogrammus hexagrammus.* Length 120 mm. Japan (Jordan and Snyder, 1902a).

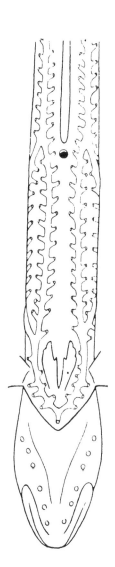

73   Figure 53.   Seismosensory system of *Ernogrammus hexagrammus,* ventral view (Makushok, 1961).

developed, open exteriorly, through constant number of pores in most cases: 2 nasal, 7 interorbital, 7 postorbital, 4 occipital, 6 suborbital, 6 preopercular, and 4 mandibular.[60] Seismosensory papillae of trunk open in 4 form of middle and upper branches, and only origin of upper branch included in short canal opening exteriorly through pores. Vertebrae 57-71, of which 15-17 precaudal. Dorsal fin with stiff spinules, which usually carry cirri at origin of fin. One poorly developed spinule located at origin of anal fin. Principal rays of caudal fin 14-15. Membrane of anal fin touches caudal fin; membrane of dorsal fin continues slightly onto base. Pectoral fins large, with 13-15 rays. Pelvic fins well developed (I 4 or I 3).

Distribution: Most species distributed in northern part of the Pacific Ocean.[61] Found near coasts among outgrowths of algae on stony bottom, and in depths of 50 to 90 m and more. Feed on minute invertebrates (mollusks, polychaetes, and hydroids).

Four genera. Three known from the Sea of Japan.

5. Genus *Bryozoichthys* Whitley, 1931

Bryolophus Jordan and Snyder, Proc. U.S. Nat. Mus., 25, 1903: 617 (type: *B. lysimus* Jordan and Snyder, 1903). Non *Bryolophus* Ehrenberg, 1839.

Bryozoichthys Whitley, Austr. Zool., 6, 1931: 334. Substitute for *Bryolophus* Jordan and Snyder. Makushok, Stichaeoidea..., 1958: 60.

Body elongate, compressed laterally. Mouth small. Branchiostegal membranes form fold across isthmus. Teeth on jaws form narrow band; teeth not present on vomer and palatines.[62] Body covered with minute scales, head naked, without scales. Lateral line represented by short row of pores above pectoral fin. Interorbital space and occiput with dermal processes. Dorsal fin commences above gill opening, consists of only spiny rays; pelvic fins jugular; caudal fin well developed (Jordan and Snyder, 1903b).

According to our observations, the main features of this genus should be the conical shape of the multiserial teeth on the jaws, which are grouped in a narrow band, presence of minute teeth on vomer and palatines, and nasal tubules which attenuate apically but open through comparatively broad pores.

1 species. Known from the Sea of Japan, Sea of Okhotsk, and Bering Sea.

[60]*Soldatovia* has 5 postorbital and 3 mandibular pores.

[61]Along the Asian coast—*Chirolophis japonicus*, *Ch. snyderi*, *Ch. otohime*, *Ch. saitone*, *Ch. wui*, and *Soldatovia polyactocephala;* along the American coast—*Ch. decoratus*, *Ch. nugator*, and *Ch. tarsodes*. *Ch. ascanii* is endemic to the Atlantic.

[62]According to Taranetz (1937b: 151), Andriyashev (1937: 328), and our observations (No. 39952), teeth are present on the vomer and palatines.

1. *Bryozoichthys lysimus* (Jordan and Snyder, 1903) (Figure 54)

Bryolophus lysimus Jordan and Snyder, Proc. U. S. Nat. Mus., 1903: 617, fig. 3 (Unalaska Island). Andriyashev, Issled. Morei SSSR, 25, 1937: 328. Shmidt, Ryby Okhotskogo Morya, 1950: 73.

Bryozoichthys lysimus, Whitley, Austr. Zool., 6, 1931: 3 (Unalaska Island). Taranetz, Kratkii Opredelitel'..., 1937: 153. Makushok, Stichaeoidea..., 1958: 60.

12438. Sea of Okhotsk, Aniva Bay. September 28, 1901. P.Yu. Shmidt. 1 specimen.

19122. Sea of Okhotsk, St. Iona Island. June 26, 1914. GEVO. 1 specimen.

30605. Bering Sea, St. Matthew Island. 1932. A.P. Andriyashev. 1 specimen.

39952. Sea of Japan, Tatar Strait near Nevel'sk. August 3, 1947. G.U. Lindberg. 1 specimen.

In our specimens D LXI-LXIV; A I 48; P 15.[63]

As percentage of total length: head length 13-17%, body depth at origin of anal fin 11-12.5%, and length of pectoral fin 10-12%. As percentage of head length: diameter of eye 25-30.5%, snout length 17.5-22.0%, length of pelvic fins 29-33%, and interorbital space 9-10%. Snout short and blunt. Maxilla extends beyond vertical from anterior margin of eye; lower jaw protrudes slightly forward; teeth on jaws relatively small and conical, arranged in several rows. Minute teeth on vomer and palatines. Gill rakers about 15, short. Nasal tubules open through broad orifices. Branchiostegal membranes do not continue downward in front, broadly fused, and not attached to isthmus. Very thin dark stripe extends along margin of branchial membranes. Pectoral fins relatively large, their length equal to length of postorbital space; dark rounded spot at base of fin continues onto rays. Dorsal fin with dark spots between 4th and 7th rays; anterior 3 rays with branched cirri. Anal fin originates at vertical from 15th dorsal ray. Dorsal and anal fins connected with base of caudal fin through membrane. Lateral line rudimentary, represented by 5 pores above base of pectoral fin. Upper surface of head, from nostrils to occiput and up to origin of dorsal fin with dermal tentacles, some of which bifurcate at tip. One unbranched barbel located above and beside nostril; unpaired cirrus occurs in center of upper surface of snout and is distinctly branched. One unbranched or slightly branched process located on anterior and posterior margin of orbit, pair of unbranched cirri in center of interorbital space, and 1 unbranched cirrus located between posterior pair of supraorbital cirrose processes. Further back, up to origin of dorsal fin, 18 unbranched or branched cirri occur. Chin with pair of barbels. Upper part of

[63] Andriyashev, 1937: D LXV, A 48, P 15; Makushok, 1958: D LXIII-LXVI, A I 50, P 14.

operculum without cirri. Cirri not found on branchiostegal membranes and sides of head. Body covered with minute scales.

Description of our specimens (length slightly more than 100 mm) similar to characters of type specimen, except that in our specimens teeth present on vomer and palatines and gill rakers short (not long).

We agree with Andriyashev (1937), who concluded that if the type specimen of *B. lysimus* actually has no teeth on the vomer and palatines, then the specimens in our collection (Nos. 19122, 30605, 39952) should be placed in a separate genus and species.

76   The fish in our collections were caught from depths of 80–240 m. One of the beam trawls, the one in which specimen no. 39952 was hauled in, contained *Heliometra glacialis* f. *maxima* and *Ophiura sarsi*, as well as *Aspidophoroides*.

Length 100 mm.

Distribution: In the Sea of Japan known from Tatar Strait near Nevel'sk (No. 39952). In the Sea of Okhotsk found in Aniva Bay (No. 12438) and near St. Iona Island (no. 19122). In the Bering Sea found off St. Matthew Island (Andriyashev, 1937: 329). Described on the basis of a specimen from the Bering Sea (Unalaska Island).

6. Genus *Chirolophis* Swainson, 1839

Chirolophis Swainson, Nat. Hist. Fish., Amph. and Rept., 2, 1839: 275 (type: *Blennius palmicornis* Yarrell = *Blennius ascanii* Walbaum). Shmidt, Ryby Okhotskogo Morya, 1950: 71. Andriyashev, Ryby Severnykh SSSR, 1954: 233.

Bryostemma Jordan and Starks, Proc. Calif. Acad. Sci., 5, 1895: 841 [type: *Blennius polyactocephalus* (non Pallas) = *Bryostemma decoratum* Jordan and Snyder]. Jordan and Snyder, Proc. U.S. Nat. Mus., 25, 1902: 464. Soldatov and Lindberg, Obzor..., 1930: 438. Taranetz, Kratkii Opredelitel'..., 1937: 151 (differentiation from *Bryozoichthys* and *Soldatovia*). Matsubara, Fish Morphol. and Hierar., 1955: 759.

Azuma Jordan and Snyder, Proc. U. S. Nat. Mus., 25, 1902: 463 (type: *A. emmnion* Jordan and Snyder). Soldatov and Lindberg, Obzor..., 1930: 438. Taranetz, Kratkii Opredelitel'..., 1937: 153. Matsubara, Fish Morphol. and Hierar., 1955: 759. Fowler, Synopsis..., 1958; 254.

Body moderately elongate, compressed laterally, covered with minute scales. Head short, compressed laterally, without scales[64]; dorsally covered with simple or branched cirrose processes or tentacles, of which usually 2 supraorbital pairs better developed.[65] Lateral line reduced to 3–15 distinct

---

[64]Except for *Ch. japonicus* Herzenstein and *Ch. wui* Wang and Wang, in which the head is partly covered with scales.

[65]In *Ch. snyderi* Taranetz and *Ch. japonicus* Herzenstein these cirrose processes are highly dendroid.

pores (upper branch) above pectoral fin. Gill openings do not continue forward, branchiostegal membranes fused, form broad fold across isthmus. Teeth on jaws arranged in single row (or in 2 alternate rows), their flat tips forming continuous cutting edge.[66] Teeth not present on vomer and palatines. Upper margin of operculum posteriorly connected with body through free skin fold, forming unique siphon for release of water from gill chamber. Dorsal fin long, with 50-62 spiny rays. One short spine located at origin of anal fin. Pectoral fins large, more than half head length. Pelvic fins I 3-4, without true spiny rays (Andriyashev, 1954). Number of pores in seismosensory canal of head constant in all members of the genus: nasal 2, interorbital 7, postorbital 7, suborbital 6, occipital 4, preopercular 6, and mandibular 4 (Makushok, 1961b: 232). Structural features of gill filaments in species of this genus described by Shmidt (1950: 72).

9 species. 4 known from the Sea of Japan and one from adjacent waters.

Key to Species of Genus Chirolophis

1 (4). Anal fin with I 36-40 rays; dorsal fin with 51-56 rays.
2 (3). Cutaneous processes present only on upper part of head, slightly branched. Head naked. D LI; A I 36..........................
.... 1. **Ch. saitone** Jordan and Snyder.
3 (2). Cutaneous processes numerous, present not only on upper surface of head, but also on anterior rays of dorsal fin and throughout opercular margin; processes sometimes dendritic. Head partly covered with scales. D LVI; A I 40 ....
.... 2. [**Ch. wui** Wang and Wang].
4 (1). Anal fin with I 45-47 rays; dorsal fin with 59-62 rays.
5 (6). Cutaneous processes present only on upper part of head; 4-5 anterior rays of dorsal fin with little cutaneous growths. A I 45-46; D LX-LXII................. 3. **Ch. otohime** Jordan and Snyder.
6 (5). Cutaneous processes present on entire head and anterior rays of dorsal fin, processes fleshy and dendritic.
7 (8). Head naked. Anterior part of dorsum between occiput, dorsal fin, and lateral line, as well as membranes of fins also without scales. D LIX-LX; A I 45...................... 4. **Ch. snyderi** Taranetz.
8 (7). Head and entire body covered with scales. Scales also present on membranes of dorsal, anal, and pectoral fins. D LXI-LXII; A I 45-47................................ 5. **Ch. japonicus** Herzenstein.

1. *Chirolophis saitone* (Jordan and Snyder, 1903) (Figure 55)
Bryostemma saitone Jordan and Snyder, Proc. U.S. Nat. Mus., 25, 1903:

---

[66]Teeth probably adapted for clipping invertebrates dwelling on the bottom (Makushok, 1958: 16).

84

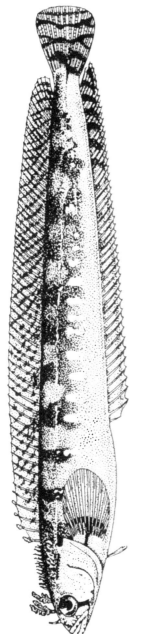

75

Figure 54. *Bryozoichthys lysimus*. Length 100 mm. Unalaska Island (Jordan and Snyder, 1903).

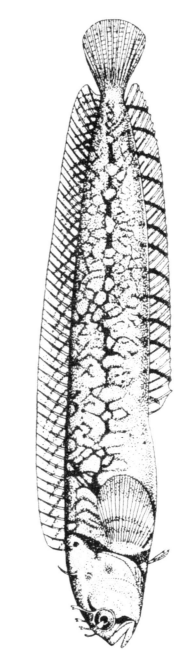

78

Figure 55. *Chirolophis saitone*. Length 95 mm. Japan (Jordan and Snyder, 1902a).

467, fig. 13 (Aomori). Taranetz, Kratkii Opredelitel'..., 1937: 153. Matsubara, Fish Morphol. and Hierar., 1955: 759.

Chirolophis saitone, Makushok, Stichaeoidea..., 1958: 61.

D LI; A I 36 (Makushok, 1958).

Eyes large, set in antero-upper part of head; snout short, suborbital space narrow; lower jaw slightly longer than upper; maxilla extends beyond vertical with anterior margin of pupil. Teeth small, close-set, arranged in two rows in anterior part of mouth. Gill rakers short and pointed. Head naked, body covered with minute cycloid scales. Anterior nostrils with long tubules. Interorbital space and occiput with long, branched tentacles; similar processes located on upper margin of orbit, slightly longer than diameter of eye. Dorsal fin originates above upper margin of gill slit, bears only highly sinuous rays about 2.25 times in head length.

Color light olive-green, with minute indistinct chocolate-brown spots; dark chocolate-brown spots occur along sides of eye and body near base of dorsal fin; several small spots situated along body and several minute spots at base of anal fin, which continue as dark stripes onto fin; abdomen light-colored. Differs from the closely related species, *Ch. otohime*, in less bright coloration of body and shorter anal fin (Jordan and Snyder, 1903a).

Length 95 mm (Jordan and Snyder, 1903a).

Distribution: Described from Mutsu Bay (Aomori). Not found at other places in the Sea of Japan.

2. [*Chirolophis wui* (Wang and Wang, 1935)] (Figure 56)

Azuma wui Wang and Wang, Contrib. Biol. Lab. Sci. Soc. China, 11, 6, 1935: 210, fig. 36 (Chefoo). Fowler, Synopsis..., 1958: 256.

Chirolophis wui, Makushok, Stichaeoidea..., 1958: 61.

D LVI; A I 40; P 15; V I 3 (Makushok, 1958).

Body slightly elongate, significantly compressed laterally throughout length; dorsal profile commencing from occiput slightly raised, ventral profile more or less flat. Head relatively small, short, blunt, and compressed to some degree. Eyes moderate in size, set in anterior upper part of head; interorbital space narrow, slightly convex, its width much less than diameter of eye; snout conical, its tip broadly rounded. Nostrils on each side considerably separate; anterior nostril small and located in front of eye. Mouth almost horizontal, not very large; jaws equal in length, but sometimes lower jaw protrudes slightly forward; maxilla reaches vertical from posterior margin of eye. Teeth small, arranged in two rows. Tongue very thick, its tip broadly rounded. Gill rakers thick and short, 14 on lower part of first arch. Scales very small, cycloid, and barely embedded in skin throughout entire body; chin, opercle, and posterior part of preorbital covered with minute scales; remaining part of head naked.

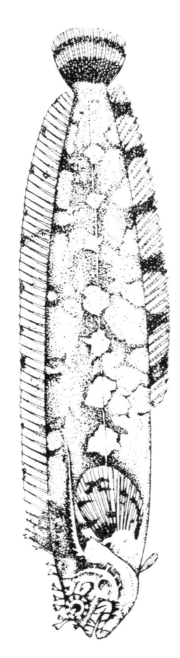

Figure 56. *Chirolophis wui.* Standard length 193 mm. Yellow Sea (Wang and Wang, 1935).

78

Membranes of dorsal, anal, and pelvic fins with scales which continue almost up to margin of fin; bases of pectoral and caudal fins also covered with scales. Top of head, occiput, margin of operculum, and anterior rays of dorsal fin with cutaneous processes, which are broad and fleshy at base and pointed or branched at tips. Dorsal fin originates almost at origin of pectoral fin and extends posteriorly to connect with caudal fin in upper part of its base; dorsal fin with only spiny rays. Anal fin commences under 17th ray of dorsal fin and posteriorly also connects with base of caudal fin, where its membrane forms narrow deep notch. Caudal fin rounded. Pectoral fin 1.2 times in head length, middle rays longest. Pelvic fins jugular.

Color of fish preserved in formalin slightly pinkish. Body with 9 dark spots which continue onto rays of dorsal fin, and 7 indistinct spots above anal fin which continue onto fin in form of irregular blotches. Caudal fin with 2 black vertical stripes, remaining part of fin white. Pectoral and pelvic fins dark, light-colored along margin.

This species is close to *Ch. japonicus* but differs from it in larger number of rays in dorsal and anal fins (Wang and Wang, 1935).

Type specimen 193 mm long.

Distribution: Described from Yellow Sea (Chefoo). No other record of this species known to date.

3. ***Chirolophis otohime*** (Jordan and Snyder, 1902) (Figure 57)

Bryostemma otohime Jordan and Snyder, Proc. U. S. Nat. Mus., 25, 1902: 466, fig. 12 (Hakodate). Wang and Wang, Contrib. Biol. Lab. Sci. Soc. China, 11, 6, 1935: 212. Taranetz, Kratkii Opredelitel'..., 1937: 153. Matsubara, Fish Morphol. and Hierar., 1955: 759.

Chirolophis otohime, Makushok, Stichaeoidea..., 1958: 61.

Soldatovia otohime, Fowler, Synopsis..., 1958: 259.

D LVIII-LXI; A I 44-45; P 15; V 3 (Makushok, 1958).

Body deep, laterally compressed for the most part. Lower jaw protrudes slightly forward, upper jaw reaches vertical from posterior margin of pupil; teeth in anterior part of jaw arranged in 2 rows, in posterior part one row, close-set, their tips forming cutting edge. Gill rakers on first gill arch 4 + 11, short, and pointed. Body covered with minute, close-set, cycloid scales. Head naked. Upper part of head with small cutaneous processes arranged along median line.

Dorsal fin originates above upper end of gill slit, connects with base of caudal fin, and consists entirely of spiny rays; tips of anterior 4-5 rays free and with cirri. Thin membrane of dorsal fin without notch between tips of rays. Height of anterior part of dorsal fin 2.25 times in head length. Anal fin not high, its rays slightly more than diameter of eye. Membrane of this fin deeply notched between tips of rays. Membrane of dorsal and anal fins not covered with scales. Caudal fin obtusely rounded, its length 1.33 times in head length. Pectoral fins also rounded, their membranes deeply notched between tips of rays, their length 1.12 times in head length; pelvic fins 3 times in head length.

Dorsal part of body with 10-11 narrow vertical stripes, with corresponding number of large dark spots on dorsal fin; ventral part of body with 10 broad vertical stripes, with correspondingly large black spots on anal fin; dark spots on abdomen separated by white spots. Head fully covered with chocolate-brown to black stripes and spots; cutaneous processes of head with transverse stripes. Caudal fin with large black blotch in middle part; base and margin white. Pectoral and pelvic fins dark, margin white (Jordan and Snyder, 1902a).

Length 82 mm (Jordan and Snyder, 1902a).

Distribution: In the Sea of Japan known from Hakodate (Jordan and Snyder, 1902a: 467). In the Yellow Sea reported from Chefoo (Wang and Wang, 1935: 212).

4. *Chirolophis snyderi* (Taranetz, 1938)—Snyder's Prickleback
(Figure 58)

Bryostemma snyderi Taranetz, Vestn. Dal'nevost. Fil. Akad. Nauk SSSR, 28, 1938: 123, fig. 6 (western coast of Sakhalin). Matsubara, Fish Morphol. and Hierar., 1955: 759.

Chirolophis snyderi, Andriyashev, Ryby Severnykh Morei, 1954: 236, fig. 120 (synonymy, description). Makushok, Stichaeoidea..., 1958: 61.

Bryostemma polyactocephalum (non Pallas), Jordan and Evermann, Fish N. and M. Amer., 3, 1898: 2408 (partly, specimens from Petropavlovsk). Jordan and Snyder, Proc. U. S. Nat. Mus., 25, 1902: 465 (partly, specimens from Petropavlovsk). Shmidt, Ryby Vostochnykh Morei..., 1904: 170 (partly, species not of Pallas). Soldatov and Lindberg, Obzor..., 1930: 439 (partly, species not of Pallas). Andriyashev, Issled. Morei SSSR, 25, 1937: 327.

12439. Sea of Okhotsk, Aniva Bay. August 24, 1901. P.Yu. Shmidt. 1 specimen.

D LVIII-LXI; A I 43-45; P 15; V I 4; vertebrae 63-65 (16-17 + 46-49) (Makushok, 1958).

D LX[67]; A I 45; P 14; pores of lateral line 7.[68] Scales in oblique transverse row from origin of anal to dorsal fin 49. Gill rakers 5 + 13-3 + 11. Pyloric caeca 5. Teeth only on jaws, arranged in two rows; their cusps as in all other members of the genus close-set, forming continuous cutting edge (Figure 32, B). Five pairs of tentacles arranged on top of

---

[67]D LIX-LX; A I 44-47 (Andriyashev, 1954).
[68]Pores in lateral line 6-8 (Andriyashev, 1954).

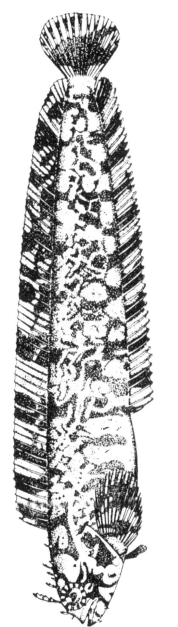

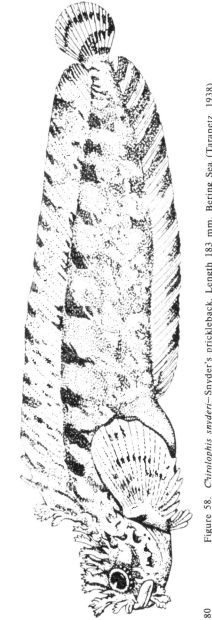

Figure 57. *Chirolophis otohime*. Length 82 mm. Japan (Jordan and Snyder, 1902a).

80

Figure 58. *Chirolophis snyderi*—Snyder's prickleback. Length 183 mm. Bering Sea (Taranetz, 1938).

80

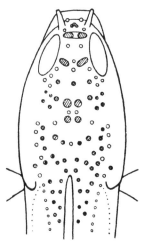

Figure 59. Arrangement of cirri (shaded) and pores of seismosensory
canal of *Chirolophis snyderi* (Makushok, 1958).

head, with additional pair of smaller ones very close-set at base. Several
tentacles along posterior margin of operculum and lower jaw, usually less
dendritic and less broad than in *Chirolophis japonicus* (*Azuma japonica*).
Pair of tentacles on chin, several scattered on other parts of head. Occiput
with several large tentacles (Figure 59). Head without scales, naked. In
lateral line first few pores directed upward and backward, other pores
arranged in tandem parallel to body axis. Row of tentacles (about 5)
parallel lateral line of pores, decreasing in size posteriorly. Another row of
4 longer tentacles located above first row. Minute tentacles located on
anterior margin of first and second rays of dorsal fin, and large tentacles at
tips of second and third rays. Body in formalin with vague dark spots
arranged in chessboard pattern. Spots of upper row continue along base of
82 dorsal fin, bifurcate, and continue onto fin. Caudal fin with two indistinct
transverse dark stripes. Anal fin with row of indistinct dark spots.[69]
Pectoral fins light-colored (Taranetz, 1938a). Color of live fish very vivid.
Body pinkish-orange with 11 short lilac-red bands, which lighten in color
on lower side and merge. Minute spots on dorsal fin. Anal fin pinkish with
11 indistinct reddish spots. Rays of caudal fin with two bright red
transverse stripes. Pelvic fins dark (Andriyashev, 1954).

Length, to 183 mm (Taranetz, 1938a).[70]

Distribution: In the Sea of Japan known from west coast of Sakhalin
(Taranetz, 1938a: 123). Sea of Okhotsk, Bering Sea (Andriyashev, 1954).

[69]First ray of this fin in form of single poorly developed spine.
[70]Probably attains much larger dimensions (Andriyashev, 1954).

5. *Chirolophis japonicus* Herzenstein, 1892—Japanese War Bonnet
(Figure 60)

Chirolophis japonicus Herzenstein, Mélanges Biologiques..., XIII, 3,
1892: 219-235 (Japan, Hakodate). Makushok, Stichaeoidea.... 1958: 61.

Azuma emmnion Jordan and Snyder, Proc. U.S. Nat. Mus., 25, 1902:
463, fig. 11 (Hakodate). Taranetz, Kratkii Opredelitel'..., 1937: 153.
Soldatov and Lindberg, Obzor..., 1930: 438.

Azuma japonica, Taranetz, Vestn. Dal'nevost. Fil. Akad. Nauk SSSR,
28, 1938: 121. Fowler, Synopsis..., 1958: 255, fig. 19.

20696. Vladivostok (market). 1908. N.A. Pal'chevskii. 1 specimen.

22186. Vladivostok (market). 1927. E.P. Rutenberg. 1 specimen.

26075. Peter the Great Bay. September 1, 1934. ZIN. 1 specimen.

26076. Peter the Great Bay. September 26, 1934. ZIN. 1 specimen.

26295. Peter the Great Bay. October, 1934. ZIN. 1 specimen.

D LXI-LXII; A I 46-47; P 13-15 (Soldatov and Lindberg, 1930).[71]

Head small, short, and blunt. Jaws equal in length, lower jaw
sometimes protrudes slightly forward; maxilla extends up to vertical with
posterior margin of pupil. Gill rakers short, thick, pointed at ends, 6 + 11
on first gill arch.

83   Body uniformly chocolate-brown to black; 10 blackish spots size of eye
located on upper part of dorsal fin; 11 or 12 indistinct broad vertical stripes
in lower half of body, 10 of which located above anal fin and continue onto
fin.

Caudal fin with 2 broad vertical blackish stripes; interval between
stripes and margin white. Pelvic fins blackish, with white edging. Head
mottled, chin and throat white (Jordan and Snyder, 1903a).

Length to 440 mm (Soldatov and Lindberg, 1930).

Distribution: In the Sea of Japan known from Peter the Great Bay
(Taranetz, 1938a: 121); near Won'san and Pusan (Mori and Uchida, 1934:
21); Hakodate (Jordan and Snyder, 1903a: 464); Sado Island (Honma,
1963: 21); Toyama Bay (Katoh et al., 1956: 322); San'in region (Mori,
1956a: 21). In the Yellow Sea found near Mokpo (Mori and Uchida, 1934:
21) and in Gulf of Chihli (Bohai) (Zhang et al., 1955: 171). Along the
Pacific coast of Japan reported from Miyako (Matsubara, 1955: 759).

7. Genus *Soldatovia* Taranetz, 1937

Soldatovia Taranetz, Kratkii Opredelitel'..., 1937: 152 (type: *Blennius*
polyactocephalus Pallas, 1811). Matsubara, Fish Morphol. and Hierar.,
1955: 759. Makushok, Stichaeoidea..., 1958: 118. Fowler, Synopsis...,
1958: 258.

[71]D LX-LXIII; A I 45-47; principal rays C 6-7 + (7) 8 (9); P 13-14; V I 4; vertebrae 63-65
(16-17 + 46-49) (Makushok, 1958).

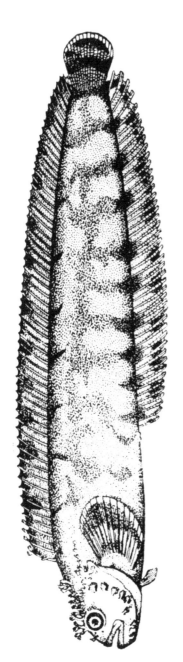

82

Figure 60. *Chirolophis japonicus*—Japanese war bonnet. Length 250 mm. Hakodate (Jordan and Snyder, 1902).

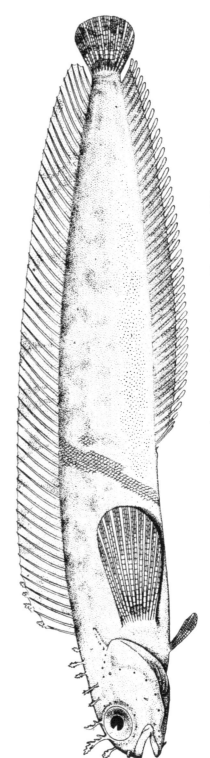

83

Figure 61. *Soldatovia polyactocephala*—Soldatov's prickleback. Standard length 75 mm. No. 26241. Aniva Bay.

Body elongate, compressed laterally, covered with small scales. Head naked. Eyes set high, near profile of head. Snout slopes downward from eye toward mouth; mouth located below eyes. Branchiostegal membranes do not continue downward toward front, forming broad fold across isthmus. Teeth on jaws arranged with alternating bases in two close-set rows, their flat cusps convergent, forming single row. Teeth absent on vomer and palatines. Notch present on upper posterior margin of operculum. Nasal tubules attenuate toward upper side and open through small orifice. Two pairs of cirri present above eyes. Top of head behind eyes and in front of dorsum with two parallel rows of cirri. Lateral line represented by 3–4 pores above base of pectoral fin. Dorsal fin with only spiny rays. V I 3 (Taranetz, see Makushok, 1958: 118).

84 1. *Soldatovia polyactocephala* (Pallas, 1811)—Soldatov's Prickleback (Figure 61)

Blennius polyactocephalus Pallas, Zool. Rosso-Asiat., 3, 1811: 179 (Kamchatka). Jordan and Evermann, Fish N. and M. Amer., 3, 1898: 2408; 4, 1900, fig. 828 (partly, specimens not from Petropavlovsk). Jordan and Snyder, Proc. U.S. Nat. Mus., 25, 1902: 465 (partly, specimens not from Petropavlovsk). Shmidt, Ryby Vostochnykh Morei..., 1904: 170 (partly, species of Pallas). Soldatov and Lindberg, Obzor..., 1930: 439 (partly, species of Pallas).

Chirolophis polyactocephalus, Shmidt, Ryby Okhotskogo Morya, 1950: 71.

Soldatovia polyactocephala, Taranetz, Kratkii Opredelitel'..., 1937: 153. Matsubara, Fish Morphol. and Hierar., 1955: 759. Fowler, Synopsis..., 1958: 259. Makushok, Stichaeoidea..., 1958: 119.

26241. Sea of Okhotsk, Cape Aniva. June 28, 1932. A.Ya. Taranetz. 1 specimen.

D LII-LX; A I 41–43; P 14–15.[72] Longitudinal rows of scales above base of anal fin 38–46; *l. l.* 3 (in 3 specimens) or 4 (in 1 specimen). Vertebrae 63 (based on 1 specimen). Pyloric caeca 3. Gill rakers 4–8 + 2–9 (based on one specimen). Head length 15.8–17.3%, body depth at origin of anal fin 14.8–16.2% and least body depth 4.7–5.1% of standard length. Diameter of eye 23.1–28.2%, length of lower jaw 47–51%, and length of pectoral fin 95–112% of head length. Above eyes, 7 pairs of cirri located on occiput and behind occiput on dorsum (Figure 62). Number and arrangement of cirri constant; shape and size highly variable. First pair of cirri equal to or longer than second pair. Several cirri present on anterior part of dorsal fin. Usually two simple barbels present on chin, sometimes also occur on operculum or along lower posterior margin of slit. Upper pores on head

[72]D LV; A I 42; principal rays C 7 + 8; P 14–15; V I 4; vertebrae 60 (15 + 45) (Makushok, 1958: based on one specimen).

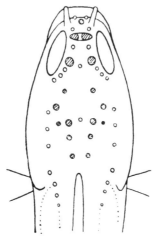

Figure 62. Arrangement of cirri (shaded) and pores of seismosensory canals of head in *Soldatovia polyactocephala* (Makushok, 1958).

very small. Body light-colored, brownish, with irregular and indistinct dark spots. Several brighter spots located along base of dorsal fin. Dark vertical stripe above eye. Light-colored triangular spot situated anterior to stripe and covers apex of snout and lips. Dorsal, pectoral, and caudal fins light-colored, with vague dark spots and stripes. Tips of rays of anal fin white, rest of fin gray. Belly and lower surface of head light-colored (Taranetz; see Makushok, 1958: 119). Dwells on pebbled bottom. Caught from depth of 56 m.

In our specimen (length 92 mm): D LX; A 42; P 14; ratio of length of head to standard length 17.1. In all other characters conforms to above description of species. This specimen was caught in the littoral zone at a depth of 40 m.

Length, to 415 mm (Soldatov and Lindberg, 1930).

Distribution: In the Sea of Japan known from Peter the Great Bay and near Aomori (Soldatov and Lindberg, 1930: 439); from Hakodate (Shmidt, 1950: 71); and Oshoro Bay (Kobayashi, 1962: 258). Eastern Sakhalin near Aniva Bay; Kamchatka (Taranetz, 1937b: 153).

3. Subfamily Lumpeninae

Makushok, Stichaeoidea..., 1958: 84.

Body very moderately to highly elongate; head region with scales only on cheeks (only in *Lumpenella* is entire head covered with scales). Head large, without cutaneous processes. Mouth medium or large (*Leptoclinus*). Teeth on vomer and palatines present or absent. Gill openings continue forward; branchiostegal membranes narrowly attached to isthmus. Siphon of operculum absent (upper notch present: Figure 27, B). Branchiostegal rays 6. Postorbital, occipital, and suborbital canals completely reduced, and mandibular canal partly reduced; seismosensory papillae open (Figure 63). Remaining canals of head open through fixed number of pores:[73] nasal 2, interorbital 1, postorbital 1, preopercular 5, and mandibular 2 (1 in *Anisarchus*). Lateral line in form of middle branch of open seismosensory papillae. Vertebrae 60–94, precaudal 23–30. Spiny rays of dorsal fin long, stiff, equal in thickness. Anal fin origin with one poorly developed spiny ray or 2–5 spines. Caudal fin either rounded, oval-pointed, or truncate, with 12 (*Anisarchus*) or 13 principal rays. Membranes of dorsal and anal fins only touch base of caudal fin and only in *Anisarchus medius* are they usually completely fused (without notch) with base. Pectoral fins large, with 12–16 rays. Pelvic fins well developed (I 3), their soft rays usually unbranched. Soft rays of anal, caudal (principal), and pectoral fins with 2–3 or more branches

Separation of caudal fin from dorsal and anal fins, reduction of seismosensory canals of head, loss of siphon on operculum, larger size of gill openings, and elongate body, "independent location of caudal fin"* are characters of specialization which, in the opinion of Makushok (1958: 54), are present in members of the subfamily Lumpeninae which have acquired great mobility.

6 genera. 5 known from the Sea of Japan.

Distribution: Northern part of the Pacific and Atlantic oceans and north Arctic Ocean. Most members of Lumpeninae dwell on the continental slope; usually found on a silty bottom at depths of 10 to 150 to 200 m or more. Only *Lumpenella longirostris* moves into the bathyal zone (400–500 m). Feed on minute benthic invertebrates: polychaetes, crustaceans, and bivalve mollusks.

8. Genus *Lumpenus* Reinhardt, 1836

86

Lumpenus Reinhardt, Dansk. Vidensk. Selsk. Nat.-Math. Afhandl., 6, 1836: 110 (type: *Blennius lumpenus* Fabricius = *L. fabricii* Reinhardt). Makushok, Stichaeoidea..., 1958: 68, 87.

Makushok (1958) separated from the composite genus *Lumpenus* Reinhardt the following independent genera: *Anisarchus* Gill (type *L. medius* Reinhardt) and *Acantholumpenus* Makushok (type: *L. mackayi* Gilbert = *L. fowleri* Jordan and Snyder = *Blennius anguillaris* Pallas).

The characters of the genus *Lumpenus* presented below were defined

[73] Only in some specimens of *Lumpenella longirostris* are individual preopercular pores fused; in *Acantholumpenus mackayi* all interorbital pores fused.

*Present in Lindberg but not in Makushok (1958: 55); says the same thing as first phrase of the sentence—Editor.

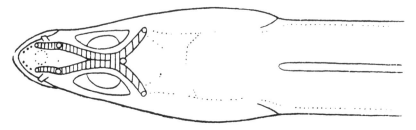

Figure 63. Seismosensory system of head in *Lumpenus fabricii* (dorsal view) (Makushok, 1958).

while taking into account these changes and are based on the key to the subfamilies and genera of the family Stichaeidae compiled by Makushok (1958: 72). Couplets 20 to 25 have also been considered.[74]

Head without cutaneous processes in form of barbels and cirri (if processes present, in form of snout-occipital crest). Pectoral fins large (less than 2 times in head length), with at least 12 (12-21) rays. Pelvic fins present. Gill openings continue forward ventrally; branchiostegal membranes narrowly attached to isthmus. Gill openings do not reach forward to vertical from posterior margin of eye. Postorbital, occipital, and suborbital canals of head completely reduced (Figure 63). Mandibular pores not more than two. Outermost ray of pelvic fins and first ray of anal fin in form of poorly developed spine. Teeth not present on vomer.[75] Lower rays of pectoral fins shorter than middle ones. Oral valves (palatine and mandibular) present. Mandibular pores two. Caudal fin free, not connected with dorsal and anal fins. Principal rays of caudal fin 13 (6 + 7). Scales cover entire body.

Three species. Two known from the Sea of Japan.

Key to Species of Genus Lumpenus[76]

1 (2). D LXIV-LXXI; A I 45-50. Base of dorsal fin usually with dark stripe on back .... 1. **L. sagitta** Wilimovsky.
2 (1). D LXI-LXV; A O-I 39-43. Dark stripe not present on back along base of dorsal fin .... 2. **L. fabricii** Reinhardt.

1. ***Lumpenus sagitta*** Wilimovsky, 1956—Sagittate Eelblenny (Figure 64)
Lumpenus sagitta Wilimovsky (= *L. anguillaris* auct., non *Blennius*

---

[74]In delineating characters of the genus *Lumpenus*, only features of external structure were considered.

[75]If teeth present on palatines, always arranged in single row. Makushok (1958: 17) has reported that the presence of teeth on the palatines and the vomer is usually typical of least specialized species. This is very helpful in determining the extent of affinity between species.

[76]From Taranetz, 1937b: 157.

anguillaris Pallas, 1811), Stanf. Ichthyol. Bull., 7, 2, 1956: 23 (new name in place of *Leptogunellus gracilis* Ayres).

Leptogunellus gracilis Ayres, Proc. Calif. Acad. Sci., 1, 1855: 26 (California).

Leptogunellus lampetraeformis Kobayashi and Ueno (non Walbaum, 1792), Bull. Fac. Fish. Hokkaido Univ., VI, 4, 1956: 239.

Lumpenus anguillaris, non Pallas, Soldatov and Lindberg, Obzor..., 1930: 470.

40199. Sea of Okhotsk, Terpenia Gulf. October 3, 1949. KSE. 1 specimen.

40313. Sea of Okhotsk, Terpenia Gulf. September 15, 1947. KSE. 3 specimens.

40314. South Kuril Strait. September 18, 1949. KSE. 1 specimen.

40315. Sea of Okhotsk, Aniva Bay. August 17, 1947. KSE. 1 specimen.

40316. Sea of Okhotsk, Terpenia Gulf. September 12, 1947. KSE. 1 specimen.

40317. Peter the Great Bay. September 1, 1954. Legeza and Dorofeeva. 1 specimen.

40324. Eastern Sakhalin coast. September 11, 1947. KSE. 1 specimen.

40344. La Perouse Strait. August 24, 1948. KSE. 5 specimens.

40345. Tatar Strait. September 24, 1948. KSE. 1 specimen.

87 D (LXIV) LXV-LXXI; A I, 45-50; C 6 + 7; P 15-16; V I, 3; vertebrae 75-80 (26-28 + 46-54) (Makushok, 1958).

Specimens of the Kuril-Sakhalin expedition,[77] with D LXIV-LXXI (more often LXV-LXVII); A I, 45-50 (often 45). Fish caught mostly from depths of 1 to 70 m, at water temperatures near bottom ranging from 7-16°C, from a sandy bottom with an admixture of silt, pebbles, and stones. These fish were caught incidentally in catches of flounders and walleye pollock. Other incidental fishes: *Triglops jordani, Podothecus gilberti, Lycodes palearis, Gymnocantus herzensteini, G. detrisus, Glyptocephalus stelleri, Myoxocephalus jaok,* and *M. polyacanthocephalus*; among the invertebrates found often were *Cucumaria japonica, Asterias amurensis, Ophiura sarsi*, rarely *Gorgonocephalus* and *Chiridota*; among nudibranch mollusks a small number of *Yoldia* and rarely *Leda*; among crustaceans *Erimacrus isenbeckii, Paralithodes platypus,* and *Hyas coarctatus*; sometimes many polychaetes were found on the bottom. Red alga was prolific.

Color not retained to the same degree in all fish, but characteristic dark stripe along base of dorsal fin was distinct, as well as thin dark stripe along outer margin of this fin. Dark transverse stripes retained on dorsal fin in some specimens. Upper half of body pigmented, lower half light-colored. Pigmented spots on sides of body with indistinct contours. Dark spot

[77]Of 140 specimens, 68 were examined.

present in upper part of pectoral fin, continues onto body in region of humeral band. Caudal fin with dark transverse stripes.

Length of our specimens ranged from 68 to 290 mm.

Distribution: In the Sea of Japan known from Peter the Great Bay (Taranetz, 1937: 157); near mouth of Tuman-gang River, Tatar Strait, and liman of Amur River (Shmidt, 1950: 78); La Perouse Strait (No. 40344); off Hakodate (Taranetz, 1937b: 157); Sado Island (Honma, 1963: 21); Toyama Bay (Katoh et al., 1956: 322). In the Sea of Okhotsk specimens of KSE were collected near the eastern and southeastern coasts of Sakhalin, South Kuril Strait, and off Šiaškotan Island. Northern part of the Pacific Ocean and south to San Francisco and Japan (Taranetz, 1937b: 157).

2. *Lumpenus fabricii* Reinhardt, 1836—Slender Eelblenny (Figure 65)

Lumpenus fabricii Reinhardt, Dansk. Vidensk. Selsk. Nat.-Math. Afhandl., 6, 1836: 110 (Greenland). Andriyashev, Ryby Severnykh Morei SSSR, 1954: 244, figs. 126, 127. Makushok, Stichaeoidea..., 1958: 61.

Gunellus fabricii Cuvier and Valenciennes, Hist. Nat. Poiss., 11, 1836: 431 (Greenland).

Lumpenus fabricii, Soldatov and Lindberg, Obzor..., 1930: 471. Rass, Acta Zool., 17, 1936: 395, figs. 1–16 (age-dependent variability of major characters). Taranetz, Kratkii Opredelitel'..., 1937: 157. Shmidt, Ryby Okhotskogo Morya, 1950: 79. Matsubara, Fish Morphol. and Hierar., 1955: 769. Fowler, Synopsis..., 1958: 279.

20340. Tatar Strait. June 24, 1910. Derbek. 1 specimen.

40284. Sea of Okhotsk, Terpenia Gulf. October 2, 1949. KSE. 1 specimen.

D (LXI) LXIII–LXV (LXVI); A I, 40–43[78]; P 15–16 (17); V I, 3; vertebrae 70–73. Head low, pointed toward snout. Mouth horizontal, upper jaw distinctly protrudes forward in relation to lower one. Lower lip interrupted anteriorly, lobes present but poorly defined. Rays of dorsal fin gradually shorten anteriorly with the first 1–2 rays barely perceptible and not connected by a common membrane. Anal fin equal in height throughout length, its posterior rays not elongate. Posterior rays of dorsal and anal fins do not continue onto caudal fin, but connect with its base (Andriyashev, 1954).

Color yellowish. About 6 dark spots located along base of dorsal fin and series of minute midbody spots. Dorsal fin with scattered spots; caudal fin with 4 dark transverse stripes; remaining fins pale yellow (Soldatov and Lindberg, 1930). Similar morphological characters and age-dependent variability of major characters reported by Rass (1936).

Seismosensory system of this species depicted in Figure 33.

[78]Soldatov and Lindberg (1930) reported D LVIII–LX; A 35–38. In our specimen (No. 40284) from Terpenia Gulf we found D LXIII; A I 38.

Information on biology of this eelblenny given by Andriyashev (1954). Length, to 365 mm.

Distribution: In the Sea of Japan known from the northern part of Tatar Strait and liman of Amur River (Soldatov and Lindberg, 1930: 471). In the Sea of Okhotsk found in Terpenia Gulf (No. 40284). Arctic Sea, Bering Sea (Taranetz, 1937b: 157).

9. Genus *Anisarchus* Gill, 1864

Anisarchus Gill, Proc. Acad. Nat. Sci., Philad., 16: 210 (type: *Clinus medius* Reinhardt). Makushok, Stichaeoidea..., 1958: 61, 71.

This genus differs from the closely related genus *Lumpenus* Reinhardt in that the gill openings continue beyond a vertical line with the middle of the eye, oral valves rudimentary, mandibular pore one, principal rays of caudal fin 12 (6 + 6), anterior part of skull slopes sharply, mesethmoid very broad, and scales on cheeks highly reduced (Makushok, 1958: 71).

Two species. Both known from the Sea of Japan.

Key to Species of Genus Anisarchus[79]

1 (2). Dorsal fin with more than 57 spiny rays. Anal fin with more than 35 soft rays. Posterior rays of dorsal and anal fins distinctly continue beyond base of caudal fin. Diameter of eyes equal to or less than length of snout .... 1. **A. medius** (Reinhardt).

2 (1). Dorsal fin with less than 57 spiny rays (54–56). Anal fin with 35 or less spiny* rays (33–35). Posterior rays of dorsal and anal fins do not continue beyond base of caudal fin. Diameter of eyes more than length of snout .... 2. **A. macrops** (Matsubara and Ochiai).

1. ***Anisarchus medius*** (Reinhardt, 1838)—Stout Eelblenny (Figure 66)
Clinus medius Reinhardt, Dansk. Vidensk. Selsk. Nat.-Math. Afhandl., 7, 1838: 114, 121, 194 (Greenland).

Lumpenus medius, Shmidt, Ryby Vostochnykh Morei..., 1904: 186 (comparison of Pacific and Atlantic specimens). Soldatov and Lindberg, Obzor..., 1930: 471. Taranetz, Kratkii Opredelitel'..., 1937: 157. Shmidt, Ryby Okhotskogo Morya, 1950: 80. Andriyashev, Ryby Severnykh Morei SSSR, 1954: 243, fig. 124 (synonymy and description). Matsubara, Fish Morphol. and Hierar., 1955: 769. Fowler, Synopsis..., 1958: 280 (synonymy and description).

Anisarchus medius, Makushok, Stichaeoidea..., 1958: 61
D LVIII–LXIII; A I, 37–42; P 14–15; V I, 3; C 6 + 6 (Makushok, 1958).

---

[79]From Matsubara (1955: 769), with modifications.

*Obvious error in Russian text. In description of the species, soft rays are indicated— General Editor.

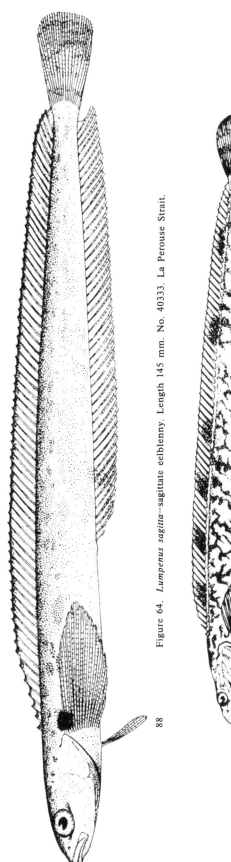

Figure 64. *Lumpenus sagitta*—sagittate eelblenny. Length 145 mm. No. 40333. La Perouse Strait.

88

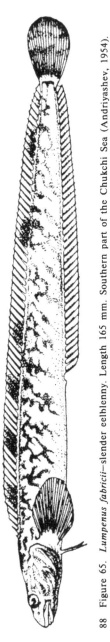

Figure 65. *Lumpenus fabricii*—slender eelblenny. Length 165 mm. Southern part of the Chukchi Sea (Andriyashev, 1954).

88

Figure 66. *Anisarchus medius*—stout eelblenny. Length 134 mm. Southern part of the Chukchi Sea (Andriyashev, 1954).

88

In the Pacific specimens examined by Makushok, vertebrae numbered
90 65-70 (23-25 + 43-50). Snout blunt, not pointed. Body comparatively
deep, height at origin of dorsal fin more than 8% of total length (Andriyashev, 1954). Mandibular seismosensory canal highly reduced (Figure
67) (Makushok, 1958).

Shmidt (1904: 186) compared specimens of *A. medius* from the
Atlantic and Pacific oceans and concluded that they were completely
identical. Color of live fish probably variable. Shmidt (1904) reported the
body as reddish, while Andriyashev (1954) reported it as light yellow with
indistinct yellowish spots.

In the northern seas, according to Andriyashev (1954), found at depths
of 10-15 to 150 m, often 30-100 m, almost exclusively on a silty bottom. It
prefers lower benthopelagic temperatures (mostly subzero), but is also
known from warmer waters with temperatures up to 3 to 5°C (Spitsbergen)
and above (White Sea). Not found in less saline marine waters, preferring
salinity above 30°/₀₀. Feeds on minute benthic animals (polychaetes,
bivalve mollusks, and crustaceans).

Specimens collected during KSE (33) from the Sea of Japan in the
region of southern Primor'e (No. 40261), Tatar Strait (Nos. 40270, 40287);
Sea of Okhotsk in the region of Terpenia Gulf (Nos. 40267, 40275, 40276);
southeastern coast of Sakhalin and Aniva Bay (Nos. 40266, 40268, 40269,
40271, 40272, 40273, 40274); eastern coast of Sakhalin (Nos. 40262, 40265,
40277, 40278, 40279); and Pacific coast of Iturup Island (No. 40321): D
LVIII-LXIII; A I-II, 36-46; diameter of eye less than length of
snout. Pigmentation preserved poorly. Traces of roundish dark spots

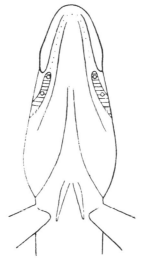

Figure 67. *Anisarchus medius.* Reduction of anterior part of mandibular
seismosensory canal (Makushok, 1961).

visible on sides of body and dorsal fin; black spot located in anterior part of dorsal fin between first and second rays.

Our specimens were found in catches of flounder (*Hippoglossoides elassodon robustus*) form depths of 14 to 300 m, where a silty bottom predominated, sometimes with an admixture of sand, stones, and clay, containing a large number of mollusks (*Leda*). The by-catches also included *Cardium groenlandicum, Yoldia hyperborea, Ophiura sarsi, Gorgonocephalus, Chionocoetes opilio,* and rarely sponges (*Geodia*). Among fishes, in addition to *Anisarchus medius,* these were also caught: *Icelus spiniger intermedius, Gymnocanthus* sp., *Stichaeus nozawae, Lycodes tanakae, Eumicrotremus birulai, Podothecus gilberti, Allolepis hollandi,* and *Dasycottus setiger.*

Length, to 246 mm.

Distribution: In the Sea of Japan known in southern Primor'e (KSE) and Tatar Strait near Cape Syurkum (Soldatov and Lindberg, 1930: 471). Found in the Sea of Okhotsk in Terpenia Gulf and Aniva Bay (KSE). Pacific coast of Iturup Island (KSE). Along the coasts of Japan, found in Volcano Bay, off Hokkaido Island (Sato and Kobayashi, 1956: 14). Arctic Ocean and Far East seas, circumpolar (Andriyashev, 1954: 242).

2. ***Anisarchus macrops*** (Matsubara and Ochiai, 1952) (Figure 68)

Lumpenus medius, Taranetz, Kratkii Opredelitel'..., 1937: 157 (indications of variability).

Lumpenus macrops Matsubara and Ochiai, Japan. J. Ichthyol., 2, 4–5, 1952: 206, fig. 1 (Sea of Japan). Matsubara, Fish Morphol. and Hierar., 1955: 769. Matsubara and Ochiai, Fig. and Descr., Fishes of Japan, 59, 1958: 1236, pl. 237, fig. 597.

Anisarchus macrops, Makushok, Stichaeoidea ... 1958: 61.

40263. Tatar Strait. August 17, 1949. KSE. 1 specimen.

40264. Tatar Strait. September 30, 1948. KSE. 1 specimen.

40311. Southern Primor'e. March 3, 1912. Lyaskovskii. 1 specimen.

D LVI; A I 35; P 13[80]; V I 3; branchiostegal rays 6; longitudinal series of 123 scales along median line of body; gill rakers on first gill arch 5 + 14. Specimens preserved in formalin light yellowish-chocolate-brown, with 11 dark chocolate-brown indistinct blotches located on sides of body; first blotch occurs immediately behind head, last one at base of caudal fin; smaller spots located between larger blotches (from fifth to tenth); minute vague spots present in preocular region; apex of snout blackish. Caudal fin with 4 broad chocolate-brown transverse stripes. Dorsal fin with 11 large, vague, elongate, light chocolate-brown spots, first notably darker than the

[80]D LII–LVI; A I, 32–35; V I, 3; vertebrae 61 (Makushok, 1958). The author notes that a reduction in number of rays in the pectoral fins is observed in *Anisarchus macrops* (less than 12 to 13 rays) and considers this phenomenon a secondary one.

others. Pectoral, pelvic, and anal fins light-colored (Matsubara and Ochiai, 1952: 206). Biology unknown Most probably feeds on different types of algae (Abe, 1958: 110).

Two of our specimens were caught in Tatar Strait from depths of 255–263 m in sandy silt, with D LVI, A I, 35–36. Dorsal fin with 11 dark blotches, the first very vivid. Last rays of dorsal fin do not continue beyond base of caudal fin and last rays of anal fin slightly continue beyond base of caudal fin. Upper profile of head slopes abruptly from eyes downward. 4 transverse dark stripes well expressed on caudal fin.

Lenght, to 170 mm (Abe, 1958).

Distribution: In the Sea of Japan known from Tatar Strait and southern Primor'e (KSE), off Sado Island in Toyama Bay near Ishikawa and Fukui Prefectures (Matsubara and Ochiai, 1952: 209). Possibly widely distributed along the coast of Japan (Abe, 1958: 110).

10. Genus *Leptoclinus* Gill, 1864

Leptoclinus Gill, Proc. Acad. Nat. Sci., Philad., 1861: 45 (type: *Lumpenus aculeatus* Reinhardt). Jordan and Snyder, Proc. U.S. Nat. Mus., 25, 1902: 498. Soldatov and Lindberg, Obzor ..., 1930: 469. Taranetz, Kratkii Opredelitel' ..., 1937: 157. Shmidt, Ryby Okhotskogo Morya, 1950: 81. Andriyashev, Ryby Severnykh Morei SSSR, 1954: 248. Matsubara, Fish Morphol. and Hierar., 1955: 768. Makushok, Stichaeoidea ..., 1958:61.

92   Close to the genus *Lumpenus,* but teeth present not only on palatines but also on vomer. Upper jaw not protractile. Lower rays of pectoral fin elongate. Caudal fin notched at end (Andriyashev, 1954).

One amphiboreal species with, according to Andriyashev (1937), two subspecies: one found in the Atlantic Ocean and the other in the Pacific Ocean and also known from the Sea of Japan.

Key to Subspecies of Leptoclinus maculatus[81]

1 (2). D LXI–LXIV; A I, 37–40; vertebrae 70–72..................
.... 1. **L. m. diaphanocarus** (Schmidt).
2 (1). D LVII–LXI; A I, 34–37; vertebrae 66–69.....................
.... [**L. m. maculatus** (Fries, 1837)].[82]

1. *Leptoclinus maculatus diaphanocarus* (Schmidt, 1904) (Figure 69)

Plectobranchus diaphanocarus, Shmidt, Ryby Vostochnykh Morei ..., 1904: 182 (eastern Sakhalin).

Leptoclinus maculatus, Matsubara, Fish Morphol. and Hierar., 1955: 768.

---

[81]From Andriyashev (1954: 248).
[82]Distribution: northern part of the Atlantic Ocean (Andriyashev, 1954: 249).

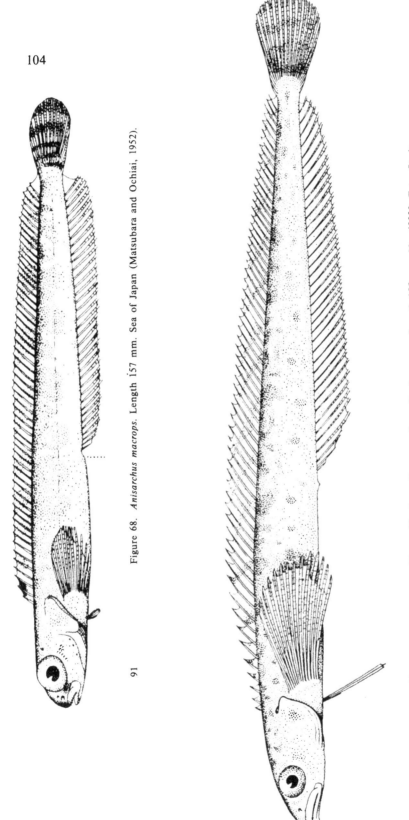

Figure 68. *Anisarchus macrops*. Length 157 mm. Sea of Japan (Matsubara and Ochiai, 1952).

91

Figure 69. *Leptoclinus maculatus diaphanocarus*. Length 155 mm. No. 40202. Tatar Strait.

92

Leptoclinus maculatus diaphanocarus, Taranetz, Kratkii Opredelitel' ..., 1937: 157. Shmidt, Ryby Okhotskogo Morya 1950: 81. Andriyashev, Ryby Severnykh Morei SSSR, 1954: 248, fig. 129, 1 (radiograph).

D. LXI-LXIII; A II, 37-39; P 15; C 13-15; V I, 3; branchiostegal rays 6 (Shmidt, 1904).

D LXI-LXIV; A I-II, 37-40; vertebrae 70-72 (Andriyashev, 1954).

Head compressed laterally, pointed anteriorly, and slightly flat dorsally. Skull bones transparent. Eyes large, round; snout lengt'i equal to diameter of eye. Mouth large, posterior end of maxillary reaching vertical from first third of pupil. Teeth of upper and lower jaws in 2 rows (anterior row consists of larger canine-shaped teeth). Body highly compressed laterally, 93 covered with minute, indistinct caducous scales. Scales also present on cheeks. Vent almost midbody. Anterior 6-7 spines in dorsal fin free but each connected by triangular membrane to (dorsum); first spine located on vertical with posterior margin of opercle.[83] Color of specimens preserved in alcohol yellowish, live fish probably almost transparent; brown pigment spots present on some specimens. Caudal fin grayish; others transparent (Shmidt, 1904: 183).

Our specimens (22; KSE, Nos. 40200-40210) from Tatar Strait, Aniva Bay, Terpenia Gulf, and southeastern coast of Sakhalin characterized by: D LXI-LXIV; A I-II, 37-39; P 15; ratio of diameter of eye to head length 3.3 to 3.8; ratio of snout length to head length 4.4 to 6.0. Round dark spots observed on many preserved specimens, arranged along sides of body and on dorsal fin. Length of our specimens 114 to 163 mm.

Biology not studied. It is known that the specimens collected by A.P. Andriyashev were obtained from depths of 8-68 m, with a stony-pebbled bottom and temperature close to zero.

KSE specimens were collected from depths of 33 to 187 m, from sandy bottom (shallow depths) and silted sand or argillaceous silt (greater depths); sometimes bottom with admixture of stones and pebbles. In some habitats bottom water temperature varied from −1.8 to + 1.9°C. This species was usually encountered incidentally in catches of flounders (*Hippoglossoides elassodon robustus, Acanthopsetta nadeshnyi, Limanda aspera*) together with *Myoxocephalus, Podothecus gilberti, Artediellus dydymovi schmidti, Triglops jordani, Lycodes uschakovi, L. tanakae, L. raridens, Lumpenus medius, Icelus spiniger, Percis japonica*; among invertebrates, catches included a large number of *Gorgonocephalus caryi,* many *Cucumaria japonica* and mollusks (*Leda*), and far fewer *Ctenodiscus crispatus, Pandalus borealis eous, P. goniurus,* and *Chionocoetus opilio.* Length, to 175 mm (Andriyashev, 1954).

Distribution: In the Sea of Japan known in Tatar Strait (Taranetz,

[83]In our specimen (No. 40202) located on vertical line with origin of gill slit.

1937b: 157 and specimens of KSE). In the Sea of Okhotsk found off the southeastern coast of Sakhalin (KSE), Aniva Bay (KSE), Terpenia Gulf (Shmidt, 1950: 81 and KSE), and off eastern coast of Sakhalin (Shmidt, 1904: 182). Lake Notoro, northern coast of Hokkaido (Hikita, 1952: 12). Bering Sea, north up to Anadyr Gulf. Possibly region of Bering Strait (Andriyashev, 1954: 248).

11. Genus *Acantholumpenus* Makuschok, 1958

Acantholumpenus Makushok, Stichaeoidea ..., 1958: 61, 72, 87 (type: *Lumpenus mackayi* Gilbert = *L. fowleri* Jordan and Snyder = *Blennius anguillaris* Pallas).

This genus is distinguished by a spiny outer ray in the pelvic fins and spiny anterior rays (at least 2) in the anal fin, absence of teeth on vomer, and location of jaw teeth in 2–4 central rows (Figure 70, A, B). Scales on
94 head present only on cheeks. Palatines with teeth. Oral valves present. Posterior precaudal vertebrae without parapophyseal connections. Skull low and narrow (Makushok, 1958).

One species. Also known from the Sea of Japan.

1. *Acantholumpenus mackayi* (Gilbert, 1893) (Figure 71)

Lumpenus mackayi Gilbert, Rept. U.S. Fish Comm., 1893 (1895): 450, pl. 32 (Bristol Bay). Jordan and Evermann, Fish N. and M. Amer., 3, 1898: 2436; 4, fig. 839. Soldatov and Lindberg, Obzor ..., 1930: 472. Shmidt, Ryby Okhotskogo Morya, 1950: 78.

Lumpenus fowleri Jordan and Snyder, Proc. U.S. Nat. Mus., 25, 1902: 500, fig. 28 (Kushiro).

Lumpenella mackayi, Taranetz, Vestn. Dal'nevost. Fil. Akad. Nauk SSSR, 13, 1935: 97; Kratkii Opredelitel' ..., 1937: 157. Matsubara, Fish Morphol. and Hierar., 1955: 768.

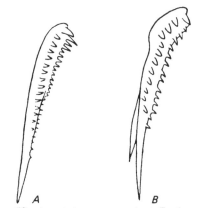

Figure 70. *Acantholumpenus mackayi.* Teeth on jaws.
A—upper jaw; B—lower jaw.

Lumpenus anguillaris, Tanaka, Figs. and Descr., Fishes of Japan, XII, 1913: 211, pl. 57, fig. 3; 2113, pl. 59, fig. 220. Fowler, Synopsis ..., 1958: 279.

Blennius anguillaris Pallas, Zoogr. Rosso-Asiat., 3, 1811: 176, pl. 42, fig. 3 (Kamchatka, America, and Islands).

Acantholumpenus mackayi, Makushok, Stichaeoidea ..., 1958: 61.

D LXIX–LXXV; A II, 41–47 (Soldatov and Lindberg, 1930)[84]

Characters given in description of genus.

Coloration: Continuous dark stripe located under dorsal fin and interrupted stripe in form of punctation along middle of side of body; indistinct row of spots in-between punctation. Sexual dimorphism well expressed. Males with longer caudal fin, notably broadened in middle and sharply pointed toward end. Caudal fin of females not only shorter and narrower, but also truncate at end, with rounded corners (Soldatov and Lindberg, 1930).

Length of our few specimens (5; nos. 40195–40198; 40308–40309), 320 to 457 mm. Collected from Peter the Great Bay, Aniva Bay, Terpenia Gulf, and Šiaškotan Island. D LXXIII–L XXV; A II, 44–45. Body color preserved: dark stripe continues beyond base of dorsal fin; elongate dark spots along middle of side of body form punctate stripe; sometimes third short stripe visible between these two stripes commencing near upper margin of gill opening and extending backward without reaching base of caudal fin. Dark spots discernible between dark stripes. Minute blackish spots present along outer margin of dorsal fin.

Length, to 582 mm (Soldatov and Lindberg, 1930).

Distribution: In the sea of Japan known from Peter the Great Bay, Tatar Strait, and liman of Amur River (Soldatov and Lindberg, 1930: 472); Pusan (Okada and Matsubara, 1938: 404); Olga Bay (Popov, 1933: 15); noted for Hakodate (Jordan and Snyder, 1902a: 501); Sado Island (Honma, 1952: 226); Niigata (Tanka, 1913: 210); Toyama Bay (Katayama, 1940: 25). In the Sea of Okhotsk found in Terpenia Gulf, Aniva Bay, and off Šiaškotan Island (KSE). Along the Pacific coast of Japan reported from Lake Notoro (Hikita, 1952: 3) to Muroran (Snyder, 1912: 449) Pacific coast of America from Alaska to San Francisco (Jordan and Snyder, 1902a: 500).

12. Genus *Lumpenella* Hubbs, 1927

Lumpenella Hubbs, Pap. Mich. Acad. Sci., 7, 1927: 378 (type: *Lumpenus longirostris* Evermann and Goldsborough). Soldatov and Lindberg, Obzor ..., 1930: 473. Makushok, Stichaeoidea ..., 1958: 61, 72.

Body elongate, covered with small cycloid scales. Lateral line absent.

---

[84]D LXVIII–LXXV; A II, 41–47; principal rays C 6 + 7; P 14–15; V I, 3; vertebrae 76–80 (27–29 + 49–52) (Makushok, 1958).

Head pointed; snout produced somewhat beyond small horizontal mouth. Eyes elliptical. Dorsal fin consists entirely of hard and fairly high spines. Anal fin with 3–5 (usually 5) spines and about 40 soft rays. Pelvic fins small, I 2.[85] This genus is close to the genus *Lumpenus* but differs from it in shape of heal, snout, and eyes, and large number of spines in anal fin (Hubbs, 1927). Head entirely covered with scales. Teeth of upper and lower jaws arranged in 2–4 central rows. Palatine and vomerine teeth absent. Oral valves (palatine and mandibular) completely reduced. Caudal fin oval, truncate. Skull high and broad (Makushok, 1958).

One species. Also known from the Sea of Japan.

1. ***Lumpenella longirostris*** (Evermann and Goldsborough, 1907)— Longsnout Prickleback (Figure 72)

Lumpenus longirostris Evermann and Goldsborough, Bull. U.S. Bur. Fish., 26, 1907: 340, fig. 115 (Aleutian Islands).

Lumpenella longirostris, Hubbs, Pap. Mich. Acad. Sci., 7, 1927: 378. Soldatov and Lindberg, Obzor .., 1930: 474. Taranetz, Kratkii Opredelitel' ..., 1937: 157. Shmidt, Ryby Okhotskogo Morya, 1950: 80, pl. IV, fig. 2.

Lumpenella nigricans Matsubara and Ochiai, Japan J. Ichthyol., 2, 4–5, 1952: 210, fig. 2 (Kushiro). Matsubara, Fish Morphol. and Hierar., 1955: 768.

D LXV–LXVI; A II–V, 39–42; P 13–14; V I, 2–3.[86] Branchiostegal rays 6 Scales in longitudinal row 190. Gill rakers on first gill arch 4 + 16 (Soldatov and Lindberg, 1930: 474; Matsubara and Ochiai, 1952: 210).

96 Body and fins of specimens preserved in formalin dark chocolate-brown but belly, lips, branchiostegal membranes, and fins (except dorsal) black (Matsubara and Ochiai, 1952).

P.Yu. Shmidt described the coloration of his live specimens as dark chocolate-brown with a light-colored pattern. In one specimen 8 vague dark spots distinguishable under base of dorsal fin. Head blackish toward anterior end; oral, gill, and belly cavities black. Belly, pectoral fins, anal and caudal fins black; dorsal fin blackened along margin.

Caught at great depths, up to 600 m, at low temperatures but not below 0°C (Shmidt, 1950: 81).

Our specimen (No. 40289) from the Sea of Okhotsk (51° 12.8′ N and 155° 27.3′ E) was caught at a depth of 820 m with bottom water temperature −2.3°C. Length of fish 340 mm, D LXV; AII, 42; P 13; V I, 3.

Length, to 370 mm (Matsubara and Ochiai, 1952: 210).

Distribution: In the Sea of Japan recorded as *L. nigricans* near Sado Island (Honma, 1963: 210). Along the Pacific coast of Japan reported

[85]V I, 3 (Matsubara and Ochiai, 1952).

[86]D LXV–LXVII; A II–III, 40–42; principal rays C 6 + 7; P 13–14; V I, 3; vertebrae 71–74 (24–25 + 47–49) (Makushok, 1958).

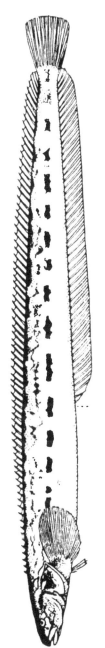

94

Figure 71. *Acantholumpenus mackayi*. Length 315 mm. Hokkaido (Jordan and Snyder, 1902).

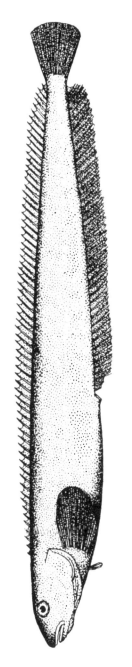

95

Figure 72. *Lumpenella longirostris*—longsnout prickleback. Length 374 mm. Hokkaido (Matsubara and Ochiai, 1952).

from Kushiro, Hokkaido (Matsubara, 1955: 768); also known southward almost to Kuji (Nos. 41542-41544). Sea of Okhotsk, eastern coast of Sakhalin, St. Iona Island, southeastern Alaska (Shmidt, 1950: 81). Our specimen was caught in the Sea of Okhotsk far away in Cape Lopatka. Bering Sea (Taranetz, 1937b: 157). Described from the Aleutian Islands.

4. Subfamily Opisthocentrinae

Makushok, Stichaeoidea ..., 1958: 90.

Body laterally compressed, low or relatively deep (*Ascoldia, Opisthocentrus*), and covered with overlapping scales. Head without cutaneous processes; scales cover entire head (*Ascoldia, Opisthocentrus ocellatus,* and *O. zonope*), or present only on cheeks (*Lumpenopsis pavlenkoi* and *Plectobranchus*[87]), or completely absent (*Kasatkia, Opisthocentrus dybowskii*). Mouth small. Teeth present only on vomer.[88]

Branchiostegal membranes broadly fused and not attached to isthums. Branchiostegal rays 5 or 6 (*Lumpenopsis, Allolumpenus*[89]). Seismosensory canals of head, except subocular one, well developed and open through constant number of pores. Lateral line of body in form of middle and upper branches of open seismosensory papillae. Vertebrae 53-72, precaudal 17-24. Spines of dorsal fin long, hard, and equal in thickness, or anterior ones slender and flexible, gradually thickening toward tail where posteriormost convert into hard spines (*Ascoldia, Opisthocentrus*). Anal fin origin with 1 (*Allolumpenus*) or 2 spines. Membranes of dorsal and anal fins only slightly touch base of caudal fin (in *Ascoldia* and *Opisthocentrus* dorsal fin slightly overlaps base of caudal fin). Principal rays of caudal fin 13-14 (*Lumpenopsis, Kasatkia*) or 14-15 (*Ascoldia, Opisthocentrus*). Pectoral fins large, with 12-21 rays. Pelvic fins well developed, with 1 spine and 3 unbranched rays, or rudimentary (*Ascoldia*), or absent (*Opisthocentrus* and *Kasatkia*).

Distributed in the northern part of the Pacific Ocean. Littoral fish; mode of life not known (Makushok, 1958).

6 genera, 4 known from the Sea of Japan.

97   ### 13. Genus *Lumpenopsis* Soldatov, 1915

Lumpenopsis, Soldatov, Ezhegodn. Zool. Muzeya Rossiisk. Akad. Nauk, 20, 1915: 635 (type: *L. pavlenkoi* Soldatov). Soldatov and Lindberg, Obzor ..., 1930: 474. Makushok, Stichaeoidea ..., 1958: 72.

Body highly elongate, compressed laterally, covered with minute scales; lateral line absent. Head long, compressed laterally, snout

[87]Distributed from California north to British Columbia.
[88]Present on vomer and palatines in *Plectobranchus.*
[89]Distributed along Pacific coast of Canada (Departure Bay).

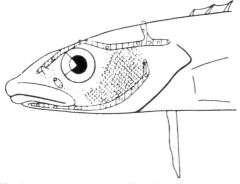

97   Figure 73. Seismosensory system of head in *Lumpenopsis pavlenkoi*,
lateral view (Makushok, 1958).

elongate. Cirri or tentacles absent on head. Eyes moderate in size, set high
on head. Teeth arranged in narrow bands on jaws. Teeth present on
vomer. Dorsal fin composed of a large number of sharp spines. Caudal fin
long. Pectoral fins large, more than half head length, middle rays longest.
Pelvic fins present, I 3 (Soldatov and Lindberg, 1930). Gill openings do
not continue ventrally toward front; branchiostegal membranes broadly
fused and not attached to isthmus. Branchiostegal rays 6. Postorbital and
occipital canals of head normally developed and subocular canal, even if
reduced, only partly so. Number of pores in canals of head constant and
constitute diagnostic character for indentification of genus: nasal 2,
interorbital 1, postorbital 5, suborbital 2, occipital 3, preopercular 5, and
mandibular 3 (Figure 73). Anal fin origin with 2 spines, its soft rays
bifurcate. Membranes of dorsal and anal fins touch only base of caudal fin.
Opercular siphon replaced by postero-upper notch (Makushok, 1958).

In the opinion of V.M. Makushok, several morphological characters of
this genus bring it closer to the subfamily Lumpeninae (presence of scales
on cheeks, 6 branchiostegal rays, 5 subopercular pores, and 1 interorbital
pore); another series of characters is typical of the subfamily Opisthocentrinae (broad fusion of branchiostegal membranes, which are not
attached to isthmus, presence of postorbital and occipital seismosensory
canals, and presence of ocelli on dorsal fin).

Two species. Both known from the Sea of Japan.

· *Key to Species of Genus Lumpenopsis*[90]

1 (2). Dorsal fin with 42–49 spines.[91] Body pale yellow, with 12 chocolate-

[90]From Matsubara, 1955: 768.

[91]An unfortunate misprint was incorporated in the diagnosis of *Lumpenopsis*
(*Leptoclinus*) *triocellatus* given by Matsubara in the key for species (1955: 768, couplet a[2]): D
LXII–LXIX. The figure published by the author shows D XLIX. It should read D XLII–
XLIX.

brown to gray spots randomly arranged on sides. Dorsal fin with 3 round black spots greater in size than pupil (in posterior half of fin) .... 1. **L. triocellatus** Matsubara.

2 (1). Dorsal fin with 47 spines. Body light chocolate-brown. 6 trapezoidal spots on dorsum with base directed toward median line of body side. Dorsal fin with 5 rounded dark spots arranged at base of fin in-between apices of trapezoidal spots of back ....

.... 2. **L. pavlenkoi** Soldatov.

1. *Lumpenopsis triocellatus* (Matsubara, 1943)—Three-spotted Eelblenny (Figure 74).

Leptoclinus triocellatus Matsubara, J. Sigenkagaku Kenkyusyo, 1, 1943: 37, fig. 1, pl. 1 (Tsurga Bay). Matsubara, Fish Morphol. and Hierar., 1955: 768, fig. 283.

Lumpenopsis triocellatus, Makushok, Stichaeoidea ..., 1958: 61, 92 (explanation erroneously attributed to genus *Leptoclinus*).

D XLII-XLIX; A II, 24-31; P 13; V I, 3 (Makushok, 1958).

Characters given in description of genus and in key to species.

Like Makushok, neither could we find the first description of this species (Matsubara, 1943); hence we reproduce the explanation given by V.M. Makushok regarding the affinity of *Leptoclinus triocellatus* Matsubara to the genus *Lumpenopsis*:

"Presence of postorbital and occipital canals in *Leptoclinus triocellatus* Matsubara, which open respectively through five and three pores, two pores on suborbital bone (Fig. 75), two well-developed spines at anal fin origin, 13 rays in pectoral fin, 49 spines in dorsal fin, oval-rounded caudal fin, and finally, 'ocelli' on dorsal fin—these justify its affinity to the genus *Lumpenopsis* ... This species differs from *L. pavlenkoi* in the presence of two anterior 'ocelli' on dorsal fin and slight elongation of unbranched lower rays of pectoral fins".

Length of specimen depicted in Matsubara's figure (Matsubara, 1955, pl. 82, fig. 283) about 110 mm.

Distribution: In the Sea of Japan known from off Sado Island (Honma, 1958: 21). Along the Pacific coast of Japan reported from Tsuruga Bay (Matsubara 1955: 768).

2. *Lumpenopsis pavlenkoi* Soldatov, 1915—Pavlenko Eelblenny (Figure 76)

Lumpenopsis pavlenkoi Soldatov, Ezhegodn. Zool. Muzeya Rossiisk Akad. Nauk, 20, 1915: 636, drawing (Peter the Great Bay). Soldatov and Lindberg, Obzor ..., 1930: 474. Matsubara, Fish Morphol. and Hierar., 1955: 767. Makushok, Stichaeoidea ..., 1958: 61, 92 (comparative notes).

18859. Peter the Great Bay, Cape Gamov. October 16, 1912. DVE. 1 specimen.

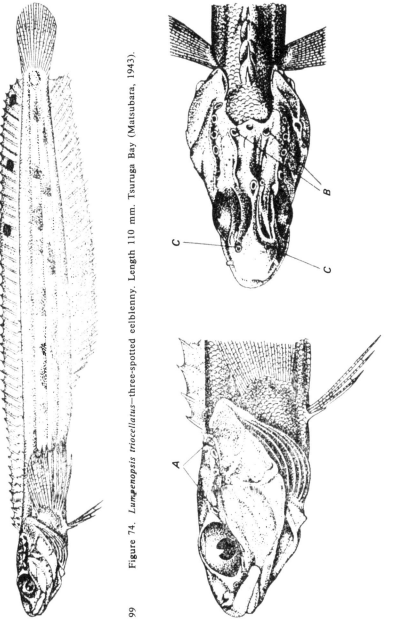

99   Figure 74. *Lumpenopsis triocellatus*—three-spotted eelblenny. Length 110 mm. Tsuruga Bay (Matsubara, 1943).

99   Figure 75 *Lumpenopsis triocellatus*. Pores of seismosensory canals of head (Matsubara, 1955). A—postorbital; B—occipital; C—preorbital.

114

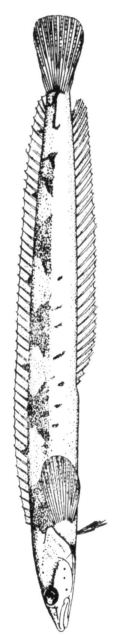

99

Figure 76. *Lumpenopsis pavlenkoi*—Pavlenko's eelblenny. Length 72 mm. Peter the Great Bay (Soldatov, 1915).

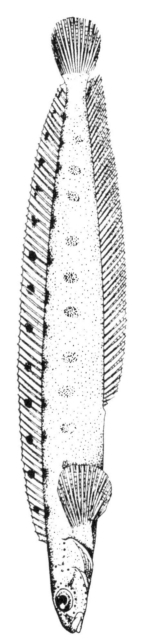

100

Figure 77. *Kasatkia memorabilis.* Length 86 mm. Tatar Strait (Soldatov and Pavlenko, 1915).

25976. Southern Primor'e. September 23, 1934. ZIN. 1 specimen.
D XLVII; A II, 30; V I, 3.[92]

Head naked; small scales present only on cheeks (Figure 73). Mouth moderate. Lower jaw short, maxilla reaches vertical with anterior margin of pupil. Rays of dorsal fin very low. Dorsal fin originates at vertical with base of pectoral fin. Rays of anal fin higher than rays of dorsal fin; anal fin originates slightly ahead of vertical with midpoint of standard length. Caudal fin long but shorter than head length, slightly rounded at end. Pelvic fins inserted ahead of base of pectoral fin, long and narrow, with three unbranched soft rays (Soldatov, 1915). Specimens examined by Soldatov were caught in Peter the Great Bay at a depth of 30 m and in Tatar Strait at a depth of 30–40 m.

99 Length 72 mm (Soldatov and Lindberg, 1930).

Distribution: In the Sea of Japan known from Peter the Great Bay and Tatar Strait (Soldatov, 1915: 636).

14. Genus *Kasatkia* Soldatov and Pavlenko, 1915

Kasatkia Soldatov and Lindberg, Ezhegodn. Zool. Muzeya Rossiisk. Akad. Nauk, 20, 1915: 638 (type: *K. memorabilis* Soldatov and Pavlenko). Soldatov and Lindberg, Obzor 1930: 444. Taranetz, Kratkii Opredelitel' ..., 1937: 154. Matsubara, Fish Morphol. and Hierar., 1955: 760. Fowler, Synopsis ..., 1958: 275. Makushok, Stichaeoidea ..., 1958: 61, 74.

100 Body elongate, compressed laterally, and covered with smooth scales except for head. Pectoral fins large, rounded, more than half head length. Branchiostegal membranes fused and not attached to isthmus. Dorsal fin with hard spines, devoid of flexible rays. Anal fin with 2 spines. Pelvic fins absent (Soldatov and Pavlenko, 1915: 638). Branchiostegal rays 5. Interbranchial pores 5, preopercular pores 6. Suborbital canals continuous (Figure 43, A). Dorsal fin with 18 rays. Soft rays of anal fin bifurcate. Membranes of dorsal and anal fins touch only base of caudal fin (Makushok, 1958).

One species. Known from the Sea of Japan.

1. *Kasatkia memorabilis* Soldatov and Pavlenko, 1915 (Figure 77)

Kasatkia memorabilis Soldatov and Pavlenko, Ezhegodn. Zool. Muzeya Rossiisk Akad. Nauk, 20, 1915: 639, drawing (Peter the Great Bay). Soldatov and Lindberg, Obzor..., 1930: 444. Taranetz, Kratkii Opredeli-tel' ..., 1937: 154. Matsubara, Fish Morphol. and Hierar., 1955: 760. Fowler, Synopsis..., 1958: 276, fig. 24. Makushok, Stichaeoidea ..., 1958: 61.

18977. 42°31′ N, 131°14′ E. April 7, 1913. DVE. 1 specimen.

[92]D XLVIII; A II, 31; V I, 3 (Makushok, 1958).

116

26043. Sea of Japan. 1934. ZIN. 1 specimen.

D LXIII-LXV; A II, 45-47; P 18.[93]

Head elongate. Eyes relatively large. Mouth small, slightly oblique. Maxilla reaches vertical with anterior margin of eye. Teeth on jaws small and pointed; slightly smaller teeth present on vomer, no teeth on palatines. Dorsal fin originates slightly anterior to vertical with base of pectoral fin. First and last rays of dorsal fin reduced. Dorsal fin originates slightly anterior to vertical with base of pectoral fin. First and last rays of dorsal fin reduced. Anal fin originates closer to tip of snout than to base of caudal fin. Caudal and pectoral fins rounded. Pelvic fins absent. Vent situated immediately anterior to origin of fin. Color of specimens preserved in alcohol yellow. Dorsal fin with 12-14 dark spots; series of dull spots located on dorsum along base of dorsal fin; 12-13 indistinct longitudinal spots on sides of body. Dark oblique stripe continues onto cheeks from eye toward margin of operculum (Soldatov and Lindberg, 1930). Number of pores in seismosensory canals of head constant: 2 nasal, 4 interorbital, 7 postorbital, 7 suborbital, 5 occipital, 6 preopercular, and 4 mandibular (Makushok, 1961b).

101   Length, to 96 mm (Soldatov and Lindberg, 1930)

Distribution: In the Sea of Japan known from Peter the Great Bay and Tatar Strait (Taranetz, 1937b: 154).

15. Genus *Ascoldia* Pavlenko, 1910

Ascoldia Pavlenko, Ryby Zaliva Petr Velikii, 1910: 50 (type: *A. variegata* Pavlenko). Soldatov, Sb. v Chest' Knipovicha, 1927: 399. Soldatov and Lindberg, Obzor ..., 1930: 440. Taranetz, Kratkii Opredelitel' ..., 1937: 153. Shmidt, Ryby Okhotskogo Morya, 1950: 76. Matsubara, Fish Morphol. and Hierar., 1955: 760. Fowler, Synopsis..., 1958: 268. Makushok, Stichaeoidea ...,1958: 61.

Body elongate, compressed laterally, covered with small cycloid scales; head covered with very small scales. Mouth small, with fleshy lips. Jaws and vomer with small conical teeth, palatines without teeth. Lower jaw slightly reduced. Dorsal fin high.[94] Pectoral fins long, equal in length to head. Pelvic fins present but reduced to a single spine and 2-3 rudimentary rays. Gill openings do not continue far forward; branchiostegal membranes broadly fused and not attached to isthmus (Soldatov and Lindberg, 1930). Pores of seismosensory canals well developed on head: nasal 2, interorbital 5, postorbital 7, suborbital 3 + 2,[95] occipital 3, pre-

[93]D LXV-LXVII; A II, 47-49; principal rays C 7-8 + 7; P 17-18; vertebrae 70-72 (18 + 52-54) (Makushok, 1958).

[94]Anterior rays of fin thin and flexible, thicken gradually posteriorly and turn stiff, and often last rays convert into stout spines (Makushok, 1958).

[95]Suborbital canal interrupted in middle; 3 pores in lower part and 2 in upper.

opercular 6, and mandibular 4 (Makushok, 1961: 232). Body without lateral line.

One species and 1 subspecies. Former known from the Sea of Japan, and latter from adjacent waters.

Key to Subspecies of Ascoldia variegata

1 (2). Anal fin with 2 spines and 38–39 soft rays; V I, 2. Sea of Japan.
.......................1. **Ascoldia variegata variegata** Pavlenko.
2 (1). Anal fin with 2 spines concealed in skin and 34–36 soft rays; V I, 3. Sea of Okhotsk .... 1a. **Ascoldia variegata knipowitschi** Soldatov.

1. *Ascoldia variegata variegata* Pavlenko, 1910—Pavlenko's Red Prickleback (Figure 78)

Ascoldia variegata Pavlenko, Ryby Zaliva Petr Velikii, 1910: 50, fig. 9 (Askold Island). Soldatov and Lindberg, Obzor..., 1930: 440. Makushok, Stichaeoidea..., 1958: 61.

Ascoldia variegata variegata, Taranetz, Kratkii Opredelitel'..., 1937: 153. Matsubara, Fish Morphol. and Hierar., 1955: 760. Fowler, Synopsis..., 1958: 268, fig. 22.

40310. Peter the Great Bay. July 10, 1949. G.U. Lindberg. 1 specimen. D LVIII–LX; A II, 38–39; P 21; C 15; V I, 2 (Pavlenko, 1910).

Head 5 times and depth 4 times in standard length; depth of caudal peduncle 2 times in head; eyes 6.9 times in head length in adults and 4.6 times in young fish; interorbital space 4 times in head; pelvic fins 2 times in longitudinal diameter of eye; pectoral fins 1.5 times and snout 4 times in head length. Interorbital space very broad, convex, covered with
103 small scales. Mouth small; jaws and vomer with minute, barely discernible teeth. Entire body and head covered with cycloid scales. Pectoral fins well developed, pelvic fins present but very small, with one spine and 2 rudimentary rays.

Color of live fish red. Large number of diffuse greenish-yellow spots scattered on red background of body and dorsal fin. Number of spots on dorsal fin 8–9, more numerous on body (Pavlenko, 1910).

Length, to 430 mm.

Distribution: In the Sea of Japan known from Peter the Great Bay and off Askold Island (Pavlenko, 1910: 51).

1a. *Ascoldia variegata knipowitschi* Soldatov, 1927—Knipovich's Red Prickleback (Figure 79)

Ascoldia variegata knipowitschi Soldatov, Sb. v Chest' Knipovicha, 1927: 399, fig. 1 (Abrek Inlet, Shantar Islands). Soldatov and Lindberg, Obzor..., 1930: 440. Taranetz, Kratkii Opredelitel'..., 1937: 153. Shmidt, Ryby Okhotskogo Morya, 1950: 76. Matsubara, Fish Morphol. and Hierar., 1955: 760. Fowler, Synopsis..., 1958: 268.

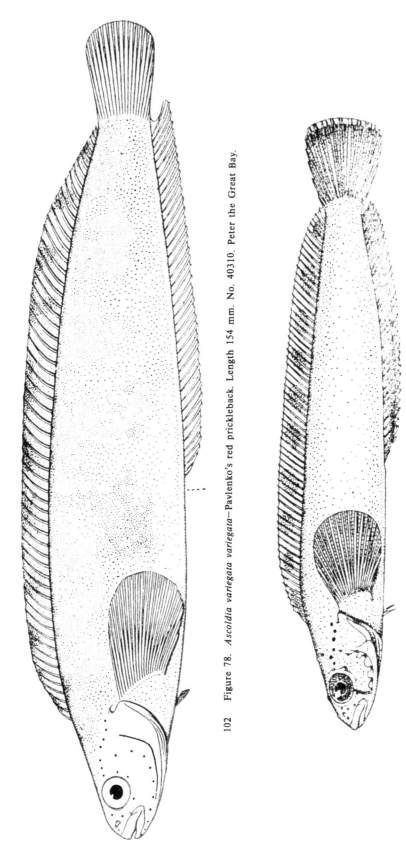

102　Figure 78.　*Ascoldia variegata variegata*—Pavlenko's red prickleback. Length 154 mm. No. 40310. Peter the Great Bay.

102　Figure 79.　*Ascoldia variegata knipowitschi*—Knipovich's red prickleback. Length 97 mm. Shantar Islands (Soldatov, 1927).

40374. Sea of Japan. Tatar Strait. August 12, 1949. KSE. 1 specimen.
41614. Sea of Okhotsk, Aniva Bay. December 27, 1972. KSE. 2 specimens.

D LVII-LIX; A II, 34-36; P 20-22; V I, 3.

Head 21.6-25.0%, body depth 15.8-21.1%, depth of caudal peduncle 7.7-8.2%, eyes 5.0-6.0%, interorbital space 4.8-5.4%, snout 5.0%, pectoral fins 17.0-17.4% and pelvic fins 3.5-3.8% of standard length. Two rays anterior to anal fin concealed in skin (Soldatov and Lindberg, 1930).

Interorbital space broad, convex, covered with minuscule scales. Eyes large. Black stripe continues from margin of eye downward. Dorsal fin originates anterior to vertical with posterior margin of operculum, consists of spines; anterior and posterior rays slightly shorter than middle ones, and membrane of last spine reaches base of upper rays of caudal fin. Almost all rays of dorsal fin stiff. Anal fin origin anterior to vertical at middle of standard length. Membrane of last ray of anal fin does not reach base of lower rays of caudal fin; membrane of anal fin deeply notched between tips of rays. Pectoral fins rounded, longer than half head length; caudal equal to pectoral fin in length and slightly rounded along posterior margin. Pelvic fins small, less than diameter of orbit.

Body color of specimens preserved in alcohol uniformly yellowish, without pattern; belly pale. Anal fin with narrow dark stripe along outer margin of fin. Dorsal fin with 9-10 dark spots. Body color in live fish light chocolate-brown, not red as in *A. v. variegata* Pavlenko (Soldatov, 1927).

Ascoldia variegata is extremely close to species of the genus *Opisthocentrus*, from which it differs only in the presence of pelvic fins. It differs further from *Opisthocentrus dybowskii* in the presence of scales on the head, and from *O. ocellatus* and *O. zonope* by the presence of 5 interorbital pores (Makushok, 1958).

Our specimens (41) varied in length from 55 to 118 mm; A II 35-36[96]; V I, 3; D LIX-LXI; ratio of head length to standard length 4-4.6. These fish were caught by KSE from Terpenia Gulf, Aniva Bay, and coastal waters of Šiaškotan Island in depths ranging from 1.5 to 60 m,[97] with fairly variable 104 bottom (stones and sand, sandy silt, silted sand, rocks, rocks and sand, rocks and shells, broken shells). Catch also included cod, flounder, and *Myoxocephalus*.

Judging from the incidentals in the catches, it may be stated that red algae,[98] laminaria, sponges, and bryozoans also occurred in the habitats of *Ascoldia variegata knipowitschi*. Clumps of *Zostera marina* and *Z. nana*, red seaweed, and hydroids were likewise present. Among invertebrates,

[96]In one specimen, A II, 37.
[97]In one case bottom temperature 0.2°C.
[98]Also *Suberites*, *Agarum*, and *Ectocarpus*.

Asteria amurensis, Ophiura sarsi, and white and chocolate-brown *Cucumaria* were encountered.

Length, to 440 mm (No. 41614).

Distribution: In the Sea of Japan known from Tatar Strait (No. 40374). Described from Abrek Inlet (Shantar Island) and found off Ola Island (Sea of Okhotsk). Our specimens were found in Aniva Bay, Terpenia Gulf, and coastal waters of Šiaškotan Island (Nos. 40360, 40361–40373), and from the southwestern coast of Paramoshir Island (No. 41614).

16. Genus Opisthocentrus Kner, 1868

Opisthocentrus Kner, Sitzb. Acad. Wiss., 58, 1868: 49 (type: *Centronotus quinquemaculatus* Kner). Soldatov and Lindberg, Obzor..., 1930: 443,. Taranetz, Kratkii Opredelitel'..., 1937: 153. Makushok, Stichaeoidea..., 1958: 61.

Pholidapus Bean and Bean, Proc. U.S. Nat. Mus., 19, 1896: 398 (type: *P. grebnitzkii*). Soldatov and Lindberg, Obzor..., 1930: 445. Taranetz, Kratkii Opredelitel'..., 1937: 154. Fowler, Synopsis..., 1958: 274.

Abryois Jordan and Snyder, Proc. U.S. Nat. Mus., 25, 1902: 486, fig. 22 (type: *A. azumae*). Matsubara, Fish Morphol. and Hierar., 1955: 761.

This genus differs from other related genera in absence of pelvic fins and medial interruption of suborbital seismosensory canal. All rays of dorsal fin spiny, anteriorly slender and flexible, gradually thickening and becoming stiff toward posterior end, and often convert into stout spines near caudal peduncle. Genus characterized by slope of 45° of pectoral fins and reduction in strength of spines of anal fin. Teeth on upper jaw arranged in 2–4 rows, lower with 1 row. Number of pores of seismosensory canals of head: nasal 2, interorbital 3 or 5, postorbital 7, suborbital 3 in lower and 2 in upper part of medially interrupted canal, occipital 3 or 6, preopercular 6, and mandibular 4 (Makushok, 1958).

3 species. All found in the Sea of Japan.

Key to Species of Genus Opisthocentrus

1 (4). Head covered with scales. 3 interorbital pores in seismosensory canal. Posterior maxillary teeth not larger than anterior ones. Height of middle rays of dorsal fin equal to or more than half body depth.

2 (3). D LVI–LXIII. Head with diffuse transverse stripes Dorsal fin with 5–6 ocellar spots..................... 1. **O. ocellatus** (Tilesius).

3 (2). D XLVIII–LV. Head with sharply delineated transverse stripes. Dorsal fin with less than 6 ocellar spots (usually 4)...........
.... 2. **O. zonope** Jordan and Snyder.

4 (1). Head naked. 5 interorbital pores in seismosensory canal. Posterior

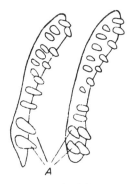

Figure 80. Maxillary teeth of *Opisthocentrus dybowskii*
(Makushok, 1958).
A—large teeth.

maxillary teeth larger than anterior ones (Figure 80).[99] Height of
middle rays of dorsal fin less than half body depth............
.... 3. **O. dybowskii** Steindachner.

1. ***Opisthocentrus ocellatus*** (Tilesius, 1811)—Ocellate Blenny (Figure 81)
Ophidium ocellatum Tilesius, Mem. Acad. Imp., St. Petersburg, 2, 1811:
237, pl. 8, fig. 2 (Kamchatka).

Opisthocentrus ocellatus, Shmidt, Ryby Vostochnykh Morei..., 1904:
180 (synonymy, description). Soldatov and Lindberg, Obzor..., 1930: 443
(synonymy, description). Taranetz, Kratkii Opredelitel'..., 1937: 154.
Shmidt, Ryby Okhotskogo Morya, 1950: 76. Matsubara, Fish Morphol.
and Hierar., 1955: 760, fig. 279. Fowler, Synopsis..., 1958: 271. Abe, Enc.
Zool., 2, Fishes, 1958: 112, fig. 330 (color diagram). Makushok,
Stichaeoidea..., 1958: 61.

Opisthocentrus ochotensis Ueno, Japan. J. Ichthyol., 3 (4-5), 1954: 102-
106, figs. 1-4 (male).

D LVI-LXIII; A II, 31-42.[100]

In addition to differences mentioned in the key to species, this species
is characterized by the absence of a stripe continuing from the origin of
the dorsal fin through the base of the pectoral (Soldatov and Lindberg,
1930). P.Yu. Shmidt noted that the available description required
supplementary information, namely, oral cavity with palatine and
mandibular membranes. Upper lip continuous, lower lip interrupted in
middle. Upper part of gill opening with siphon formed by fleshy margin of
operculum and very short fleshy fold attached to trunk (Shmidt, 1950).

[99]This peculiarity is an exception in the subfamily Opisthocentrinae.
[100]D LVIII-LXII; A II, 37-39; principal rays C 7-8 + 6-7; P 20-21; vertebrae 63-67 (22-
23 + 40-44) (Makushok, 1958).

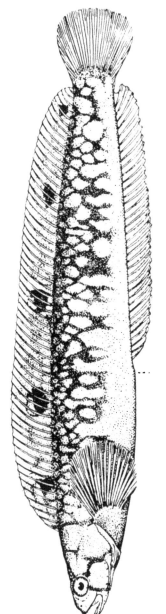

105   Figure 81.   *Opisthocentrus ocellatus*—ocellate blenny. Length 160 mm (Jordan and Snyder, 1902).

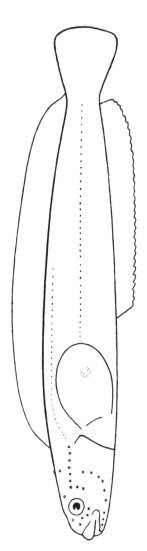

106   Figure 82.   Seismosensory system in *Opisthocentrus ocellatus* (Makushok, 1958).

Color of live fish reddish; reticulate gray spots on sides continue onto dorsal fin on which ocellar spots additionally located.[101] Eye intercepted by brown stripe that continues from occiput toward chin; similar brown stripe originates from posterior corner of eye and extends toward angle of operculum (Shmidt, 1904).

Some males with flexible spines in anterior part of dorsal fin, differing from hard ones, which can be greatly elongate. *O. ocellatus* has upper branch of open seismosensory papillae situated parallel to median line and usually interrupted in middle of trunk (Figure 82) (Makushok, 1958).

106   Our specimens (41; Nos. 31586-31589, 40281; 40304-40393) 34 to 167 mm long, D LVI-LXI, with 5 to 6 rounded dark spots on dorsal fin and 3-4 indistinct dark stripes on head (well developed in large specimens) directed from eye downward toward interbranchial space and lower angle of operculum and upward toward occipital region. In smaller fish, 34 to 52 mm, these stripes almost indiscernible. Color also retained in some specimens in form of dark network. Fish were caught in Tatar Strait near Moneron Island, in Aniva Bay, and off Šiaškotan Island from depths of 1 to 68 m, from a stony or silted sand bottom. They were encountered in catches mainly comprising flounders and walleye pollock. Occurrence of red algae, laminaria, sponges, bryozoans, ascidians, and hydroids was incidental. Catch likewise included *Cucumaria japonica, Asterias amurensis, Triglops jordani, Podothecus gilberti,* and *Lycodes palearis fasciatus.*

Length, to 171 mm (Soldatov and Lindberg, 1930).

Distribution: In the Sea of Japan from Peter the Great Bay north to Tatar Strait (Soldatov and Lindberg, 1930: 444); off Moneron Island (No. 13022); Primor'e indicated for Olga Bay and Soviet Gavan (Popov, 1933: 149). Reported from Oshoro Bay (Kobayashi, 1962: 258), Otaru, Hakodate, Aomori (Shmidt, 1950: 76), Toyama Bay (Katayama, 1940: 24), Sado Island (Honma, 1963: 6). Found near the Korean Peninsula, Sonen Inlet, Port of Shestakov (Shmidt, 1950: 76). Widely represented in the Sea of Okhotsk from liman of Amur River to western coast of Kamchatka, Terpenia Gulf, Aniva Bay, and farther up to Šiaškotan Island (KSE). In the north found to the Bering Sea (Andriyashev, 1939b: 84). Along the Pacific coast of Japan found in Lake Notoro (Hikita, 1952: 13), near Nemuro (Franz, 1910: 85), in Volcano Bay (Sato and Kobayashi, 1956: 13), near Muroran, and reported from off Nagasaki (Shmidt, 1950: 76), which needs confirmation.

---

[101]Number of ocelli on dorsal fin probably a variable character since different researchers report different numbers. Abe (1958, Fig. 330) depicts 6; Jordan and Snyder (1902, Fig. 20) and Steindachner (1880) have published figures with 5. Shmidt (1904) indicated 4-5 for fish from the Gulf of Busse; our specimens had 5-6.

2. **Opisthocentrus zonope** Jordan and Snyder, 1902—Girdled Blenny (Figure 83)

Opisthocentrus zonope Jordan and Snyder, Proc. U.S. Nat. Mus., 25, 1902: 485, fig. 21 (Muroran, Hokkaido). Soldatov and Lindberg, Obzor..., 1930: 444. Taranetz, Kratkii Opredelitel'..., 1937: 253. Matsubara, Fish Morphol. and Hierar., 1955: 760. Fowler, Synopsis..., 1958: 272, fig. 23.

18691. Peter the Great Bay. October 13, 1912. DVE 2 specimens.

18692. Ussuri Bay. October 1, 1912. DVE. 2 specimens.

18975. Peter the Great Bay. September 25, 1911. DVE. 2 specimens.

37297. Sea of Japan, Ussuri Gulf. 1962. ZIN. 1 specimen.

40280. Sea of Japan. August 28, 1970. E. Tsimbalyuk. 1 specimen.

40560. Vladivostok. May 25, 1947. E.P. Rutenberg. KSE. 2 specimens.

107   D LI; A II, 33[102] (Jordan and Snyder, 1902a).

Description of species based on characters given in genus and in key to species.

Color slightly olive-green, body sides with distinct randomly arranged stripes, spots, and lines, which in some specimens form reticulate pattern; head with sharply demarcated narrow dark stripes; narrow stripes extend 108 from origin of dorsal fin downward and intercept base of pectoral fin; base of caudal fin with narrow vertical dark stripe (Jordan and Snyder, 1902a).

Our specimens (10): D L-LV, fin with 4-5 dark rounded spots. Other characters accord with description of species.

Length to 118 mm (Soldatov and Lindberg, 1930).

Distribution: In the Sea of Japan known from Peter the Great Bay, in Primor'e reported from Olga Bay and Soviet Gavan (Soldatov and Lindberg, 1930: 444). Found in Oshoro Bay (Kabayashi, 1962: 258), near Otaru (Jordan and Snyder, 1902a: 486).

3. **Opisthocentrus dybowskii** (Steindachner, 1880)—Dybowskii's Blenny (Figure 84)

Centronotus dybowskii Steindachner, Sitzb. Akad. Wiss., 82, 1880: 259 (Strelok Bay).

Abryois azumae Jordan and Snyder, Proc. U. S. Nat. Mus., 25, 1902: 488, fig. 22 (Muroran). Matsubara, Fish Morphol. and Hierar., 1955: 761.

Pholidapus dybowskii, Jordan and Snyder, Proc. U. S. Nat. Mus., 25, 1902: 488. Shmidt, Ryby Vostochnykh Morei..., 1904: 148 (synonymy, description). Soldatov and Lindberg, Obzor..., 1930: 447 (synonymy, description). Taranetz, Kratkii Opredelitel'..., 1937: 153. Shmidt, Ryby Okhotskogo Morya, 1950: 77. Fowler, Synopsis..., 1958: 275.

---

[102]D XLVIII-L; A II, XXXIV-XXXVII*; principal rays C 7-8 + 6-7; P 20-21; vertebrae · 56-58 (19-20 + 36-38) (Makushok, 1958).

*Apparent misprint in Russian original as notation conveys spiny rays when, in fact, the second part of the anal fin contains soft rays—General Editor.

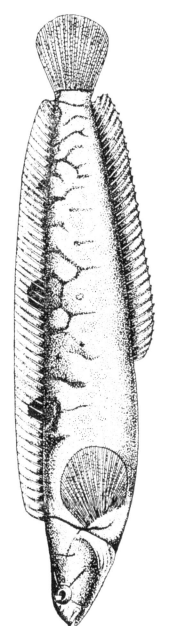

Figure 83. *Opisthocentrus zonope*—girdled blenny. Length 125 mm. Hokkaido (Jordan and Snyder, 1902).

107

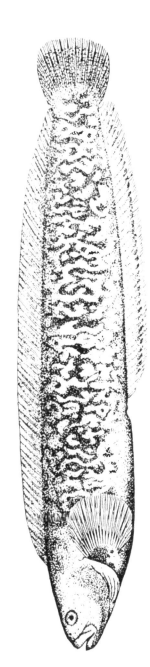

Figure 84. *Opisthocentrus dybowskii*—Dybowski's blenny. Length 400 mm. Hokkaido (Jordan and Snyder, 1902).

107

Opisthocentrus dybowskii, Makushok, Stichaeoidea..., 1958: 61.
D LVII–LXIV; A II, 36–44.[103]

More oblong and broader body than in closely related species, covered with small smooth scales, head naked. Mouth small, horizontal or slightly oblique. Teeth on jaws and vomer; palatines without teeth. One or two highly conical teeth on premaxilla behind narrow band of teeth. Posterior rays of dorsal fin thicker and slightly lower than remaining ones. Pectoral fin longer than caudal. Anal fin lower than dorsal fin. Body color variable. Number of ocelli usually not more than four, sometimes completely absent (Soldatov and Lindberg, 1930: 447).

Our specimens (130; Nos. 40562–40574) ranged in length from 14 to 430 mm; D LX–LXIV; A II, 38–41; dorsal fin with 1 to 4 rounded dark spots usually located in anterior part of fin. Fish collected by KSE in Tatar Strait, off southwestern coast of Sakhalin, in Aniva Bay, and off Šiaškotan Island, from silty bottom at shallow depths (up to 2 m), usually with laminaria, Zosteraceae, and algae.

Length, to 460 mm (Soldatov and Lindberg, 1930).

Distribution: Widely represented in the Sea of Japan: off Wonsan (Taranetz, 1937b: 154), in Peter the Great Bay and north to Tatar Strait (Soldatov and Lindberg, 1930: 448). Known near Otaru (Jordan, Tanaka and Snyder, 1913: 392). In the Sea of Okhotsk found in Aniva Bay, also Terpenia Gulf, north to Avachinskaya Inlet (Shmidt, 1950: 77). Southward found to Kuril Islands (Taranetz, 1937b: 154), Šiaškotan Island (KSE). Sea of Okhotsk and Pacific coast of Hokkaido, and reported from Lake Notoro (Hikita, 1952: 12) and Volcano Bay (Shmidt, 1950: 77). Farther south known from Sagami Bay (Franz, 1910: 85).

5. Subfamily Alectriinae

Makushok, Stichaeoidea..., 1958: 96.

Body moderately elongate. Embedded scales retained only in posterior part of body or completely absent (*Pseudalectrias*). Head with snout-occipital fleshy crest. Mouth relatively large (extends up to vertical posterior to eye or beyond it as in *Alectrias alectrolophus benjamini*). Teeth present on vomer and palatines (absent on palatines in *Pseudalectrias*). Branchiostegal membranes broadly fused and not attached to isthmus, or broadly fused with it (*Anoplarchus*). Branchiostegal rays 5. Seismosensory canals of head normally developed, opening exteriorly through constant number of pores: nasal 2, interorbital 4, suborbital 1, postorbital 7, preopercular 6, and mandibular 4[104] (Figure 85, A, B).

109

---

[103]D LXI–LXV; A II, 38–40; principal rays C 7–8 + 7; P 18–19; vertebrae 67–70 (23–24 + 43–46) (Makushok, 1958).

[104]Numbers differ in *Pseudalectrias* (Figure 86, A, B).

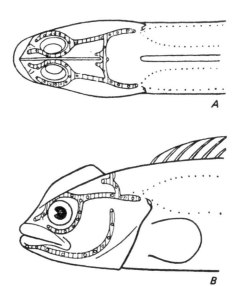

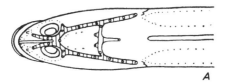

109   Figure 85. Seismosensory system of head in *Alectrias alectrolophus*
(Makushok, 1958).

A—dorsal view; B—lateral view.

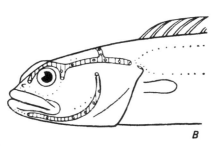

109   Figure 86. Seismosensory system of head in *Pseudalectrias tarsovi*.
A—dorsal view; B—lateral view.

Lateral line of body in form of middle and upper branches of free seismosensory papillae. Vertebrae 62-69, precaudal 16-21. Spiny rays of dorsal fin slender and flexible in anterior part, gradually becoming thicker and stiffer posteriorly. One undeveloped spine located at origin of anal fin. Principal rays of caudal fin 12-14. Membranes of unpaired fins usually fused with base of caudal fin. Pectoral fins small or highly reduced (*Pseudalectrias*), with 8-10 rays. Pelvic fins absent.

Loss of one ray of branchiostegal membranes (5 versus 6), reduction of scales, unique structure of rays of dorsal fin, loss of pelvic fins, and partial reduction of pectoral fins, are traits of specialization which, in the opinion of Makushok (1958: 54), have developed as a result of adaptation of movement in the supralittoral zone.

Distributed in the northern part of the Pacific Ocean. Inhabitants of pebbled-gravelly shores in the littoral zone. Parents protect their eggs. Feed on minute benthic invertebrates (polychaetes, mollusks, and crustaceans).

3 genera. Two known from the Sea of Japan.

110   **17. Genus *Alectrias* Jordan and Evermann, 1898**

Alectrias Jordan and Evermann, Bull. U. S. Nat. Mus., 47, 1898: 2869 [type: *Anoplarchus alectrolophus* (Pallas) = *Blennius alectrolophus* Pallas]. Jordan and Snyder, Proc. U. S. Nat. Mus., 1902: 475. Shmidt, Ryby Vostochnykh Morei..., 1904: 176. Hubbs, Pap. Michigan Acad. Sci., 7, 1927: 371. Lindberg, Genera and Species of Fishes of the Family Blenniidae, 1938: 499. Shmidt, Ryby Okhotskogo Morya, 1950: 69. Matsubara, Fish Morphol. and Hierar., 1955: 763. Andriyashev, Ryby Severnykh Morei SSSR, 1954: 238. Fowler, Synopsis..., 1958: 286. Makushok, Stichaeoidea..., 1958: 61.

Alectridium Gilbert and Burke, Bull. Bur. Fish., 30, 1910: 87 (type: *A. aurantiacum*). Soldatov and Lindberg, Obzor..., 1930: 459. Taranetz, Kratkii Opredelitel'..., 1937: 155. Shmidt, Ryby Okhotskogo Morya, 1950: 70. Matsubara, Fish Morphol. and Hierar., 1955: 764. Fowler, Synopsis..., 1958: 288.

An addition should be made to the description of the subfamily, namely, that in members of the genus *Alectrias* the upper and lower jaw teeth are arranged in 2-4 central rows (Figure 87, A, B), and the occipital pit is absent. Branchiostegal rays 5 (Makushok, 1958).

3 species. 2 known from the Sea of Japan, and 1 from adjacent waters.

Key to Species of Genus Alectrias[105]

1 (2). Cirri present above eyes. P 8-9. Length of pectoral fins 6% of standard length. Crest on head high.... 1. **A. cirratus** (Lindberg).

[105]From Lindberg (1938), with additions (Taranetz, 1937b).

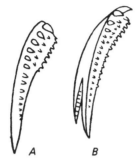

Figure 87. Jaw teeth of *Alectrias alectrolophus* (Makushok, 1938).
A—upper jaw; B—lower jaw.

2 (1). Cirri not present above eyes.
3 (4). Crest on head low, with process in interorbital space resembling cockscomb. P 10. Length of pectoral fins 8% of standard length.
.... 2. [**A. gallinus** (Lindberg)].
4 (3). Crest on head without process. P 11. Length of pectoral fins 5% of standard length.
5 (6). D LX-LXVI; A 41-45[106]...
.... 3. **A. alectrolophus alectrolophus** (Pallas).
6 (5). D LIII-LXI; A 36-41[107] ....
.... 3a. **A. a. benjamini** Jordan and Snyder.

1. *Alectrias cirratus* (Lindberg, 1938) (Figure 88)
Alectridium cirratum Lindberg, O Rodakh i Vidakh Ryby Sem. Blenniidae, 1938: 505, fig. 4 (Peter the Great Bay). Matsubara, Fish Morphol. and Hierar., 1955: 764.
Alectrias cirratus, Makushok, Stichaeoidea..., 1948: 61, 99 (comparative notes).
18856. Vladimir Bay. September 17, 1913. DVE. 1 specimen.
25160. Peter the Great Bay. August 17, 1937. A.I. Savinov. 1 specimen. D LX; A XLII; P 8-9.[108]
Body and head compressed laterally, crest on head well developed,
111 high, slightly dentate on upper side, not smooth, extending from snout to midpoint of occiput; distance from pore behind crest to origin of dorsal fin 2.5% of standard length. Well-developed lobate cirrus located above each eye, situated on upper rim of eyeball. Upper jaw extends to vertical with posterior margin of eye. Posterior part of body covered with small rounded scales embedded in skin and adjacent (not overlapping) (Lindberg, 1938).

[106]Body depth 7 3/5 in standard length (Fowler, 1958: 286).
[107]Body depth 5 4/5 in standard length (Fowler, 1958: 286).
[108]D LVII-LX; A I XLI-XLIV; C 6-7 + 6 (7); vertebrae 61-65 (16-18 + 45-47) (Makushok, 1958).

Length, to 97 mm.

Distribution: In the Sea of Japan known from Peter the Great Bay and Vladimir Bay.

2. [*Alectrias gallinus* (Lindberg, 1938)] (Figure 89)

Alectridium gallinum Lindberg, O Rodakh i Vidakh Ryb Sem. Blennii-dae, 1938: 506, fig. 5 (Sea of Okhotsk, Ukoi). Shmidt, Ryby Okhotskogo Morya, 1950: 70. Matsubara, Fish Morphol. and Hierar., 1955: 764. Fowler, Synopsis..., 1958: 288.

Alectrias gallinus, Makushok, Stichaeoidea..., 1958: 61, 99 (comparative notes).

40358. Sea of Okhotsk, Terpenia Gulf. September 8, 1947. KSE. 2 specimens.

40359. Sea of Okhotsk, Terpenia Gulf. October 1, 1947. KSE. 1 specimen.

D LXI; A I, 44; P 10.[109]

Body and head compressed laterally. Crest on head moderately developed, with unique process between eyes giving crest appearance of a cockscomb. Upper margin. of crest, excluding process, smooth or edentate. Crest extends from apex of snout almost to origin of dorsal fin; distance from pore behind crest up to origin of dorsal fin 1.2% of standard length. Cirri absent above eyes. Upper jaw extends to vertical with posterior margin of pupil. Pigmented spots arranged beyond midpoint of posterior part of body. Pectoral fins relatively long (1/2 head length) (Lindberg, 1938).

Three specimens of this species from the collection of KSE ranged in length from 78 to 98 mm. D LXI; A I, 44; P 10; length of pectoral fins 8.3-8.8% of standard length; ratio of body depth to this length 7.0-7.8. Fish caught at a depth of 48 m with a bottom temperature of 2.3°C. Bottom in region of catches comprised pebbles, gravel, and silted green sand. Incidental catch comprised numerous hydroids, sponges, bryozoans, and exoskeletal parts of the crab *Paralithodes camtschaticus*; among fishes only *Melletes papilio* was found.

Length, to 98 mm.

Distribution: Not known from the Sea of Japan. Sea of Okhotsk, near Cape Ukoi, Erineiskii Gulf, and Tanisk Inlet (Lindberg, 1938: 507). Specimens of KSE from Terpenia Gulf (Nos. 40358, 40359).

3. *Alectrias alectrolophus alectrolophus* (Pallas, 1811)—Stone Cockscomb (Figure 90)

Blennius alectrolophus Pallas, Zoogr. Rosso-Asiat., 3, 1811: 174 (Penzha Gulf).

---

[109]D LXI-LXIII; A I, 44; C 6-7 + 6 (7); vertebrae 65-67 (18-19 + 47-48) (Makushok, 1958).

Alectrias alectrolophus, Jordan and Evermann, Fish N. and M. Amer., 1898: 2869. Shmidt, Ryby Vostochnykh Morei..., 1904: 174 (partly, specimens of the Sea of Okhotsk). Andriyashev, Ryby Severnykh Morei SSSR, 1954: 238, fig. 122. Makushok, Stichaeoidea..., 1958: 61, 99 (comparative notes).

Alectrias alectrolophus alectrolophus Hubbs, Pap. Michigan Acad. Sci., 7, 1927: 371 (synonymy). Lindberg, O Rodakh i Vidakh Ryb Sem. Blenniidae, 1938: 502, fig. 3. Shmidt, Ryby Okhotskogo Morya, 1950: 70. Matsubara, Fish Morphol. and Hierar., 1955: 763.

Anoplarchus alectrolophus, Jordan and Evermann, Fish N. and M. Amer., 1898: 2421. Soldatov and Lindberg, Obzor..., 1930: 459 [partly, specimens from stations of TONS (5) and DVE (46, 161)].

Anoplarchus alectrolophus alectrolophus, Taranetz, Kratkii Opredelitel'..., 1937: 155.

Alectridium aurantiacum Gilbert and Burke, Bull. Bur. Fish., 30, 1910: 87, fig. 31. Soldatov and Lindberg, Obzor..., 1930: 459. Taranetz, Kratkii Opredelitel'..., 1937, 155. Shmidt, Ryby Okhotskogo Morya, 1950: 70. Matsubara, Fish Morphol. and Hierar., 1955: 764.

443   *Addendum**: Pinchuk (1974) confirmed on the basis of new studies the conclusion of Peden (1967) regarding the independent status of the species *Alectridium aurantiacum* Gilbert and Burke, 1942.

112   D LXI-LXII (LIX-LXIV); A 42-44; P 10.[110]

Crest on head well developed, with significant height in occipital region; distance from pore behind crest to origin of dorsal fin 2.2-2.3% of standard length. Lateral line continuous along side, well developed; dorsal line of pores also well defined (Lindberg, 1938).

113   Upper part of gill opening with siphon formed from dermal outgrowth of operculum as well as short dermal fold on trunk, which does not reach base of pectoral fin. Lower lip interrupted for short distance. Palatine and mandibular membranes present in oral cavity (Shmidt, 1950).

Seismosensory system of head in this species does not differ from that in other species of *Alectrias* (Makushok, 1958)

Color of live fish highly variable, from dull gray or almost black to bright pattern of spots. Sinuous line usually extends along dorsum, separating dark portion from light-colored spots that continue onto dorsal fin (Andriyashev, 1954).

Our specimens (50, of which 20 examined) from KSE ranged in length

[110]D LVIII-LXIV; A I, 42-46; C 6-7 + 6-7; P 9-10; vertebrae 62-69 (16-19 + 45-50) (Makushok, 1958).

*The original Russian book contains addenda and corrections to the text on p. 443. These addenda and corrections have now been included in the flow of the text at the appropriate places with indication of the type of change and page number of the original book in the left-hand margin—General Editor.

132

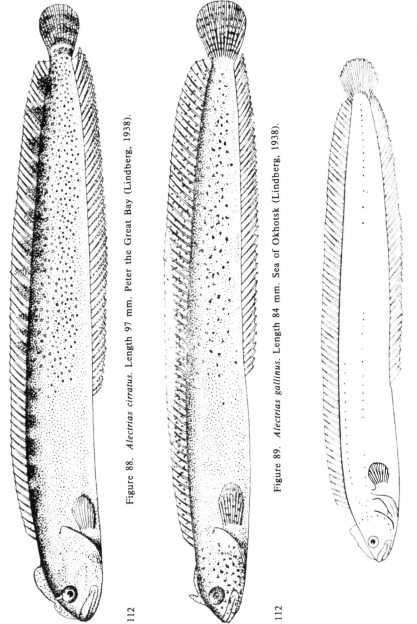

Figure 88. *Alectrias cirratus*. Length 97 mm. Peter the Great Bay (Lindberg, 1938).

112

Figure 89. *Alectrias gallinus*. Length 84 mm. Sea of Okhotsk (Lindberg, 1938).

112

Figure 90. *Alectrias alectrolophus alectrolophus*—stone cockscomb. Length 100 mm. Sea of Okhotsk (Andriyashev, 1954).

112

from 46 to 110 mm. D LXI-LXVI; A 42-45; P 10; ratio of body depth to standard length 7.1-8.5 times of length of pectoral fin 5-7.2. Fish caught in Aniva Bay, near Šiaškotan and Kunashir Islands, at depths of 2.5 to 50 m. Bottom stony or with broken shells and gravel.

Length, to 128 mm (Shmidt, 1904).

Distribution: In the Sea of Japan known in De-Kastri Bay; not found farther south. Northern part of the Sea of Okhotsk, Bering Sea, and farther east to west coast of Alaska (Lindberg, 1938: 503). Like *Alectridium aurantiacum,* found near Aniva Bay (Shmidt, 1950: 70 and collections of KSE, No. 31699). Known from near Šiaškotan and Kunashir Islands (KSE, Nos. 40395-40399; 40551-40559).

3a. *Alectrias alectrolopus benjamini* Jordan and Snyder, 1902
(Figure 91)

Alectrias benjamini Jordan and Snyder, Proc. U.S. Nat. Mus., 25, 1902: 475, fig. 16 (Hakodate). Fowler, Synopsis..., 1958: 287.

Alectrias alectrolophus benjamini Hubbs, Pap. Michigan Acad. Sci., 1927: 372. Lindberg, O Rodakh i Vidakh Ryb Sem. Blenniidae, 1938: 503. Shmidt, Ryby Okhotskogo Morya, 1950: 70. Matsubara, Fish Morphol. and Hierar., 1955: 763.

Alectrias alectrolophus, Shmidt, Ryby Vostochnykh Morei..., 1904: 176 (partly, specimens from Vladivostok). Pavlenko, Ryby Zaliva Petr Velikii, 1910: 70.

Anoplarchus alectrolophus, Soldatov and Lindberg, Obzor..., 1930: 459 (partly, stations of DVE 207, 210, 256, 331; TONS 40, 140, 172, 273, 289).

18066. Primor'e. Preobrazheniya Inlet. October 6, 1908. Derbek. 1 specimen.

20472. Amur Bay. July 19, 1898. M. Yankovskii. 1 specimen.

25973. Sea of Japan. October 7, 1934. ZIN. 1 specimen.

34340. West coast of Sakhalin. August 20, 1933. A. Kuznetsov. 1 specimen.

D LV; A 1, 41.

Body depth equal to head length. Head large. Mouth oblique. Maxilla 114 extends beyond vertical with posterior margin of eye. Jaws equal in length. Interorbital space convex. Teeth minute, pointed, arranged on jaws in narrow band, outer teeth largest; vomer and palatines with narrow bands of minute teeth. Gill rakers about 12; pseudobranchs large. Head with distinct crest extending from tip of snout to occiput; maximum height of crest slightly less than diameter of eye. Head naked, without dermal cirri. Dorsal fin originates at vertical with base of pectoral fin, continues to caudal fin, and fuses with latter; fin membrane unnotched between tips of spines. Caudal fin rounded posteriorly and its length half head length; pectoral fins also rounded, 2.5 times in the head length.

Color of specimens preserved in alcohol yellowish-olive-green, darker on upper side than on lower. Several whitish spots, longer than eye, arranged along dorsum; background of dorsum between these spots darker, light-colored spots themselves mottled black, and several smaller spots extend along median line of body side. Cheeks, chin, and throat marked with minute black dots; crest with 4 dark vertical stripes; anal fin with alternate white and black spots near base. Caudal fin with indistinct dark- and light-colored vertical stripes. Pectoral fins light-colored, with a few dark stripes (Jordan and Snyder, 1902a).

Length, to 95 mm (Jordan and Snyder, 1902a).

Distribution: In the Sea of Japan known from Poset Bay (Taranetz, 1937b: 155); Peter the Great Bay, Olga Bay, Vladimir Bay (Soldatov and Lindberg, 1930: 459). In the north found to Aleksandrovsk-Sakhalin (Taranetz, 1937a: 39); described from Hakodate, reported from Sado Islands (Honma, 1963: 21); Gulf of Chihli (Bohai) in Yellow Sea (Zhang et al., 1955: 173); and Chefoo (Wang and Wang, 1935: 217). In the Sea of Okhotsk found in Aniva Bay (Shmidt, 1950: 70). Reported from Aikawa and Muroran (Snyder, 1912: 449).

18. Genus *Pseudalectrias* Lindberg, 1938

Pseudalectrias Lindberg, O Rodakh i Vidakh Ryb Sem. Blenniidae, 1938: 507 (type: *Alectrias tarasovi* Popov). Matsubara, Fish Morphol. and Hierar., 1955: 764. Fowler, Synopsis..., 1958: 290. Makushok, Stichaeoidea..., 1958: 61, 100 (comparative notes).

Pseudalectrias, in which branchiostegal membranes are not attached to the isthmus and broadly fused with each other, differs from *Alectrias* in these characters: complete reduction of scale cover; absence of palatine teeth; reduction of pyloric caeca and dermal crest; reduction of pectoral fins and primary elements of shoulder girdle; reduction in number of preopercular, mandibular, and suborbital pores; and laterally compressed skull (Makushok, 1958: 100).

One species. Known only only from the Sea of Japan.

1. *Pseudalectrias tarasovi* (Popov, 1933) (Figure 92)
   Alectrias tarasovi Popov, Issled. Morei SSSR, 19, 1933: 150 (De-Kastri).
   Pseudalectrias tarasovi, Lindberg, O Rodakh i Vidakh Ryb Sem. Blenniidae, 1938: 507, fig. 6 (Petrov Island, Sea of Japan). Matsubara, Fish Morphol. and Hierar., 1955: 764. Makushok, Stichaeoidea..., 1958: 61. Fowler, Synopsis ..., 1958: 290, fig. 28.
   25308. Sea of Japan, Petrov Island. September 12, 1934. ZIN. 2 specimens.

443 *Addendum*: Pinchuk (1974), based on an examination of 13 specimens,

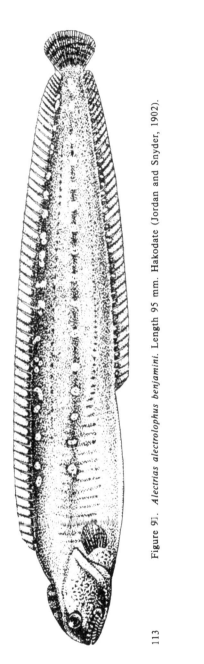

113

Figure 91. *Alectrias alectrolophus benjamini*. Length 95 mm. Hakodate (Jordan and Snyder, 1902).

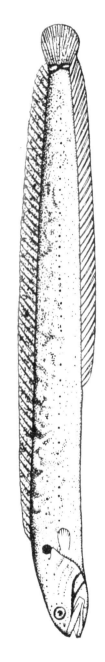

115

Figure 92. *Pseudalectrias tarasovi*. Length 117 mm. Tatar Strait (Popov, 1933).

has provided further data on color in vivo, range of dimensions (up to 133 mm), and distribution (Šiaškotan Island).

D XXVI-XXXV (61); A 43 (Lindberg, 1938).[111]

115   Body entirely naked, moderately elongate, compressed laterally, and uniformly and gradually attenuate toward caudal fin. Least body depth near base of last spine on dorsal fin 2.5 times in maximum depth. Membranes of dorsal and anal fins fused with caudal fin. Dorsal fin with two types of rays: about 25-26 rays in anterior part of fin soft and flexible, simple, and nonsegmented; remaining 35 rays stiff and spiny. Rays of anal fin soft and unbranched. Makushok (1958: 33) has reported that only a small number of rays of this fin have retained traces of branching, the others have completely lost this character and converted into simple, unbranched soft rays. Head moderate in size, about 7 times in standard length, and without scales. Crest poorly developed, located only in region of snout and interorbital space, and does not fuse on occiput. Eyes small, about 5.5 times in head length. Mouth large, slightly oblique, and conical. Posterior end of upper jaw distinctly continues beyond posterior margin of eye (Lindberg, 1938). Maxillary teeth arranged in 2-4 rows, and teeth on dentary in 1 row. Size of teeth reduces from front to back (as in Opisthocentrinae, Stichaeus, and Xiphisterinae). Occipital pit present. Pores of seismosensory canals of head (Figure 86, A, B): nasal 1, interorbital 3, postorbital 6, suborbital 1, occipital 4, preopercular 5, mandibular 2 (Makushok, 1961b).

Body color of preserved specimens dark chocolate-brown, with no sharply expressed pattern; dark spot present near upper margin of gill opening; 2 white spots located near base of caudal fin. Oblique light-colored stripe distinct on head (Lindberg, 1938).

The specimen described by A.P. Popov was caught in May under stones in De-Kastri Bay near Bazal'tovogo Island.

Length, to 117.6 mm (Popov, 1933a).

Distribution: Known only from the Sea of Japan, De-Kastri Bay, and off Petrov Island (Lindberg, 1938: 509).

6. Subfamily Xiphisterinae

Makushok, Stichaeoidea..., 1958: 100.

Body elongate, compressed laterally, covered with minute imbricate scales. Head naked, with well developed snout-occipital fleshy crest in Cebidichthys[112]; in some members of Dictyosoma rudiment of this struc-

---

[111]D LXI-LXII; A I, 43-44; principal rays C 6 + 6; P 9-10; vertebrae 64-65 (20-21 + 44-45) (Makushok, 1958).

[112]Distributed along coast of California (Norman, 1957: 467). Makushok (1957) thinks that the fleshy crest on head is associated with some peculiarities of life in the supralittoral zone. This structure is seen in many littoral members of Blenniidae and in species of Neozoarces.

ture evident, while in *Phytichthys*[113] and *Xiphister*[114] it is totally absent.
16 Mouth small. Teeth present on vomer and palatines (*Cebidichthys.*
Dictyosoma) or absent. Seismosensory canals of head well developed, only
some open exteriorly through constant number of pores: nasal 2, interorbital 5, preopercular 6 or 7 (*Dictyosoma*), mandibular 4 (*Dictyosoma*) or
3. Number of other pores highly variable. Seismosensory canals of trunk
well developed; one (upper) canal present in *Cebidichthys,* complex
network of anastomoses in *Dictyosoma,* and 4 canals on each side
with regular branches terminating blindly in *Phytichthys* and *Xiphister.*
Vertebrae 68-81: precaudal 19-31. Dorsal fin with only short spines
(*Phytichthys, Xiphister*) or consistently elongate spines in anterior part, or
soft branched rays in posterior part (*Cebidichthys, Dictyosoma*). Anal fin
origin with 1 undeveloped spine (*Xiphister*), 2 (*Dictyosoma*), or 2-3
spines (*Phytichthys*). Principal rays of caudal fin 13-14 (*Dictyosoma*).
or 12 (*Xiphister*). Pectoral fins small (*Cebidichthys, Dictyosoma*),
considerably reduced (*Phytichthys*), or rudimentary (*Xiphister*), with
10-12 rays. Pelvic fins absent. As in the subfamily Alectriinae, with
adaptation to the supralittoral zone, body elongation is observed in
members of the subfamily Xiphisterinae, which is related to serpentine
movements, reduction of paired fins, and loss of pelvic fins. Probably the
complexity of the seismosensory system is also a specialization.

Distribution: Northern part of the Pacific Ocean along the American
coast. Only the genus *Dictyosoma* is represented along the Asian coast,
and is endemic to the Sea of Japan. Dwell in the littoral zone in clumps
of algae. Parents protect their eggs (Makushok, 1958).

4 genera. 1 known from the Sea of Japan.

19. Genus *Dictyosoma* Schlegel, 1846

Dictyosoma Schlegel, Fauna Japonica, Poiss., 1846: 139 (type: *D.
buergeri* Van der Hoeven). Jordan and Snyder, Proc. U.S. Nat. Mus., 1902:
481. Soldatov and Lindberg, Obzor..., 1930: 448 Taranetz, Kratkii
Opredelitel'..., 1937: 154. Matsubara, Fish Morphol. and Hierar., 1955:
761. Fowler, Synopsis..., 1958: 269. Makushok, Stichaeoidea..., 1958:
61, 104 (comparative notes).

Characters of the genus given in description of the subfamily.

It should be noted that the genus *Dictyosoma* is characterized by:
doubling of row of pores of some seismosensory canals of head (posterior
part of supraorbital, postorbital, occipital, and posterior part of suborbital
canals) (Figure 93); increase in number of preopercular pores (7 versus
6); and development of complex network of seismosensory canals of trunk

---

[113]Distributed from Alaska to California, especially the Aleutian Islands (Norman, 1958:
464).

[114]Distributed from Alaska to California (Norman, 1957: 464).

138

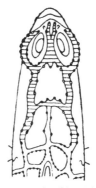

Figure 93. Seismosensory canals of head in *Dictyosoma buergeri*
(Makushok, 1958).

(Figure 94). Makushok (1958: 15) considered these structural peculiarities
an expression of specialization of the genus.

Membranes of dorsal and anal fine fused for most part with base of
caudal fin, but usually with fairly deep notch.

1 species. Known from the Sea of Japan.

1. *Dictyosoma buergeri* Van der Hoeven, 1850 (Figure 95)

Dictyosoma Schlegel, Fauna Japonica, Poiss., 1845: 139, pl. 73, fig.
3 (color figure) (Nagasaki).

Dictysoma buergeri Van der Hoeven, Handbuch der Dierkunde, 1850:
347. Jordan and Snyder, Proc. U.S. Nat. Mus., 1902: 482. Soldatov and
Lindberg, Obzor..., 1930: 449. Taranetz, Kratkii Opredelitel'..., 1937:

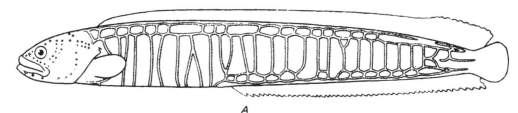

A

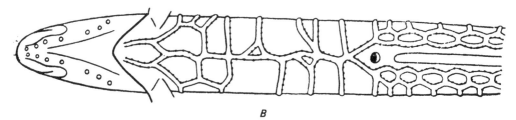

B

Figure 94. Seismosensory canals of trunk in *Dictyosoma buergeri*
(Makushok, 1958)

A—lateral view; B—ventral view.

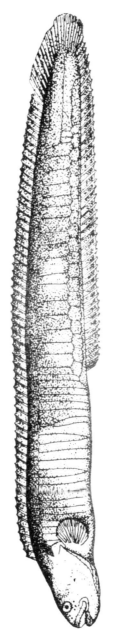

Figure 55. *Dictyosoma buergeri*. Length 290 mm. Japan (Temminck and Schlegel, 1845).

117

154. Matsubara, Fish Morphol. and Hierar., 1955: 761. Fowler, Synopsis..., 1958: 269. Abe, Enc. Zool., 2., Fishes, 1958: 113, fig. 333 (color figure). Makushok, Stichaeoidea..., 1958: 61.

Dictysoma temminckii Bleeker, Verhand. Batavia. Genootsch., 25, 1853: 42 (Japan).

D LVIII 9[115]; A II, 43; P 10; C 10; branchiostegal rays 6 (Temminck and Schlegel, 1845: 139).

Head 6.5 times, body depth 7.5 times in standard length. Eyes 6.5 times, interorbital space 13 times, and snout 4.33 times in head length. Gill rakers 2 + 10 (Soldatov and Lindberg, 1930). Dorsal fin with only 119 spines in anterior part, only last few rays of fin soft. Transition between these sharp, distinct. Spines gradually increase in height, but soft rays are slightly greater in height than posteriormost spines, and do not decrease in length toward caudal fin (as happens in most fish), but rather, elongate. At origin of anal fin first spine particularly well developed. Pores of seismosensory canals of head: nasal 2, interorbital 11-13, postorbital 12-13, suborbital 12-15, occipital 4-6, preopercular 7, and mandibular 4 (Makushok, 1958).

Length, to 375 mm (Soldatov and Lindberg, 1930).

Distribution: In the Sea of Japan known from Pusan (Shmidt, 1904: 368); off Hakodate (Jordan and Snyder, 1902a: 449); Aomori (Okada and Matsubara, 1938: 401); Sado Island (Honma, 1963: 21); Toyama Bay (Katayama, 1940: 25); San'in region (Mori, 1956a: 21). Reported for Cheju-do Island (Uchida and Yabe, 1939: 13) and Chefoo (Wang and Wang, 1935: 318). Pacific coast of Japan known all along Honshu (Okada and Matsubara, 1938: 401). Nagasaki (Shmidt, 1931b: 148).

[Subfamily Azygopterinae]

Makushok, Stichaeoidea..., 1958: 104.

Body elongate, highly compressed laterally, covered with extremely small imbricate scales. Head naked, without fleshy processes. Vomer and palatines without teeth. Branchiostegal membranes broadly fused and not attached to isthmus. Branchiostegal rays 6. Seismosensory canals of head well developed, open exteriorly through constant number of pores (Figure 96, A-C): nasal 2, interorbital 1, postorbital 3, occipital 1, suborbital 6, preopercular 3, and mandibular 3. Trunk with middle branch of open seismosensory papillae. Vertebrae 110-113, precaudal 42-45. Dorsal fin with short spines. Anal fin origin with one small spine, other rays soft, unbranched. Highly reduced caudal fin almost completely fused with membranes of dorsal and anal fins; 10 rays present, of which 6 principal (Figure 97). Pectoral and pelvic fins absent.

[115]D LII-LVIII, 7-10 (Abe, 1958). D LIII-LIX 7-11; A II, 40-44; principal rays C 6-7 + 6-7; P 12; vertebrae (63) 68-72 [(19) 20-22 + (44) 47-50] (Makushok, 1958).

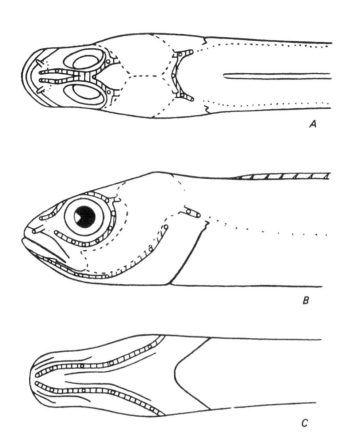

Figure 96. Seismosensory system of head in *Azygopterus corallinus*
(Makushok, 1958).

A—dorsal view; B—lateral view; C—ventral view.

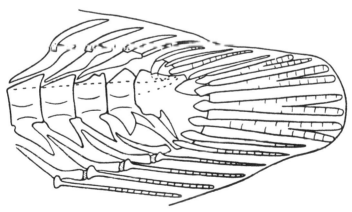

Figure 97. Caudal fin of *Azygopterus corallinus* (Makushok, 1958).

Mode of life not known.

1 genus. Not known from the Sea of Japan. Known off Kuril Islands.

[Genus *Azygopterus* Andriashev and Makuschok, 1955]

Azygopterus Andriyashev and Makushok, Vopr. Ikhtiologii, 3, 1955: 50, figs. 1-2 (type: *A. corallinus* Andriashev and Makuschok).

Description of genus given in characters of the subfamily.

1 species. Not known from the Sea of Japan.

[*Azygopterus corallinus* Andriashev and Makuschok, 1955] (Figure 98)

Azygopterus corallinus, Andriyashev and Makushok, Vopr. Ikhtiologii, 3, 1955: 50, figs. 1-2 (Kuril Islands).

D CVI; A I, 64; vertebrae 111 (45 + 66).[116]

Species described on the basis of a specimen (1 of 5) caught from depths of 100-150 m off the Kuril Islands. Most probably this fish is adapted to life in outgrowths [of hydrocorals], leading to elongation of its body, confluence of the dorsal and anal fins, reduction of caudal fin, and complete loss of pectoral and pelvic fins. Specimens of *A. corallinus* differ in color. Those caught along with hydrocorallines were uniformly pale pink, while those found among brittle stars had a large number of light-colored transverse stripes, which made them resemble brittle stars somewhat (Andriyashev and Makushok, 1955).

Length 96.5 mm (Andriyashev and Makushok, 1955).

Distribution: Not known from the Sea of Japan. Described from the Kuril Islands. Nadezhda Strait.

120   7. Subfamily Eulophiinae

Makushok, Stichaeoidea..., 1958: 107.

Body significantly elongate, rounded anteriorly (in cross section), naked. Head without fleshy processes. Mouth small. Vertebrae at least 130, of which precaudal at least 30. Posterior part of dorsal fin with soft rays. Large spine occurs at origin of anal fin. Caudal fin greatly reduced, entirely confluent with dorsal and anal fins. Pectoral fins small. Soft rays of all fins unbranched. Pelvic fins absent.

Mode of life not known.

One genus. Known from the Sea of Japan.

20. Genus *Eulophias* Smith, 1902

Eulophias H.M. Smith, Bull. U.S. Fish. Comm., 1902: 93 (type: *E. tanneri* Smith). Jordan and Snyder, Proc. U.S. Nat. Mus., 25, 1902: 477, fig.

---

[116]D CV-CVII; A I, 63-64; principal rays of C 3 + 3; vertebrae 110-113 (42-45 + 68-72) (Makushok, 1958).

17. Soldatov and Lindberg, Obzor..., 193C: 460. Taranetz, Kratkii Opredelitel'..., 1937: 155. Makushok, Sticha oidea..., 1958: 61.

Description of genus as given in characters of the subfamily.

2 species. 1 known from the Sea of Japan, the other from the Pacific coast of Japan.

Key to Species of Genus Eulophias[117]

1 (2). Anal fin with 1 spine and 75 soft rays. Caudal fin with 7 soft rays. Length of pectoral fins almost 3 times in head length ....
.... 1. **E. tanneri** Smith.
2 (1). Anal fin with 1 spine and 95 soft rays. Caudal fin with 10 soft rays. Length of pectoral fins 3.6 times in head length ....
.... **[E. owasii** Okada and Suzuki, 1954].[118]

1. *Eulophias tanneri* Smith, 1902 (Figure 99)

Eulophias tanneri H.M. Smith, Bull. U.S. Fish. Comm., 1902: 94 (Tsuruga). Matsubara, Fish Morphol. and Hierar., 1955: 756. Soldatov and Lindberg, Obzor..., 1930: 461. Makushok, Stichaeoidea..., 1958: 61.

121   D CXXI 13; A I, 75.

Body eel-like, cylindrical in anterior part, compressed laterally in posterior part, and pointed at end. Maximum body depth 5% and head length 12% of standard length. Eyes large, about 33% of head length; snout short, equal to half eye length. Caudal fin distinct, but confluent with dorsal and anal fins. Pectoral fins short, narrow, pointed, less than half head length (Soldatov and Lindberg, 1930). Length of only specimen examined 50 mm.

Distribution:‘In the Sea of Japan known from Peter the Great Bay (Soldatov and Lindberg, 1930: 461). Along the Pacific coast of Japan reported from Sagami Bay (Matsubara, 1955: 756). Described from Tsuruga Bay.

CLII. Family ZOARCIDAE—Eelpouts

Body notably elongate, sometimes eellike, covered with minute, contiguous, cycloid scales, or scaleless. Dorsal and anal fins long, border body, completely confluent with caudal fin. Fins with only soft, segmented or branched rays; rarely posterior part of dorsal fin with short spiny rays (Zoarcinae, Neozoarcinae). Pectoral fins well developed, pelvic fins rudimentary and jugular, or absent. Size of gill openings highly variable, usually in form of small vertical slit, but sometimes continue far forward (Lycogramminae) or, contrarily, reduced (Gymnelinae); in *Lycozoarces* branchiostegal membranes form fold across isthmus.

[117]From Matsubara (1955: 756).
[118]Distributed along the Pacific coast of Japan (Mie Prefecture) (Matsubara, 1955: 756).

144

120

Figure 98. *Azygopterus corallinus*. Length 96 mm. Kuril Islands (Andriyashev and Makushok, 1955).

120

Figure 99. *Eulophias tanneri*. Length 45 mm. Japan (Jordan and Snyder, 1902).

Branchiostegal rays 5-7. Pseudobranchs present. Mouth nonprotruding, inferior, or terminal. Teeth present on jaws; sometimes absent on vomer and palatines. Opercle without spines. Swim bladder absent. Pyloric ceca usually rudimentary, not more than three. Parietals divided by supraoccipital. Wings of parasphenoid contiguous with nondivergent process of frontals. Opisthotics small. Suborbitals membranous. Each radial of dorsal and anal fins corresponds to neural or hemal process of vertebrae. Number of vertebrae varies from 66 (*Bothrocarina*) to 139 (*Lycenchelys, Zoarces*). Body of vertebra symmetrical (Lycodinae, Lycogramminae) or middle strand shifted toward anterior margin of vertebra. Benthic fishes; large eggs deposited in small numbers on bottom. Larvae do not pass through planktonic stage (Andriyashev, 1954).

Among the 14 genera of this family known from the Sea of Japan, radiographs were taken to 10 (*Lycozoarces, Krusensterniella, Zoarces, Neozoarces, Lycodes, Bilabria, Davidojordania, Gymnelopsis, Allolepis,* and *Lycogramma*).

In our opinion, it is essential to determine the number of rays of the unpaired fins—dorsal, caudal, and anal—from radiographs of all members of the eelpout family. Many ichthyologists in counting the rays of this group of fishes have included some rays of the upper half of the caudal fin in the number of rays of the dorsal fin, and some rays of the lower half in the rays of the anal fin. In counting vertebrae we have not taken into account here the last one (with a urostyle). By the term "lateral line" we mean the series of freely located neuromasts, and not the pores of the seismosensory canal.

More than 30 genera are known in the northern part of the Pacific Ocean, as well as in the Arctic, Atlantic, and Antarctic Oceans; these fishes are known from the littoral to the profundal zones (Andriyashev, 1954). Species of 14 genera have been recorded from the Sea of Japan.

122   *Key to Genera of Family Zoarcidae*[119]

1 (2). Branchiostegal membranes broadly connected, from transverse fold across isthmus. Pelvic fins present (Lycozoarcinae)......
....1. **Lycozoarces** Popov.

2 (1). Branchiostegal membranes more or less attached to isthmus, do not form transverse fold across it.

3 (10). Dorsal fin with spiny rays.

4 (7). Short spiny rays occur in posterior part of dorsal fin (Zoarcinae).

5 (6). Pelvic fins absent. Teeth present on vomer and palatines. ..
.... 2. **Krusensterniella** Schmidt.

[119]From Taranetz (1937b: 159) and Matsubara (1955: 770), with additions and modifications applicable to specimens from the Sea of Japan.

6 (5). Pelvic fins present. Teeth absent on vomer and palatines...
.... 3. **Zoarces** Cuvier.

7 (4). Short spiny rays occur in anterior part of dorsal fin (Neozoarcinae).

8 (9). Head with fleshy process...... 4. **Neozoarces** Steindachner.[120]

9 (8). Head without fleshy process.................................
.... 5. **Zoarchias** Jordan and Snyder.[121]

10 (3). Dorsal fin without spiny rays.

11 (22). Pelvic fins present, albeit very small, in adult fish.

12 (17). Chin with pair of crests with one row of large pores or a stretched dermal fold at base (Lycodinae).

13 (16). Teeth present on palatines, sometimes absent on vomer. Mental crests usually not fused anteriorly.

14 (15). Palatine membrane absent. Length of anal fin more than one-third total length. .... 6. **Lycodes** Reinhardt.

15 (14). Narrow palatine membrane present. Length of anal fin not more than one-third total length .... [**Lycenchelys** Gill, 1884].[122]

16 (13). Teeth absent on vomer and palatines. Mental crests fused anteriorly........ 7. **Petroschmidtia** Taranetz and Andriashev.

17 (12). Chin without crests or stretched dermal fold (Hadropareinae).

18 (19). Head and anterior part of body roundish in cross section. Cheeks highly inflated. Body naked ....
.... [**Hadropareia** Schmidt, 1904].[123]

19 (18). Head and anterior part of body compressed laterally. Cheeks not highly inflated. Body with scales, sometimes few in number.

20 (21). Teeth absent on palatines. Upper lip with notch in fornt and attached to tip of snout.................. 8. **Bilabria** Schmidt.

21 (20). Teeth present on palatines. Upper lip entire, without notch, not attached to tip of snout .... 9. **Davidojordania** Popov.

123 22 (11). Pelvic fins absent.

23 (28). Membrane of palatines well developed. Slit of gill openings relatively small, slopes below upper margin of base of pectoral

[120]Makushok (1961c), on the basis of a careful analysis of the morphological peculiarities of the genera *Neozoarces* and *Zoarchias*, established their close affinity with the family Zoarcidae, and assigned these genera to the subfamily Neozoarcinae in this family. The close affinity between Neozoarcinae and Zoarcidae is confirmed by similarity of structure of the mouthparts, digestive tract, gill apparatus, seismosensory system, vertebrae and related elements, soft fin rays, tail with supportive skeleton, skull, etc. Honma and Sugihara (1963) and Honma and Kitami (1970) have included the genus *Zoarchias* in the family Cebidichthyidae.

[121]See footnote No. 120.

[122]Distributed mainly in deep seas of the northern hemisphere and the Antarctic Ocean (Andriyashev, 1954: 307). Not found in the Sea of Japan.

[123]Reported from the northwestern part of the Sea of Okhotsk; described from Shantar Islands.

fin, but does not continue beyond lower margin of base. Lateral line entire, not interrupted (Gymnelinae).

24 (25). Body naked. Teeth present on vomer and palatines. Lower lip interrupted in front........... [**Gymnelis** Reinhardt, 1836].[124]

25 (24). Body with scales, often only in caudal part, in small numbers and barely discernible.

26 (27). Vomer with two canine teeth. Body with scales only in caudal part. Pectoral fins without black dots. A 67–75.............
.... 10. **Gymnelopsis** Soldatov.

27 (26). Vomer without two canine teeth. Body with scales not only in caudal part, but also on trunk and unpaired fins. A 86......
.... 11. **Gengea** Katayama.

28 (23). Palatine membrane absent or rudimentary. Slit of gill openings large, distinctly continues beyond lower margin of base of pectoral fin. Lateral line consists of two parts—upper and middle (Lycogramminae).

29 (30). Scales on body elongate, with longitudinal axis pointing various directions (Figure 144, A).... 12. **Allolepis** Jordan and Hubbs.

30 (29). Scales not elongate.

31 (32). Scales present on occiput, isthmus, and anterior half of abdomen .... 13. **Lycogramma** Gilbert.

32 (31). Scales absent on occiput, isthmus, and anterior half of abdomen .... 14. **Zestichthys** Jordan and Hubbs.

1. Genus *Lycozoarces* Popov, 1935—Eelpouts

Lycozoarces Popov, Dokl. Akad. Nauk SSSR, IV (IX), 6-7 (75-76), 1935:[125] 285 (type: *L. hubbsi* Popov). Taranetz, Kratkii Opredelitel'..., 1937: 166. Matsubara, Fish Morphol. and Hierar., 1955: 783.

Body elongate, compressed laterally. Head large, slightly compressed. Mouth opening large; corner of mouth extends beyond posterior margin of eye. Lips fleshy. Lower lip rather notably thickened anteriorly and thin posteriorly. Teeth well developed on jaws, vomer, and palatines. Gill openings broad. Branchiostegal membranes broadly connected, form broad fold across isthmus. Unpaired fins high. Body of vertebrae asymmetrical. Hard rays not present in fins. Body naked. Lateral line present (Popov, 1935).

The distinctive character of this genus is the presence of a fold across the isthmus.

In the diagnosis of the genus Popov (1935) mistakenly indicates that

[124]Distributed from Barents Sea eastward to Chukchi Sea; also reported from the northern part of the Bering Sea and the Sea of Okhotsk.

[125]Article probably delayed in publication since the description of *Lycozoarces regani* n. sp. was prepared in 1933.

148

"the corners of the mouth extend far beyond the posterior margin of the eyes." This is not a character of the genus, but merely a specific distinction, about which the author himself made a mention in his brief description of *L. regani* (Popov, 1933) and *L. hubbsi* (Popov, 1935). In the former species "the ends of the corners of the mouth do not extend beyond the vertical from the posterior margin of the eyes," while in the latter species "the upper jaw continues beyond the vertical from the posterior margin of the eyes."

124 We compared the type specimens of these species and are convinced of the correctness of this distinctive character. Because our own material was limited, we could not ascertain whether this character persists at different stages of growth nor in which sex. Possibly this morphological peculiarity disappears with age. Taranetz (1937b: 166) and Shmidt (1950: 109, pl. 10, fig. 2) consider only one species, *L. hubbsi*. But until further studies are done, we accept two species—*L. regani* and *L. hubbsi*. One is known from the Sea of Japan as well as the Sea of Okhotsk, and the other found only in the Sea of Okhotsk.

Key to Species of Genus Lycozoarces

1 (2). Maxilla does not extend beyond vertical from posterior margin of eye. Margin of fold of branchiostegal membranes without notch in interbranchial space (Figure 100, A)........ 1. **L. regani** Popov.
2 (1). Maxilla extends beyond vertical from posterior margin of eye.

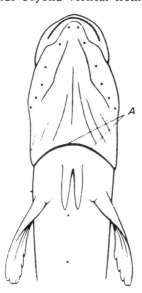

Figure 100. Head of *Lycozoarces regani*, ventral view. No. 29987.
A—margin of fold of branchiostegal membranes.

Margin of fold of branchiostegal membranes notched in interbranchial space (Figure 102, A) .... 2. [**L. hubbsi** Popov].

1. *Lycozoarces regani* Popov, 1933—Regan's Eelpout (Figure 101)

Lycozoarces regani Popov, Issled. Morei SSSR, 19, 1933: 151, fig. 2[126]
(Tatar Strait). Taranetz, Kratkii Opredelitel' ..., 1937: 166.

24833. Sea of Okhotsk, 53°40' N, 144°02' E. September 10, 1932. M.
Krivobok. 2 specimens.

29987. Sea of Japan, Tatar Strait. September 12, 1931. N.I. Tarasov.
1 speciemen.

30601. Sea of Okhotsk, 58°22.8' N, 143°06' E. August 19-21, 1932.
I.A. Polutov. 1 specimen.

33333. Sea of Okhotsk, 58°50' N, 146°18' E. July 20, 1916. GEVO.
1 specimen.

33750. Sea of Okhotsk, collection of E/S *Vityaz.* August 24, 1949.
Institute of Oceanology, Academy of Sciences of the USSR. 1 specimen.

34843. Sea of Okhotsk, 57°25' N, 141°18' E. August 14, 1910. Derbek.
1 specimen.

36979. Sea of Okhotsk. 1912. Lyaskovskii. 1 specimen.

Because the description of this species given by Popov is so laconic
(1933a), we give here a brief description on the basis of eight specimens:
D 64-69; A 49-54; P 15; C 13-14; vertebrae 66-71;[127] maxilla does not
extend beyond vertical from posterior margin of eye; margin of fold of
branchiostegal membranes across isthmus without notch (Figure 100, A).
Anterior part of dorsal fin with one or two roundish dark spots. Pectoral
fins not flabelliform and roundish along posterior margin. Head length
4.2 to 5.0 times in the total length, diameter of eye 2.8 to 3.7 times
in the head length. Width of interorbital space 2.0 to 4.0 times in the
diameter of eye. Maximum height of dorsal fin 1.4 to 2.1 times and that
of anal fin 1.5 to 2.6 times in the maximum body depth.

Length of pelvic fins 0.9 to 1.3 times in the diameter of eye.[128] Opercular
125 siphon well developed. Pores of seismosensory canals of head:
preopercular 4 (rarely 3), mandibular 3, suborbital 7, preorbital 1,
postorbital 5, interorbital 1, and occipital 3. Lateral line mediolateral,
complete, represented by about 86 freely located neuromasts (in
specimens preserved a long time, difficult to discern).

Length of our specimens ranged from 76 to 142 mm.

Distribution: In the Sea of Japan known from Tatar Strait (No. 29987).
A larger number of specimens obtained from the Sea of Okhotsk.

[126]In the key, erroneously included under no. 1.

[127]Counts of rays of fins and of vertebrae based on radiographs.

[128]Pelvic fins not shown in the drawing given by Popov (1933a), although these are well
developed in the type specimen. We have added these fins in the copy of the drawing given
here.

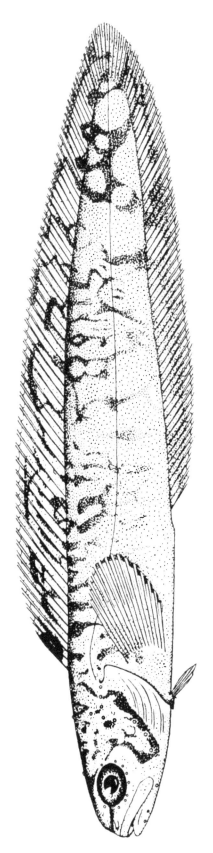

Figure 101. *Lycozoarces regani*—Regan's eelpout. Length 96 mm. Tatar Strait (Popov, 1933).

125

2. [*Lycozoarces hubbsi* Popov, 1935—Hubb's Eelpout] (Figure 102)

Lycozoarces hubbsi Popov, Dokl. Akad. Nauk SSSR, 4, 6-7, 1935: 285-286, fig. (Sea of Okhotsk). Taranetz, Kratkii Opredelitel'..., 1937: 166. Matsubara, Fish Morphol. and Hierar., 1955: 783.

26566. Sea of Okhotsk, 54°14′ N, 143°45′ E. July 12, 1928. A.M. Popov. 1 specimen.

33749. Sea of Okhotsk, 57°47.5′ N, 148°06′ E. August 19, 1932. Fishing trawler *Plastun*. 1 specimen.

Head length 4.5 times in standard length. Head slightly compressed laterally, but much broader than body. Cheeks prominent. Eyes large, about 4.5 times in the head length. Oral slit large, oblique, opens slightly below lower horizontal line of eye. Upper jaw extends beyond vertical from posterior margin of eye; length of upper jaw more than 0.5 head length. Postorbital distance equal to length of upper jaw. Both jaws almost equal anteriorly. Anterior pair of nostrils in form of tubules. Series of pores (7) very distinct under eyes. Teeth present on jaws, vomer, and palatines. Teeth in anterior part of jaws irregularly arranged, while those posteriorly arranged in regular row. Anal fin originates behind vertical from end of pectoral fins, but closer to end of head than to end to body. Pelvic fins very short and located slightly anterior to pectoral fins (Popov, 1935).

Examination of our specimens yielded the following data: D 64-67; A 49-53; P 15; C 13-14; vertebrae 65-68.[129] Maxilla extends far beyond vertical from posterior margin of eye. Margin of fold of branchiostegal membranes across isthmus with notch (Figure 102, A). Anterior part of dorsal fin without well-developed spots. Pectoral fins flabelliform. Length of head 4.0 to 4.5 times in total length; diameter of eye 4.0 to 4.1 times in the head length; width of interorbital space 2.2 to 2.4 times in the diameter of eye. Maximum height of dorsal fin 0.8 to 0.9 times and that of anal fin 1.5 to 1.6 times in the maximum body depth. Length of pelvic fins equal to diameter of eye.

Pores of seismosensory system of head sufficiently distinct and similar in number to *L. regani*. Lateral line mediolateral, complete, represented by about 80 free neuromasts.

L. hubbsi differs from *L. regani* not only in larger size of maxilla, which extends beyond vertical from posterior margin of eye, and presence of notch in branchiostegal fold across interorbital space, but also in relatively larger height of dorsal fin, relatively smaller size of eyes, and shape of pectoral fins.

Length, to 180 mm.

Distribution: Not known from the Sea of Japan. Known only from the

---

[129]Count of rays of fins and of vertebrae based on radiographs.

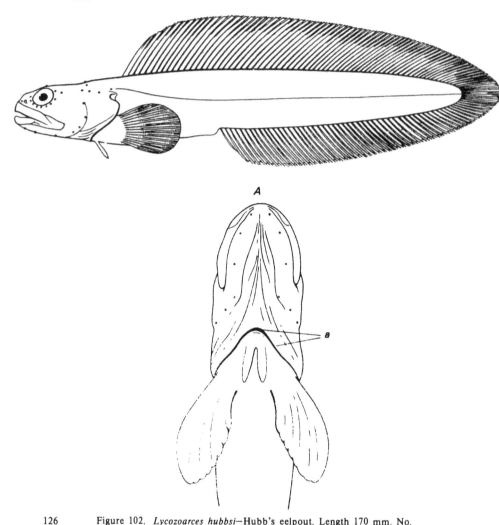

A

Figure 102. *Lycozoarces hubbsi*—Hubb's eelpout. Length 170 mm. No. 26566. Sea of Okhotsk.

A—head of same specimen, ventral view:
a—margin of fold of branchiostegal membranes.

Sea of Okhotsk. Records of distribution in Primor'e (Ueno, 1971: 87) require confirmation.

2. Genus *Krusensterniella* Schmidt, 1904

Krusensterniella Shmidt, Ryby Vostochnykh Morei..., 1904: 197 (type: *K. notabilus* Schmidt). Soldatov and Lindberg, Obzor..., 1930: 490. Taranetz, Kratkii Opredelitel'..., 1937: 161. Andriyashev, Vestn.

Dal'nevost. Fil. Akad. Nauk SSSR, 32, 1938: 117. Matsubara, Fish Morphol. and Hierar., 1955: 772. Fowler, Synopsis..., 1958: 302.

Body very elongate, its depth 13 to 16 times in its length. Scales cover entire body or only posterior part. Lateral line poorly expressed.[130] Structure of dorsal fin resembles that of *Zoarces* in few short spiny 127 rays in posterior part (2-20).[131] Pelvic fins absent. Gill openings small.[132] Lips thin, upper lip continuous, lower lip interrupted in front, with very poorly expressed anterior lobes. Palatine membrane present. Minute but distinct pores present on head. Pyloric ceca two (Andriyashev, 1938). This genus differs sharply from *Zoarces* in presence of teeth on palatines and vomer, smaller size of gill openings, and absence of pelvic fins.

Based on an examination of 13 specimens in our collection, the following data may be added: preanal distance varies from 30.0 to 34.4% of total length; pores of seismosensory system of head: preopercular 3-4, mandibular 4, suborbital 7, preorbital 1, postorbital 4, unpaired interorbital 1, and occipital commissures 3; vertebrae 99-119 (20-22 + 79-97); rays in dorsal fin 99-116 (45-60 + III-XVIII + 35-57); anal fin with 83-96 rays; skeleton of free part of caudal fin with 6 principal and 3 to 4 marginal rays (2 + 3 + 3 + 1-2).

3 species. 1 known from the Sea of Japan; 2 known from the Sea of Okhotsk.

Key to Species of Genus Krusensterniella[133]

1 (4). Spiny rays in dorsal fin less than 10. Pyloric caeca tubercular.

2 (3). Spiny rays in dorsal fin 2-3. Scales extremely sparse anterior to vertical from commencement of vent and not represented in region up to gill openings. Color monochromatic, light, or with barely discernible pattern of bands across body.....................
.... [**K. notabilis** Schmidt].

3 (2). Spiny rays in dorsal fin 5-7. Scales covered entire body up to gill openings and base of pectoral fins. Series of distinct dark spots, not larger than diameter of eye, visible along median line of body
.... 1. **K. maculata** Andriaschev.

4 (1). Spiny rays in dorsal fin more than 10. Pyloric caeca digitate...
.... 2. [**K. multispinosa** Soldatov].

---

[130]Lateral line mediolateral, incomplete, represented by about 40 free neuromasts. Pores of seismosensory system of head well developed.

[131]Makushok (1961c) rightly noted that the formula of the dorsal fin of *Krusensterniella* should be accepted with appropriate allowances, since all the rays of the dorsal fin before the stiff spines are entire (do not consist of lateral halves), i.e., these are essentially the same spines in spite of their slenderness and flexibility.

[132]Dermal fold behind and on upper side of gill openings forms siphon together with margin of operculum (Shmidt, 1950).

[133]From Andriyashev (1938), with additions.

[*Krusensterniella notabilis* Schmidt, 1904] (Figure 103).

Krusensterniella notabilis Shmidt, Ryby Vostochnykh Morei..., 1904: 198, fig. 12 (Sakhalin, northern part of eastern coast). Soldatov and Lindberg, Obzor..., 1930: 490. Taranetz, Kratkii Opredelitel'.... 1937: 161. Andriyashev, Vestn. Dal'nevost. Fil. Akad. Nauk SSSR, 32, 1938: 119. Matsubara, Fish Morphol. and Hierar., 1955: 772.

Enchelyopus elongatus Tanaka (non Kner), J. Univ. Tokyo, 3, 1, 1931: 48 (northern Japan).

13011. Northeastern coast of Sakhalin. 1899. V.K. Brazhnikov. 6 specimens.

13012. Northeastern coast of Sakhalin. 1899. V.K. Brazhnikov. 3 specimens.

D 53-57, II-III 61-63; A 98-103; *Br.* 5.

Trunk highly compressed laterally, attenuate, and thins toward posterior end; covered with minute cycloid scales, particularly dense at posterior end. Body depth 14 to 15 times in its length. Lateral line expressed only in anterior part of body. Head small, compressed laterally, its length 6.8 to 7.4 times in the standard length; forehead convex. Snout length always equal to maximum diameter of eye. Lower jaw shorter than 129 upper. Jaws, vomer, and palatines with sharp, conical, fairly large, wide-set teeth.[134] Scales absent on head. Series of large distinct pores under eye and along margin of preopercle. Anterior nostrils without tubules, posterior ones with long tubules. Dorsal fin high in anterior part; length of largest rays 1.2 times in the head length in adult fish and twice in young fish, and significantly greater than body depth. Height of soft rays in posterior part of dorsal fin 3.0 to 3.6 times in the head length.[135] Body color yellowish, with diffused dark spots forming indistinct broad transverse bands. Fins transparent, yellowish. Dorsal with large spot in anterior part[136] (Shmidt, 1904).

In our specimens preanal distance varied from 31.3 to 34.0% of total length.

Radiographs of our specimens (7 of 9) ranging in length from 82 to 182 mm yielded the following additional data: vertebrae 115-119 (25 + 90-94); D 113-116 (54-57 + III + 55-57); A 94-96; skeleton of free part of caudal fin morphologically well isolated from dorsal and anal fins, which

---

[134]Anterior part of palatines covered with numerous minute teeth, which correspond to group of minute teeth on symphysis of lower jaw. Palatine membrane broad, covering part of teeth on vomer. Gill openings extend up to midbase of pectoral fins. Gill rakers smooth, spine-shaped, 12 in number (in both series) (Andriyashev, 1938).

[135]Largest ray of anterior part of dorsal fin usually three times length of largest ray of posterior part and seven to ten times length of hard spiny rays (Andriyashev, 1938).

[136]Andriyashev (1938) has reported up to three such spots.

are very close to it, and represented by 3 upper and 3 lower principal rays and 4 marginal (2 each in upper and lower lobes of this fin).

Length, to 189 mm.

Distribution: Not known from the Sea of Japan. All known specimens caught near east coast of northern part of Sakhalin (Andriyashev, 1938: 119).

1. *Krusensterniella maculata* Andriashev, 1938[137] (Figure 104)

Krusensterniella maculata Andriashev, Vestn. Dal'nevost. Fil. Akad. Nauk SSSR, 32, 1938: 118 (Tatar Strait). Lindberg, Predvaritel'nyi Spisok..., 25, 1947: 170. Matsubara, Fish Morphol. and Hierar., 1955: 773. Fowler, Synopsis..., 1958: 303.

29989. Tatar Strait. October 13, 1933. Z.I. Kobyakova. 2 specimens.
40166. Sea of Japan, southern Primor'e. 1900. Lyaskovskii. 1 specimen.
D 49-53, V-VII 64; A ca 100; P 11-12 (Andriyashev, 1938).

This species is close to *K. notabilis* Schmidt, but differs in greater number of spiny rays (5-7 versus 2-3) in the dorsal fin, shorter predorsal distance (11.2-12.9%), shorter and higher head, much better developed scale cover, longer pectoral fins, and characteristic coloration in the form of 20 to 30 black spots and dots along median line of body, extending from apex of gill opening to tail (each spot not more than diameter of eye). A similar series of smaller dark spots extends along the dorsal fin. Black ocellar spot absent in anterior part of dorsal fin. In the type specimen anterior part of dorsal and anal fins with black edging. Head with minute but distinct, uniformly arranged pores: (3 + 4) on lower jaw and preopercle, (8) under eye, 3 across occiput, 3 on each side in longitudinal occipital series, and 1 interorbital. Number and arrangement of pores as in *K. notabilis* (Andriyashev) 1938.

Specimens of this species were caught from depths of 53 to 150 m.

Preanal distance in our specimens varied from 30.0 to 31.6% of the total length.

130   Radiographs of our specimens (3) yielded the following data: vertebrae 104 112; dorsal fin with 99-113 rays, anterior part with 45-60 rays, followed by 7 spines, then 41-50 rays; anal fin with 87-96 rays; skeleton of free part of caudal fin as in *K. notabilis*.

Length, to 144 mm.

Distribution: In the Sea of Japan known from Peter the Great Bay (Lindberg, 1947: 160) and Tatar Strait up to Pos'et (Andriyashev, 1938: 119).

2. [*Krusensterniella multispinosa* Soldatov, 1917] (Figure 105)

Krusensterniella multispinosa Soldatov, Ezhegodn. Zool. Muz. Rossiisk.

---

[137]*K. maculata* Andriashev was first mentioned by Taranetz (Kratkii Opredelitel'..., 1937b) in his key to species.

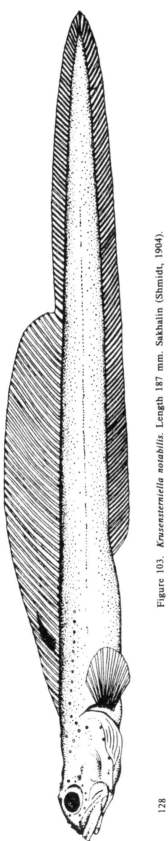

Figure 103. *Krusensterniella notabilis*. Length 187 mm. Sakhalin (Shmidt, 1904).

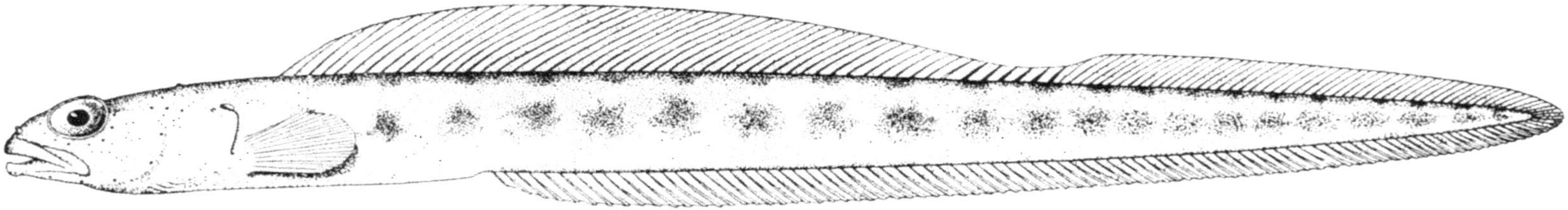

128

Figure 104. *Krusensterniella maculata*. Length 100 mm. No. 40166. Sea of Japan.

128

Akad. Nauk SSSR, 23, 1917: 159, drawing (Sea of Okhotsk, Shantar Islands). Soldatov and Lindberg, Obzor ..., 1930: 490, fig. 68. Taranetz, Kratkii Opredelitel'..., 1937: 161, fig. 99. Andriyashev, Vestn. Dal'nevost Fil. Akad. Nauk SSSR, 32 1938: 120. Shmidt, Ryby Okhotskogo Morya, 1950: 84. Matsubara, Fish Morphol. and Hierar., 1955: 772, fig. 289. Fowler, Synopsis..., 1958: 303, fig. 34.

19961. Sea of Okhotsk, Ayan Inlet. July 26, 1912. DVE. 1 specimen.

34728. Sea of Okhotsk, 50°3′ N, 144°8′ E. July 29, 1919. GEVO. 1 specimen.

34991. Sea of Okhotsk, southeast coast of Sakhalin. September 29, 1949. KSE. 1 specimen.

D 46-48, XVII-XX 37; A 71-75; P 11-12; *Br.* 5[138] (Soldatov, 1917b).

Differs from *K. notabilis* and *K. maculata* in greater number of spiny rays in dorsal fin, relatively less elongate and stouter body, presence of scales only in posterior part of caudal peduncle, greater number of flexible rays in dorsal and anal fins, and presence of not 4 but 3 pores on preopercle.[139]

Head larger than in any other species, its length 15.9% of the total length. Predorsal distance also greater than in other species (17.6% of the total length versus 14.2 to 15.3% in *K. notabilis* and 11.2 to 12.9% in *K. maculata*), preanal distance 34.1% of this length (versus 30.2 to 33.1% in *K. notabilis* and 29.1 to 30.8% in *K. maculata*). Lateral line mediolateral, rather well defined, consists of minute pores, and extends from apex of gill opening along median line of body up to origin of scaly cover, i.e., extends beyond vertical from origin of anal fin.[140] Color (in alcohol) monochromatic, light, without traces of spots and stripes (Andriyashev, 1938).

131　We may also add that the preanal distance in our specimens varied from 33.6 to 34.4% of the total length. It should be noted that in specimen No. 34991 (length 116 mm) dark, almost rectangular, spots were distinguishable on sides of body. Specimen No. 34728 had 4 pores along margin of opercle, 3 in region of lower angle of this bone and near the upper end of its outer margin. Radiographs of our specimens (3) yielded the following data: vertebrae 99-103; rays in dorsal fin 99-102 (45-47 + XVIII + 35-41); rays in anal fin 83-85; and structure of free part of caudal fin as in two previous species.

Length 105 mm (Soldatov, 1917b).

Distribution: Not known from the Sea of Japan. Two specimens caught

---

[138]According to Andriyashev (1938): D 46, XVIII 41; A 78; P 11.

[139]Other pores on head similar to those in other species (Andriyashev, 1938).

[140]In our specimens about 40 neuromasts located freely between upper end of gill opening and vertical from first fifth of anal fin; further counting in preserved specimens was difficult due to nonavailability of appropriate techniques.

from the Sea of Okhotsk at a depth of 87 m near Shantar Islands (Soldatov, 1917b) and third near the southeast coast of Sakhalin from a depth of 160 m (No. 34991).

3. Genus *Zoarces* Cuvier, 1829—Eelpouts

Zoarces Cuvier, Régne Anim., ed. 2, 2, 1829: 240 (type: *Blennius viviparus* L.). Soldatov and Lindberg, Obzor..., 1930: 488. Andriyashev, Ryby Severnykh Morei SSSR, 1954: 256.

Posterior part of dorsal fin with short spiny rays.[141] Body highly elongate, covered with minute cycloid scales. Upper jaw protrudes slightly forward in relation to lower one. Jaws with blunt conical teeth; vomer and palatines without teeth. Palatine and mandibular membranes well developed. Gill openings continue below lower margin of base of pectoral fin. Pelvic fins slightly anterior to vertical from base of pectoral fins (Andriyashev, 1954).

An examination of our specimens of species of this genus (15) yielded the following data: preanal distance from 31.7 to 39.4% of the total length. Lateral line mediolateral, complete, and well defined through free neuromasts. Pores of seismosensory system of head: preopercular 4, mandibular 4, suborbital 7, preorbital 1, postorbital 4, unpaired interorbital 1, and occipital 3.

Radiographs of our specimens of 2 species of this genus (Nos. 12397, 13004, 13008, 18054, 19123, 20352, 22871, 28014, 31669, 35564, 37098, 40824) revealed: vertebrae from 118 to 132 (precaudal 21-26, caudal 89-107); number of rays in dorsal fin (118) 122-132 (82-94 + VII-XVIII + 20-29); rays in anal fin 93-110; skeleton of free part of caudal fin with 2-3 principal rays in upper part and 4 such rays in lower part; and number of marginal rays from 2 to 3 on each side of fin.

4 species. 2 species known from the Sea of Japan.[142]

Key to Species of Genus *Zoarces*[143]

1 (2). Dorsal fin with 6 to 16 spiny rays; anterior part of fin without dark spot....................................... 1. **Z. elongatus** Kner.

2 (1). Dorsal fin with more than 16 spiny rays; anterior part of fin with dark spot .... 2. **Z. gillii** Jordan and Starks.

---

[141]These rays are much shorter than the soft rays anterior and posterior to them, and tend to form a characteristic notch in the dorsal fin of *Zoarces*. Cases of complete absence of these short spiny rays are known, in which case the notch is also totally absent (J. Schmidt, 1917; Andriyashev, 1954; Makushok, 1961c).

[142]Composition of the genus requires special analysis. Possibly, Shmidt (1904) was right in considering all the presently known species of this genus simply forms of *Z. viviparus* L.

[143]From Matsubara, 1955, with modifications.

132 1. **Zoarces elongatus** Kner, 1868—Elongate Eelpout (Figure 106)

Zoarces elongatus Kner, Sitzb. Acad. Wiss., 1868: 52, pl. 7, fig. 2 (Tatar Strait, De-Kastri Bay). Shmidt, Ryby Vostochnykh Morei..., 1904: 196. Soldatov and Lindberg, Obzor..., 1930: 488. Taranetz, Kratkii Opredelitel'..., 1937: 161. Fowler, Synopsis..., 1958: 301, fig. 32.

Zoarces viviparus elongatus, Shmidt, Ryby Okhotskogo Morya, 1960: 83.

Enchelyopus elongatus, Matsubara, Fish Morphol. and Hierar., 1955: 773, fig. 292.

D 75–94, VII–XII 18–23; A 80–100; P 17–19 (Soldatov and Lindberg, 1930).

D 80, XII 22; A over 90; C 11; V 3; P 19–20; *Br.* 6 (Kner, 1868).

Body firm, elongate, slightly attenuate posteriorly, and covered with cycloid scales. Head elongate, without scales, rounded on sides. Snout short, blunt, convex. Eyes relatively small, lips fleshy, mouth large.

Maxilla barely reaches middle of eye. Teeth on lower and upper jaws blunt, strong and broad, but thin slightly toward apices; anteriorly arranged in two rows and laterally in one row. Teeth on vomer and palatines absent. Dorsal fin originates anterior to vertical from gill opening; anterior part highest. Base and entire fin in lower part covered with sparser scales than in posterior part. Anal fin anteriorly also higher and covered with scattered scales anteriorly only at base and posteriorly up to margin of fin. Pectoral fins rounded on lower side, rays thickened. Pores well developed near eyes, on jaws, preopercle, and occiput. Lateral line forms curve in anterior part, and barely discernible in posterior part of body (Soldatov and Lindberg, 1930).

Z. elongatus Kner does not differ in body coloration from *Z. viviparus* L. Basic color greenish-gray or chocolate-brown to gray, dorsal side with series (14–16) of dark spots that continue into dorsal fin; series of similar but more diffuse spots located along lateral line (Shmidt, 1904).

In our specimens (12) preanal distance varied from 31.7 to 39.4% of the total length. Lateral line mediolateral, complete, represented by free neuromasts, numbering about 100 up to vertical from end of first third of

133 anal fin; beyond this point counting of neuromasts extremely difficult. Some neuromasts were detected at isolated places along lateral line of body up to base of caudal fin, indicating completeness of lateral line. Arrangement and number of pores of seismosensory system of head given in characters of genus. Radiographs of our specimens (12) yielded the following data: number of vertebrae (118) 122–127, with precaudal ranging from 21 to 26 and caudal 89 to 104; number of rays in dorsal fin 115–126 (82–92 + VII–XV + 22–29); rays in anal fin 93–103; skeleton of free part of caudal fin with 3–4 principal rays in upper half, 3–4 in lower half, and 4–6 marginal.

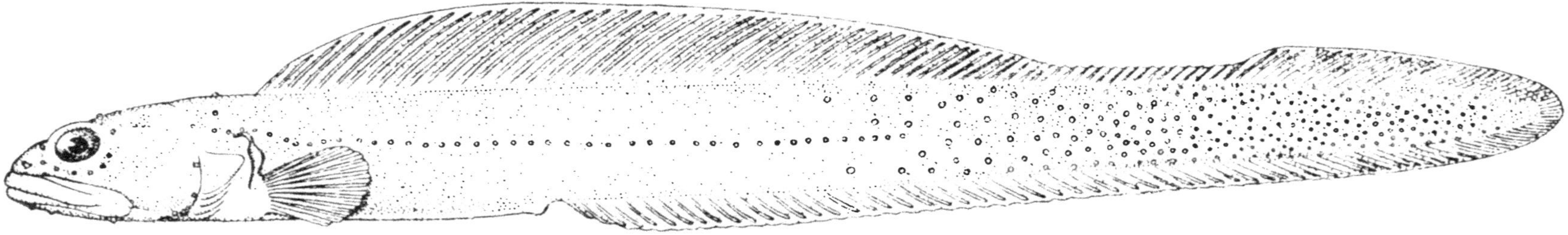

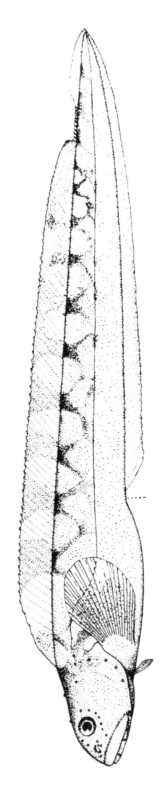

Figure 105. *Krusensterniella multispinosa*. Length 105 mm. Sea of Okhotsk (Soldatov and Lindberg, 1930).

130

Figure 106. *Zoarces elongatus*—Elongate eelpout (Matsubara, 1955).

132

Fish from the Sea of Okhotsk (Nos. 41653-41661) were caught from depths of 4 to 36 m, with bottom temperature 5 to 6°C, and bottom consisting of rocks, stones, sand, and sandy silt. At these places the biocenosis comprised *Tellina calcarea* and laminarians. The stones in these sites were covered with tubes of *Lithothamnion* and Serpulidae. Together with specimens of this species, catches comprised sponges, hydroids, and bryozoans.

Length, to 300 mm (Shmidt, 1904).

Distribution: In the Sea of Japan known from Tatar Strait and liman of Amur River (Soldatov and Lindberg, 1930: 488); Primor'e (Ueno, 1971: 85); from west (Taranetz, 1937b: 39), southwest, and east coasts of Sakhalin (recorded as *Z. viviparus*; Ueno, 1971: 85). In the Sea of Japan, coast of Hokkaido (Jordan and Fowler, 1902b: 766); Sado Island (Honma, 1952: 226); Wakasa Bay (Takegawa and Morino, 1970: 383); San'in region (Mori, 1956a: 21). In the Yellow Sea recorded from the Gulf of Chihli (Bohai) (Zhang et al., 1955: 174). In the Sea of Okhotsk found in northern part (Shmidt, 1950: 83) to southern Kuril Islands (specimens of KSE, Nos. 41653-41661). Pacific coast of Japan (Ueno, 6, 1971: 85).

2. *Zoarces gillii* Jordan and Starks, 1905—Gill's Eelpout[144]
(Figure 107)

Zoarces gillii Jordan and Starks, Proc. U.S. Nat. Mus., 28, 1905: 212, fig. 11 (Pusan). Mori, Mem. Hyogo Univ. Agric., 1952: 130. Fowler, Synopsis..., 1958: 300.

Zoarces tangwangi Wu, Contrib. Biol. Lab. Sci. Soc. China, 6, 6, 1930: 60 (China). Chu, Biol. Bull. St. John's Univ., 1, 1931: 182. Lindberg, Predvaritel'nyi Spisok..., 1947: 169.

Enchelyopus gilli, Jordan, Tanaka and Snyder, J. Coll. Sci. Univ. Tokyo, 33, 1913: 399 (Pusan, west coast of Hokkaido). Mori and Uchida, J. Chosen Nat. Hist Soc., 19, 1934: 22 (Korean Peninsula). Wang and Wang, Contrib. Biol. Lab. Sci. Soc. China, 11, 6, 1935: 221, fig. 43. Matsubara, Fish Morphol. and Hierar., 1955: 773.

22871. Sea of Japan (Pusan). March 27, 1901 P.Yu. Shmidt. 4 specimens.

134 D 84, XIX 14; A 80.

Head length 5.6 times and depth 10 times in standard length. Diameter of eye 5 times, length of maxilla 2.2 times, and length of snout 3.3 times in head length. Gill rakers short and pointed, 3 + 14 on anterior gill arch (Jordan and Starks, 1905: 212). Vertebrae 131-132 (25 + 106-107); rays of dorsal fin 131-132 (91-94 + XVII-XVIII + 20-23); and rays of anal fin 106-110. This species differs from *Z. elongatus* not only in greater number of spiny rays in dorsal fin, presence of dark spot in

[144]Chu et al. (1963: 381) consider this species a synonym of *Enchelyopus elongatus* (Kner).

162

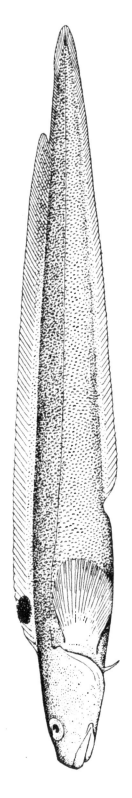

Figure 107. *Zoarces gilli*—Gill's eelpout. Length 225 mm. Pusan (Jordan and Snyder, 1905).

133

anterior part of dorsal fin, greater number of rays in anal fin, relatively large preanal distance (37.4 to 39.4%), but also broad and even interorbital space, high-set eyes, relatively short maxilla (not extending beyond vertical from posterior margin of eye), position of origin of dorsal fin (at vertical from gill opening versus anterior to this vertical in *Z. elongatus*), and coloration of body and fins.

Length, to 240 mm (Fowler, 1958).

Distribution: In the Sea of Japan known from Pusan and Wonsan (Mori, 1952: 130); coast of Hokkaido (Jordan, Tanaka and Snyder, 1913: 399); and San'in region (Mori, 1956a: 21). In the Yellow Sea found near Chefoo (Wang and Wang, 1935: 221). China (as *Z. tangwangi* Wu), recorded from Fu-chow (Wu, 1930: 60).

4. Genus *Neozoarces* Steindachner, 1880

Neozoarces Steindachner, Ichthyol. Beitr., 9, 1880: 26 (type: *N. pulcher* Steindachner). Soldatov and Lindberg, Obzor..., 1930: 461. Matsubara, Fish Morphol. and Hierar., 1955: 756.

Body elongate, laterally compressed, posteriorly pointed. Dorsal and anal fins confluent with very small caudal fin. Dorsal fin consists of two parts: longer part with low, slender, pointed spiny rays, and slightly shorter part with higher soft rays. Pectoral fins well developed. Mouth large, maxilla distinctly extends beyond vertical from posterior margin of eye. Blunt conical teeth in jaws, vomer, and palatines. Snout with fleshy process. Gill openings broad; branchiostegal membranes fused and not attached to isthmus. Scales small, deeply embedded in skin (Soldatov and Lindberg, 1930). Seismosensory canals of head with narrow lumen, open exteriorly through small number of pores arranged in single row. Number of pores (with rare individual deviations) strictly constant: nasal 2, interorbital 1, postorbital 6, occipital 3, suborbital 6, preopercular 4, and mandibular 4 (Figure 108, A). Trunk with two branches of open seismosensory pores: middle one, a continuation of postorbital canal, extends along mediolateral line of body up to base of tail; upper one interrupted in middle of body (Makushok, 1961c: 203).

Radiographs of 16 of our specimens of 2 species (Nos. 12399, 18761, 20349, 20483, 33312, 37335, 41106, and Nos. 25475, 31625, 37667, 37668) permits us to add these characters of the genus: number of vertebrae varies from 94 to 100 (precaudal 18-21, caudal 75-82); number of rays in dorsal fin 93-99 (XLII-XLVI + 49-59), of anal fin I 73-79; and skeleton of free part of caudal fin with 6-7 principal and 4-6 marginal rays.

Two species. Both known from the Sea of Japan.

164

Key to Species of Genus Neozoarces[145]

1 (2). Head not more than 6 times in standard length.[146] Reticulate
       pattern very distinct in lower half of head . . . .
       . . . . 1. **N. pulcher** Steindachner.
2 (1). Head more than 6 times in standard length. Reticulate pattern
       in lower half of head either absent of poorly developed. . . . .
       . . . . 2. **N. steindachneri** Jordan and Snyder.

1. *Neozoarces pulcher* Steindachner, 1880—Handsome Eelpout
(Figure 108)
Neozoarces pulcher Steindachner, Ichthyol. Beitr., 9, 1880: 27, pl.
6, fig. 2 (Peter the Great Bay). Shmidt, Ryby Vostochnykh Morei...,
1904: 177. Soldatov and Lindberg, Obzor..., 1930: 481. Taranetz, Kratkii
Opredelitel'..., 1937: 155. Shmidt, Ryby Okhotskogo Morya, 1950: 69.
Matsubara, Fish Morphol. and Hierar., 1955: 756. Fowler, Synopsis...,
1958: 320.

8660. Peter the Great Bay. 1888. Sliunin. 1 specimen.

12399. Sea of Okhotsk, Aniva Bay. August 24, 1901. P.Yu. Shmidt. 3
specimens.

18761. Peter the Great Bay, Strelok Strait. June 12, 1911. DVE. 1
specimen.

20349. Tatar Strait, De-Kastri Bay. August 5, 1909. Derbek. 4
specimens.

20483. Peter the Great Bay. Amur Gulf. August 7, 1898. M. Yankovskii.
4 specimens.

33312. Tatar Strait, Shirokaya Pad'. August 18 and 22 [no year]. L.
Kuznetsov. 2 specimens.

37335. Peter the Great Bay. September 30, 1907. V.K. Brazhnikov. 1
specimen.

41106. Peter the Great Bay, Amur Gulf. September 9, 1974. V.I.
Pinchuk. 1 specimen.

41107. Peter the Great Bay, Ussuriisk Gulf. September 5-7, 1972. V.I.
Pinchuk. 2 specimens.

D XLI, 50; A I, 75; P 10 (Steindachner, 1980).

Spiny rays of dorsal fin low, 1.5 to 2.0 times less in height than soft
rays (Shmidt, 1904).

Head length 5.25 to 6.3 and maximum body depth 9 to 9.5 times in total
length. Diameter of eye 4.75 to 6.0 times, snout 4.5 to 5.0 times, length of
pectoral fins 2 to 2.5 times in the head length (Steindachner, 1980).

---

[145]From Taranetz, 1937b.
[146]Possibly, 6.3 times in standard length (Steindachner, 1880).

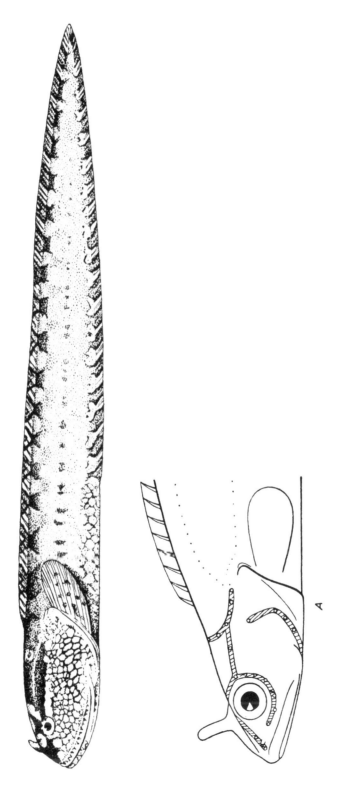

Figure 108. *Neozoarces pulcher*—handsome eelpout. Length 110 mm. Peter the Great Bay (Steindachner, 1880). A—arrangement of seismosensory canals of head.

136

Branchiostegal rays 7. Trunk with two branches of open seismosensory papillae: middle, a continuation of postorbital canal, extends along mediolateral line of body up to base of tail; and upper branch interrupted in middle part of body (Makushok, 1961c: 203). In our specimens, preanal distance varies from 34.0 to 40.6% of total length.

Radiographs of 10 specimens from 84 to 160 mm in length yielded the following data: vertebrae 96-100 (18-21 + 75-82); number of rays in dorsal fin 94-99 (XLII-XLVI + 49-59), in anal fin I 73-78; skeleton of free part of caudal fin with 6-8 principal and 4-5 marginal rays.

These are small, brightly colored fish,[147] dwelling near the coast among stones and clumps of algae. Eggs large, bright pink, and few in number. Parents probably protect their eggs (Makushok, 1961c: 207).

Length, to 160 mm.

Distribution: In·the Sea of Japan known from Hamhung-Namdo Province on the east coast of the Korean Peninsula (Mori, 1952: 127); 137 Peter the Great Bay (our specimens). In the north found to Tatar Strait (Soldatov and Lindberg, 1930: 462); along the east (Popov, 1933a: 149) and west coasts (Taranetz, 1937b: 155). In the Sea of Okhotsk known from Aniva Bay (Shmidt, 1950: 69), near the southwest and east coasts of Sakhalin and off the coast of Hokkaido (Ueno, 1971: 85).

2. *Neozoarces steindachneri* Jordan and Snyder, 1902—Steindachner's Eelpout (Figure 109)

Neozoarces steindachneri Jordan and Snyder, Proc. U.S. Nat. Mus., 25, 1902: 479, fig. 18 (Hokkaido, Otaru and Hakodate islands). Soldatov and Lindberg, Obzor..., 1930: 462. Taranetz, Kratkii Opredelitel'..., 1937: 155. Matsubara, Fish Morphol. and Hierar., 1955: 756. Fowler, Synopsis..., 1958: 321.

25475. Western Sakhalin. May 29, 1934. A.Ya. Taranetz. 3 specimens.

31625. Southern Sakhalin. August–September, 1946. KSE. 21 specimens.

37296. Sea of Japan, Ussuriisk Gulf. 1962. EZIN. 1 specimen.

37667. Western Sakhalin, Antonovo. July 29, 1963. V.G. Averintsev. 1 specimen.

37668. Sea of Okhotsk. Aniva Bay. August 16, 1963. L.L. Chislenko and A.N. Golikov. 1 specimen.

D XXXVIII, 49; A I, 72; gill rakers on first gill arch 4 + 12 (Jordan and Snyder, 1902a).

Radiographs of our 6 specimens 73 to 108 mm in length yielded the following data: number of vertebrae varies from 94 to 100, including 18–

---

[147]Shmidt (1904) mentions that the light-colored bands on the dorsal and anal fins which continue on the trunk broaden in such a way that they become clavate. Body gray on sides, with pattern of minute yellowish dots.

19 precaudal and 76-81 caudal; variation in number of vertebrae probably even greater, since Makushok (1961c: 203) found: vertebrae 106-108, precaudal 20 and caudal 86-88. Number of rays in dorsal fin varies from 94 to 99; spiny part with XLII-XLVI rays, followed by 49-59 soft rays. Anal fin I 73-78. Skeleton of free part of caudal fin with 6-8 principal and 4-5 marginal rays. Meristic characters very similar to those of *N. pulcher.*

This species differs from *N. pulcher* in relatively shorter head, shorter upper jaw, relatively smalller preanal distance (32.5 to 38.3% of total length), absence (or poor development) of pattern on lower surface of head, and more variegated body color. It dwells among algae in coastal waters.

Length, to 108 mm.

Distribution: In the Sea of Japan known from Peter the Great Bay (Soldatov and Lindberg, 1930: 462); west coast of Sakhalin (Taranetz, 1937b: 155); coast of Japan, recorded off Otaru (Jordan and Snyder, 1902a; 481). In the Sea of Okhotsk known from Aniva Bay, Lake Busse (Okada and Matsubara, 1938: 401); near east coast of Sakhalin (Ueno, 1971: 85). Recorded in the Sea of Okhotsk from coast of Hokkaido and Lake Notoro (Hikita, 1952: 12). Along the Pacific coast of Japan known from Hakodate (Jordan and Snyder. 1902a: 479).

5. Genus *Zoarchias* Jordan and Snyder, 1902—Eelpouts

Zoarchias Jordan and Snyder, Proc. U.S. Nat. Mus., 25, 1902: 480 (type: *Z. veneficus* Jordan and Snyder). Soldatov and Lindberg, Obzor..., 1930: 462.

This genus differs from *Neozoarces* in shorter spiny part of dorsal fin and greater number of rays in higher posterior part. Fleshy process on snout absent (Soldatov and Lindberg, 1930).

4 or 5 species. 2 known from the Sea of Japan.

138

Key to Species of Genus Zoarchias[148]

1 (2). Dorsal fin with 28 spiny rays; part of dorsal fin with soft rays begins far behind vertical from origin of anal fin. A I, 85-88 ....
.... 1. **Z. veneficus** Jordan and Snyder.

2 (1). Dorsal fin with 15 spiny rays; part of dorsal fin with soft rays begins slightly anterior to vertical from origin of anal fin. A I, 69.....
.... 2. **Z. uchidai** Matsubara.

1. *Zoarchias veneficus* Jordan and Snyder, 1902 (Figure 110)

Zoarchias veneficus Jordan and Snyder, Proc. U.S. Nat. Mus., 25, 1902: 480, fig. 19 (Hakodate, Muroran, Otaru). Soldatov and Lindberg,

[148]For species known from the Sea of Japan and adjacent parts of the Sea of Okhotsk and the Yellow Sea.

Obzor..., 1930: 462. Matsubara, Fish Morphol. and Hierar., 1955: 757. Fowler, Synopsis..., 1958: 322.

D XXVIII, 77; A I, 78; gill rakers on first gill arch 3 + 12 (Jordan and Snyder, 1902a).

Tomiyama (1972) researched the reason for the discrepancy in number of rays in the anal fin indicated by Jordan and Snyder in text (A I, 78) and depicted in their figure (A I, 88). After examining several specimens of *L. veneficus* from Hakodate, Tomiyama states that the correct number of rays is shown in the figure and should be retained as A I, 85–88.

Anal fin originates almost under the 18th ray of the dorsal fin. Standard length about 11 times the depth. Typical members pale chocolate-brown with dark reticulate pattern on body (Matsubara, 1955).

Close to the species *Z. major* Tomiyama, 1972,[149] from which it differs in greater number of rays in the dorsal fin (XXIX 80), shorter base of anal fin (judging from figure, begins under the 21st ray of dorsal fin), and (as pointed out by Tomiyama) broad dark spots in dorsal fin and presence of elongate lustrous black spot in anterior part.

Length 70 mm.

Distribution: In the Sea of Japan known from coasts of Hokkaido (Ueno, 1971: 85), Otaru (Jordan and Snyder, 1902a: 481); Sado Island (Honma, 1963: 31); Toyama Bay (Katayama, 1940: 25); Wakasa Bay (Takegawa and Morino, 1970: 382); San'in region (Mori, 1956a: 21). Found along the Pacific coast of Japan from Muroran (Jordan and Snyder, 1902a: 480) to Boshu Province (Okada and Matsubara, 1938: 401).

2. **Zoarchias uchidai** Matsubara 1932—Uchida's Eelpout (Figure 111)

Zoarchias uchidai Matsubara, Bull. Japan. Soc. Sci. Fish., 1, 2, 1932: 1, fig. 1 (Pusan). Matsubara, Fish Morphol. and Hierar., 1955: 757, pl. 80, fig. 275. Fowler, Synopsis..., 1958: 323.

D XV, 78; A I, 69.

Anal fin originates under 6th soft ray of dorsal fin. Eyes small, diameter about 7 times and interorbital space 10 times in head length (Matsubara, 1955). Anterior spiny rays of dorsal fin slender and highly elongate (equal to soft rays of this fin), gradually shortening posteriorly in such a way that posterior spiny rays lower than soft rays which follow them (Makushok, 1961c).

Length 120 mm (Matsubara, 1932).

Distribution: In the Sea of Japan known off Pusan (Matsubara, 1932: 1). Reported from south coast of the Korean Peninsula (Mori and Uchida, 1934: 22).

---

[149]Recorded from west coast of Kyushu Island (Ike Island.)

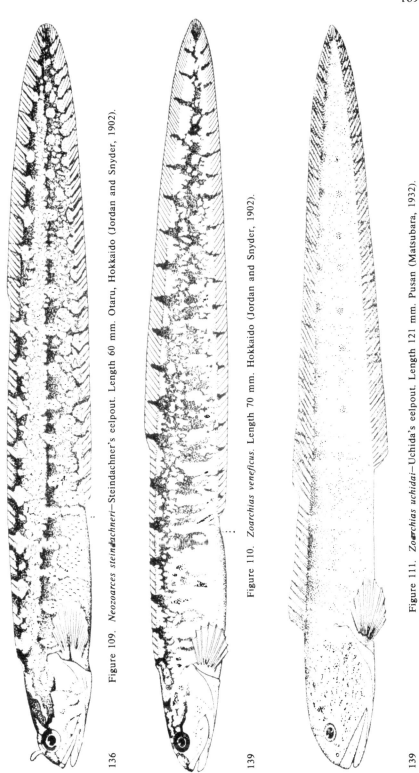

Figure 109. *Neozoarces steindachneri*—Steindachner's eelpout. Length 60 mm. Otaru, Hokkaido (Jordan and Snyder, 1902).

136

Figure 110. *Zoarchias veneficus*. Length 70 mm. Hokkaido (Jordan and Snyder, 1902).

139

Figure 111. *Zoarchias uchidai*—Uchida's eelpout. Length 121 mm. Pusan (Matsubara, 1932).

139

6. Genus *Lycodes* Reinhardt, 1838—Eelpouts

Lycodes Reinhardt, Overs. Dansk. Vidensk. Selsk. Nat.- Math. Afhandl., 7, 1838: 153 (type: *L. vahlii* Reinhardt). Jordan and Evermann, Fish. N. and M. Amer., 3, 1898: 2461. Soldatov and Lindberg, Obzor..., 1930: 492. Taranetz, Kratkii Opredelitel'..., 1937: 161. Andriyashev, Ryby Severnykh Morei SSSR, 1964: 266.

Body moderately elongate, its depth at origin of anal fin 7 to 14 times in total length (or 7 to 14% of it). Teeth on jaws, vomer, and palatines present. Mouth inferior, upper jaw protrudes forward beyond lower jaw. Mandibular and maxillary membranes completely reduced. Pair of mental crests present on lower side of head, each crest consisting of skin-covered cartilaginous overgrowth of lower margin of dentary. Neither cirri nor tentacles present on head. Lower lip broadens posteriorly (infralabial lobe) and fuses anteriorly (interlabial space). Lower jaw and upper jaw without large nostril-shaped pores. Gill openings do not continue forward, separated by isthmus; folds across latter absent. Body naked or covered
140 with minute scales separated from each other. Lateral line usually consists of minute and barely discernible free neuromasts and differs in position (Figure 112, A–F); mediolateral line (extends along median line of body); ventral (slopes down and back toward anal fin, continuing above it for greater or lesser distance); or doubled, mediolateral and ventral (bifurcates near apex of gill opening; usually anterior part of mediolateral
141 branch poorly developed). In some species lateral line transitional or ventrolateral, i.e., arcs down toward belly in anterior part of body, but at vertical from vent rises up and continues along median line of body up to tail; usually the posterior (ascending) part of arc poorly discernible, interrupted, or totally absent. In many species pores of dorsal series even wider-set and individual minute pores additionally present on head (latter do not form a true circumorbital ring). Dorsal fin without spiny rays. Pelvic fins rudimentary, jugular. Pectoral fins usually with 14–16 rays. Most species with 2 tubercular pyloric caeca; rarely caeca absent. Otoliths large. Vertebrae 87–119 with symmetrical centra (Andriyashev, 1954).

Benthic fishes, preferring silty bottom, which many burrow into. Eggs laid on bottom, large, and numerous. Emerging larvae probably do not pass through a pelagic stage. Food comprises benthic crustaceans, polychaetes, bivalve mollusks, echinoderms, and rarely fishes (Andriyashev, 1954).

About 50 species. In the Sea of Japan 12 species are known, and 1 species from adjacent waters.

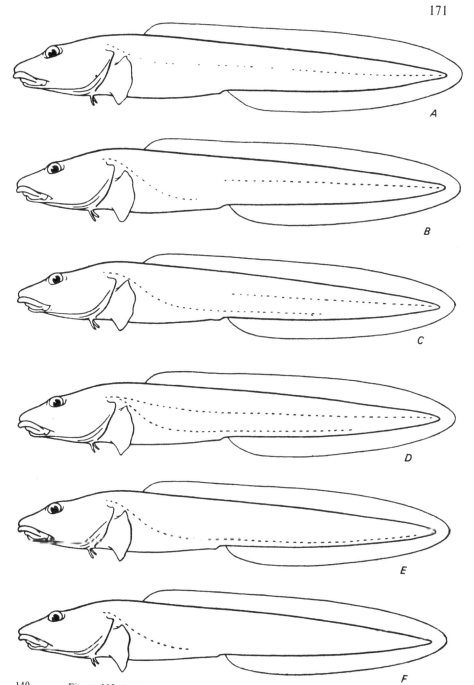

Figure 112. Major types of lateral line in *Lyodes* (Andriyashev, 1954).
A—mediolateral; B—ventrolateral; C—incomplete double; D—complete
double; E—complete ventral; F—incomplete ventral.

172

Key to Species of Genus Lycodes[150]

1 (12). Mental crests fused anteriorly.[151] Gill openings large.

2 (3). Gill openings do not reach base of pelvic fins. P 14-15; rays of lower part unbranched and relatively thick. Length of pelvic fins equal to diameter of eye or slightly more. D 79-84; A 69-73. Vertebrae 87-93. Dark pattern well expressed on sides of body. Lateral line incomplete, ventral ....

.... 1. **L. japonicus** Matsubara and Iwai.

3 (2). Gill openings reach base of pelvic fins.

4 (5). Posterior end of body with large white blotch, almost equal to head length. Several diffuse black spots located over and around this blotch. D 98-101; A 83-88; P 19-21. Vertebrae 106-111. Lateral line ventral ..... 2. **L. caudimaculatus** Matsubara.

5 (4). Posterior end of body without large white blotch.

6 (7). Upper part of body and dorsal fin with pattern in form of stripes arranged in shape of ⅄, with upper vertical part located on dorsal fin .... [**L. semenovi** Popov, 1931].

7 (6). Body and dorsal fin without such pattern.

8 (9). Body pale, without stripes, grayish-red. D 76-79; A 64-67; P 15-16. Vertebrae 84-85. Lateral line mediolateral ....
.... 3. **L. teraoi** Katayama.

9 (8). Body with light-colored transverse stripes.

10 (11). Teeth absent on vomer............. [**L. colletti** Popov, 1931].

11 (10). Teeth present on vomer. Lateral line mediolateral ....
.... 4. **L. uschakovi** Popov.

142 12 (1). Mental crests not fused anteriorly.[152] Gill openings do not continue to pelvic fins.

13 (18). Lower rays of pectoral fins elongate, form lobe.

14 (15). Dorsal fin with less than 90 rays (82-88), anal fin with 67-72, and pectoral fin with 19-20. Lateral line ventral ....
.... [**L. macrochir** Schmidt, 1950].

15 (14). Dorsal fin with more than 90 rays and anal fin with more than 70.

16 (17). P 17-19. Rays of lower part of pectoral fins unbranched, pointed

---

[150]Key based on that given by Matsubara (1955) with additions from the works of Taranetz (1937b), Matsubara and Iwai (1951), and our data. For a more complete understanding of the genus, species distributed in the northern part of the Sea of Okhotsk are included: *L. semenovi, L. macrochir, L. heinemanii, L. brunneofasciatus, L. jenseni, L. brevicauda,* and *L. colletti.*

[151]In fused mental crests the inner margins form a fold across the chin, which is very distinct in well-developed crests and poorly defined in less-developed crests.

[152]In independent mental crests the anterior ends fuse with the lower lip in the center and no fold forms across the chin (Figure 128, A, a).

at ends. Vertebrae 112–117. Interorbital space 12 to 25 times in the head length. Upper jaw distinctly protrudes forward beyond lower one. In closed mouth, teeth on premaxilla greatly exposed. Dorsal side of body grayish to chocolate-brown, underside light-colored, lateral sides without narrow pale transverse stripes. Lateral line ventral....... 5. **L. diapterus nakamurai** (Tanaka).

17 (16). P 20–23. Rays of lower part of pectoral fins unbranched. Caudal fin comparatively long, more than 0.5 diameter of eye. Vertebrae 108–111. Interorbital space 10–18 times in the head length. Upper jaw just barely protrudes beyond lower one. In closed mouth teeth almost not visible on premaxilla. Body dark chocolate-brown, usually with six to seven narrow white transverse stripes. .... [**L. hubbsi** Matsubara, 1955].[153]

18 (13). Lower rays of pectoral fin not elongate, do not form lobe.

19 (20). Body without scales, naked. Dorsal fin with less than 80 rays. Length of pelvic fins less than diameter of eye.........
.... [**L. heinemanni** Soldatov, 1916]

20 (19). Body with scales, at least in posterior part

21 (22). Lateral line ventral, passing along base of anal fin almost up to rays of caudal fin....... [**L. brunneofasciatus** Suvorov, 1935].

22 (21). Lateral line different, not ventral.

23 (34). Belly covered with scales.

24 (25). Lower 9 to 10 rays of pectoral fins form negligible lobe. D 91–106; A 76–87. Body black, without stripes and spots (at places of sloughed scales only white spots remain). Lateral line ventrolateral....... 6. **L. soldatovi** Taranetz and Andriashev.

25 (24). Lower rays of pectoral fin do not form notched lobe; fin obliquely truncated in lower part. D 79–89; A 64–72. Body relatively light-colored; seven to eight dark spots along back separated by light-colored intervals. Lateral line mediolateral.
.... 7. **L. macrolepis** Taranetz and Andriashev.

26 (27). Pelvic fins very short, less than 0.5 diameter of eye. Body light-colored, without stripes and spots. Margin of dorsal and anal fins dark. Lateral line ventrolateral.....................
.... 8. **L. brevipes ochotensis** Schmidt.

143 27 (26). Pelvic fins longer than 0.5 diameter of eye. Body with

---

[153]L. hubbsi has been recorded from the Pacific coast of Hokkaido and Honshu. Matsubara (1955: 776) indicated that a description of the species was forthcoming in the near future. None has been published to date. Neither has another species, L. sadoensis Matsubara and Honma, been described although recorded by Honma (1963: 21) from near Sado Island, and by Morino and Takegawa (1970: 383) from Wakasa Bay. We cannot consider the latter species for our waters due to the absence of a published description, and the inadequacy of the photograph published by Honma (1969: 31, Fig. 9) for writing up a description.

spots or stripes (except in *L. jenseni*). Margin of dorsal and anal fins not darker than background color of fins.

28 (29). Pectoral fins with 21 rays. Dorsal fin with light-colored Y-shaped spots, which continue onto back. Lateral line ventrolateral...

.................................9. [**L. ygreknotatus** Schmidt].

29 (28). Pectoral fin with not more than 18 rays. Dorsal fin without Y-shaped spots.

30 (31). Mental crests poorly developed, low, with rounded tips. Upper jaw short, reaches posteriorly only to anterior margin of eye. Body color monochromatic. Lateral line ventrolateral........

..................[**L. jenseni** Taranetz and Andriashev, 1935].

31 (30). Mental crests well developed, from pointed process or lobe toward front. Upper jaw extends posteriorly beyond anterior margin of eye, and protrudes beyond lower jaw.

32 (33). Pectoral fins rounded at posterior margin, longest rays in middle. P 17. Lateral line ventrolateral.....................

...........................10. **L. palearis fasciatus** (Schmidt).

33 (32). Pectoral fin truncated at posterior margin, longest rays in upper third. P 18. Lateral line ventrolateral.................

........................10a. **L. palearis schmidti** Gratzianov.

34 (23). Belly naked, or if scales present, only in posterior part.

35 (40). Preanal distance less than 50% of total length. Dorsal and anal fins at least partly covered with scales.

36 (39). Teeth on lower jaw distinctly close-set, numerous.

37 (38). Upper half of body and head chocolate-brown, with rounded light-colored spots. Lower half of body and head light-colored. Margin of dorsal and anal fins dark; fins usually with light-colored spots and oblique stripes. Lateral line medio-lateral.

........................11. **L. tanakae** Jordan and Thompson.

38 (37). Upper half of body and head with light-colored S-shaped marks, which continue onto dorsal fin. Vertical fins usually without light-colored oblique stripes. Lateral line mediolateral.......

..........12. **L. sigmatoides** Lindberg and Krasjukova n. nov.

39 (36). Teeth on lower jaw wide-set (especially in posterior part) and few. Upper half of body with about seven dark transverse stripes which continue onto dorsal fin. Apices of gill openings connected by light-colored stripe which curves forward. Lateral line mediolateral...

·..............13. **L. raridens** Taranetz and Andriashev, 1935.

40 (35). Preanal distance more than 50% total length. Dorsal and anal fins not covered with scales. Lateral line mediolateral.......

..............[**L. brevicauda** Taranetz and Andriashev, 1935].

1. *Lycodes japonicus* Matsubara and Iwai, 1951—Japanese Eelpout
(Figure 113)

Lycodes japonicus Matsubara and Iawi, Japan. J. Ichthyol., 1, 6, 1951:
368, figs 1-3 (Toyama Bay). Matsubara, Fish Morphol. and Hierar., 1955:
775.

D 79-84; A 69-73; P 14-15; vertebrae 87-93.

Six lower rays of pectoral fin simple, more or less thickened, and
membrane between rays deeply notched; lower half of fin symmetrical to
upper half. Pelvic fins relatively long, almost equal or slightly larger than
diameter of eye, and length 3.3 to 4.5 times in the head length. Pyloric
144 caeca 2, rarely 3. Head in males broader and higher than in females
(Figure 114, C, D). Eyes fairly large, their diameter slightly less or equal to
snout length. Lower inner margin of lower jaw does not form protruding
fleshy lobes. Belly covered with scales; scales continue downward and
forward, beyond base of pectoral fin. Teeth minute, sharp. On premaxillary teeth arranged in 2 rows in front and in 1 row along sides.
Vomer with 9 to 10 teeth. Lower jaw with 3 rows of teeth in anterior part
and 1 row in posterior part (Figure 114, A, B). Sides of body of adult fish
with characteristic coloration, especially pattern (Figure 113) (Matsubara
and Iwai, 1951b). Color of males and females changes with age and,
probably, in fish close to 122 mm in length stabilizes (Figure 115, A-D).

The authors of this species have provided a detailed description and
drawing (Figure 114, C, D) of the arrangement of the seismosensory pores
of the head: pores minute (visible only under lens) on operculum and
lower jaw, suborbital bones, near nasal tubules, in postorbital region,
occiput, and on surface anterior to dorsal fin. Mandibular-opercular series
consists of about 7-8 pores arranged along margin of operculum and 3-4
on lower jaw. Suborbital series comprises 7-8 pores, of which 2 or 3 on
cheek, 3 on upper lip, and 3 on anterior margin of snout. Nasal pores 6-7,
arranged from base of nasal tubules to midway between nostril and
anterior margin of eye. Post-orbital series of 5 pores begins immediately
behind eye and directed toward beginning of lateral line, 1 or 2 pores
located on occiput. Predorsal series comprises 4-5 pores[154] arranged from
origin of dorsal fin to upper end of operculum. Lateral line indistinct,
incomplete, and does not curve downward; instead it proceeds smoothly
from occiput to pectoral fin, barely extending beyond its anterior third
(Matsubara and Iwai, 1951b).

Length, to 137 mm (Matsubara and Iwai, 1951b).

Distribution: In the Sea of Japan known from Toyama Bay (Matsubara
and Iwai, 1951b: 368) and off Sado Island (Honma, 1963: 21).

---

[154]The authors of this description make no distinction between pores and free
neuromasts, referring to both as pores.

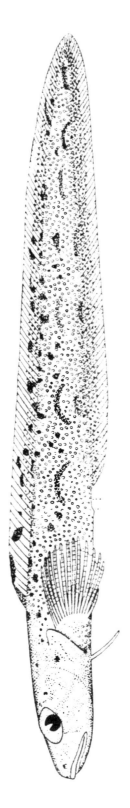

Figure 113. *Lycodes japonicus*—Japanese eelpout. Length 129 mm. Toyama Bay (Matsubara and Iwai, 1951).

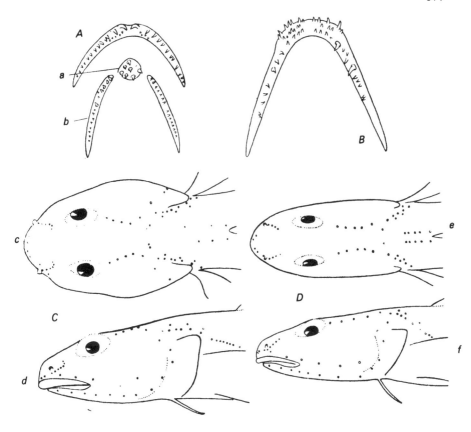

145 Figure 114. Teeth and head of *Lycodes japonicus* (Matsubara and Iwai, 1951).

A—premaxilla, a—vomer, b—palatine; B—lower jaw; C—head of male, c—dorsal view, d—lateral view; D—head of female, e—dorsal view, f—lateral view.

2. *Lycodes caudimaculatus* Matsubara, 1936—White-Blotched Tail Eelpout (Figure 116)

Lycodes caudimaculatus Matsubara, J. Imper. Fish. Inst., 31, 2, 1936: 115,.fig. 1 (Mie Prefecture, Pacific coast of Japan). Matsubara and Iwai, 145 Japan. J. Ichthyol., 1, 6, 1951: 373. Matsubara, Fish morphol. and Hierar., 1955: 775, fig. 293.

41193. Pacific coast of Japan, 37°45'8" N, 141°55'3" E. January 13, 1973. V.V. Fedorov. 3 specimens.

41194. Pacific coast of Japan, 36°58' N, 141°29' E. April 2, 1973. A.S. Sokolovskii. 1 specimen.

41195. Pacific coast of Japan, 36°55' N, 141°23' E. April 15, 1973. A.S. Sokolovskii. 3 specimens.

178

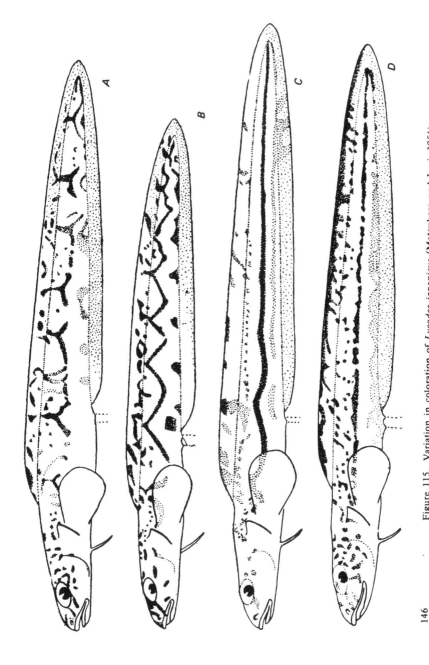

Figure 115. Variation in coloration of *Lycodes japonicus* (Matsubara and Iwai, 1951).

A—female, length 109.5 mm; B—female, length 100 mm; C—female, length 119 mm; D—male, length 116 m.

146

D 98-101; A 83-86; P 19-21; vertebrae 106-111 (Matsubara and Iwai, 1951b).

Differs from *L. japonicus* in asymmetrical pectoral fin,[155] smaller pelvic fins with length less than diameter of eye, and absence of dots on sides of body. Oval white spot and a few small dark spots located only in caudal part. Dark spots present in posterior part of dorsal and anal fins (Matsubara and Iwai, 1951b).

In 7 of our specimens, 171 to 242 mm long, mental crests were poorly developed and fused. Lateral line ventral, complete, and in form of free neuromasts (more than 200 throughout line). Pores of seismosensory
146 system of head distinct: suborbital 8, postorbital 4, preopercular 4-6, mandibular 4, and 3 neuromasts in occipital row, with 3 free neuromasts near posterior nostril. Preanal distance 33.6 to 39.9% of the total length. Radiographs of these specimens revealed that the number of vertebrae varied from 107 to 109 (precaudal 22-23, caudal 84-86); number of rays in dorsal fin from 101 to 103, of anal fin from 86 to 88; and skeleton of free part of caudal fin with 8 principal and 5 marginal rays.

Information on biology not available. Probably dwells at fairly large depth since our specimens were caught between 260 to 500 m, and places of occurrence given below in the distribution of this species are quite deep. Catches from shallow waters not known to date.
147 Length, to 242 mm.

Distribution: In the Sea of Japan known only in Wakasa Bay (Takegawa and Morino, 1970: 383). Recorded from the Pacific coast of Japan near Mie Prefecture (Matsubara, 1936), reported for Teshio, and farther south to Kochi Prefecture (Matsubara, 1955: 775).

3. *Lycodes teraoi* Katayama, 1943—Pale Eelpout (Figure 117)
Lycodes teraoi Katayama, Ann. Zool. Japon, 22, 2, 1943: 103, fig. 2 (Tsuiyama, Hyogo Prefecture). Matsubara and Iwai, Japan. J. Ichthyol., 1, 6, 1951: 373. Matsubara, Fish Morphol. and Hierar., 1955: 775. Fowler, Synopsis..., 1958: 305

D 76; A 64; P 15[156]; V 3 (Katayama, 1943).

Head length 4.76 times, depth 10.18 times, distance from tip of snout to vent 2.43 times, and distance from origin of pelvic fin to anal fin 3.62 times in the total length. Longitudinal diameter of eye 5.66 times, interorbital distance 17.0 times, snout length 3.09 times, length of maxilla 2.61 times, length of postorbital part of head 2.0 times, length of pectoral fin 2.26 times and length of pelvic fin 6.8 times in head length. Body elongate, compressed laterally, attenuate toward tip

[155]Lower rays branched and membrane between them not deeply notched; in *L. japonicus* lower rays simple and membrane between them deeply notched.
[156]D 76-79; A 64-67; P 15-16 (Matsubara, 1955).

of tail. Head depressed, interorbital space narrow and convex. Mouth moderate in size, maxilla continues to vertical from middle of eye. Upper jaw protrudes beyond lower by distance almost equal to half longitudinal diameter of eye. Teeth small but strong, present on jaws, vomer, and palatines. Tongue thick, blunt at end, and not free toward front. Gill openings large, almost equal in length to maxilla. Branchiostegal membranes widely fused with isthmus. Nostrils with small tubules. Lateral line mediolateral, continues along middle of body, without reaching end. Scales small and round, embedded in skin, and absent only on head, in middle part of belly, and at base of pectoral and pelvic fins. Dorsal fin originates almost at vertical from middle of pectoral fin. Anal fin originates almost at vertical from 15th ray of dorsal fin. Dorsal fin higher than anal. Pectoral fins short. Pelvic fins very small (Katayama, 1943).

Characterized by grayish-pink pale body devoid of stripes and spots, light-colored fins, and notably protruding upper jaw.

Length more than 300 mm (Matsubara and Iwai, 1951b).

Distribution: In the Sea of Japan known from Wakasa Bay (Takegawa and Morino, 1970: 383); off coast of Hyogo Prefecture (Matsubara, 1955: 775); Tayama (Katayama, 1952a: 5); and San'in region (Mori, 1956a: 21).

4. *Lycodes uschakovi* Popov, 1931—Ushakov's Eelpout (Figure 118)

Lycodes uschakovi Popov, Issled. Morei SSSR, 14, 1931: 141, pl. II, fig. 7 (northern part of Sea of Okhotsk). Shmidt, Ryby Okhotskogo Morya, 1950: 96. Matsubara, Fish Morphol. and Hierar., 1955: 775.

Lycodes collettii Popov, Issled. Morei SSSR, 14, 1931: 143, pl. II, fig. 6 (57°11'5" N, 148°19'5" E).

Lycodes lindbergi Popov, Issled. Morei SSSR, 14, 1931: 142, pl. II, fig. 5 (58°11'5" N, 148°19'5" E).

24834. Sakhalin, east coast. July 22, 1918. GEVO. 6 specimens.

25269. Sea of Japan, northern part. June 7, 1931. D. Okhryamkin. 3 specimens.

34845. Sakhalin, east coast. July 22, 1918. GEVO. 1 specimen.

36935. Sakhalin, east coast. July 22, 1918. GEVO. 3 specimens.

39308. Sea of Okhotsk. September 23, 1963. V.P. Shuntov. 1 specimen.

39342. Sea of Okhotsk. October 31, 1963. V.P. Shuntov. 1 specimen.

41637. Sea of Japan, Tatar Strait. August 23, 1949. KSE. 1 specimen.

41638. Sea of Okhotsk, Aniva Bay. September 24, 1947. KSE. 3 specimens.

41639. Sea of Japan, Tatar Strait. August 22, 1949. KSE. 4 specimens.

The description of this species given by Popov (1931a) on the basis of a young specimen has been rightly considered by Shmidt (1950) as insufficient and too short. Shmidt furnished a description after examining 9 adult specimens.

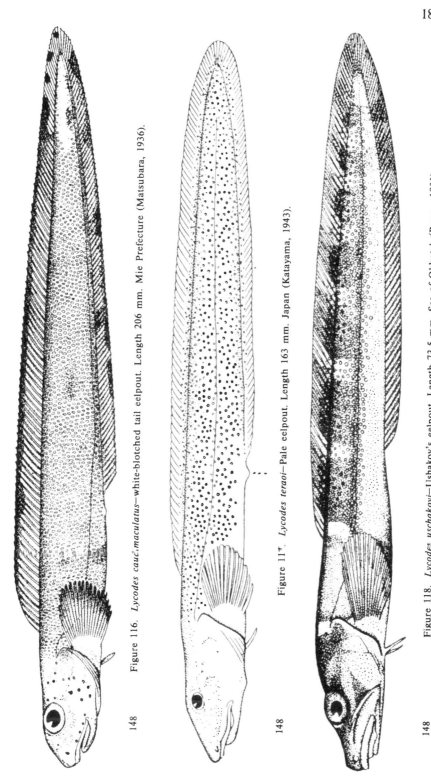

Figure 116. *Lycodes cauc.maculatus*—white-blotched tail eelpout. Length 206 mm. Mie Prefecture (Matsubara, 1936).

148

Figure 117. *Lycodes teraoi*—Pale eelpout. Length 163 mm. Japan (Katayama, 1943).

148

Figure 118. *Lycodes uschakovi*—Ushakov's eelpout. Length 73.5 mm. Sea of Okhotsk (Popov, 1931).

148

D 78–80; A 64; P 18; gill rakers on first arch 3 + 12 (Shmidt, 1950).

Body relatively long, its depth 10 times and length of head 4 to 5 times in total length. Head large, slightly depressed, occiput convex. Eyes protrude slightly above upper contour of head. Posterior end of maxilla reaches vertical from anterior margin of pupil. Lips thick; lateral lobe of lower lip broad, suspended, and lip not interrupted toward front but attenuate. Mental crests short, anterior ends fused. Teeth on upper jaw arranged in 2 rows: outer with 7–8 blunt teeth (anterior ones larger), and inner row with 1–2 teeth. Teeth on lower jaw arranged in 1 row posteriorly and 3–4 rows anteriorly. Vomer with 2–3 small blunt teeth, palatines with 4–5 arranged in single row. Trunk covered with relatively large scales, which continue on sides of body to posterior margin of pectoral fins. Naked strip extends along anterior half of dorsal fin near its base; scales on posterior end of fin continue onto base. Surface of back anterior to dorsal fin, head, abdomen, thorax, and chin naked. Color gray to chocolate-brown, back darker, and abdomen and lower part of head light-colored. Light-colored stripe present between upper corners of gill openings; sides of body with 8 to 10 transverse light-colored stripes, which continue onto dorsal fin; those in posterior part of body also continue onto anal fin. Sometimes dark spot present in anterior part of dorsal fin (Shmidt, 1950).

The specimens of KSE were caught from depths of 78 to 148 m, where the bottom water temperature ranged from +0.4 to −1.9°C. Bottom silty with admixture of sand, pebbles, and stones. Sites of catches were abundantly inhabited by *Gorgonocephalus caryi, Ctenodiscus crispatus,* many *Pandalus borealis eous*; fishes in the catches comprised many flounders (*Limanda aspera, Hippoglossoides elassodon robustus, Pleuronectes quadrituberculatus, Acanthopsetta nadeshnyi*), numerous cods, fewer eelpouts, and rarely *Lumpenus medius, Myoxocephalus polyacanthocephalus, Enophrys diceraus, Dasycottus setiger,* and *Icelus spiniger cataphractus*.

Length, to 356 mm.

Distribution: In the Sea of Japan known from northern part near Cape Peshchernyi, west coast of Tatar Strait (Shmidt, 1950: 98), and Primor'e (Lindberg, 1947: 171). In the Sea of Okhotsk known from Aniva Bay (Lindberg, 1947: 171), and north to Shantar Islands (Shmidt, 1950: 98).

5. *Lycodes diapterus nakamurai* (Tanaka, 1914)—Nakamura's Eelpout
   (Figure 119)

Furcimanus nakamurae Tanaka, Fig. and Descript., Fish of Japan, XVIII, 1914: 303, pl. 82, fig. 277[157] (Niigata).

---

[157]Misprinted as fig. 276 on p. 303 of text; should read pl. 82 and fig. 277, which were included in volume XVII.

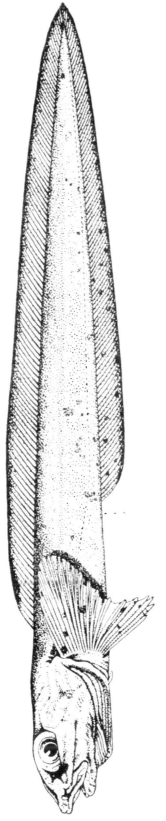

Figure 119. *L·rodes diapterus nakamurai*—Nakamura's eelpout. Length 285 mm. Niigata (Tanaka, 1914).

184

Lycodes nakamurai, Matsubara, Fish Morphol. and Hierar., 1955: 776.

150   *Lycodes diapterus nakamurae,* Taranetz, Kratkii Opredelitel'..., 1937: 164. Lindberg, Predvaritel'nyi Spisok..., 1947: 171. Shmidt, Ryby Okhotskogo Morya, 1950: 98.

24822. Sakhalin, east coast. September 11, 1932. M.N. Krivobok. 3 specimens.

26887. Sea of Okhotsk, 56°18′ N, 145°04′ E. August 28, 1914. GEVO. 4 specimens.

30966. Sea of Japan, 47°41′ N, 140°06′ E. July 21, 1932. M.I. Krivobok. 1 specimen.

41611. Sea of Japan, Tatar Strait. August 23, 1949. KSE. 2 specimens.

41612. Sea of Japan, Tatar Strait. August 23, 1949. KSE. 2 specimens.

41613. Sea of Okhotsk, Cape Aniva. September 24, 1947. KSE. 2 specimens.

D 104-107; A 94-96; P 18-19; gill rakers on first arch $9 + 13$[158]; nasal tubules less than half diameter of eye. Pectoral fins with well-developed deep notch; length of shortest ray about two-thirds length of longest ray (Shmidt, 1950).

Tanaka (1914) reported differences from the subspecies *L. diapterus diapterus* Gilbert such as shorter pectoral fins, broader interorbital space, deeper body, and longer nasal tubules. Shmidt (1950) mentioned that the only differences between *L. diapterus nakamurai* and *L. diapterus diapterus* Gilbert are smaller size of eyes and deeper notch in pectoral fins in the former. It should be noted that the latter has 8-9 transverse stripes on the body, which are absent in our subspecies.

Our subspecies differs from *L. diapterus beringi* Andriashev in deeper notch in pectoral fins, absence of transverse stripes on body, and very distinct ventral lateral line. In the 8 specimens examined by us, the mental crests were not fused and poorly developed. Preanal distance varied from 33.5 to 36.0% of total length, vertebrae from 112 to 117 $(20-22 + 92-96)$, rays in dorsal fin from 106 to 114, in anal fin 93-98. Caudal fin with 8-10 principal and 2-4 marginal rays. Specimens of our subspecies were obtained from depths of 148 to 207 m, with bottom water temperatures of $+0.9$, 2.0, and $-0.4°C$. Bottom consisted of argillaceous silt with a few pebbles.

Sites of catches were abundantly inhabited by *Gorgonocephalus caryi,* Pennatulidae with *Asteronyx loveni, Ctenodiscus crispatus,* and *Pandalus borealis* predominant; among fishes only *Sebastolobus macrochir* was recorded.

Length 273 mm (Tanaka, 1914).

152   *Distribution*: In the Sea of Japan known from Pohang (Mori, 1952: 130); in Primor'e (Lindberg, 1947: 171); near the west coast of Hokkaido

[158]$4 + 11$ (Tanaka, 1914).

(Hikita and Hirosi, 1952: 48); Toyama Bay (Katayama, 1940: 25); off Sado Island (Honma, 1952: 226); Wakasa Bay (Takegawa and Morino, 1970: 383); and San'in region (Mori, 1952: 21). Sea of Okhotsk (Nos. 24822, 26887).

In the region of the southern Kuril Islands, in the Pacific Ocean, and east of the Iturup and Zelenyi islands, eelpouts (9 specimens) from the "*diapterus*" group in the collection of KSE had a ventral lateral line. These were identified by A.P. Andriyashev in 1950 as *L. (Furcimanus) taranetzi,* sp. n. Description of the species was not published. Here we present the diagnosis and an illustration of the species kindly supplied by A.P. Andriyashev, the editor of this volume.

Lycodes (Furcimanus) taranetzi Andriashev, sp. n.–Taranetz Eelpout (Figure 119-A)

Lycodes taranetzi Lindberg, 1950: 251 (see Andriyashev in litt., nomen nudum).

443   *Addendum* : This species does not appear in the key to species of the genus *Lycodes* because the description and drawing were received from the author after this book went to press.

Holotype: No. 42247. Kuril-Sakhalin Expedition of the Institute of Zoology, Academy of Sciences of the USSR and TINRO, trawler *Toporok,* St. 101, September 14, 1949. Pacific side of Iturup Island. Depth 414 m, bottom temperature 2.3°C; caught in beam trawl along with sand and gravel, and considerable nodular material. Collected by G.U. Lindberg and M.I. Legeza. Total length 303 mm. Institute of Zoology, Academy of Sciences of the USSR–paratypes: same place, male, total length 340 mm; others juv. 187, 180, 175, 167, and 136 mm. Institute of Zoology, Academy of Sciences of the USSR, No. 42248; *Toporok,* St. 88, September 11, 1949; east of Zelenyi Island in group of southern Kuril Islands, juv. 150 mm at a depth of 382-295 m. Institute of Zoology, Academy of Sciences of the USSR, No. 42262.

Close to the American species *Lycodes diapterus* Gilbert in ventral type of lateral line, poorly expressed and anteriorly fused mental crests, smooth slender lobe in lower lip, distinctly notched pectoral fins, general coloration, and other characters. However, distinctly differs in smaller number of vertebrae (105 in holotype, 104-107 in 7 paratypes),[159] and complete absence of scales on occiput, scales not continuing in front beyond transverse occipital white stripe, and scales far from reaching minute pores of supratemporal (occipital) commissure. According to

[159]According to Schultz (1967: 5), in the holotype and paratype of *L. diapterus* Gilbert (1891) vertebrae 121 (21 + 100); in our specimen (Zoological Institute, Academy of Sciences of the USSR, No. 25869) 20 + 101.

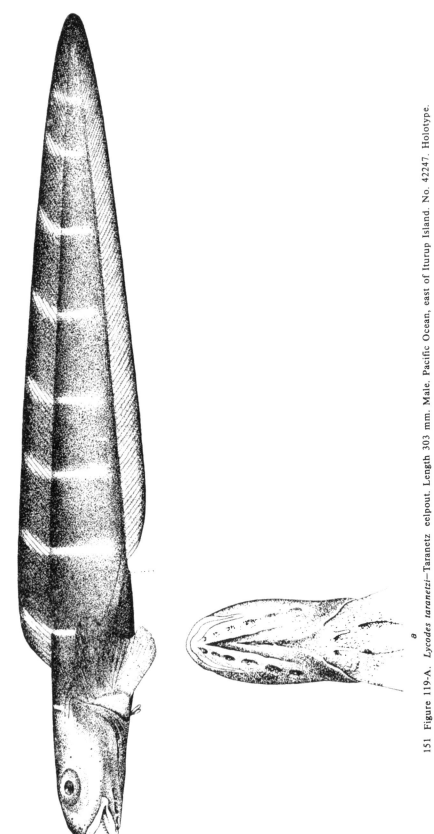

151 Figure 119-A. *Lycodes taranetzi*—Taranetz eelpout. Length 303 mm. Male. Pacific Ocean, east of Iturup Island. No. 42247. Holotype. a—head, ventral view.

Gilbert, *L. diapterus* has minute scales in larger specimens (total length 288 and 101 mm), "which continue onto occiput and cover it completely." In our specimens of *L. diapterus* from California (Santa Barbara Channel; 150 fathoms, R.L. Bolin, Zoological Institute Academy of Sciences of the USSR, No. 25869, total length 169 mm) scales cover entire occiput, and extend farther forward (at some places with intact skin) almost to level of posterior margin of lens. In addition, our specimens of *L. diapterus* differ from close forms in strong development of scales on pectoral fins as well as in details of coloration.

D + 1/2 C in holotype 107 (in paratype with total length 342 mm—108); A respectively 92 (92); P 20 (20). Head length with membrane 18.1 (18.0)%, predorsal distance 19.8 (20.2)%, preanal distance 34.3 (34.8)%, length of upper lobe P 10.2 (10.8), of lower lobe P 9.1 (8.9), and body depth at origin of anal fin 9.2 (9.1)% of the total length. As percentage of head length, longitudinal diameter of eye 27.3 (29.3), snout length from anterior margin of eye 26.4 (24.4). Scales cover not only entire body and unpaired fins, but also pectoral fins, reaching middle of their upper lobe and almost to end of lower lobe. Gill rakers $\dfrac{3+8}{1+8}$; pyl. caeca—0.

153   Body dark chocolate-brown, abdomen bluish; narrow straight white stripe continues across occiput; body with 6-8 narrow light-colored stripes, which continue onto D, broaden toward lower side, and become quite diffuse; in younger specimens these stripes are sharper. Belly chocolate-brown to black or completely black.

Species named in honor of the famous expert on fishes of the Pacific Ocean, Anatoli Yakovlevich Taranetz.

Distribution: Southern Kuril Islands (Iturup and Zelenyi) in depths ranging from 295 to 414 m.

6. **Lycodes soldatovi** Taranetz and Andriashev, 1935—Soldatov's Eelpout
(Figure 120)

Lycodes soldatovi Taranetz and Andriyashev, Zool. Anz., 112, 9-10, 1935: 246, Abb. 3 (Sea of Okhotsk, Terpenia Gulf). Taranetz, Kratkii Opredelitel'..., 1937: 163. Shmidt, Ryby Okhotskogo Morya, 1950: 90, pl. 6, fig. 2. Matsubara. Fish Morphol. and Hierar., 1955: 777.

24846. West coast of Kamchatka. 1922. I.A. Polutov. 1 specimen.

25190. Sea of Okhotsk, Cape Terpenia. July 5, 1932. S. Generozova. 2 specimens.

37963. 57°58′ N, 154°15′ E. July 30, 1963. V.P. Shuntov. 1 specimen.

37978. Bering Sea. September 7, 1963. V.V. Fedorov. 2 specimens.

37979. 51°06′ N, 156°06′ E. July 9. 1913. V.P. Shuntov. 1 specimen.

D 105-106; A 87; P 22-23 (Andriyashev and Taranetz, 1935). Differs from close species in very small eyes (11.8 to 13.9% of head length), broad interorbital space (6.2 to 6.6% of head length), long snout (31.5 to 33.4% of head length), and dark almost blackish body. Semi-deepwater species (Taranetz and Andriyashev, 1935).

In our 7 specimens ranging in length from 266 to 607 mm, mental crests were not fused, and had poorly defined lobes at anterior end. Lateral line ventrolateral, complete, fairly distinct in preserved fish. Pores of seismosensory system of head in preserved specimens almost not discernible. Preanal distance 41.3 to 49.0% of total length.[160] Radiographs showed number of vertebrae varied from 96 to 106 (precaudal 20–25, caudal 76–83); rays of dorsal fin from 91 to 100, of anal fin 76 to 86; skeleton of free part of caudal fin with 8–9 principal and 4–5 marginal rays.

Length, to 660 mm (Andriyashev and Taranetz, 1935).

Distribution: In the Sea of Japan reported from Sado Island (Katho et al., 1956: 323). Recorded from the Sea of Okhotsk, east of Cape Terpenia, and near the west coast of Kamchatka; also known from the Bering Sea (Taranetz and Andriyashev, 1935: 246).

7. *Lycodes macrolepis* Taranetz and Andriashev, 1935—Large-Scaled Eelpout (Figure 121)

Lycodes macrolepis Taranetz and Andriyashev, Zool. Anz., 112, 9–10, 1935: 251, figs. 6 and 7. Taranetz, Kratkii Opredelitel'..., 1937: 163. Shmidt, Ryby Okhotskogo Morya, 1950: 88, pl. 5, fig. 1. Matsubara, Fish Morphol. and Hierar., 1955: 777. Fowler, Synopsis..., 1958: 305, fig. 35.

24837. Sea of Okhotsk, 57°02′ N, 114°40′ E. September 5, 1932. M.N. Krivobok. 3 specimens.

25274. Sea of Okhotsk. September 2, 1932. M.N. Krivobok. 1 specimen.

29998. Sea of Okhotsk, 55°36′ N, 139°55′ E. August 29, 1932. P.Yu. Shmidt. 1 specimen.

30985. Sea of Japan, Soviet Gavan'. July 22, 1932. M.N. Krivobok. 1 specimen.

D. 83–89; A 64–72; P 20 (Taranetz and Andriyashev, 1935).

Distinguished by presence of 7–8 dark stripes on body, lateral line mediolateral, very small interorbital space (about 4% of head length), relatively small number of rays in dorsal and anal fins, and absence of lobe in lower part of pectoral fin. In our 4 specimens (length from 150 to 162 mm), mental crests were not fused, and were well developed, and preanal distance varied from 43.3 to 47.0%. On the basis of radiographs number of vertebrae varied from 86 to 90 (precaudal 19–21, caudal 67–69); rays of dorsal fin 79–83, of anal fin 66–69. Caudal fin in free part with 8–10 principal and 3–5 marginal rays.

[160]Given as 44.8 to 46.0% by Taranetz and Andriyashev (1935: 247).

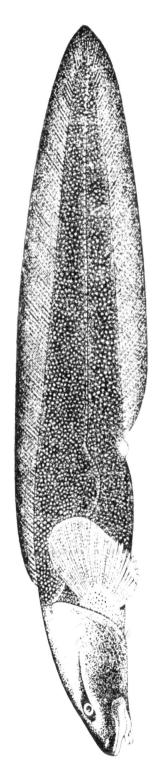

Figure 120. *Lycodes soldatovi*—Soldatov's eelpout. Length 560 mm. Terpenia Gulf (Taranetz and Andriyashev, 1935).

154

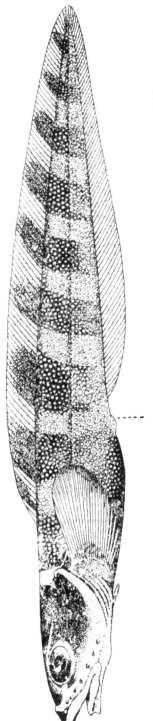

Figure 121. *Lycodes macrolepis*—large-scaled eelpout. Length 153 mm. Sea of Okhotsk (Taranetz and Andriyashev, 1935).

154

Length 161 mm (Taranetz and Andriyashev, 1935).

Distribution: In the Sea of Japan known from Tatar Strait, Soviet Gavan (Shmidt, 1950: 88); near west coast of Sakhalin, Cape Aleksandr (Taranetz and Andriyashev, 1935: 251); off Sado Islands (Honma, 1963: 21); Toyama Bay (Katho et al., 1956: 323); near coast of Tayima Province (Katayama, 1952: 5); and San'in region (Mori, 1956a: 22). Sea of Okhotsk (Shmidt, 1950: 88).

8. *Lycodes brevipes ochotensis* Schmidt, 1950—Short-finned eelpout
(Figure 122)

Lycodes brevipes ochotensis Shmidt (east coast of Sakhalin) in: Taranetz Kratkii Opredelitel'..., 1937: 163 (without description). Shmidt, Ryby Okhotskogo Morya, 1950: 89 (description).

Lycodes brevipes, Matsubara, Fish Morphol. and Hierar., 1955: 777.

25271. Sea of Okhotsk, 53°17' N, 144°10' E. 1932. S. Generozova. 1 specimen.

36177. Sea of Okhotsk, 51°58' N, 144°27' E. August 19, 1959. A.P. Nikolaev. 2 specimens.

D 87; A 76; P 20. Lateral line ventrolateral (Shmidt, 1950). Close to typical form, known from southeastern part of the Bering Sea. Differs from typical form in smaller size of eyes and monochromatic chocolate-brown body devoid of stripes and spots. Differs further from description of subspecies *L. b. diapteroides* reported from the northeastern part of the Bering Sea by Taranetz and Andriyashev in 1937 in smaller number of rays in dorsal and anal fins (87 and 76 versus 99–101 and 83–85), relatively larger head length, incomplete lateral line (in *L. b. diapteroides* reaches only to anal opening), and smaller number and different arrangement of teeth on jaws, vomer, and palatines.

Examination of our 3 specimens of *L. b. ochotensis* (length 281 to 342 mm) revealed variation in the preanal distance (percentage of total length) from 39.1 to 40.0 and unfused mental crests. Pores of seismo-sensory system of head well defined: suborbital 8, postorbital 4, mandibular 4, preorbital 2, occipital 3, and margin of preopercle 4. Number of vertebrae in radiographs of all specimens 98 (precaudal 20–21, caudal 77–78); rays of dorsal fin 93, of anal fin 78. Skeleton of free part of caudal fin with 8–10 principal and 3–4 marginal rays.

Length 277 mm (Shmidt, 1950).

Distribution: In the Sea of Japan known only from the coast of Hokkaido (Ueno, 1971: 86). In the Sea of Okhotsk recorded from the east coast of Sakhalin (Shmidt, 1950: 90) and Okhotsk Sea coast of Hokkaido (*Oshoro Maru* cruise, 1969: 390).

9. [*Lycodes ygreknotatus* Schmidt, 1950]—Y-barred eelpout (Figure 123)

Lycodes ygreknotatus Shmidt (Shantar Islands) in: Taranetz, Kratkii

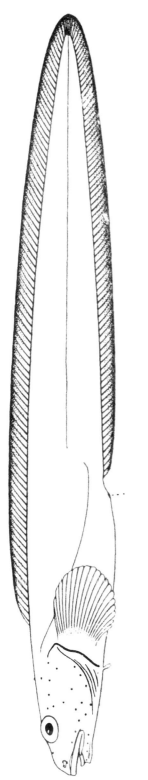

Figure 122. *Lycodes brevipes ochotensis*. Short-finned eelpout. Length 345 mm. No. 36177. Sea of Okhotsk.

156

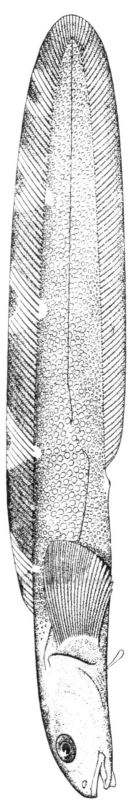

191

Figure 123. *Lycodes ygreknotatus*. Y-barred eelpout. Length 191 mm. No. 24843. Sea of Okhotsk.

156

Opredelitel'..., 1937: 163 (without description). Shmidt, Ryby Okhotskogo Morya, 1950: 92 (description). Matsubara, Fish Morphol. and Hierar., 1955: 777.

24843. Sea of Okhotsk. August 28, 1932. M.N. Krivobok. 1 specimen. 37966. Sea of Okhotsk, 56°24' N, 145°45' E. August 27, 1963. V.P. Shuntov. 2 specimens.

157 D 88; A 77; P 21.

Body elongate, depth 9 times and head length 5 times in total length. Eyes large, their diameter slightly less than snout length. Interorbital space narrow. Bony crest well developed on occiput. Upper jaw longer than lower one. Maxilla reaches vertical from anterior margin of eye. Mental crests well developed, with rounded and anteriorly protruding ends of lobes (not fused). Teeth on upper jaw 11 and arranged in single row; on lower jaw arranged in 3 rows in front, 2 rows in middle part of jaw, and in 1 row (16 teeth) in posterior part. Vomer and palatines with teeth. Dorsal and anal fins covered with scales at base, number increasing posteriorly. Lateral line complete, ventrolateral. Pectoral fins rounded, do not reach vertical from vent by distance equal to twice diameter of eye; rays in lower lobe protrude slightly at ends, all branched. Length of pelvic fins equal to 0.5 diameter of eye. Color chocolate-brown; belly, thorax, and chin gray. Dorsal fin smoky with black edging; 5 Y-shaped, milk-white blotches located on fin which continue onto sides of body. White patches also present above gill opening. Pectoral fins dark gray (Shmidt, 1950).

This detailed description can be supplemented with characters of the preanal distance, which in our specimens (length from 192 to 228 mm) varied from 42 to 46%, and with results of analysis of radiographs. Number of vertebrae varied insignificantly, 92–93 (precaudal 20–21, caudal 72); rays of dorsal fin 87–88, of anal fin 73–74; and free part of caudal fin with 8 principal and 4 marginal rays.

Length 228 mm.

Distribution: Reported from the Sea of Japan by just one author, in Primor'e (Ueno, 1971: 86). This find needs verification since to date we know only of a specimen from the sea of Okhotsk near Shantar Islands (Shmidt, 1950: 93) and from the northern part of this sea (No. 37966). No other records have been published.

10. *Lycodes palearis fasciatus* (Schmidt, 1904)—Crested Striped Eelpout (Figure 124)

Lycenchelys fasciatus Shmidt, Ryby Vostochnykh Morei..., 1904: 203, pl. 6 (Aniva Bay). Soldatov and Lindberg, Obzor..., 1930: 498.

Lycodes palearis fasciatus, Taranetz and Andriyashev, Zool. Anz., 1 12, 112, 9-10, 1935: 253 (notes on classification of *L. palearis*). Taranetz, Kratkii Opredelitel'..., 1937: 163 (northern part of the Sea of Japan, Aniva Bay).

Lycodes fasciatus, Fowler, Synopsis ..., 1958: 308, fig. 37.

13092. Sea of Okhotsk, Aniva Bay. August 23, 1901. P.Yu. Shmidt. 1 specimen.

13093. Sea of Okhotsk, Aniva Bay. August 23, 1901. P.Yu. Shmidt. 1 specimen.

24840. Sea of Japan, Silant'ev Bay, June 29, 1931. D. Okhryamkin. 1 specimen.

25272. Sea of Japan, Jigit Bay. July 1, 1932. M. Krivobok. 2 specimens.

29102. Sea of Okhotsk, 50°34′ N, 155°46′ E. August 4, 1932. S. Generozova. 4 specimens.

32554. Sea of Okhotsk, 52°30′ N, 167°47′ E. August 4, 1932. S. Generozova. 1 specimen.

41621-41632. Sea of Okhotsk, Sakhalin Island. 1947-1949. KSE. 42 specimens.

D 103-105; A 86-88; P 17 (Shmidt, 1904).

Close to typical form; differs from it in relatively longer pelvic fins (0.5 diameter of eye versus 1.5 to 1.3), relatively larger size of eyes (4.0 times in the head length versus 5 to 6 times), and different body coloration (light gray instead of chocolate-brown to olive-green).

Blackish dark spots and bands present on light gray background of body. Head with series of irregularly diffuse patches under eye and on operculum and parietals. Occiput with large spot. First light-colored transverse stripe located slightly anterior to dorsal fin, followed by series of light-colored stripes (14 to 15) which reach middle of body and continue in upper part of dorsal fin. Oblong black stripe occurs on upper margin of anterior part of dorsal fin. In young specimens light-colored stripes continue to lower body surface (Shmidt, 1904).

Shmidt (1950) provided a description of this subspecies and considered it a synonym of *L. palearis brashnikowi.* We do not agree with him since *L. palearis brashnikowi* differs sharply from *L. palearis fasciatus* and can be considered an independent subspecies and, possibly, even

158 species (see synonymy of *L. p. schmidti* Gratzianov — *L. p. brashnikowi* Sold., as interpreted by Taranetz and Andriyashev, 1935).

Furthermore, in the 5 specimens of this species examined by us (length 45 to 367 mm), the mental crests were not fused. Not all the pores of the seismosensory system were discernible on the head; we detected only 7 suborbital, 4 mandibular, preopercular, and postorbital, and 1 preorbital. Preanal distance was 38.3 to 42.2% of total length. Radiographs showed that the number of vertebrae varied from 101 to 104 (precaudal 23-28, caudal 78-81) and a lesser number (than indicated by Shmidt) of rays in the dorsal (96-98) and anal fins (80-82).

Specimens of KSE were caught in depths of 25 to 187 m with bottom water temperature ranging from +0.2 to −0.3 to −1.7°C. Bottom silty, rarely sandy and pebbled. Incidentals in the catch comprised: *Asterias amurensis, Yoldia hyperborea, Leda,* rarely *Chionoecetes opilio* and *Cucumaria japonica,* and rarer still *Chiridota* and *Gorgonocephalus.* In all cases in which specimens of our species were obtained from silt, *Macoma,* Amphipoda, and Maldanidae were also recovered. Among fishes, the more frequently found were: *Podothecus gilberti, Hippoglossoides elassodon robustus,* cods, and walleye pollock, rarely *Lumpenus gracilis* and *L. medius,* and rarer still *Artedeillus dydymovi schmidti, Triglops jordani, Davidojordania brachyrhyncha,* and *Percis japonica.*

Length 367 mm.

Distribution: In the Sea of Japan known from the northern part (Taranetz, 1937b: 163). In the Sea of Okhotsk described from Aniva Bay; catches known from the southeast coast of Sakhalin (KSE) and off Hokkaido (*Oshoro Maru* Cruise, 1969: 390). The latter record requires confirmation.

10a. **Lycodes palearis schmidti** Gratzianov, 1907—Schmidt's Crested
    Eelpout (Figure 125)

Lycodes sp. Shmidt, Ryby Vostochnykh Morei..., 1904: 200, fig. 13 (east coast of Sakhalin, Cape Senyavin, No. 13010, two specimens).

Lycodes schmidti, Gratsianov, Tr. Otd. Ikhtiologii Russkogo Obshch. Akklimatizatsii Zhivotnykh i Rastenii, IV, 1907: 426, 430 (from Shmidt, 1904: 200, specimens not examined).

Lycodes brashnikovi Soldatov, Ezegodn. Zool. Muzeya Rossiisk. Akad. Nauk, 22, 1917: 112, fig. 1 (Sea of Japan, 42°18′ N, 130°46′ E, including No. 13010, two specimens). Soldatov and Lindberg, Obzor..., 1930: 495, fig. 71.

Lycodes palearis brashnikovi, Taranetz and Andriyashev, Zool. Anz., 1, 12, 112, 9–10, 1935: 253 (remarks on classification of *L. palearis*). Taranetz, Kratkii Opredelitel'..., 1937: 163. Andriyashev, Issled. Morei SSSR, 25, 1937: 332.

Lycodes perspecillum (non Kröyer) Tanaka, Ann. Zool. Japan, 1908: 252. Shmidt, Ryby Okhotskogo Morya, 1950: 93.

13010. Sakhalin, Cape Senyavin. 1899. V. Brazhnikov. 2 specimens.

19167. Sea of Okhotsk, 53°17′ N, 154°47′ E. August 5, 1907. Smirnov 3 specimens.

34671. Sea of Okhotsk, 58°38′ N, 152°45′ E. August 22, 1912. Derbek. 1 specimen.

D 95; A 78; head 4.5 (4.2) times in length (Gratsianov, 1907: 426, see Shmidt, 1904: 200).

Differs from the typical form in lower number of rays in dorsal and anal fins (91–98 and 71–83[161] versus 105 and 90), almost entirely covered

---

[161]Rays counted in radiographs of 5 specimens.

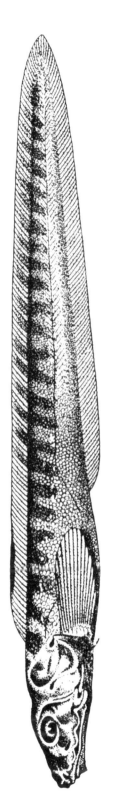

156

Figure 124. *Lycodes palearis fasciatus*—Crested striped eelpout. Length 125 mm. Aniva Bay (Shmidt, 1904).

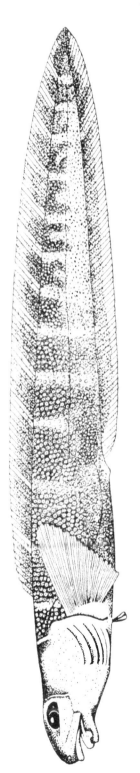

159

Figure 125. *Lycodes palearis schmidti* Gratzianov–Shmidt's crested eelpout. Length 262 mm. Sea of Japan (Soldatov, 1917).

with scales, and truncate posterior margin of pectoral fin with longest rays in upper third versus middle part in *L. p. palearis*. Preanal distance in our specimens 40.0 to 46.8% of total length.

Length 262 mm (Soldatov, 1917a).

Distribution: In the Sea of Japan known from Peter the Great Bay 160 (Soldatov, 1917a: 114). Reported from Sado Island as *L. perspecillum* Kröyer (Katho et al., 1956: 323). In the Sea of Okhotsk known off east coast of Sakhalin (Shmidt, 1904: 200).

11. *Lycodes tanakae* Jordan and Thompson, 1914—Tanaka's Eelpout
(Figure 126)

Lycodes tanakae Jordan and Thompson, Mem. Carnegie Mus., 6, 1914: 299, pl. 37, fig. 2 (Noto Peninsula, Honshu). Matsubara, Fish Morphol. and Hierar., 1955: 778.

300016. Cape Chanadedan (Cape Peshchurov), east coast of the Korean Peninsula. June 17, 1931. D.I. Okhryamkin. 1 specimen.

41633. Sea of Okhotsk, Sakhalin Island, south of Cape Senyavin. October 4, 1949. KSE. 2 specimens.

41634. Sea of Okhotsk, Sakhalin Island, south of Cape Senyavin. October 4, 1949. KSE. 2 specimens.

41635. Sea of Okhotsk, Sakhalin Island, south of Cape Senyavin. October 4, 1949. KSE. 1 specimen.

41636. Sea of Okhotsk, Sakhalin Island, Mordvinov Gulf. September 2, 1947. KSE. 1 specimen.

D 97; A 76; P 20; V 3. Head length 4.4 times and body depth 8.75 times in total length. Body depth 2 times, length of maxilla 2.6 times, diameter of eye 8.0 times, snout length 3.5 times, and length of pectoral fin 1.6 times in head length. Pelvic fin 1.33 times in eye diameter. Snout blunt at tip. Preanal distance 2.14 times in the total length. Dorsal fin originates at vertical from middle of pectoral fin. Anal fin originates at vertical from 24th ray of dorsal fin. Teeth on upper jaw strong, arranged in 1 row, slightly bent at tips, and continue only to midway between snout and posterior end of upper jaw. Teeth of lower jaw close-set, large, strong, and arranged in 1 row along sides of jaw; minute teeth in anterior part of jaw arranged in 2 distinct rows. Vomer with 2 strong teeth. Color mainly chocolate-brown, with rounded white spots in upper half of body sides and on dorsal fin. Abdomen and thorax light-colored. Dorsal and anal fins dark (Jordan and Thompson, 1914).

Our examination of 7 specimens (ranging in length from 208 to 433 mm) yielded the following additional data: mental crests not fused and well developed. Lateral line mediolateral. Length of head 4.0–4.9, body depth 6.4–9.5 times in total length. Body depth 1.6–2.0, length of maxilla 2.2–2.6, diameter of eye 5.2–7.5, length of snout 3.0–3.9, and length of pectoral fin 1.5–1.7 times in head length. Pelvic fin 1.2–1.6 times in eye

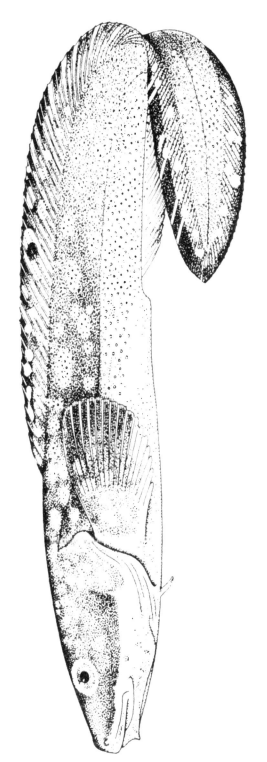

Figure 126. *Lycodes tanakae*—Tanaka's eelpout. Length 160 mm. Sea of Japan. Honshu (Jordan and Thompson, 1914).

159

diameter. Preanal distance 2.0–2.4 times in total length. Dorsal fin originates at vertical between 0.2 to 0.5 length of pectoral fin. Anal fin originates at vertical between 20th to 24th ray of dorsal fin. Width of interorbital distance 13.0–19.0 times in head length. D 93–97; A 72–76; vertebrae 98–101 (27–29 + 70–73); skeleton of free part of caudal fin with 6–8 principal and 3–5 marginal rays.[162] Color of our fish did not fully correspond to description of type specimen, differing in shape and size of light-colored spots in upper half of body and dorsal fin.

Specimens of KSE were caught at depths of 56 to 96 m, with bottom water temperature ranging from −1.2 to −1.8°C. Bottom sandy, rarely silted sand. Catches of our species included the following: *Ophiura sarsi, Leda, Yoldia, Lycodes raridens, Lumpenus medius,* and *Eumicrotremus birulai.*

Length 880 m (Andriyashev, 1955: 393).

Distribution: In the Sea of Japan known from off Wonsan (Mori, 1952: 130); Cape Chanadedan (No. 30016); off south coast of Sakhalin (Nos. 41633, 41634, 41635); off northwest coast of Hokkaido (Hikita and Hirosi, 1952: 49); off Sado Island (Honma, 1952: 226); in Toyama Bay (Katayama, 161 1940: 25); Wakasa Bay (Takegawa and Morino, 1970: 383); and San'in region (Mori, 1956a: 21). In the Sea of Okhotsk known from Mordvinov Gulf (No. 41636), near Sakhalin (Nos. 41633–41635), and off coast of Hokkaido (*Oshoro Maru* Cruise, 1969: 361).

12. *Lycodes sigmatoides* nomen novum—Sigmoid Eelpout (Figure 127)

Lycodes schmidti Soldatov (nomen praeoccupatum, Gratsianov, 1907: 430), Ezhegodn. Zool. Muzeya Rossiisk. Akad. Nauk, 22, 1917: 115, fig. 2 (partly, western Sakhalin, Ognevo). Soldatov and Lindberg, Obzor..., 1930: 496 (partly).

Lycodes tanakae (non Jordan and Thompson), Taranetz, Kratkii Opredelitel'..., 1937: 163. Lindberg, Predvaritel'nyi Spisok..., 1947: 171 (partly); Issled. Dal'nevost Morei SSSR, 1959: 251 (partly).

19165. Tatar Strait. July 5, 1913. DVE. 3 specimens.

19166. Tatar Strait, Cape Syurkum. August 25, 1911. DVE. 1 specimen.

29098. Sea of Japan. August 13, 1932. M. Krivobok. 1 specimen.

29104. Tatar Strait. June 26, 1910. F.A. Derbek. 1 specimen.

30017. Sea of Japan, 50°45′ N, 141° E. June 17, 1910. F.A. Derbek. 2 specimens.

30019. Tatar Strait. August 5, 1932. M. Krivobok. 1 specimen.

D 92–96; A 72–83; P 18–20. Head length 21.3 to 24.1%, anterodorsal distance 26.6 to 27.6%, preanal distance 44.3 to 50.0%, length to pelvic fin 17.7 to 20.2%, length of snout 6.0 to 7.5%, length of pelvic fin 2.1 to 2.9%, length of pectoral fin 14.3 to 15.0%, longitudinal diameter of eye 2.6

---

[162]Counts of rays in fins and of vertebrae based on radiographs of 7 specimens.

to 4.0%, body depth before origin of dorsal fin 13.2 to 15.0%, body depth before origin of anal fin 11.2 to 12.6%, and length of gill opening 9.0 to 12.0% of total length. Teeth sharp, canine-shaped, arranged in single row on upper jaw and on palatines. Teeth not present on symphysis of upper jaw. Vomer with 3-5 teeth, 1-2 enlarged. Teeth on lower jaw minute, in anterior part arranged in 3-4 rows, those on sides strong, sharp, and canine-shaped. Characteristic light-colored S-shaped spots located on back, which continue onto dorsal fin. Back and dorsal fin without dark oblique stripes. Margins of dorsal and anal fins light-colored. Upper half of head dark, on sides without pattern of spots and lines. Occiput with a few spots (Soldatov, 1917a).

In the 7 specimens examined by us (length from 290 to 552 mm) the preanal distance varied from 44.1 to 47.7% of total length. Radiographs showed vertebrae 99-106 (precaudal 26-27, caudal 73-80); number of rays in dorsal fin from 93 to 96, of anal fin from 75 to 77; skeleton of free part of caudal fin with 8 principal and 4-5 marginal rays.

Length, to 556 mm (Soldatov, 1917a).

Distribution: In the Sea of Japan known from Tatar Strait (our specimens) and off the southwest coast of Sakhalin (Lindberg, 1959: 251). In the Sea of Okhotsk reported from Aniva Bay (Lindberg, 1959: 251). Recorded from the eastern coast of Sakhalin (Cape Rymniv, 50°15' N). In the eastern part of the Sea of Okhotsk replaced by the closely related species, *L. brevicauda* Taranetz and Andriashev, 1935

13. *Lycodes raridens* Taranetz and Andriashev, 1937—Few-Toothed Eelpout (Figure 128)

Lycodes sp. Shmidt, Ryby Vostochnykh Morei..., 1904: 199 (Sakhalin, east coast, Cape Rymnik, No. 13009).

Lycodes schmidti Soldatov, Ezhegodn. Zool. Muzeya Rossiisk. Akad. Nauk, 22, 1917: 115 (partly, only No. 13009).[163]

163   *Lycodes raridens* Taranetz and Andriashev. In: Andriyashev, Issled. Morei SSSR, 25, 1937: 335, fig. 15 (southern part of the Anadyr Gulf). Taranetz, Kratkii Opredelitel'..., 1937: 162. Shmidt, Ryby Okhotskogo Morya, 1950: 87 (partly). Andriyashev, Ryby Severnykh Morei SSSR, 1954: 288, fig. 160, 161.

Lycodes paucidens[164] Taranetz and Andriashev. In: Taranetz, Kratkii Opredelitel'..., 1937: 163

[163]Identification of one of the syntypes (No. 13009) was done by A.Ya. Taranetz and A.P. Andriyashev in 1934 (A.A.).

[164]In the *Keys to...*, *Lycodes raridens* Taranetz and Andriashev was erroneously printed under the name *Lycodes paucidens* Taranetz and Andriashev (*L. paucidens* = *L. raridens*). The first description should be considered invalid since the work of A.P. Andriyashev was published earlier (1937, sent to press December 27, 1936) than the *Kratkii Opredelitel'*..., by A.Ya. Taranetz (1937, sent to press August 31, 1937) (A.A.).

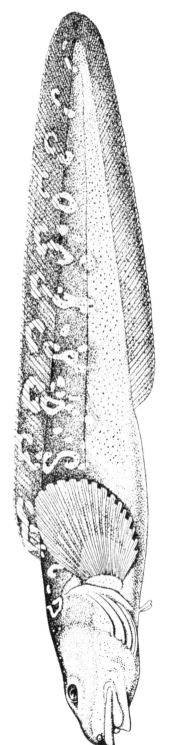

Figure 127. *Lycodes sigmatoides*—sigmoid eelpout. Length 149 mm. East coast of Sakhalin (Soldatov, 1917).

162

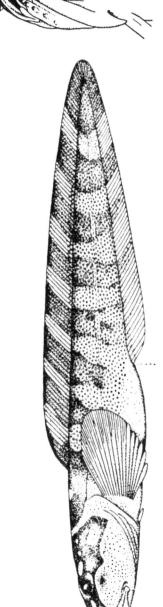

Figure 128. *Lycodes raridens*—few-toothed eelpout. Length 120 mm. Anadyr Gulf (Andriyashev, 1964). A—head of same specimen, ventral view; a—mental crests.

162

13009. Sea of Okhotsk, Sakhalin, Cape Rymnik. 1899. V. Brazhnikov. 1 specimen.

30010. Sea of Okhotsk, Sakhalin. July 4, 1932. S. Generozova. 2 specimens.

36178. Sea of Okhotsk, 51°21′ N, 144°11′ E. August 20, 1959. A.P. Nikolaev. 1 specimen.

36179. Sea of Okhotsk, 51°41′ N, 144°10′ E. August 20, 1959. A.P. Nikolaev. 1 specimen.

36936. Sea of Okhotsk, 46°39′ N, 143°43′ E. July 5, 1960. V.V. Barsukov. 1 specimen.

41615. Sea of Okhotsk, Terpenia Peninsula. September 9, 1947. KSE. 1 specimen.

41616. Sea of Okhotsk, Sakhalin coast, Cape Svobodnyi. September 4, 1947. KSE. 1 specimen.

41617. Sea of Okhotsk, east coast of Sakhalin. October 2, 1949. Makarov. KSE. 2 specimens.

41618. Sea of Japan, Tatar Strait, west coast of Sakhalin. October 8, 1949. KSE. 2 specimens.

41619. Sea of Okhotsk, east coast of Sakhalin, Cape Ostryi. October 4, 1949. KSE. 4 specimens.

41620. Sea of Okhotsk, east coast of Sakhalin, Cape Ostryi. October 4, 1949. KSE. 1 specimen.

D 83-93; A 72-76; P 18-19; gill rakers on first gill arch $\frac{3 + 10\text{-}12}{0 + 13\text{-}14}$ Head length 19.5 to 24.1%, predorsal distance 24.8 to 28.7%, preanal distance 45.6 to 48.0% length of pectoral fin 13.0 to 14.7%, body depth at vertical with origin of dorsal fin 9 to 11%, body depth at vertical with origin of anal fin 8.8 to 10.8%, and interorbital distance very broad—1.7 to 1.9% of total length. Snout length 27.0 to 31.7%, length of pelvic fin 11.4 to 13.0%, longitudinal diameter of eye 14.6 to 19.3%, and interorbital distance 7.4 to 8.3% of head length.

Teeth on jaws very characteristic for this species—sparse, few, with spaces in-between. On lower jaw 3-4 wide-set large teeth arranged posteriorly in 1 row, and smaller teeth arranged in 2 rows anteriorly. On upper jaw 6-9 small teeth arranged in 1 row, sometimes with 1-2 additional teeth located at inner margin toward the front. Palatines with 6-8 sharp teeth arranged in 1 row; vomer with 2-4 teeth. Scales minute, cover entire posterior part of body. Scales absent on belly in specimens from the Bering and Chukchi Seas, usually present in specimens from the Sea of Okhotsk.[165] Scales also present on dorsal and anal fins. Lateral line single, simple, mediolateral, and extends along median line of body to tail.

[165]Usually in posterior part (Shmidt, 1950: 87).

Head of young specimens dark, light-colored spots usually located behind each eye; number of light-colored spots increases with age and, through fusion, form complex reticulate light-colored pattern in larger specimens. Lower margin dark on sides of head, with distinct pattern rising under eyes and descending on cheeks, operculum, and toward snout; dark spot present under each eye, separated from remaining dark region by light-colored up-curved stripe. Upper posterior parts of operculum connected through distinct light-colored stripe directed in front in form of semicircle. Upper part of body with about 7 dark transverse spots which continue in form of broad stripe onto dorsal fin and 164 terminate at margin of fin in darker color, giving impression of continuous dark stripe edging dorsal fin. Posterior part of body with dark marbled spots that also continue onto anal fin; this pattern is more complex in adult specimens, forming an intricate reticulate pattern in upper part of body. Lower part of body, belly, and lower surface of head yellowish-white; sometimes belly and anal fin darker (Andriyashev, 1937).

Specimens of KSE (11) caught from depths of 53 to 187 m with bottom water temperature ranging from +1.4 and +3.5 to −0.7°C. Bottom sandy-argillaceous, rarely with silt. Incidentals in the catch comprised large pink sea anemones, *Hyas coarctatus, Macoma calcarea,* and *Gorgonocephalus*; among fishes mainly walleye pollock and flounders were caught (*Acanthopsetta nadeshnyi*), and a small number of *Enophrys dicereus, Gymnocanthus* sp., *Leptoclinus maculatus,* and *Crystallias matsushimae.*

Length, to 700 mm (Andriyashev, 1955: 393).

Distribution: In the Sea of Japan known from Tatar Strait (No. 41618). In the Sea of Okhotsk represented from the east coast of Sakhalin (Nos. 30010, 41615, 41616, 41617, 41619, 41620) to the west coast of Kamchatka (Andriyashev, 1954: 290). Bering and Chukchi seas (Andriyashev, 1954: 290).

[Genus *Lycenchelys* Gill, 1884—Wolf Eelpouts]

Lycenchelys Gill, Proc. Acad. Nat. Sci. Philad., 1884 (type: *Lycodes muraena* Collett). Shmidt, Ryby Okhotskogo Morya, 1950: 106. Andriyashev, Ryby Severnykh Morei SSSR, 1954: 307. Andriyashev, Tr. Zool. Inst. Akad. Nauk USSR, 18, 1955: 350.

Close to the genus *Lycodes* Reinhardt, but mandibular and maxillary branches of canals on lateral line usually open through series of large nostril-like pores arranged on lower jaw to lower side of posterior margin of preoperculum, and above upper jaw almost in straight line from tip of snout to anterior margin of preoperculum, without forming circumorbital ring of round minute pores (Figure 129). Characteristic mental crests of *Lycodes* not developed, converted into low, bifurcate raised portions,

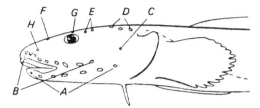

Figure 129. Diagram of seismosensory system of head in *Lycenchelys* (Andriyashev, 1955).

A—pores of mandibular series; B—pores of maxillary series; C—preopercular pore; D—occipital pores; E—postorbital pores; F—preorbital pore (unpaired); G—interorbital pore (unpaired); H—pore behind tubular nostril.

fused with each other anteriorly and with space between lips. Palatine-maxillary membrane present but narrow. Mouth inferior. Teeth present on jaws, vomer, and palatines. Teeth on lower jaw arranged in patch of few irregular rows in front; canine-shaped teeth absent. Branchiostegal rays 6. Body more elongate than in *Lycodes,* its depth about 4 to 6 (8)% of total length. Anterior part of body short, preanal distance less than 1/3 of total length. Body covered with dense, minute scales embedded in skin. Lateral line indistinct, usually ventral; rarely mediolateral branch present. Dorsal fin originates above pectoral fin or slightly behind end of it. Vertebrae 111–136. Body color in most species monochromatic, dark to bluish-black in lower part of head and belly. Body of species from less deep waters light-colored, often spotted. Deep sea, silt-loving forms, with length up to 35 cm. Eggs laid on bottom, young not planktonic (Andriyashev, 1955). About 30 species.

Members of this genus are not found in the Sea of Japan. In the Sea of Okhotsk two species are known: *L. rassi* Andriashev, 1955 and *L. hippopotamus* Schmidt, 1950. Both species are known only from the central part of this sea, east of Sakhalin.

7. Genus *Petroschmidtia* Taranetz and Andriashev, 1934

Petroschmidtia, Taranetz and Andriyashev, Dokl. Akad. Nauk SSSR, 11, 1934: 506 (type: *P. albonatata* Taranetz and Andriashev). Taranetz, Kratkii Opredelitel'..., 1937: 161. Shmidt, Ryby Okhotskogo Morya, 1950: 107. Matsubara, Fish Morphol. and Hierar., 1955: 774.

165 Body moderately elongate, covered with fairly large scales. Teeth on jaws minute, none significantly enlarged. Vomer and palatines without teeth. Mental crests formed by inner margin of dentaries strong, high, entirely fused anteriorly and covered with skin. Upper lip thin and continuous; lower lip thickened in posterior part. Large nostril-like pores (as in *Lycenchelys* Gill) absent; very minute pores occur on head and

poorly developed mucous cavities above upper jaw and on lower jaw. Gill openings very broad, entend downward to base of pelvic fins. Palatine membranes not developed. Pectoral fins rounded, without notch; pelvic fins well developed.

Species of this genus dwell on silty bottom and feed on minute animals (Cumacea, Ostracoda, etc.). Because of these ecological peculiarities, it is necessary to review the series of features distinguishing this genus from all known representatives of the family Zoarcidae, namely: complete reduction of teeth on vomer and palatines; strong development of fused mental crests in form of plough; numerous minute teeth on jaws; small mouth; relatively thin and nontubercular gill rakers without spinules; absence of pyloric caeca; and large gill openings (Taranetz and Andriyashev, 1934).

Two species. One known only from the Sea of Japan, the other from the Sea of Okhotsk.

Key to Species of Genus Petroschmidtia[166]

1 (2). Interorbital space broad (17.7 to 26.4% of the head length). Head length 19.6 to 25.0% of total length. D 88-95; P 18-20. Dorsal fin with 4 to 10 dark spots (in young fish with white margin).[167]
.... 1. **P. toyamensis** Katayama.
2 (1). Interorbital space narrow (about 2.7 to 3.7% of the head length). Head length 17.7 to 21.2% of total length. D 95-98; P 17-18. Dorsal fin with 4 to 7 white spots that continue onto back...........
.... 2. **P. albonotata** Taranetz and Andriashev.[168]

1. **Petroschmidtia toyamensis** Katayama, 1941 (Figure 130)

Petroschmidtia toyamensis Katayama, Zool. Mag., 53, 12, 1941: 593 (Toyama Bay). Matsubara, Fish Morphol. and Hierar., 1955: 774, fig. 290.

D about 97; A about 86; P 19; V 2.

Head length 4.55, body depth 8.0, distance from origin of pelvic fins to anal fins 3.9 times in total length. Maximum eye diameter 6.1, snout 3.04, maxilla 2.4, distance between eyes 6.7, length of pectoral fin 2.1, and length of pelvic fin 6.7 times in head length.

Body elongate, laterally compressed, attenuate toward tail. Head also compressed, lateral sides vertical, width less than maximum depth.

[166]Matsubara, 1955: 774.

[167]In specimens with a length of 305 mm these spots were absent, and margins of D and A were dark (Katayama, 1941, Figure 1) (see Figure 130).

[168]This species was earlier known only from the northern part of the Sea of Okhotsk, but the Japanese ichthyologist Ueno (1971: 86) has reported its occurrence in the Sea of Japan along the coast of Hokkaido. Thus we must consider it a species of the marine area under study. However, additional confirmation for this region is not known.

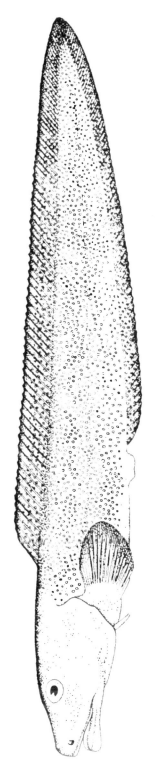

Figure 130. *Petroschmidtia toyamensis.* Length 305 mm. Toyama Bay (Katayama, 1941).

Mouth moderate in size. Maxilla extends up to vertical from middle of eye. Lower jaw slightly shorter than upper, mental crests strong and high, fused anteriorly. Nostrils with small fleshy tubules. Lateral line curves
166 above pectoral fin and continues farther horizontally along medial line of body to tail. Head, occiput, and bases of pectoral fins naked; trunk, tail, and fins covered with scales embedded in skin. Dorsal fin originates slightly posterior to vertical with base of pectoral fin. Anal fin originates under vertical with the 14th ray of dorsal fin. Dorsal fin higher than anal fin. Pectoral fins short. Pelvic fins very short, almost equal to diameter of eye. Body color in specimens preserved in formalin chocolate-brown to gray, head darker; dorsal and anal fins, howerver, dark along margins; pectoral fins dark (Katayama, 1941). Body color changes with age (Figure 131, A-C).

Length, up to 305 mm.

Distribution: In the Sea of Japan known off the coast of Hokkaido (Ueno, 1971: 86); Sado Island (Honma, 1963: 21); Toyama Bay (Matsubara, 1955: 774); Wakasa Bay (Takegawa and Morino, 1970: 383); littoral zone of Hyogo Prefecture (Matsubara, 1955: 774); entire coast of Tajima Strait (Katayama, 1952: 5); and San'in region (Mori, 1956a: 21). Along the Pacific coast of Japan reported from Yamato Province (Mori, 1956a: 29).

2. *Petroschmidtia albonotata* Taranetz and Andriashev, 1934
(Figure 132)

Petroschmidtia albonotata Taranetz and Andriyashev, Dokl. Akad. Nauk SSSR, 11, 8, 1934: 507, figs. 1-2 (Sea of Okhotsk). Taranetz, Kratkii Opredelitel'..., 1937: 161. Shmidt, Ryby Okhotskogo Morya, 1950: 107. Matsubara, Fish Morphol. and Hierar., 1955: 774.

24606. Sea of Okhotsk. September 6, 1932. P.Yu. Shmidt. 2 specimens.
24607. Sea of Okhotsk. September 6, 1932. P.Yu. Shmidt. 3 specimens.
34421. Sakhalin Island, southeastern coast. 1947. KSE. 1 specimen.
37976. Sea of Okhotsk. August 3, 1963. V.P. Shuntov. 3 specimens.
37977. Sea of Okhotsk. July 19, 1963. V.P. Shuntov. 3 specimens.
D 95-98; A 80-84; P 17-18.

Head 17.7 to 21.2% of standard length; diameter of eye 22.1 to 28.4% of the length, less than that of snout, length of which reaches 25.7 to 28.4% of head length (eyes of smaller specimen equal to snout); interorbital space very narrow (2.7 to 3.7% of head length). Teeth on jaws minute, sharp, not enlarged in front. Teeth on upper jaw arranged in 2 rows (outer row with 18, inner row with 9 to 10), with 2-3 teeth in-
168 between them in anterior part; all teeth fine, slightly retrorse. Gill rakers on first gill arch not tubercular, same as in *Lycodes,* compressed laterally, and totally without spines in inner as well as outer row of rakers. Mental

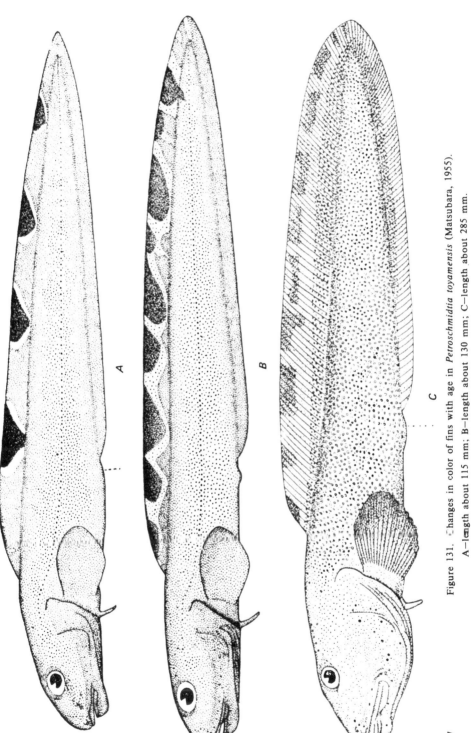

Figure 131. Changes in color of fins with age in *Petroschmidtia toyamensis* (Matsubara, 1955).
A—length about 115 mm; B—length about 130 mm; C—length about 285 mm.

167

crests well developed, thick, high, covered with skin and fused anteriorly. Upper jaw short, hardly reaches vertical with middle of eye, slightly protrudes beyond lower jaw. Minute pores and indistinct mucous cavities developed on head. Gill openings very large, reach base of pelvic fins. Body moderately elongate, compressed laterally. Scales fairly large, separate, and cover entire body except for part of belly behind pelvic fins, narrow strip under these fins, and triangular space from origin of dorsal fin to line connecting apices of gill openings. Head and occiput naked. Individual scales continue on anterior part at base of dorsal fin; posteriorly number of scales increases and cover up to three-fourths height. Anterior half of anal fin naked, posterior half with scales over one-half to three-fourths height of fin. Lateral line mediolateral, without curves, and distinct along body to tail. Dorsal fin fairly high, anal fin distinctly low. Body color uniform, gray to olive-green, slightly lighter on lower side. Scales lighter in color than surrounding skin. Dorsal fin dark, with 4-7 light-colored longitudinal or roundish spots, which continue onto back in form of short, narrow, light-colored stripes. Usually an oblong white spot present on caudal fin.

Food mainly consists of minute crustaceans (Cumacea as well as Ostracoda, minute Isopoda and Gammaridae). A characteristic feature is the absence in their gut of Decapoda, Mollusca, Polychaeta, and other larger organisms, which are common in the food of species of *Lycodes* Reinh. (Taranetz and Andriyashev, 1934).

An examination of 10 of our specimens (length 205 to 390 mm) yielded the following additional data: Pores of seismosensory canals of head difficult to distinguish, but lateral line well expressed and represented by freely located neuromasts numbering about 100. Preanal distance varied 169 from 39.4 to 44.3% of total length. Analysis of radiographs showed some variations in meristic characters. Number of vertebrae varied from 95 to 103 (precaudal 21-22, caudal 74-82); number of rays in dorsal fin from 89 to 98, in anal fin from 74 to 82. Skeleton of free part of caudal fin with 10 principal and 2-4 procurrent rays. Bivalve molluscs were seen in radiographs of the gut of some fish (specimen No. 37977), and since they were sufficiently numerous one may assume that they constitute some part of the food of *Petroschmidtia albonotata*.

Length, to 390 mm

Distribution: In the Sea of Japan reported from the coast of Hokkaido (Ueno, 1971: 86). Known from the northern part of the Sea of Okhotsk (Taranetz and Andriyashev, 1934: 507) and off the southeastern coast of Sakhalin (No. 34421).

[Genus *Hadropareia* Schmidt, 1904 – Stout Pouts]

Hadropareia, Shmidt, Ryby Vostochnykh Morei..., 1904: 204 (type:

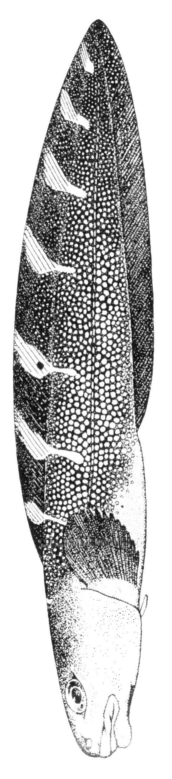

Figure 132. *Petrosomidtia albonotata*. Length 241 mm. Sea of Okhotsk (Taranetz and Andriyashev, 1934).

168

210

H. middendorffii Schmidt). Soldatov and Lindberg, Obzor..., 1930: 492.

Body narrow and long, entirely naked. Large head characterized by strong development of cheeks and notable dilation in preopercular region. Teeth slightly conical (blunt at end) on jaws and vomer; palatines without teeth. Lateral line in form of a few pores in anterior part of body. Branchiostegal membranes attached to isthmus. Gill openings small. Mouth large, maxilla extends distinctly beyond vertical with posterior margin of eye. Pelvic fins small. This genus differs from other genera of the family Zoarcidae in absence of teeth on palatines and presence of teeth on vomer (Soldatov and Lindberg, 1930). Oral cavity with palatine and mandibular membranes (Shmidt, 1950).

To date one species has been described—*H. middendorffii* Schmidt, 1904, which is known from the northwestern part of the Sea of Okhotsk.

8. Genus *Bilabria* Schmidt, 1936—Double-Lips

Bilabria Shmidt, Dokl. Akad. Nauk SSSR, 1, 2, 1936: 93-96 (type: *Lycenchelys ornatus* Soldatov). Shmidt, Ryby Okhotskogo Morya, 1950: 113.

Body elongate, compressed laterally, and covered with scales. Head small, with almost vertical cheeks. Mouth horizontal; upper jaw longer than lower one. Upper lip consists of two separate folds, anteriorly attached to skin of snout (Figure 133); lower lip also split, anteriorly wide. Teeth on jaws and vomer, but none on palatines. Gill openings large, with skin fold behind. Vertical fins confluent. Lateral line straight (medio-lateral). This genus is close to the genus *Davidojordania* Popov, but differs in absence of teeth on palatines, structure of upper lip, and presence of lateral line (Shmidt, 1936a).

One species. Known from the Sea of Japan and the Sea of Okhotsk.

1. *Bilabria ornata* (Soldatov, 1917)—Ornate Double-Lip (Figure 134)
Lycenchelys ornatus Soldatov, Ezhegodn. Zool. Muzeya Rossiisk, Akad. Nauk, 23, 1917: 162, fig. 2 (Tatar Strait).

Figure 133. Head of *Bilabria ornata*, anterior view (Shmidt, 1950).

Bilabria ornata, Shmidt, Dokl. Akad. Nauk SSSR, 1, 2, 1936: 94 (Tatar Strait and Aniva Bay). Taranetz, Kratkii Opredelitel'..., 1937: 164. Shmidt, Ryby Okhotskogo Morya, 1950: 114, fig. 9. Fowler, Snyopsis..., 1958: 316, fig. 40.

13089. Sea of Okhotsk, Aniva Bay. August 28, 1901. P.Yu. Shmidt. 1 specimen.

170   D 110; A 93; P 16.

Body oblong, gradually attenuate toward tail, and covered with roundish, separate scales uniformly arranged throughout body. Trunk roundish on lower side, caudal part compressed laterally. Head comparatively short, flat on occiput, rounded on sides; snout slopes downward abruptly and much larger than longitudinal diameter of eye. Eyes raised, close-set on top of head. Upper jaw protrudes over lower one. Maxilla extends beyond middle of eye. Teeth on upper jaw arranged in 2 rows: those of outer row large, blunt at tip; those of posterior row small, also blunt at tip; at posterior ends of upper jaw teeth arranged in 1 row only. Vomer with just 3 blunt teeth. Palatines without teeth. Teeth of lower jaw blunt, arranged in 2 rows but at posterior ends of jaw also arranged in 1 row. Dorsal fin almost twice height of anal fin. Pectoral fins rounded, without elongate lower part; dark spots present on bases. Pelvic fins small, almost equal to diameter of eye. Gill openings equal to snout length (Soldatov and Lindberg, 1930).

Preanal distance 32.3% of total length. Lateral line mediolateral, difficult to discern. Pores of seismosensory system of head: mandibular 4, preopercular 4, subopercular 7, preorbital 1, postorbital 4, interorbital 1, and occipital 3.

Radiographs of our specimen showed: vertebrae 117 (23 precaudal and 94 caudal); number of rays of dorsal fin 117, of anal fin 98. Skeleton of free part of caudal fin with 6 principal rays; number of procurrent rays difficult to determine.

Body color (preserved in alcohol) chocolate-brown; entire body covered with minute dark spots. Irregularly shaped dark spots distinct on dorsal and anal fins.

Length 166 mm (Soldatov and Lindberg, 1930).

Distribution: In the Sea of Japan known from Tatar Strait, near Grossevich (Soldatov, 1917c: 162). In the Sea of Okhotsk reported from Aniva Bay (Shmidt, 1936a: 94) and east coast of Sakhalin (Ueno, 1971: 86).

9. Genus *Davidojordania* Popov, 1931—Jordan's Eelpouts

Davidojordania Popov, Dokl. Akad. Nauk SSSR, 1931: 212 (type: *Lycenchelys lacertinus* Pavlenko). Shmidt, Dokl. Akad. Nauk SSSR, 1, 2,

171 1936: 93 (comparison with the genus *Bilabria*). Taranetz, Kratkii
Opredelitel'..., 1937: 164. Lindberg, Predvarital'nyi Spisok..., 1947: 170.
Shmidt, Ryby Okhotskogo Morya, 1950: 111 (distribution). Matsubara,
Fish Morphol. and Hierar., 1955: 780. Fowler, Synopsis..., 1958: 309.

Body elongate and compressed laterally. Body depth 16 to 17 times in
length. Head length 6 to 7 times in length. Length of pectoral fin slightly
more than postorbital distance. Eye diameter slightly less than length
of snout,[169] but much more than width of interorbital space. Corners
of mouth reach posterior margin of eye. Length of pelvic fin equal to
diameter of eye. Upper jaw slightly longer than lower one. Lower lip very
thin posteriorly and uniformly thickened anteriorly. Upper lip fairly
thick, gradually thickening toward front. Mental crests on chin absent.
Teeth on jaws conical. Teeth present on vomer and palatines. Dorsal fin
originates at vertical with base of pectoral fin. Dorsal and anal fins
consist of soft rays. Pectoral fins do not reach origin of anal fin. Dorsal
and anal fins confluent with caudal fin. Body covered with minute scales.
This genus differs from the genera *Lycenchelys* and *Lycodes* in absence
of nasal pores under upper lip and distinct mental crests on chin, presence
of minute round pores under eye and on occiput, and almost equal length
of both jaws (Popov, 1931c). It differs from the genus *Bilabria* in absence
of lateral line, presence of teeth on palatines, and ordinary upper lip
(Shmidt, 1936a).

An examination of our species of this genus yielded the following
additional data: preanal distance varied from 30.0 to 38.1% of total length;
number of vertebrae from 98 to 120 (precaudal 20–23, caudal 77–97);
number of rays in dorsal fin 97 to 118, in anal fin 81 to 102. Skeleton of free
part of caudal fin with 6 principal and 6–7 marginal rays[170]

It is not correct to state that the lateral line is absent. It is better
to consider it incomplete, since in *D. jordaniana* it is expressed by
7 to 8 free neuromasts, and in *D. lacertina* by 14. The seismosensory
system of the head comprises these pores: mandibular 4, preopercular 4,
postorbital 4, suborbital 6, preorbital 1, and occipital 3.

5 species, all known from the Sea of Japan.

Key to Species of Genus Davidojordania[171]

1 (6). Snout length equal to or more than diameter of eye.
2 (5). Head large; its length 5.6 to 6.0 times in total length.

---

[169]Diameter of eye more than snout length in *D. spilota* and *D. brachyrhyncha*. Also, in
these species corners of mouth do not reach vertical with posterior margin of eye, and jaws
are equal in length.

[170]Counts of vertebrae and rays of fins based on radiographs of 12 specimens of species of
this genus.

[171]From Fowler, 1958: 310.

3 (4). Postorbital distance 1.6 times in head length. Gill opening barely reaches half base of pectoral fin. Body depth slightly less than 11 times in its length . . . . 1. **D. jordaniana** Popov.

4 (3). Postorbital distance half head length. Gill opening almost reaches level of lower margin of base of pectoral fin. Body depth 13 times in its length . . . . 2. **D. poecilimon** (Jordan and Fowler).

172 5 (2). Head small, its length 7.0 to 7.6 times in total length . . . .
. . . . 3. **D. lacertina** (Pavlenko).

6 (1). Snout length in specimens of corresponding size less than diameter of eye.

7 (8). Body depth 9.2 times in total length; D 83-93; A 68-70[172]; P 12-15. . . . . 4. **D. brachyrhyncha** (Schmidt).

8 (7). Body depth 11.4 times in total length; D 70; A 64; P 17 . . . . . .
. . . . 5. **D. spilota** (Fowler).

1. *Davidojordania jordaniana* Popov, 1936 (Figure 135)

Davidojordania jordaniana Popov. In: Shmidt, Dokl. Akad. Nauk SSSR, 1, 2 1936: 95 (Sea of Japan, Tatar Strait). Taranetz, Kratkii Opredelitel'. . ., 1937: 165. Soldatov and Lindberg, Predvaritel'nyi Spisok. . ., 1947: 170. Matsubara, Fish Morphol. and Hierar., 1955: 780. Fowler, Synopsis. . ., 1958: 310.

23946. Tatar Strait. June 14, 1912. F.A. Derbek. 2 specimens.
23947. Tatar Strait. June 14, 1912. F.A. Derbek. 3 specimens.
D 90; A 80; P 16; C 10.

Body compressed laterally, elongate; its depth 11.2 times and head length 5.6 times in total length. Snout length equal to diameter of eye; postorbital length 1.6 times in head length. Upper jaw reaches posterior margin of eye. Upper lip continuous; lower lip interrupted and each half broadens anteriorly. Minute teeth on jaws, vomer, and palatines, teeth on jaws arranged in 2-3 rows toward front, and in 1 row at back. Palatine membrane present. Gill opening reaches almost half base of pectoral fin and with skin fold behind it. Dorsal fin originates behind base of pectoral fin, anteriorly low, thereafter its height equals diameter of eye. Anal fin lower than dorsal fin. Caudal fin short. All 3 fins confluent. Pectoral fins relatively large; their length 8.3 times in total length. Length of pelvic fins equal to diameter of eye; they extend beyond base of pectoral fins by about 1/3 their length. Entire body covered with cycloid scales. Color yellowish to chocolate-brown; sides with 3 rows of dark chocolate-brown spots; dark spots also located on upper surface of head. Dorsal fin with brown tinge and 3-5 rounded dark spots with diffuse margins (Shmidt, 1936a).

In our 5 specimens (length 90 to 116 mm) short lateral line well

[172]Radiographs of our specimens showed a greater number of rays: D 102-106; A 92-96.

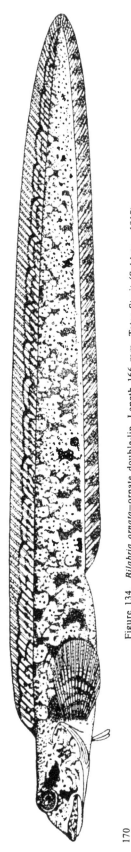

Figure 134. *Bilabria ornata*—ornate double-lip. Length 166 mm. Tatar Strait (Soldatov, 1917).

170

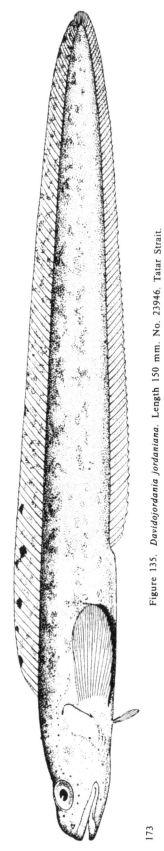

Figure 135. *Davidojordania jordaniana*. Length 150 mm. No. 23946. Tatar Strait.

173

developed in form of 7–8 free neuromasts beginning from last pore of postorbital seismosensory canal of head and extending slightly down and back. Pores of seismosensory system of head well defined, their number indicated in characters of genus. Preanal distance varied from 35.1 to 38.1% of total length. Radiographs showed insignificant variation in' number of vertebrae, from 98 to 99, and number of rays of dorsal fin—97 to 98; anal fin with 81 rays. Skeleton of free part of caudal fin with 6 principal and 7 marginal rays.

Length 116 mm.

Distribution: In the Sea of Japan known from Tatar Strait (Taranetz, 1937b: 165) and near Primor'e (Ueno, 1971: 86).

2. ***Davidojordania poecilimon*** (Jordan and Fowler, 1902) (Figure 136)

Lycenchelys poecilimon Jordan and Fowler, Proc. U.S. Nat. Mus., 25, 1902, 748, fig. 2 (Sendai Bay or Matsushima Bay, off Kinkazan Island).

174 *Davidojordania poecilimon*, Taranetz, Kratkii Opredelitel'..., 1937: 165. Lindberg, Predvaritel'nyi Spisok..., 1947: 170. Matsubara, Fish Morphol. and Hierar., 1955: 780. Fowler, Synopsis..., 1958: 310, fig. 38.

20138. Sea of Japan, Cape Gamov. October 16, 1912. DVE. 1 specimen.

29988. Sea of Japan, Ussuriisk Gulf. September 3, 1925. 2 specimens.

D 107; A 90; P 17.

Head length almost 6 times, body depth 13 times in total length; body depth slightly less than half head length, [?—Ed.] length more than 4.5 times, diameter of eye almost 4 times, length of upper jaw 2 times, pectoral fin slightly less than 2 times in head length. Head length slightly less than half snout-to-vent length; snout-to-vent length almost half length of caudal part of body.

Body elongate, its depth more or less equal throughout length, but maximum depth occurs immediately behind head and gradually reduces posteriorly. Body compressed laterally and width of head more than maximum width of body. Head elongate oblong; snout slightly blunt or entirely blunt and convex; eyes large, elongate; lips moderately thick, mouth large; maxilla extends slightly beyond vertical from posterior margin of eyes. Teeth on jaws arranged in 1 row; those of upper jaw stronger. Tongue thick, rounded, relatively long, and not free anteriorly. Gill opening large, located on side of head; branchiostegal membranes broadly attached to isthmus. Head naked, as is occiput and region around pelvic and pectoral fins. Body covered with minute, roundish cycloid scales; greater part of base of vertical fins covered with minute scales, especially in posterior part.

Color of preserved specimens light chocolate-brown, with 11 H-shaped dark chocolate-brown marks along sides of body, of which 3 at posterior end actually dark transverse stripes. These markings always continue

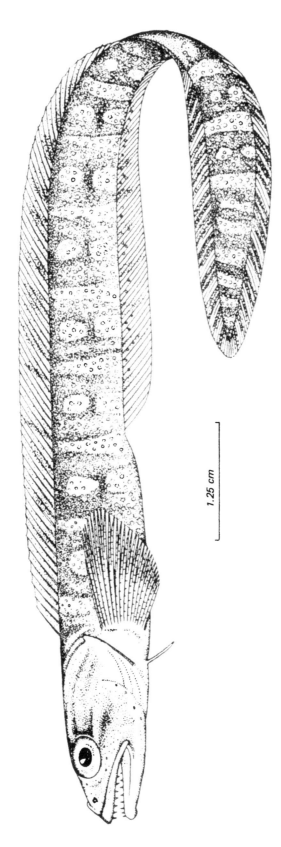

Figure 136. *Davidojordania poecilimon.* Matsushima (Jordan and Fowler, 1902).

onto vertical fins. Upper surface of head chocolate-brown with inter-secting isolated dark stripes. Lower surface of trunk, pectoral and pelvic fins, and anterior part of anal fin light-colored (Jordan and Fowler, 1902b: 749).

Radiographs of a specimen 147 mm long (No. 29988) additionally revealed: vertebrae 117 (22 precaudal and 95 caudal), 115 rays in dorsal fin, and 96 in anal fin. Preanal distance 30% of total length.

Length 150 mm (Jordan and Fowler, 1902b).

Distribution: In the Sea of Japan known from Peter the Great Bay (Nos. 20138, 29988); off Sado Island (Honma, 1963: 8); Niigata (Lindberg, 1947: 170); Toyama Bay (Katayama, 1940: 25); Wakasa Bay (Takegawa and Morino, 1970: 383); and San'in region (Mori, 1956a: 22). Along the Pacific coast of Japan known from Hokkaido (Ueno, 1971: 86) to Matsushima Bay (Jordan and Fowler, 1902b: 749).

3. *Davidojordania lacertina* (Pavlenko, 1910)—Lizard-headed Eelpout (Figure 137)

Lycenchelys lacertinus Pavlenko, Ryby Zaliva Petr Velikii, 1910: 53, figs. 10, 11 (Peter the Great Bay). Soldatov and Lindberg, Obzor..., 1930: 499.

Davidojordania lacertina, Shmidt, Dokl. Akad. Nauk SSSR, 1, 2, 1936: 95 (Vladamir Bay). Taranetz, Kratkii Opredelitel'..., 1937: 165. Lindberg, Predvaritel'nyi Spisok..., 11947: 170. Matsubara, Fish Morphol. and Hierar., 1955: 780. Fowler, Synopsis..., 1958: 311.

20139. Peter the Great Bay, Askold Island. April 5, 1913. DVE. 1 specimen.

20145. Peter the Great Bay, Sibiryakov Island. October 11, 1912. DVE. 1 specimen.

175 D 102; A 90; P 15; V 4. Head length about 9 times[173] and body depth 12.5 times in total length. Pelvic fin almost equal to diameter of eye; interorbital distance 5.5 times in head length. Lateral line absent. Head similar in shape to head of lizard, naked. Snout length much greater than longitudinal diameter of eye. End of upper jaw at level of posterior margin of pupil. Teeth on jaws arranged in two rows, in front in three rows, conical, sharp; anterior row with 8 teeth, posterior row with 15 to 17 very large teeth, and middle row with 15 minute teeth. Vomer with 5 teeth arranged in form of arch. Palatines with 4 to 5 minute teeth each. Branchiostegal membranes broadly attached to isthmus. Length of gill opening greater than snout length.

[173]Shmidt (1936a: 95) reported that the head length is not 9.0 but 7.0 to 7.3 times in the total length; this corresponds to the data obtained while studying our specimen depicted in Figure 137 (specimen No. 20139).

Color very characteristic: body along dorsal fin covered with numerous dark spots, on lower side of tail up to pectoral fin with isolated groups of large black spots in which 3 fuse and form an angle (such groups of spots about 18). Dorsal and anal fins with numerous dark transverse stripes (Pavlenko, 1910).

An examination of our specimens 69 to 170 mm in length yielded the following additional data: Lateral line short, begins at large pore of postorbital canal of head and extends in form of 14 to 15 free neuromasts in anterior part of body sides. Head with distinct pores in seismosensory canals, equal to number given in characters of genus. Radiographs (specimen Nos. 20139, 20145) showed: vertebrae 119-120 (precaudal 22-23, caudal 97); rays of dorsal fin 117-118, of anal fin 99-102; skeleton of free part of caudal fin with 6 principal and 6 procurrent rays.

Length 150 mm (Pavlenko, 1910).

Distribution: In the Sea of Japan known from Peter the Great Bay (Pavlenko, 1910: 53) and Vladamir Bay (Shmidt, 1936a: 96).

4. *Davidojordania brachyrhyncha* (Schmidt, 1904)—Jordan's Short-Snouted Eelpout (Figure 138)

Lycenchelys brachyrhynchus Shmidt, Ryby Vostochnykh Morei..., 1904: 201, pl. 6, fig. 3 (Sea of Okhotsk, Aniva Bay). Soldatov and Lindberg, Obzor..., 1930: 497.

Hadropareia brachyrhynchus, Popov, Dokl. Akad. Nauk SSSR, 1931: 211 (Aniva Bay, No. 13091).

Davidojordania brachyrhyncha, Shmidt, Dokl. Akad. Nauk SSSR. 1, 2, 177 1936: 94. Taranetz, Kratkii Opredelitel'..., 1937: 164. Shmidt, Ryby Okhotskogo Morya, 1950: 111. Matsubara, Fish Morphol. and Hierar., 1955: 780 Fowler, Synopsis..., 1958: 312.

13091. Sea of Okhotsk, Aniva Bay. August 18, 1901. P.Yu. Shmidt. 3 specimens.

19618. Sea of Okhotsk, Alexander Bay. July 15, 1911. DVE. 2 specimens.

D 85-93; A 68-70; P 12-15. Head length 7.2 times in total length. Snout length equal to about 0.6 to 0.7 times the eye diameter. Length of postorbital part of head 1.75 times in the head length. Upper lip interrupted; lower lip double, with lobes broadening anteriorly. Gill openings reach level of one-third base of pectoral fin. Pectoral fins short, their length 11 times in body length. Pelvic fins extend beyond base of pectoral fins by half their length; length of pelvic fins equal to diameter of eye. Body sides and belly covered with cycloid, barely sessile scales; scales absent on thorax and anterior to pectoral fins (Shmidt, 1936a).

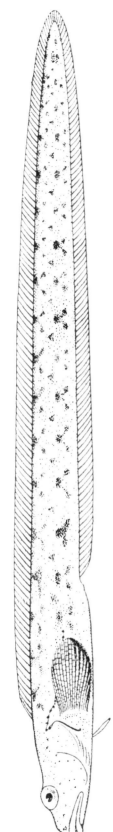

Figure 137. *Davidojordania lacertina*—lizard–headed eelpout. Length 175 mm. No. 20139. Peter the Great Bay.

175

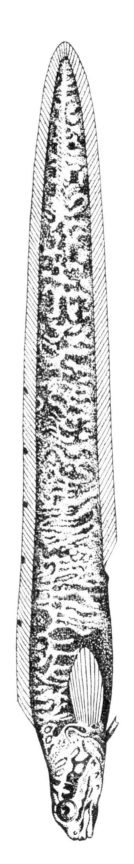

Figure 138. *Davidojordania brachyrhyncha*—Jordan's short-snouted eelpout. Length 112 mm. Aniva Bay (Shmidt, 1904).

176

Body reddish or brownish. Entire trunk and head with pattern of minute, irregularly arranged brown spots that continue onto unpaired fins. Light-colored stripe continues through lower part of eye toward lower corner of operculum and fuses toward front on snout. Occiput with 4 light-colored stripes, forming X-shaped pattern. Dorsal fin with 2–5 rounded black spots located in anterior part (young specimen with circular spots and haze on dorsal fin). Lower side of body yellowish. Pectoral fins reddish, without spots (Shmidt, 1904).

An examination of our specimens 74 to 120 mm in length yielded the following additional data; preanal distance varied from 31.1 to 34.8% of the total length. Radiographs showed slight variation in number of vertebrae, from 104 to 108 (precaudal 20–21, caudal 83–87); rays of dorsal fin from 102 to 106, of anal fin from 92 to 96. Number of rays in fins much higher than indicated by Fowler (1958: 310).

Length 121 mm (Shmidt, 1904).

Distribution: In the Sea of Japan known from Tatar Strait (Taranetz, 1937b: 164). In the Sea of Okhotsk mainly found in northwestern part (Shmidt, 1950: 113) and south to Aniva Bay (Shmidt, 1904: 204).

5. **Davidojordania spilota** (Fowler, 1943)—Jordan's Slender Eelpout
   (Figure 139)
   Lycenchelys spilotus Fowler, New Philippine Fishes, 1943: 89, fig. 24 (Niigata).
   Davidojordania spilotus Fowler, Synopsis ..., 1958: 31 (east coast of Sea of Japan).

D 70; A 64.

Body depth 11.4 times and head length 6 times in total length. Snout 5.2 times in head length, convex. Eyes larger than snout length and interorbital space. Mouth low, horizontal, jaws equal in length. Maxilla almost reaches vertical with middle of eye, its length 3 times in the head length. Interorbital space narrow, 4 times in the eye diameter, concave. Scales present only on trunk and caudal part of body; anterior part of back along base of dorsal fin, strip in middle of belly, and head naked (Figure 140, A, B).

Color of preserved fish slightly chocolate-brown: dark square spots located along sides of body, and elongate dark chocolate-brown patches along base of dorsal fin. Two dark patches present on operculum. Lower surface of head and body whitish (Fowler, 1943).

Length 70 mm (Fowler, 1943).

Distribution: In the Sea of Japan known near Niigata (Fowler, 1943: 90). Other records not known.

10. Genus *Gymnelopsis* Soldatov, 1917

Gymnelopsis Soldatov, Ezhegodn. Zool. Muzeya Rossiisk. Akad. Nauk,

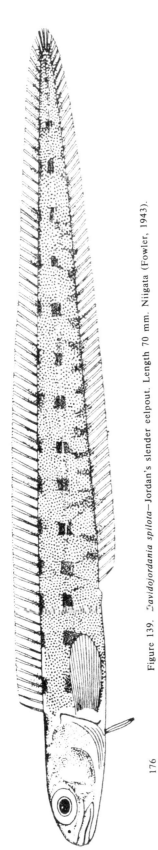

Figure 139. *Davidojordania spilota*—Jordan's slender eelpout. Length 70 mm. Niigata (Fowler, 1943).

176

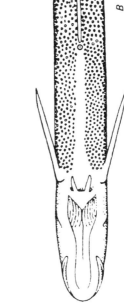

221

Figure 140. *Davidojordania spilota*. Scaleless parts (Fowler, 1943).
A—dorsal view; B—ventral view.

176

23, 1917: 160 (type: *G. ocellatus* Soldatov). Soldatov and Lindberg, Obzor..., 1930: 504. Taranetz, Kratkii Opredelitel'..., 1937: 165. Matsubara, Fish Morphol. and Hierar., 1955: 781.

Body elongate, compressed lateally, attenuate posteriorly, and only tail covered with minute round cycloid scales. Minute teeth present on jaws and palatines. Jaws almost equal in length. Gill openings small, located along sides of head. Branchiostegal membranes broadly attached to isthmus. Large pores along eyes and on sides of head. Pores of lateral line discernible only in anterior part of body. Pelvic fins absent. Dorsal and anal fin confluent with caudal fin and consist only of soft rays (Soldatov and Lindberg, 1913).

Two species. One species known from the Sea of Japan, the other from both the Sea of Okhotsk and the Sea of Japan.

Key to Species of Genus Gymnelopsis

1 (2). Vomer with 2 pairs of large canine-shaped teeth. Head without row of pores under eye.......... [**G. ocellatus** Soldatov, 1917].
2 (1). Vomer without 2 pairs of canine-shaped teeth. Head with well-developed row of pores under eye.... 1. **G. brashnikovi** Soldatov.

[*Gymnelopsis ocellatus* Soldatov, 1917—Ocellate Gymnelopsis]
(Figure 141)

Gymnelopsis ocellatus Soldatov, Ezhegodn. Zool. Muzeya Rossiisk. Akad. Nauk SSSR, 23, 1917: 161, fig. 1 (northern part of Sea of Okhotsk). Soldatov and Lindberg, Obzor..., 1930: 504, fig. 73. Taranetz, Kratkii Opredelitel'..., 1937: 165. Shmidt, Ryby Okhotskogo Morya, 1950: 124. Matsubara, Fish Morphol. and Hierar., 1955: 781, fig. 298.

Gymnelopsis ocellatus güntheri Popov. In: Shmidt, Ryby Okhotskogo Morya, 1950: 125.

20167. Sea of Okhotsk, between Ayan and Prokof'ev Island. July 26, 1912. DVE. 1 specimen.

25256. Sea of Okhotsk, Tauisk Inlet. August 18, 1930. P.V. Ushakov. 2 specimens.

D 91-97; A 67-75; P 11; *Br*. 6.[174] Head and trunk without scales; minute scales present only in caudal part of body. Head compressed laterally, oblong, constricted dorsally; profile of head convex. Mouth moderate in size. Teeth small and conical on jaws and palatines. Vomer with 2 large canine-shaped teeth. Eyes large, set high in anterior part of head. Lips not fleshy. Upper jaw does not protrude over lower one. Gill

---

[174]Radiographs of our 3 specimens (Nos. 20167, 25256) showed: D 104-108; A 86-90; vertebrae 106-110 (precaudal 21-23, caudal 85-88); canine teeth on vomer well developed; origin of dorsal fin at vertical of 4th to 5th vertebra; and skeleton of free part of caudal fin consists of 6 principal and 4 procurrent rays.

openings small. Branchiostegal membranes broadly attached to isthmus. Pectoral fins relatively broad and rounded. Color of specimens preserved in alcohol, light chocolate-brown with dark indistinct stripes and large number of minute white spots along back. Dorsal fin with 3-7 well-developed ocelli[175] (Soldatov, 1917c).

Buccal cavity with palatine and mandibular membranes. Well-developed fold behind operculum slopes down to beginning of base of pectoral fin, and together with fleshy margin of operculum forms respiratory siphon (Shmidt, 1950). Preanal distance varies from 33.0 to 34.4% of total length.

Length 117 mm.

Distribution: Not known in the Sea of Japan. In the Sea of Okhotsk known north of Shantar Island between Ayan and Prokof'ev Island (Soldatov, 1917c: 161).

1. **Gymnelopsis brashnikovi** Soldatov, 1917—Brazhnikov's Gymnolepsis (Figure 142)

Gymnelopsis brashnikovi Soldatov, Ezhegodn. Zool. Muzeya Rossiisk. Akad. Nauk, 23, 1917: 162 (east Sakhalin coast). Soldatov and Lindberg, 179 Obzor..., 1930: 505. Taranetz, Vestn. Dal'nevost Fil. Akad. Nauk SSSR, 13, 1935: 98.

13029. East coast of Sakhalin near Cape Eustaphia. July 3, 1899. V. Brazhnikov. 1 specimen.

D 110; A 78; P 11; Br. 6.[176] This species differs from *G. ocellatus* in better developed scales on caudal part of body, relatively smaller [*sic*] number of rays in dorsal fin, absence of rounded dark spots on dorsal fin, and absence of canine teeth on vomer (Soldatov, 1917c).

Taranetz (1935: 98) and Shmidt (1950: 124) do not consider the species *G. brashnikovi* described by Soldatov (1917c) an independent one because of its extreme similarity to *G. ocellatus*; they therefore relegate *G. brashnikovi* to the synonymy of the latter. Neither author paid sufficient attention to the presence of 2 pairs of canine teeth on the vomer of *G. ocellatus*, absence of numerous seismosensory pores under the eye and closer position of the dorsal fin to the head, the first ray of which (judging from radiographs) is located at the vertical of the 4th to 5th vertebra in *G. ocellatus*. Our radiographs revealed that *G. brashnikovi* has almost 180 the same number of rays in the unpaired fins and vertebrae as *G. ocellatus* but differs from the latter in absence of canine teeth on the vomer, greater distance between occiput and origin of dorsal fin, the first ray of which in *G. brashnikovi* is located at the vertical of the 8th to

---

[175]These spots may be faint or totally absent.

[176]Radiographs of specimen No. 13029 showed: D 104-105; A 89; P 11; vertebrae 109-110; vomer without canine teeth; origin of dorsal fin at vertical of 8th to 9th vertebra.

Figure 141. *Gymnelopsis ocellatus*—Ocellate gymnelopsis. Length 120 mm. Northern part of the Sea of Okhotsk (Soldatov, 1917).

[An error has occurred in this drawing or in the text. Upper jaw protrudes in relation to lower, but in the description of the holotype it is mentioned that the upper jaw does not protrude over the lower one—Editor.]

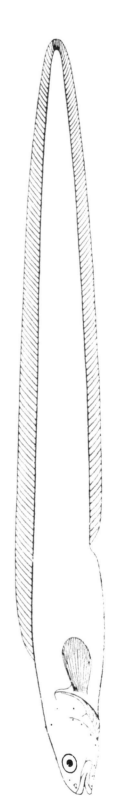

Figure 142. *Gymnelopsis brashnikovi*—Brazhnikov's gymnelopsis. Length 97 mm. No. 13209. Coast of east Sakhalin.

9th vertebra, notably smaller preanal distance (31.6% of total length), and presence of well-defined series of large rounded pores under the eyes.[177] Furthermore, as pointed out by Soldatov (1917c), the scale cover in *G. brashnikovi* is better developed in the caudal part of the body and no rounded dark spots occur on the dorsal fin. These differences provide a basis for considering *G. brashnikovi* an independent species, as done by V.K. Soldatov.

Length 100 mm (Soldatov, 1917c).

Distribution: In the Sea of Japan known from Pohang (Mori, 1952: 130). In the Sea of Okhotsk caught off the east coast of Sakhalin near Cape Eustaphia (Soldatov, 1917c: 162).

11. Genus *Gengea* Katayama, 1941

Gengea Katayama, Zool. Mag., Tokyo, 53, 12, 1941: 591 (type: *G. japonica* Katayama). Katayama, Ann. Zool. Japon., 22, 2, 1943: 101. Matsubara, Fish Morphol. and Hierar., 1955: 781. Fowler, Synopsis..., 1958: 319.

Body oblong and compressed laterally. Mouth terminal; teeth present on jaws, vomer, and palatines. Gill openings relatively small. Lateral line present along middle of anterior part of body, and absent in posterior part. Scales small and rounded, embedded in skin, cover body except for head, ventral surface, and bases of pectoral fins. Origin of dorsal fin slightly ahead of vertical from origin of anal fin. Pelvic fins absent (Katayama, 1943).

One species. Known from the Sea of Japan.

1. *Gengea japonica* Katayama, 1941 (Figure 143)

Gengea japonica Katayama, Zool. Mag., Tokyo, 53, 12, 1941: 591 (Toyama Bay). Katayama, Ann. Zool. Japon., 22, 2, 1943: 101. Matsubara, Fish Morphol. and Hierar., 1955: 781. Fowler, Synopsis..., 1958: 319.

D 93; A 89; P 11. Head 6.48, body depth 11.88, and preanal distance 3.24 times in total length. Eye diameter 4.10, interorbital distance 9.42, snout 5.50, length of upper jaw 2.27, postorbital part of head 1.73, and length of pectoral fin 2.27 times in head length (Katayama, 1943).

181 Differs from *Gymnelopsis ocellatus* Soldatov in the following characters; 1) vomer with 2 canine teeth; 2) scales present on trunk, caudal part of body, and unpaired fins; 3) origin of dorsal fin much behind head; 4) pectoral fin with black spot; and 5) larger number of rays in anal fin (89 versus 67–75 in *G. ocellatus*) (Katayama, 1943). The last distinguishing feature mentioned by Katayama was not confirmed by our radiographs of *G. ocellatus,* in which A 86–90.

Length 214 mm (Katayama, 1943).

[177]Suborbital pores 6. Head with other well-defined pores: mandibular 3, preopercular 3, postorbital 4, occipital 3, and interorbital 1.

Distribution: In the Sea of Japan known from near Niigata (Matsubara, 1955: 781); Sado Island (Honma, 1963: 8); Toyama Bay (Katayama, 1941: 591); near Fukui Prefecture (Matsubara, 1955: 781); Wakasa Bay (Takegawa and Morino, 1970: 383); off the coast of Hyogo Prefecture (Katayama, 1943: 10); and San'in region (Mori, 1956a: 22).

12. Genus *Allolepis* Jordan and Hubbs, 1925

Allolepis Jordan and Hubbs, Mem. Carnegie Mus., 10, 2, 1925: 322 (type: *A. hollandi* Jordan and Hubbs). Taranetz, Kratkii Opredelitel'..., 1937: 165. Lindberg, Predvaritel'nyi Spisok..., 1947: 170. Matsubara, Fish Morphol. and Hierar., 1955: 782. Fowler, Synopsis..., 1958: 317. Lindberg et al., Issled. Dal'nevost Morei SSSR, 6, 2, 1959: 251.

Body elongate, gradually attenuate toward caudal end. Dorsal fin consists only of soft rays. Pelvic fins absent. Pectoral fins usual in shape. Gill openings broad but do not continue far forward ventrally. Branchiostegal membranes attached along sides of isthmus. Gill rakers reduced to very short blunt tubercles. Teeth on premaxilla form thin outer row, better developed in anterior part. Teeth on lower jaw arranged in broad band with slightly enlarged teeth in outer row. Teeth present on vomer and palatines. Head behind eyes covered with rounded scales. Scales on body elongate but do not overlap (Jordan and Hubbs, 1925).

Two species. Both recorded from the Sea of Japan.

1. *Allolepis hollandi* Jordan and Hubbs, 1925 (Figure 144)

Allolepis hollandi Jordan and Hubbs, Mem. Carnegie Mus., 1925: 323, pl. 12, fig. 2 (Fukui). Taranetz, Kratkii Opredilitel'..., 1937: 165. Lindberg, Predvaritel'nyi Spisok..., 1947: 171. Matsubara, Fish Morphol. and Hierar., 1955: 782. Fowler, Synopsis..., 1958: 317, fig. 41. Lindberg et al., Issled. Dal'nevost. Morei SSSR, 6, 2, 1959: 251.

Lycogramma crystallonota Schmidt. In: Popov, Issled. Morei SSSR, 1933: 151.

24499. Sea of Japan. October 4, 1931. D.I. Okhryamkin. 3 specimens.

32227. Sea of Japan. 1933. N.M. Somova. 1 specimen.

33081. Sea of Okhotsk. August 27, 1949. P.Yu. Shmidt. 3 specimens.

33091. Sea of Japan. August 10, 1933. N. Spasskii. 2 specimens.

37967. Sea of Japan, Peter the Great Bay. September 1, 1965. I.I. Serobaba. 3 specimens.

D 115; P 17. Head and snout-to-vent length together 1.95 times in the body length up to base of caudal fin; head 6.15 and depth 9.0 times in the same length. Arrangement of scales on body very characteristic (Figure 144, A). Color light, pinkish-chocolate-brown, darker along base of dorsal fin, at tip of snout, and on opercles. Vertical fins blackish along margin (Jordan and Hubbs, 1925).

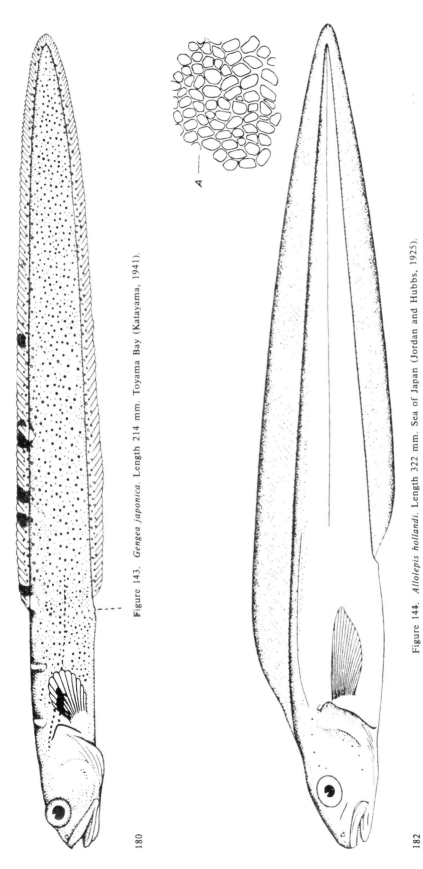

Figure 143. *Gengea japonica*. Length 214 mm. Toyama Bay (Katayama, 1941).

Figure 144. *Allolepis hollandi*. Length 322 mm. Sea of Japan (Jordan and Hubbs, 1925). A—arrangement of scales.

180

182

182   Examination of our specimens (length 113 to 250 mm) and radiographs of 9 specimens revealed significant variation in meristic and morphological characters. Number of vertebrae varied from 115 to 125 (precaudal 17-19, caudal 97-107); number of rays in dorsal fin from 110 to 122, in anal fin from 90 to 110; skeleton of free part of caudal fin with 8-10 principal and 2 procurrent rays. Preanal distance varied in the range of 32.8 to 36.6% of total length. Lateral line developed by free neuromasts in two branches: upper one begins at upper end of gill opening and, rising upward, passes near base of dorsal fin with vertical of end of first quarter of dorsal fin; other branch begins at vertical with end of upper branch and passes along median line of body to caudal fin (lateral line not immediately perceptible, requires careful scrutiny; hence Jordan and Hubbs erroneously state that it is reduced to an indistinct strip). Pores of seismosensory system of head: suborbital 8, postorbital 4, preorbital 1, unpaired interorbital 1, preopercular 3, and mandibular 4. Pores of occipital series not observed by us (in fixed material).

Length 322 mm (Jordan and Hubbs, 1925).

Distribution: In the Sea of Japan known from Peter the Great Bay (our specimens); Wonsan (Mori, 1952: 131); northern part of Tatar Strait
183 (Taranetz, 1937b: 165), off coast of Hokkaido (Ueno, 1971: 87); Sado Island (Honma, 1963: 21); Toyama Bay (Katayama, 1948: 25); off coast of Fukui (Jordan and Hubbs, 1925: 323); Wakasa Bay (Takegawa and Morino, 1970: 343); and San'in region (Mori, 1956a: 22). In the Sea of Okhotsk found off the northeast coast of Sakhalin (Shmidt, 1950: 120) and in Aniva Bay (Ueno, 1971: 87).

2. *Allolepis nazumi* Mori, 1956[178]

Allolepis nazumi Mori, Sci. Rep. Hyogo Univ. Agric., 2, 2; Nat. Sci., 1956: 29 (Yamato Bank, central part of the Sea of Japan). Fowler, Synopsis..., 1958: 318.

D 110; A 88-91; P 14: gill rakers 2-3 + 12, short. Lateral line very indistinct. Body depth 7.67 to 8.10 times in total length. Head and body compressed laterally throughout length. Snout protrudes forward, 3.1 times in head length; eyes 4.2 to 4.8 times in the same length. Maxilla reaches vertical with anterior margin of eye. Upper jaw longer than lower one. Teeth present on jaws, vomer, and palatines. Head naked except for part behind eyes, with very minute and rounded scales embedded in skin. Body without scales on occiput, in anterior half of abdomen, and on strips

---

[178]Since the author's description and drawing of this species were not available to us, we present its characters as listed by Fowler (1958), and tentatively include it in this genus as an independent species. No doubt a very careful examination will place this species only as a synonym of *A. hollandi*.

behind bases of pectoral fins. Mainly these characters distinguish *A. nazumi* from *A. hollandi* Jordan and Hubbs (Fowler, 1958).

Length 298 mm.

Distribution: In the Sea of Japan reported from Yamato Bank.

13. Genus *Lycogramma* Gilbert, 1915

Lycogramma Gilbert, Proc. U.S. Nat. Mus., 48, 1905: 364 (type: *Maynea brunnea* Bean). Jordan and Hubbs, Mem. Carnegie Mus., 10, 2, 1925: 320. Matsubara, Fish Morphol. and Hierar., 1955: 783.

Deep sea eelpout, without pelvic fins and with broad gill openings that continue forward ventrally onto throat. Branchiostegal membranes narrowly separated from each other anteriorly. Bones of head perforated, with deeply embedded seismosensory canals. Body covered with scales. 184 Lateral line very distinct, consists of two parts: anterior part located high along body side, parallel to back, and ends above vertical line behind anal opening (at a distance almost equal to eye diameter); posterior part originates below and slightly ahead of end of anterior part of lateral line and continues to end of body along middle part of side (Gilbert, 1905).

1. *Lycogramma zesta* (Jordan and Fowler, 1902) (Figure 145)

Bothracara zesta Jordan and Fowler, Proc. U.S. Nat. Mus., 25, 1902: 749, fig. 3 (Sagami Bay). Jordan and Starks, Bull. U.S. Fish Comm., 22, 1902 (1904): 601.

Lycogramma zesta, Jordan and Hubbs, Mem. Carnegie Mus., 10, 2, 1925: 320. Matsubara, Fish Morphol. and Hierar., 1955: 783.

D 112; A 92; P 17; *Br.* 6. Head 5 times in total length; longitudinal diameter of eye 6.5 times in head length and 2 times in snout length. Length of pectoral fin half head length.

Body elongate, highly compressed laterally, attenuate toward posterior end, and covered with very minute rounded cycloid scales. Head broad, its width about half head length. Dorsal, anal, and caudal fins confluent. Dorsal fin originates slightly behind vertical from base of pectoral fin Pectoral fins broad, with pointed tip.

Color of fish monochromatic, chocolate-brown, without spots (Jordan and Fowler, 1902b).

Length 482 mm (Jordan and Fowler, 1902b).

Distribution: In the Sea of Japan known off the northwest coast of Hokkaido (Hikita and Hirosi, 1952: 51); Toyama Bay (Katayama, 1940: 25); and San'in region (Yanai, 1905: 22). Along the Pacific coast of Japan known from Sagami Bay (Jordan and Fowler, 1902b: 749).

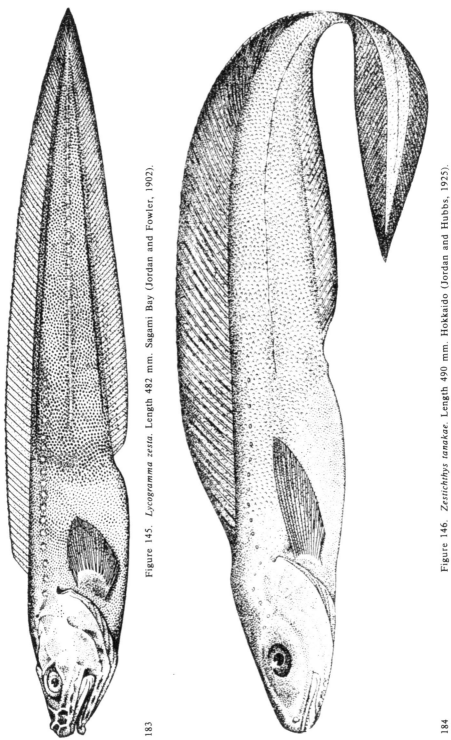

Figure 145. *Lycogramma zesta*. Length 482 mm. Sagami Bay (Jordan and Fowler, 1902).

Figure 146. *Zestichthys tanakae*. Length 490 mm. Hokkaido (Jordan and Hubbs, 1925).

183

184

14. Genus *Zestichthys* Jordan and Hubbs, 1925

Zestichthys Jordan and Hubbs, Mem. Carnegie Mus., 10, 2, 1925: 321 (type: *Z. tanakae* Jordan and Hubbs).

This genus is very close to the genus *Lycogramma*, but differs from it in lack of scales throughout entire head, on occiput, on surface behind pectoral fins, and in anterior half of abdomen. Lateral line very poorly developed; upper line very short, main line probably medial, without clearly defined pores. Teeth on jaws, vomer, and palatines minute and arranged in bands. Head compressed laterally (Jordan and Hubbs, 1925).

1. *Zestichthys tanakae* Jordan and Hubbs, 1925 (Figure 146)

Zestichthys tanakae Jordan and Hubbs, Mem. Carnegie Mus., 10, 2, 1925: 321, pl. 12, fig. 1 (Hokkaido, Kushiro). Matsubara, Fish Morphol. and Hierar., 1955: 783, fig. 302.

D about 112; A difficult to count precisely; P 14. Head and trunk 1.7 times in the body length up to the caudal fin; head 5.85 times and body depth 10.8 times in the same length. Longitudinal diameter of eye 5.65 times in head length. Head and body compressed laterally. Head very soft and with many long stripes of indefinite shape. Maxilla reaches vertical with anterior margin of eye (in closed mouth). Gill rakers on first gill arch $3 + 13 = 16$, short. Body, except for head, occiput, and anterior half of abdomen, covered with very small random scales, partly embedded in skin. Fins partly covered with scales. Main lateral line median in position, begins at vertical with anal opening and expressed only in form of "poreless fold"; upper branch of lateral line in form of series of light-colored minute spots extending short distance backward from upper corner of gill opening, apparently without pores. Vertical fins completely confluent. Pelvic fins absent. Body color very light chocolate-brown, darkening toward abdomen. Fins dark, especially along margin, with darker color in caudal part of body (Jordan and Hubbs, 1925).

Length 490 mm (Jordan and Hubbs, 1925).

Distribution: In the Sea of Japan known along coast of Hokkaido (Ueno, 1971: 87) and off Sado Island (Katoh et al., 1956: 323). Found along the Pacific coast of Japan (Matsubara, 1955: 783).

3. Suborder Ophidioidei

Fins, even pelvic fins, without spiny rays. Pelvic fins, if present, jugular or mental, represented by 1-2 filamentous cirrose rays. Vent behind pectoral fins; if, however, pelvic fins absent, vent located near base of pectoral fins. Caudal fin absent or fused with dorsal fin or anal fin and, if distinguishable, then pointed or rounded, but not bifurcate. Body oblong or elongate. Operculum usually with spines. Rays of dorsal and anal fins more numerous than vertebrae.

Osteological characters of the suborder and individual families given by Gosline (1960: 373-381; 1968: 1-78) and Nielsen (1969: 10-11). McAllister (1968: 114) separated the suborder Ophidioidei into an independent order, Ophidiiformes.

5-6 families, 3 found in the Sea of Japan.

Key to Families of Suborder Ophidioidei[1]

1 (4). Vent on belly behind pectoral fins. Scales usually present. Pelvic fins developed. Mouth usually protractile.

2 (3). Pelvic fins, if present, jugular, rarely under eyes, but in this case pectoral fins almost reach vent. Arrangement of scales usual and not at a right angle in relation to each other. Branchiostegal membranes usually separate and not attached to isthmus..................................... CLIII. **Brotulidae**

3 (2). Pelvic fins under anterior margin of eyes, almost on chin between right and left halves of lower jaw, and with mental barbels. Pectoral fins short, not more than half distance to vent. Arrangement of scales sometimes unusual (Figure 147); scales situated at a right angle in relation to each other, as in fresh-water eels of the genus *Anguilla*. Branchiostegal membranes almost separate, slightly attached to isthmus. .... CLIV. **Ophidiidae**.

4 (1). Vent on throat near base of pectoral fins. Scales absent.

5 (8). Pelvic fins absent.

6 (7). Palatines with teeth. Body and tail compressed laterally; body usually not very elongate..................... CLV. **Carapidae**.

---

[1]Nielsen (1969: 57) recorded the occurrence of members of the genus *Barathronus* (Aphyonidae) in Sagami Bay (Pacific coast of Japan). The family Aphyonidae differs from other families of the suborder Ophidioidei (Nielsen, 1969: 10) in a larger number of caudal vertebrae (31-48 versus 9-23) and the absence of scales, pyloric caeca, and spines on the opercle. Nielsen relegated the family Pyramodontidae to a subfamily of Carapidae.

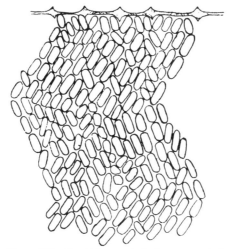

186   Figure 147. *Otophidium taylori*. Arrangement of scales.

7 (6). Palatines without teeth. Body and tail cylindrical; body highly elongate...................................... **[Disparichthyidae]**.[2]

8 (5). Pelvic fins present, cirrose, jugular, but distinctly behind vertical from eye.................................. **[Pyramodontyidae]**.[3]

CLIII. Family BROTULIDAE—Brotulas

Body elongate, compressed laterally, covered with minute cylindrical scales, sometimes naked. Dorsal and anal fins well developed, long, and usually more or less confluent with reduced caudal fin; rarely latter free. Pelvic fins, if present, jugular. Vent usually at some distance from 187 head. Mouth usually protractile. Gill openings broad. Branchiostegal membranes separate and not attached to isthmus. Opercle usually with spine (Weber and Beaufort, 1951: 398).

Marine, coastal, and deep sea fishes. About 80 genera, of which 20 known from off the coasts of Japan, including the Sea of Japan.

Key to Genera of Family Brotulidae

1 (2). Snout and lower jaw with 6 long fleshy barbels each........... .... 1. **Brotula** Cuvier.

2 (1). Snout and lower jaw without barbels.

3 (6). Base of pelvic fins near vertical with eye.

4 (5). Preopercle without spines, opercle with 1 spine. Rays of pelvic fins not bifurcate at end.................... 2. **Sirembo** Bleeker.

---

[2]New Guinea, Tahiti, Cuba.
[3]Pacific coast of Japan (J. Smith, 1955b: 546).

5 (4). Preopercle with 3 spines, opercle with 1 spine. Rays of pelvic fins bifurcate at end. .... 3. **Hoplobrotula** Gill.

6 (3). Base of pelvic fins far behind vertical with eye. Preopercle with 2 spines. Rays of pelvic fins bifurcate at end. Dorsal fin origin behind vertical with upper angle of gill opening. Body and head covered with scales. Lateral line distinct in anterior part of body. First gill arch with at least 5 long pointed rakers (excluding rudimentary ones). Pectoral fins without filamentous rays. .... 4. **Neobythites** Goode and Bean.

1. Genus *Brotula* Cuvier, 1829—Brotulas

Brotula Cuvier, Regne Animal, 2 ed. 2, 1829: 335 (type: *Enchelyopus barbatus* Schneider). Beaufort and Chapman, Fish. Indo-Austral. Arch., IX, 1951: 403.

Body elongate. Head, tail, and body covered with minute cycloid scales. Lateral line usually compete, passing slightly above median line of body. Mouth fairly large. Bands of minute teeth on jaws, vomer, and palatines. Upper and lower jaws with barbels. Branchiostegal membranes separate and not attached to isthmus. Pseudobranchs rudimentary. Eyes oblong, moderate in size. Dorsal and anal fins confluent with caudal fin, with large number of rays (more than 150), and covered with minute thin scales. Pectoral fins roundish. Pelvic fins jugular and each consists of two close-set rays (Beaufort and Chapman, 1951).

Few species, dwelling at moderate depths in the coastal waters of the Atlantic, Indian, and Pacific oceans.

1. ***Brotula multibarbata*** Temminck and Schlegel, 1842—Multi-whiskered Brotula (Figure 148)

Brotula multibarbata Temminck and Schlegel, Fauna Japonica, Poiss., 1842: 251, pl. 111, fig. 2 (Nagasaki). Hubbs, Copeia, 3, 1944: 170 (detailed list of synonyms). Kamohara, Rep. USA Mar. Biol. Sta., 1 (2), 1954: 2, fig. 1. Abe, Enc. Zool., 2, fishes, 1958: 155, fig. 458 (color figure).

Brotula japonica Steindachner and Döderlein, Fische Japan, Beiträge, 4, 1887: 24 (Tokyo).

7654. Nagasaki. 1884. Polyakova. 1 specimen.

22872. Nagasaki. 1901. P.Yu. Shmidt. 1 specimen.

22873. Kagoshima. 1901. P.Yu. Shmidt. 1 specimen.

D ca 120; A ca 85; P 22; *l. l.* ca 150, 22/48; gill rakers on first arch 4 + 18 (3 well developed) (Kamohara, 1954: 2).

189 In our specimens 400 and 423 mm in length D> 100; A > 80; P 20–24; V 2; gill rakers 5 + 1 + 15 (3 well developed).

Length, to 600 mm (Abe, 1958).

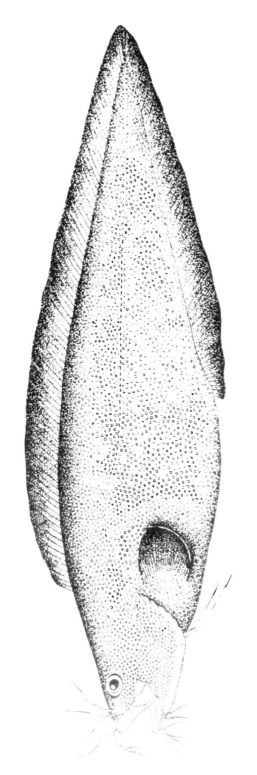

Figure 148. *Brotula multibarbata*—multiwhiskered brotula. Length 330 mm. Japan (Temminck and Schlegel, 1842).

236

Distribution: In the Sea of Japan found off Sado Island (Katoh et al., 1956: 323) and in Toyama Bay (Katayama, 1940: 28). Pacific coast of Japan from Tiba Prefecture southward (Matsubara, 1955: 803). Eastern part of the Pacific and Indian oceans (Hubbs, 1944: 165).

2. Genus *Sirembo* Bleeker, 1858—Sirembo

Sirembo Bleeker, Act. Soc. Sci. Indo-Néerl., 3, Japan, 4, 1858: 22 (type: *Brotula imberbis* Temminck and Schlegel). Jordan and Fowler, Proc. U.S. Nat. Mus., 25, 1902: 756.

Body moderately elongate, covered with very minute scales. Lateral line simple, anteriorly well developed, posteriorly indistinguishable. Eyes moderate in size. Dorsal and anal fins confluent with caudal fin. Each pelvic fin reduced to 1 ray, its base located almost on vertical from eye. Preopercle without spines, opercle with one spine (Jordan and Fowler, 1902b: 756).

Coastal fishes of the Pacific coast of Japan as well as the Sea of Japan, and usually the China and South China seas. 2 species known; 1 found in the waters under survey here.

1. *Sirembo imberbis* (Temminck and Schlegel, 1842)—Nonwhiskered Sirembo (Figure 149)

Brotula imberbis Temminck and Schlegel, Fauna Japonica, Poiss. (1842) 1846: 253, pl. 111, fig. 3 (Omura Bay, Nagasaki).

Sirembo imberbis Bleeker, Act. Soc. Inod-Néerl., 3, Japan, 4, 1858: 22. Jordan and Fowler, Proc. U.S. Nat. Mus., 25, 1902: 757. Shmidt, Tr. Tikhookeansk Kom. Akad. Nauk SSSR, 1931: 150. Kamohara, Rep. USA Mar. Biol Sta., 1 (2), 1954: 8.

22874. Nagasaki. February 12, 1901. P.Yu. Shmidt. 1 specimen.
22875. Kagoshima. February 18—March 9, 1901. P.Yu. Shmidt. 1 specimen.
22876. Tokyo. March 25, 1901. P.Yu. Shmidt. 1 specimen.
23122. Obama. March 28, 1903. N. Grebnitskii. 1 specimen.

D ca 90; A ca 70; P 22; V 1; *l. l.* 110. 10/20; gill rakers on first arch 3-5 + 8-16 (4 well developed) (Kamohara, 1954).

Counting rays was extremely difficult in our 3 specimens; hence these counts lay no claim to precision and are slightly higher than the number reported by Kamohara: D 87-100; A 72-80; P 24-26; V 1; gill rakers 4-5 + 8-10.

In other respects our specimens conform to the description of this species.

Characters of the species given in description of genus. Differs from *S. marmoratum* (Goode and Bean, 1895), found off the Philippines and China within the limits of the South China Sea, in absence of longitudinal dark narrow stripes on body and head and spots along margins of dorsal and anal fins.

Length, to 230 mm (Jordan and Fowler, 1902b).

Distribution: In the Sea of Japan reported from Pusan (Mori, 1952: 131); Sado Island (Honma, 1952: 225); Toyama Bay (Katayama, 1940: 25); and San'in region (Mori, 1956a: 22). Along the Pacific coast of Japan found from Tokyo southward (Matsubara, 1955: 803). East China and South China seas (Zhu et al., 1963: 384).

190   **3. Genus *Hoplobrotula* Gill, 1863—Hoplobrotulas**

Hoplobrotula Gill, Proc. Acad. Nat. Sci., Philad., 1863: 253 (type: *Brotula armata* Temminck and Schlegel). Jordan and Fowler, Proc. U.S. Nat. Mus., 25, 1902: 760.

This genus differs from the genus *Sirembo* in presence of strong spines on preopercle and terminal bifurcation of ray of pelvic fin. Differs from the genus *Neobythites* in attachment of pelvic fins near vertical from eye and larger number of spines on preopercle (3 versus 2).

1 species found on the coasts of Japan and China. Sea of Japan.

1. *Hoplobrotula armata* (Temminck and Schlegel, 1847)—
Armored Hoplobrotula (Figure 150)

Brotula armata Temminck and Schlegel, Fauna Japonica, Poiss., 1847: 255 (Nagasaki).

Sirembo armata, Steindachner and Döderlein, Fische Japan, Beiträge, 4, 1887: 24.

Hoplobrotula armata, Jordan and Snyder, Proc. U.S. Nat. Mus., 1900: 767, pl. 38. Jordan and Fowler, Proc. U.S. Nat. Mus., 25, 1902: 760. Shmidt, Tr. Tikhookeansk Kom. Akad. Nauk SSSR, 1931: 150. Abe, Enc. Zool., 2, Fishes, 1958: 155, fig. 458 (color figure).

22877. Kagoshima. March 20, 1901. P.Yu. Shmidt. 2 specimens.

22878. Kagoshima. February 25, 1901. P.Yu. Shmidt. 1 specimen.

22879. Nagasaki. February 18–March 9, 1901. P.Yu. Shmidt. 2 specimens.

D 79–86; A 61–74; P 21–23; V 1, forked; vertebrae 44 (Abe, 1958: 155); gill rakers 5 + 16; *l. l.* 112, 9/27 (Jordan and Fowler, 1902b: 760).

Scales on body fairly large, also on operculum, and along sides of upper part of head; upper surface of head and anterior part naked. Upper posterior part of maxilla not covered by preorbital. Pelvic fins with 1 ray each, divided into 2 long branches, inner one much longer than outer. Branchiostegal rays 8.

Length of our specimens 600 mm.

Distribution: In the Sea of Japan reported from Bohai (Mori, 1952: 131); Sado Island (Honma, 1952: 226); Niigata (Matsubara, 1955: 803);

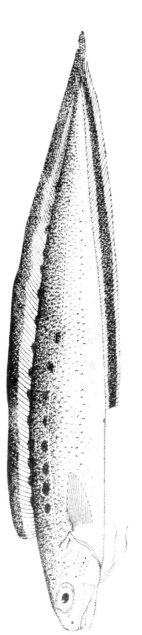

188

Figure 149. *Sirembo imberbis*—nonwhiskered sirembo. Length 175 mm. Japan (Temminck and Schlegel, 1842).

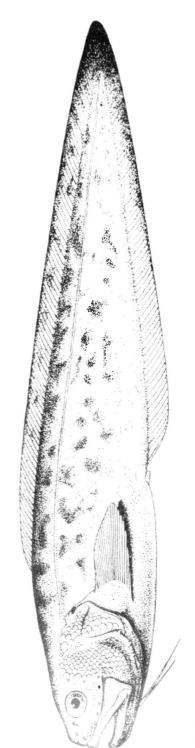

190

Figure 150. *Hoplobrotula armata*—armored hoplobrotula. Length 307 mm. **No. 22879. Nagasaki.**

Toyama Bay (Katayama, 1940: 25); Wakasa Bay (Takegawa and Morino, 1970: 383); Miyazu (Jordan and Hubbs, 1925: 325); and San'in region (Mori, 1956a; 22). Known from off Cheju-do Island (Mori, 1952: 131) and from the Gulf of Chihli (Bohai) in the Yellow Sea (Zhang et al., 1955: 176). 191 Along the Pacific coast of Japan found from Tiba Prefecture south to Nagasaki. Coast of China south to Hainan (Zhu et al., 1962: 716; 1963: 383).

4. Genus *Neobythites* Goode and Bean, 1866

Neobythites Goode and Bean, Proc. U.S. Nat. Mus., 8, 1886: 600 (type: *N. gilli* Goode and Bean). Beaufort and Chapman, Fish. Indo-Austral. Arch., IX, 1951: 415.

Watasea Jordan and Snyder, Proc. U.S. Nat. Mus., 23, 1901: 765 (type: *W. sivicola* Jordan and Snyder).

Body elongate, tail pointed. Head, body and tail covered with minute scales. Lateral line terminates in caudal part of body at some distance from caudal fin. Eyes moderate in size; snout short, roundish, sometimes protruding slightly. Barbels absent. Mouth fairly broad. Bands of very minute teeth present on jaws, vomer, and palatines. Opercle with spine; preopercle sometimes with 2 weak spines. Gill rakers well developed. Branchiostegal rays 8. Pseudobranchs present but few. Dorsal and anal fins more or less confluent with caudal fin. Each pelvic fin consists of 2 rays partly or entirely separate throughout their length (Beaufort and Chapman, 1951: 415).

The pelvic fins of *Neobythites,* unlike those in *Brotula, Sirembo,* and *Hoplobrotula,* are attached far behind a vertical with the eye, as in many other genera from Japan.

About 10 species in the Atlantic, Indian, and Pacific oceans, found at great depths and near the coast. 3 species known from the Pacific coast of Japan, 1 of which found in the Sea of Japan.

1. *Neobythites sivicolus* (Jordan and Snyder, 1901)—White-spotted Brotula (Figure 151)

Watasea sivicola Jordan and Snyder, Proc. U.S. Nat. Mus., 23, 1901: 765, pl. 37 (Misaki, Iokogama). Jordan and Fowler, Proc. U.S. Nat. Mus., 25, 1902: 759 (description of type from Iokogama). Jordan, Tanaka and Snyder, Catalogue Fishes Japan, 1913: 404, fig. 376.[4] Shmidt, Tr. Tikhookeansk Kom. Akad. Nauk SSSR, 1931: 150.
192 *Neobythites sivicolus* Matsubara, Fish Morphol. and Hierar., 1955: 797, fig 305. Fowler, Synopsis..., 1958: 334, fig. 46.

22880. Nagasaki. February 18, 1901. P.Yu. Shmidt. 2 specimens.

---

[4]Coloration in drawing similar to *N. fasciatus* (see Matsubara, 1955: 797).

22881. Pusan. March 26, 1901. P.Yu. Shmidt. 1 specimen.

D 93-96; A 74; *l. l.* ca 100.

Differs from other Japanese species, as pointed out by Matsubara, in absence of dark ocellate spot with white border on dorsal fin, which is typical for *N. nigromaculatus* Kamohara, and dark spots on sides of body and dorsal fin, which are typical for *N. fasciatus* Smith and Radcliffe. It differs from the latter species in smaller number of rays in anal fin (74-75 versus 88-90) and position of vent, location of which from tip of snout constitutes 43 to 44% (versus 30%) of standard length.

Our specimens, 178 to 203 mm in length, conform to the description of the species.

Length 234 mm (Jordan and Fowler, 1902b).

Distribution: In the Sea of Japan known from Pohang (Mori, 1952: 131); Pusan (Shmidt, 1931b: 150); Sado Island (Honma, 1963: 21); Toyama Bay (Katayama, 1940: 26); Miyazu (Jordan and Hubbs, 1925: 325); San'in region (Mori, 1956a; 22); Cheju-do Island (Mori, 1952: 131). Found throughout the Pacific coast of Japan (Matsubara, 1955: 797).

CLIV. Family OPHIDIIDAE—Cusk Eels

Body elongate, compressed laterally, more or less eel-like, usually covered with minute oval scales not arranged in regular pattern; instead scales imbricate and form oblique rows in which groups of scales located perpendicular to each other (Figure 147). Head fairly large. Lower jaw shorter than upper. Both jaws and usually vomer and palatines with teeth. Premaxilla protractile. Gill openings very broad. Branchiostegal membranes separate and only slightly attached to isthmus behind base of pelvic fins. Pseudobranchs small, gill arches 4; slit present behind last arch. Vent more or less far from head. Dorsal and anal fins not high, without spiny rays, fused around tail. Pelvic fins inserted under eyes and usually resemble long barbel divided in two. Swim bladder and pyloric caeca present (Jordan and Fowler, 1902b: 751).

About 10 genera found in tropical and temperate seas, especially near America. 1 genus found off the coast of Japan, also represented in the Sea of Japan.

1. Genus *Otophidium* Gill, 1885—Cusk Eels

Otophidium Gill. In: Jordan, Cat. Fish. North Amer., 1885: 126 (type: *Genypterus omostigma* Jordan and Gilbert).

Head naked. Scales on body rudimentary, poorly embedded in skin. Swim bladder short, thick, with large opening posteriorly. Opercle with spine concealed in skin.

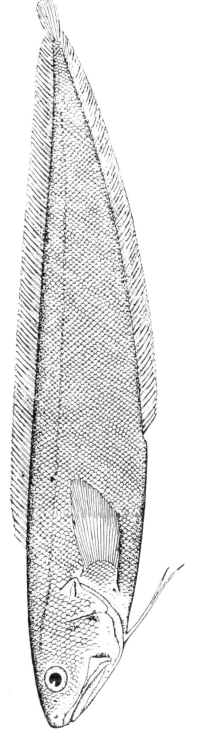

Figure 151. *Neobythites sivicolus*—white-spotted brotula. Length 190 mm. Japan (Matsubara, 1955).

242

Several species. 1 species off the coast of Japan, also known from the Sea of Japan.

1. **Otophidium asiro** Jordan and Fowler, 1902—Japanese Cusk Eel (Figure 152)

Otophidium asiro Jordan and Fowler, Proc. U.S. Nat. Mus., 25, 1902: 752, fig. 4 (Misaki).

193   D 155; A 125; P 25; V 2.

Body elongate, fairly deep, compressed laterally, with pointed tail. Head compressed laterally, about equal in length to anterior part of body. Snout rather bluntly rounded. Eyes large; posterior margin of eye nearer to tip of snout than to gill opening. Maxilla extends slightly beyond vertical with posterior margin of eye; posterior part broadens considerably, constituting about 1/3 bone length. Nostrils rather small, located in front of eyes. Teeth on jaws in form of broad bands, in front of which 1 row of large teeth occurs; vomer and palatines with conical teeth. Tongue fairly thin, pointed, and attached to lower surface of oral cavity. Gill openings very large. Branchiostegal membranes slightly attached to isthmus. Pseudobranchs with 1 small filament; gill rakers round, 3 + 3 on first arch. Spine of opercle covered with skin. Head naked; body covered with more or less minute oblong and cycloid scales. Dorsal, anal, and caudal fins confluent, caudal fin pointed. Dorsal fin originates above posterior part of pectoral fin, which is rather small; tip of pectoral fin pointed. Pelvic fins inserted ahead of vertical through middle of eye and usually consist of 2 filamentous rays, longer one $1\frac{1}{2}$ times in head length. Lateral line above median line of body, almost on back, and extends parallel to upper profile of back up to base of caudal fin. Swim bladder much thicker and short, with large opening. Color of specimens preserved in alcohol uniformly chocolate-brown; margins of vertical fins blackish-chocolate-brown (Jordan and Fowler, 1902b: 752). Absent in our collection.

Length 204 mm (Jordan and Fowler, 1902b).

Distribution: In the Sea of Japan reported from San'in region (Morei, 1956a: 22). Pacific coast of Japan—Misaki, Kumano, Koti (Matsubara, 1955: 803).

194   CLV. **Family CARAPIDAE (FIERASFERIDAE)—Pearlfishes**

Body elongate, thin, compressed laterally or cylindrical, naked, with pointed caudal part. Head short and usually the broadest and deepest part of body. Snout bluntly rounded. Eyes large, oval, located in upper part of sides of head. Posterior nostrils in form of crescent-shaped slit immediately in front of eye; anterior nostrils roundish, situated at tip of small papilla near middle of snout. Mouth large, usually oblique; lower jaw shorter than upper. Maxilla reaches or extends beyond posterior margin of orbit. Teeth on jaws, vomer, and palatines; usually larger on vomer, smallest on upper jaw. Tongue smooth, pointed, with free tip. Branchiostegal membranes slightly attached to isthmus. Gill openings broad. Branchiostegal rays 6 or 7. Upper posterior margin of opercle elongate, resembles small cusp protruding downward over base of pectoral fin. Vent in adult fishes shifted far forward. Dorsal and anal fins long, low, originate immediately behind head and continue up to end of body where they become confluent. Caudal fin absent. Neither pelvic fins nor their girdles present. Pectoral girdles always present, but pectoral fins may be reduced or even absent. Sex cannot be determined on the basis of external appearance. Eggs pelagic and probably pass through planktonic stage in their development; larvae benthic (Arnold, 1956: 259).

Atlantic, Indian, and Pacific oceans. 3 genera. 2 genera off Pacific coast of Japan, one of which has been indicated for the Sea of Japan.

Key to Genera and Subgenera of Family Carapidae (Adults Only)[5]

1 (4). Maxilla concealed in skin. Teeth on jaws and palatines arranged in 1 row.............................. 1. **Encheliophis** Müller.
2 (3). Pectoral fins present........ Subgenus **Jordanicus** Gilbert, 1905.
3 (2). Pectoral fins absent...... [Subgenus **Encheliophis** Müller, 1842].
4 (1). Maxilla well developed (not concealed in skin). Teeth on jaws and palatines arranged in several rows in form of band. Anterior teeth not canines. Body cylindrical or slightly compressed laterally; maximum body depth usually in region of head..............
.... [**Carapus** (**Carapus**) Rafinesque, 1810[6]].

1. Genus *Encheliophis* Müller, 1842

Encheliophis Müller, Ber. Verh. Preuss. Akad., 1842: 205 (type: *E. vermicularis* Müller). Smith, J. Ann. Mag. Nat. Hist. (12) 8, 1955: 415.

Jordanicus Gilbert, Bull. U.S. Fish. Comm., 23 (2) (1903) 1905: 656 (type: *Fierasfer umbratilis* Jordan and Evermann). Matsubara, Japan. J. Ichthyol, 3, 1, 1953: 31. Smith, J Ann. Mag. Nat. Hist. (12) 8, 1955: 403.

Encheliophiops Reid, Rep. Allan Hancock Pacific Exp., 9, 1940: 47 (type: *E. hancock* Reid).

Encheliophis (subgenus *Jordanicus*), Arnold, Bull. Brit. Mus. (Nat. Hist.), 4, 6, 1956: 259, 295.

Differs from the genus *Carapus* Rafinesque, 1910 in that the maxilla of *Encheliophis* is covered with skin, and the teeth on the jaws and

---

[5]Arnold, 1956: 259 (in part, only for the Pacific Ocean).

[6]2 species reported off the Pacific coast of Japan and 1 species off the northern island of Ryukyu (Kamohara and Yamakawa, 1965: 26).

244

195 palatines are arranged in 1 row and not several. The subgenus *Jordanicus* differs from the subgenus *Encheliophis* in that it has at least a small pectoral fin (undeveloped in *Encheliophis*) and larger number of branchiostegal rays (7 versus 6).

The subgenus includes 2 species; 1 species known from the Sea of Japan and off the Pacific coast of Japan.

1. *Encheliophis (Jordanicus) sagamianus* (Tanaka, 1908)—Japanese Pearlfish (Figure 153)

Carapus sagamianus Tanaka, Ann. Zool. Japan., 7, 1, 1908: 40 (Sagami); Fig. and Descr., Fishes of Japan, II, 1911: 26, pl. 7, fig. 23.

Carapus sagamius Franz, Abhandl. Math.-Phys. Klasse, Akad. Wiss., 4, Suppl., 1, 1910: 31, pl. 5, fig. 25 (Uraga and Misaki, Sagami Province).

Jordanicus sagamianus, Jordan and Hubbs, Mem. Carnegie Mus., 10, 2, 1925: 323 (Misaki). Matsubara, Japan. J. Ichthyol., 3, 1, 1953: 31.

Encheliophis (Jordanicus) sagamianus, Arnold, Bull. Brit. Mus. (Nat. Hist.), 4, 6, 1956: 301.

Head length 9.75 times and body depth at origin of anal fin 14 times in total length. Eye diameter 4.66, interorbital space 4.75, snout length 4.50, and length of upper jaw 2.25 times in head length.

Body long and slender, eel-like, but compressed laterally, and terminating in long slender caudal end. Head rather small, slightly larger in depth than in width. Eyes moderate in size, interorbital space broad and convex. Snout appears pointed in lateral view, broadly rounded in dorsal view. Mouth semi-inferior; upper jaw protrudes beyond lower jaw. Maxilla extends to vertical with posterior margin of eye. Teeth on jaws minute, sharp, and arranged in 1 row; teeth on palatines slightly larger and also arranged in 1 row; teeth on vomer larger than on palatines and arranged in narrow band of 4 longitudinal rows. Dorsal fin originates behind pectoral fins; distance from lower end of base of pectoral fin to origin of dorsal fin equal to distance from former to snout tip; height of dorsal fin, measured along ray at end of first third of fin equal to diameter of eye. Anal fin originates behind base of pectoral fin, almost at vertical from midpoint of its length; height of anal fin measured along ray in highest part of fin equal to length of postorbital part of head. Pectoral fins small, inserted below median line of body side, with pointed tip. Caudal fin small and confluent with dorsal and anal fins. Pelvic fins absent. Body naked; anterior part with series of pores 196 passing in upper half of body parallel to dorsum and forming smooth arch, extending in posterior part along middle of body. Color of fish preserved in formalin light chocolate-brown, with numerous minute dark spots; all fins light-colored, without spots (Tanaka, 1911: 26).

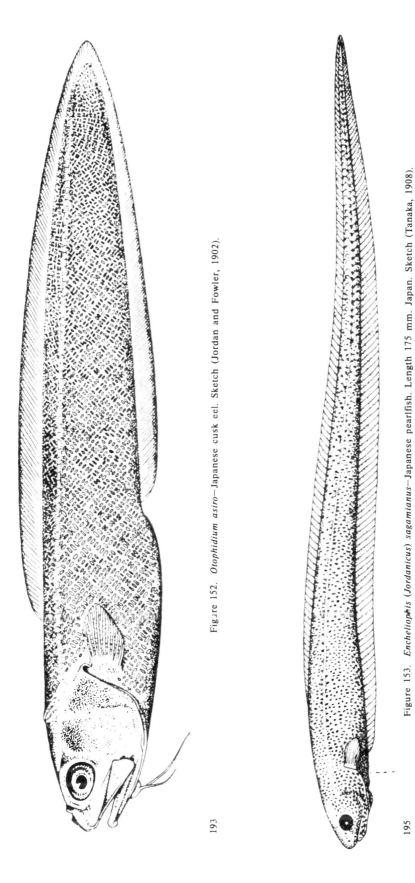

Figure 152. *Otophidium asiro*—Japanese cusk eel. Sketch (Jordan and Fowler, 1902).

193

Figure 153. *Encheliophis (Jordanicus) sagamianus*—Japanese pearlfish. Length 175 mm. Japan. Sketch (Tanaka, 1908).

195

Species differs from *Encheliophis* (*Jordanicus*) *gracilis* (Bleeker, 1856)[7] in that teeth on vomer arranged not in 1 short median row, but in band of about 4 rows.

Length, to 175 mm (Tanaka, 1908).

Distribution: In the Sea of Japan reported from Toyama Bay (Katayama, 1940: 25). Along the Pacific coast of Japan known from Sagami southward (Matsubara, 1955: 804), where it lives inside the body cavity of littoral holothurian *Holothuria monocaria* (Lesson) (Tanaka, 1911: 28).

[7]Smith (1955a: 404) considered *Carapus sagamianus* Tanaka a synonym of *Jordanicus gracilis*, but Arnold (1956: 301) recognized it as an independent species.

4. Suborder Ammodytoidei

196 Body elongate; vent behind midpoint of body. Dorsal and anal fins without spiny rays, and even without thin flexible rays. Caudal fin bifurcate or deeply forked, and well separated from dorsal and anal fins. Pelvic fins usually absent, but if present, small, jugular, with one spine and three soft rays. Head oblong; snout slightly elongate, pointed; lower jaw protrudes forward notably. Lateral line present. Gill openings normal, not reduced. Scales, if present, cycloid. Swim bladder absent (Berg, 1940: 318).

These are small marine fishes dwelling along sandy coasts and in coastal waters of the Atlantic, Indian, and Pacific oceans, as well as in the Barents, Kara, East Siberian, and Chukchi seas in the Northern Arctic Ocean.

Three families. Two known from the Sea of Japan.

Key to Families of Suborder Ammodytoidei

1 (4). Dorsal fin long; its base much longer than base of anal fin. Lateral line passes along back. Branchiostegal membranes separate. Body covered with cycloid scales of moderate size, but often on sides of body below lateral line scales detectable only in dry specimens examined under binocular microscope. Branchiostegal rays 7–8. Pectoral fins with 13 rays; caudal fin with 15 principal rays.

2 (3). Rays of dorsal fin about 40. Dermal folds (keels) not present on sides of belly. Oblique dermal folds not present on sides of body. Caudal peduncle deep and long, almost equal to head length. . . . . **[Bleekeriidae]**.

3 (2). Rays of dorsal fin about 60 (51–64). Dermal folds (keels) present on both sides of belly. Oblique dermal folds present on sides of body. Caudal peduncle low and very short . . . .
. . . . CLVI. **Ammodytidae**.

4 (1). Dorsal fin short, its base equal to length of base of anal fin. Lateral line passes along median line of body side. Branchiostegal membranes fused but not attached to isthmus. Body entirely

naked. Branchiostegal rays 4. Pectoral fins with 9 rays; caudal fins with 13 principal rays.................. CLVII. **Hypoptychidae.**[1]

CLVI. Family AMMODYTIDAE—Sand Lances

Body elongate and compressed laterally. Long dorsal fin with about 60 (50-64) rays, anal fin with about 30 (24-36) rays; number of vertebrae (60-78) slightly more than number of rays of dorsal fin. Caudal peduncle very short and low. Skin with large number of oblique folds directed downward toward caudal end. Skin between folds covered with minute, deeply embedded, cycloid scales. Lateral line continues along upper margin of body sides. Two dermal folds (keels) extend along both sides of belly from throat to midpoint of base of anal fin and even further. Median dermal fold also present on belly but poorly expressed.

About five genera found in temperate waters of all oceans. One genus in waters of Japan, also known from the Sea of Japan.

1. Genus *Ammodytes* Linné, 1758—Sand Lances

Ammodytes Linné, Syst. Nat., ed. 10, pt. 1, 1758: 247 (type: *A. tobianus* L.). Andriyashev, Ryby Severnykh Morei SSSR, 1954: 316.

Body elongate, compressed laterally. Head elongate and pointed. Premaxilla protractile. Jaws and vomer (Figure 154, A, B) without teeth. Vertebrae 60-78. Dorsal fin long, with large number of soft unbranched rays (51-68), originates above pectoral fin and continues almost up to base of caudal fin, but not confluent with it. Anal fin much shorter than half length of dorsal fin, with 24-36 rays similar to dorsal ones, and also not confluent with caudal fin. Pectoral fins rather long and narrow; their length more than three times width at base and greater than length of lower jaw. Pelvic fins absent. Skin with large number of oblique folds extending from back downward toward caudal end; skin between folds covered mainly with cycloid scales. Lateral line passes along upper margin of body side. One dermal fold present on each side of belly, which extends from throat almost (sometimes) to base of caudal fin. Sometimes dermal fold present along abdominal surface of body.

---

[1]Gosline (1963: 100) indicated the following as distinctive features of the family Hypoptychidae: jaws relatively equal in length ("jaws subequal"), firm attachment ("firmly attached") of ascending processes on premaxillae and presence of teeth on premaxillae. However, in the large number of specimens of *Hypoptychus dybowskii* examined by us we found the following: lower jaw protrudes notably beyond upper, although shorter than in *Ammodytes*; ascending processes of premaxillae not firmly attached, and bones shifted well forward; teeth on premaxillae detected with difficulty on paired and dry bones and only under high magnification of binocular microscope discernible as minute tubercles.

It should be noted that the postlarval stage of development in *Ammodytes* is very similar to that in adult forms of *Hypoptychus dybowskii* (Einarsson, 1951).

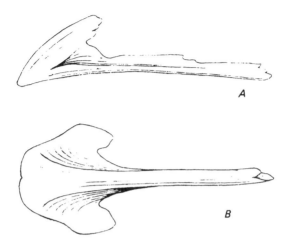

Figure 154. *Ammodytes hexapterus.* No. 34102. Mednyi Island. Shape of vomer.

A–lateral view; B–ventral view.

198

About 10 species described (for details, see Richards et al., 1963: 367–374), but classification still controversial, especially for species in the northern part of the Pacific Ocean, from where 5 species have been described: *A. hexapterus* Pallas, 1831; *A. personatus* Girard, 1859; *A. alascanus* Cope, 1873; *A. aleutiensis* Duncker and Mohr, 1939; and *A. japonicus* Duncker and Mohr, 1939.

198   As observed by Richards et al. (1953), *A. hexapterus* Pallas, 1831 is recognized as a circumpolar and highly variable species, which differs from *A. dubius* Reinhardt, 1838, found in the northern part of the Atlantic Ocean off the North American coast, and described on the basis of specimens from Greenland, in greater body depth and distribution not at the coast of freshened waters, but in much deeper seas and in marine waters with normal salinity. Meristic characters are very similar in the two species, and even differences in body depth are relatively minor. As indicated by Lindberg (1937), *A. hexapterus* is widely distributed in the Pacific Ocean. It differs from *A. personatus,* with which the northwestern Pacific sand lances were confused, solely in smaller number of vertebrae (60–66 versus 67–72). But now that the number of vertebrae in *A. hexapterus* (Richards et al., 1963: 376) has been established as 61–73, the difference between these two species in terms of this character disappears. Hence only one species should be recognized in the northwestern part of the Pacific Ocean–*A. hexapterus.* The position of *A. japonicus* Duncker and Mohr, 1939 remains unclear since these authors reported a large number of oblique body folds (pl. str. 165–188), while in other species, except for *A. alascanus* (pl. str. 182), this number does not exceed 169 (*A. aleutiensis* D. and M.; the authors included Japanese sand lances, *A.*

personatus, as a synonym of this species). Furthermore, in *A. japonicus* the lateral line originates (according to these authors) not above the pectoral fin, but at the vertical from the midpoint of the opercle. An examination of our specimens from the Sea of Okhotsk and the Sea of Japan off the south coast of Sakhalin (5 specimens) and from Obama (Japan) and the Yellow Sea (5 specimens), showed that the lateral line originates in them above the operculum, and the number of folds (pl. str.) along the sides of the body varies from 143 to 169.

The foregoing indicates that a special revision of sand lances from the northern part of the Pacific Ocean is required. At present the highly variable species—*A. hexapterus* Pallas, 1811—should be recognized from these waters, including the Yellow Sea.

1. **Ammodytes hexapterus** Pallas, 1811—Pacific Sand Lance (Figure 155)
Ammodytes hexapterus Pallas, Zoogr. Rosso-Asiat., 3, 1811: 226 (northern Kuril Islands). Lindberg, Vestn. Dal'nevost Fil. Akad. Nauk SSSR, 27, 1937: 85. Richards, Perlmutter and McAneny, Copeia, 2, 1963: 358.

Ammodytes personatus Girard, Proc. Acad. Nat. Sci. Philad., 8, 1856: 137 (Cape Flattery, Washington State). Abe, Enc. Zool., 2, Pisces, 1958: 151, fig. 447.

Ammodytes alascanus Cope, Proc. Amer. Philad. Soc., 13, 1873: 30 (Sitka, Alaska).

Ammodytes tobianus (non Linné) Shmidt, Ryby Vostochnykh Morei..., 1904: 209. Lindberg and Dul'keit, Izv. Tikho-okeansk. Nauchno-Prom. St., 3, 1, 1929: 562.

Ammodytes personatus, Lindberg, Vestn. Dal'nevost Fil. Akad. Nauk SSSR, 27, 1937: 90, fig. 4 (Obama, near Nagasaki).

199 *Ammodytes aleutensis* Duncker and Mohr, 1939: 20 (Unalaska, Aleutian Island).

Ammodytes japonicus Duncker and Mohr, 1939: 20 (Otaka on Takaido).

Ammodytes hexapterus hexapterus, Andriyashev, Ryby Severnykh Morei SSSR, 1954: 321 (Anadyr Gulf; vertebrae in 17 specimens, 67–70).

D 51–62; A 23–33; V 61–73 (Richards et al., 1963).

A 31 (29–34); vert. 65 (61–69) from 202 specimens (Ishigaki and Kaga, 1957: 13).

Counting of vertebrae from radiographs showed the following: from Anadyr Gulf 67, 68, 70, 71; Alaska 70, 71; Puget Sound 69, 69, 69; and Gulf of Mordvinov, Sea of Okhotsk 67. In specimens from Japan (Obama) the number of vertebrae was 62. Information on biology is available in works by Kitakata (1957) and Kobayashi (1961c).

Length, to 180 mm.

Distribution: Found throughout the Sea of Japan. Yellow Sea: Bohai (Zhang et al., 1955: 177); Tsingtao (Wang and Wang, 1935: 293). Okhotsk and Bering seas, Bristol Bay, south to California. Found throughout Japan (Matsubara, 1955: 720). In the Arctic Ocean adults recovered from the Chukchi Sea (Kolyuchinskaya Inlet) and larvae from the East Siberian Sea near the mouth of the Kolyma River (Andriyashev, 1954: 322).

[Family BLEEKERIIDAE—Japanese Sand Lances]

Tropical and subtropical fishes. Body fairly elongate, compressed laterally. Fairly long dorsal fin with about 40 rays; anal fin with about 15 rays. Number of vertebrae almost corresponds to number of rays in dorsal fin. Caudal peduncle deep and long, almost equal to head length. Oblique dermal folds not present along sides of body. Body covered with scales arranged in distinct rows and directed downward toward caudal end on sides. Lateral line passes near base of dorsal fin and only on caudal peduncle curves downward and terminates near midpoint of base of caudal fin. Dermal folds (keels) not present along sides of abdomen, and median abdominal fold also absent. Pelvic fins absent (*Bleekeria* Günther, 1862)[2] or present and jugular (*Embolichthys* Jordan, 1903).

Two to three genera known in the Indian Ocean, western part of the Pacific Ocean, and the Atlantic Ocean (the Lesser Antilles). One genus near the Pacific coast of Japan and Taiwan (China) represented by a single species—*Embolichthys mitsukurii* (Jordan and Evermann, 1903) (Figure 156). Not found in the Sea of Japan.

The species *Ammodytes septipinnis* described by Pallas (1811) and later (Bean, in: Jordan and Evermann, *Fishes N. and M. Amer.*, 1898: 2841) included in the new genus *Rhynchias,* has never been found subsequently. In all probability this species does not belong to Ammodytoidei.

CLVII. Family HYPOPTYCHIDAE—Shortfin Sand Lances

Differs from other families of the suborder Ammodytoidei in short dorsal fin, length of which is equal to length of anal fin; two fins similar in shape and position; anterior rays of both fins much higher than posterior rays. Body naked. Lateral line passes along middle of sides of body. Branchiostegal membranes broadly connected but not attached to isthmus. Very distinct and almost transparent dermal fold passes along middle of belly from base of pectoral fins to vent.

Osteological characters of the family furnished by Gosline (1963).

One genus in the northern part of the Sea of Japan and southern part of the Sea of Okhotsk, as well as off the Pacific coast of northern Japan.

[2]There are 3 specimens of *Bleekeria viridianguilla* (Fowler) in the collection of the Institute of Zoology, Academy of Sciences of the USSR registered as No. 36539 (coast of Hainan Island, November-December, 1959, B.E. Bykhovskii and L.F. Nagibina). A figure of this species is presented in an article by Zhu and associates (1962: 721, fig. 534).

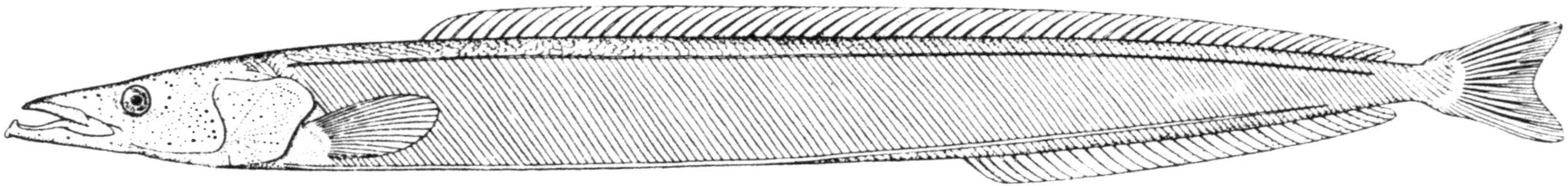

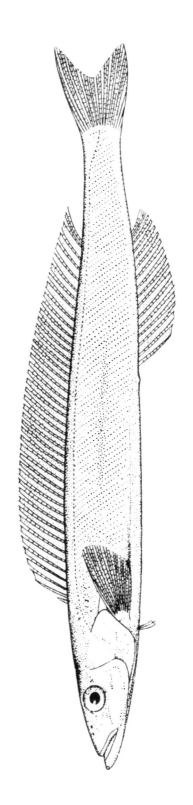

Figure 155. *Ammodytes hexapterus*—Pacific sand lance. Length 20 mm [sic]. No. 34102. Mednyi Island.

Figure 156. *Embolichthys mitsukurii*. Length 115 mm. Japan (Matsubara, 1955).

1. Genus *Hypoptychus* Steindachner, 1880

Hypoptychus Steindachner, Sitzb. Akad. Wiss., 82, 1, 1880 (1881): 257 (*Hypoptychus dybowskii* Steindachner). Fowler, Synopsis..., 12, 1-2, 1959: 72.

Characters of this monotypic genus given in description of family.

One species, distributed in the Sea of Japan, southern part of the Sea of Okhotsk, off Shikotan Island, and off the north coast of Japan in the Pacific Ocean.

1. *Hypoptychus dybowskii* Steindachner, 1880—Shortfin Sand Lance (Figure 157)

Hypoptychus dybowskii Steindachner, Sitzb. Akad. Wiss., 82, 1, 1880 (1881): 257, pl. 2, fig. 2[3] (Peter the Great Bay). Fowler, Synopsis..., 12, 1-2, 1959: 73, fig. 50 (from Steindachner).

Hypoptychus steindachneri Franz, 1910, Abhandl. Acad. Wiss. Math. Phys. Klasse, 1910: 8, pl. 5, fig. 28.

D 19-21; A 19-21; P 9.

Body elongate, thins slightly toward tail and head. Snout pointed, oral opening almost horizontal, lower jaw distinctly protrudes beyond upper one. Eyes fairly large, slightly smaller than snout length, upper margin of eye at level of upper profile of head.

Examination of our specimens (27) from Peter the Great Bay, Tatar Strait (Soviet Gavan and Kholmsk), Aniva Bay, and Shikotan Island showed that they conform to the characters of this species given by Steindachner (Fowler, 1959: 73). The only differences include: number of rays of dorsal fin [19 (3), 20 (18), 21 (5) versus 20]; number of rays of anal fin [19 (3), 20 (16), 21 (4) versus 20][4]; size of eyes (3.7-4.0 versus 3.4-3.75); and size of snout[5] (3.1-3.4 versus 3.0) in relation to head length. These ratios in fish from different regions are given in Table 2.

TABLE 2

Place of collection	No. of specimens	Total length	D	A	Ratio to head length	
					Eye diameter	Snout length
Peter the Great Bay	9	61-88	20-21	20-21	3.8	3.4
Tatar Strait	5	65-95	19-20	19-20	3.7	3.1
Aniva Bay	1	77	20	20	4.0	3.2
Shikotan Island	12	66-92	20	19	3.9	3.4

[3]Misprinted as "Figure 3" in the text of Steindachner's article.
[4]Abe (1958) reported: D 20-21; A 18-20.
[5]Snout length measured from tip of upper jaw to vertical from anterior margin of eye.

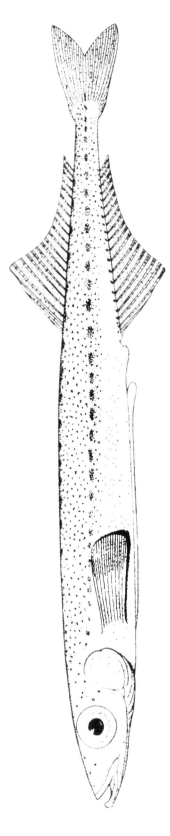

Figure 157. *Hypoptychus dybowskii*—shortfin sand lance. Length 94 mm. No. 40043. Peter the Great Bay.

Body pigmentation in our fish much more vivid than described by Steindachner. Pigmented spots occur not only in anterior part of back and sides of body, but further, beyond origin of dorsal fin, densely covering dorsal and lateral sides of caudal fin (see figure by Fowler, 1959, fig. 50).

202   Length, to 95 mm.

Distribution: In the Sea of Japan known from northern parts, along continental coast of Soviet Gavan (Soldatov and Lindberg, 1930: 509), Olga Bay (Popov, 1933a: 140), and Peter the Great Bay where it is common (described from Strelok Bay). Also reported slightly north of Wonsan (Mori, 1952: 132); found by us along island coast near Kholmsk; reported from Oshoro Bay near Otaru (Kobayashi, 1962: 258); and off Sado Island (Honma, 1963: 20). In the Sea of Okhotsk we confirmed its occurrence in Aniva Bay and near Cape Korsakov. Found by us off Shikotan Island; along the Pacific coast of Japan indicated for Muroran (Jordan and Tanaka, 1927: 391) and Sagami Bay (Franz, 1910: 8).

5. Suborder Callionymoidei[1] – Dragonets

202 The flatness of the head and trunk of these fishes and the nature and location of the broad pelvic and pectoral fins recall fishes of the order Cottiformes. The suborders differ in the absence of the characteristic suborbital stay of Cottiformes in the suborder Callionymoidei. Pelvic fins inserted ahead of pectoral fins, with 1 very short spine and 5 soft rays. Body naked. Mesethmoid (Figure 158, A) large, forming interorbital septum and replacing orbitosphenoid. Vertebrae 21. Ribs absent (Berg, 1940: 319).

203 2 families distributed in the Atlantic, Indian, and Pacific oceans. Both families represented off the coast of Japan and one (Callionymidae) also known from the Sea of Japan.

Key to Families of Suborder Callionymoidei[2]

1 (2). Preopercle with long bony process in form of spine (Figure 159, A-C). Opercle and subopercle not rudimentary. Lateral line present. Gill openings very small .... **Callionymidae**.
2 (1). Preopercle without bony spiny process. Opercle and subopercle rudimentary, each represented by straight prickly spine (Figure 160). Lateral line absent. Gill openings moderate in size. ......
.... **[Draconettidae]**.

[Family DRACONETTIDAE]

Close to Callionymidae but differs in cephalic structures: preopercle with smooth margins, without processes, and opercle and subopercle reduced and each represented by an almost straight, simple, pointed spine (Figure 160). Gill openings much broader than in Callionymidae, and branchiostegal membranes widely attached to isthmus. Lateral line absent (Jordan and Fowler, 1903).[3]

2 genera. 1 genus known from the Pacific coast of Japan.

[1]Gosline (A reinterpretation of the teleostean fish order Gobiesociformes, *Proc. Calif. Acad. Sci.*, 4th series, 38, 19, 363-382, 7 figs.) includes the suborder Callionymoidei, with its 2 families, Callionymidae and Draconettidae, as a suborder in the order Gobiesociformes.
[2]From Matsubara, 1955: 710.
[3]Diagnosis of family from Briggs and Berry (1959).

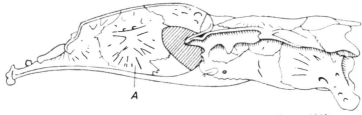

Figure 158. *Callionymus* sp. Lateral view of skull (Berg, 1940).
A—mesethmoid.

[Genus *Draconetta* Jordan and Fowler, 1903]

Draconetta Jordan and Fowler, Proc. U.S. Nat. Mus., 25, 1903: 939 (type: *D. xenica* Jordan and Fowler).

Characters of the genus given in description of family.

2 species; both known from the Pacific coast of Japan, but not reported for the Sea of Japan.

CLVIII. Family CALLIONYMIDAE—Dragonets

Head and body flat and caudal part of body compressed laterally, or

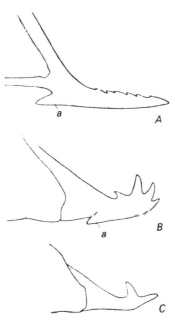

Figure 159. Processes of preopercle (Matsubara, 1955).
A—*Callionymus japonicus*; B—*C. valenciennesi*; C—*Synchiropus allivelis*;
a—spine directed forward.

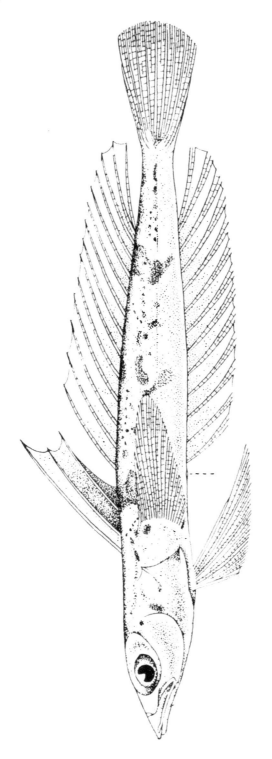

Figure 160. *Draconetta xenica.* Length 95 mm. Japan (Matsubara, 1955).

head and body more or less cylindrical. Scales absent. Lateral line single, complete, continues onto head; on upper surface of head and dorsal surface of caudal peduncle lateral line from each side of body joined together through transverse processes. Dorsal fins usually 2, rarely 1; first fin short, with 2-4 thin spines; second fin long, with simple unbranched or branched rays. Anal fin moderate in length, its base usually shorter than base of second dorsal fin. Pelvic fins inserted before pectoral fins, wide-set, large, with 1 very short spine and 5 soft rays; last ray usually connected through membrane with spine of base of pectoral fin. Pectoral fins large, rounded. Mouth protractile, small, horizontal or almost horizontal. 205 Branchiostegal membranes entirely fused with isthmus. Gill openings almost entirely covered with skin and hence acquire shape of small slits near upper margin of opercles. Preopercle with well-developed bony process, partially covered with skin; lower margin often with small spine near base directed forward in form of cusp (Figure 159, A, B); one or several odontoid spines on upper side of process closer to its pointed apex. Jaws with several rows of pointed teeth. Palatines without teeth. Vertebrae few—21 (Beaufort and Chapman, 1951: 50).

About 10 genera[4] in temperate and warm waters of the Atlantic, Indian, and Pacific oceans, but not found near the Pacific coast of America. 4 genera represented in waters of Japan, 3 from the Sea of Japan.

Key to Genera of Family Callionymidae

1 (2). One dorsal fin (Figure 161); rays soft. Lateral line one. Pelvic and pectoral fins not connected through dermal membrane. Gill opening located near upper margin of operculum, lateral. Opercular dermal valve present, formed by outgrown branchiostegal membrane supported by elongate rays. Preopercular bony process with 3-4 odontoid spines along upper margin; lower margin without spine directed forward in form of cusp.*.......
.... 1. **Dracule** Snyder.

2 (1). Two dorsal fins.

3 (4). Opercular dermal valve present, formed by outgrown branchiostegal membrane supported by elongate rays. Posterior margin of upper jaw with conical process distinctly protruding out near corner of mouth (Figure 162). Longitudinal dermal fold present, inclined upward, and passes below lateral line form top of pectoral fin to base of caudal fin..
.... [**Calymmichthys** Jordan and Thompson, 1914].[5]

[4]Schultz and Woods, 1948: 419.

*Does not correspond with description of genus or species or Fig. 161—General Editor.

[5]1 species, *Calymmichthys xenicus* Jordan and Thompson, 1914, described from Sagami Bay (Pacific coast of Honshu Island).

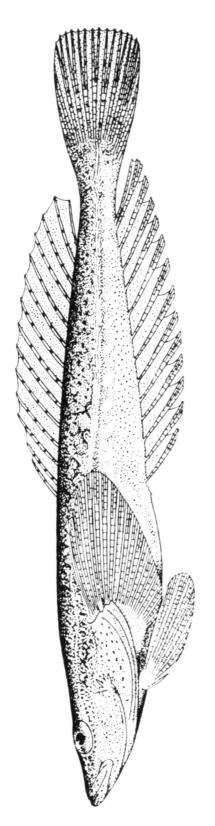

Figure 161. *Draculo mirabilis.* Standard length 35 mm. Japan (Snyder, 1912).

4 (3). Opercular dermal valve formed by outgrown branchiostegal membranes supported by elongate rays absent. Conical process of upper jaw absent. Longitudinal dermal fold passing below lateral line from top of pectoral fin to base of caudal fin absent. Gill opening located either on upper side or lateral side of head.

5 (6). Spine directed forward in from of cusp present (Figure 159, A, B, a) on lower margin of bony process of preopercle. Head and body extremely flat. Soft rays of second dorsal fin simple, only last ray branched. Gill opening located on dorsal side of head ....
.... 2. **Callionymus** Linné.

6 (5). Spine directed forward in form of cusp (Figure 159, C) on lower margin of bony process of preopercle absent. Head and body more or less cylindrical. Soft rays of second dorsal fin branched except for first short ray.* Gill opening slightly shifted to lateral side of head .... 3. **Synchiropus** Gill.

1. Genus *Draculo* Snyder, 1911

207

Draculo Snyder, Proc. U.S. Nat. Mus., 40, 1911: 545 (type: *D. mirabilis* Snyder).

Dorsal fin with 13-14 soft but unbranched rays; first dorsal fin consisting of weak spiny rays absent. Anal fin similar to dorsal fin in size and location, with 13-14 rays. Pelvic and pectoral fins not connected through membrane. Preopercular bony process with three to four odontoid spines along upper margin and one spine directed forward in shape of cusp near base of process on lower margin.

Opercular dermal valve present, formed by outgrown branchiostegal membranes supported by elongate rays, and continues backward up to base of pectoral fins. Gill opening in form of narrow slit located laterally on head near upper end of base of valve.

1 species along the Pacific coast of Japan, in the Sea of Japan (Pos'et), and in the Yellow Sea.

1. ***Draculo mirabilis*** Snyder, 1911—Weever (Figure 161)

Draculo mirabilis Snyder, Proc. U.S. Nat. Mus., 40, 1911: 545 (Tomakomai, Hokkaido). Snyder, Proc. U.S. Nat. Mus., 42, 1912: 447, pl. 61, fig. 2. Lindberg, Tr. Zool. Inst. Akad. Nauk SSSR, 18, 1955: 385–388. Zhang et al., Ryby Zaliva Bokhai..., 1955: 179, fig. 114. Arai, Japan. J. Ichthyol., 18, 1, 1971: 33–35, fig. 5.

32193. Pos'et, in Peter the Great Bay. December 6, 1948. O.B. Mokievskii. 1 specimen.

Since this species is very rare, we provide a description of our specimen.

*In *S. ijimai* only (see Figure 185)—General Editor.

D 13; A 14; P 18; V I, 5.

Head and anterior part of trunk flat and broad; body depth slightly less than width; caudal part of body compressed laterally, especially in region of caudal peduncle; height of latter 3 times width. Skin thin, smooth, especially on head. Margins of nostrils slightly raised. Mouth semisuperior, small; posterior margin of upper jaw does not reach vertical from anterior margin of eye; both sides of lower lip with group of fairly long papillae originating from center, which resemble teeth or festoons of fimbria directed not downward but upward in direction of upper jaw. Teeth small, arranged in narrow bands on jaws; vomer and palatines without teeth. Gill openings very narrow, located laterally on head near upper end of base of transparent dermal process, i.e., opercular valve supported by rays of branchiostegal membranes which extend to base of pectoral fin. Preopercular bony process with 4 large odontoid spines on upper margin, and 1 small spine directed forward on lower margin. Lateral line mediolateral, with sparse wide-set pores, and distinctly visible throughout length from occiput to caudal fin. Line slopes downward above middle of pectoral fins and continues further along middle of body. Short branches proceed from main lateral line; one of the larger branches originates near base of dorsal fin and is directed backward and upward on dorsal surface of caudal peduncle, where it fuses with branch of opposite side; two other branches fuse on occiput; one branch on each side originates near head with another branch proceeding from it toward margin of operculum.

Dorsal and anal fins originate at vertical from immediately behind vent; predorsal distance slightly less than postdorsal. Middle rays of both fins almost equal in height, constituting 1/3 head length; posterior rays of anal fin when folded extend slightly beyond base of caudal fin. Caudal fin slightly rounded, and not pointed as indicated by Snyder, but otherwise corresponds to his figure (Snyder 1912, pl. 61, fig. 2). Pectoral fins pointed, continue beyond vertical from origin of dorsal and anal fins. Longest ray of pelvic fin the penultimate (fourth branched ray), and not the last ray as mentioned by Snyder; fin free, fused with neither pectoral fin nor skin of trunk.

In our specimen preserved in alcohol, the basic background color was chocolate-brown, on which minute dark spots were distinctly visible and scattered along dorsal surface of body, tail, and rays (upper) of dorsal, caudal, and partly pectoral fins. Description of larvae and young fish already published (Kobayashi and Abe, 1963).

Standard length of our specimen 36 mm.

Distribution: In the Sea of Japan reported from Pos'et (Lindberg, 1955: 386), where it was caught in the sandy littoral zone. In the Yellow Sea reported for the southwest coast of the Korean Peninsula, Cholla Pukdo

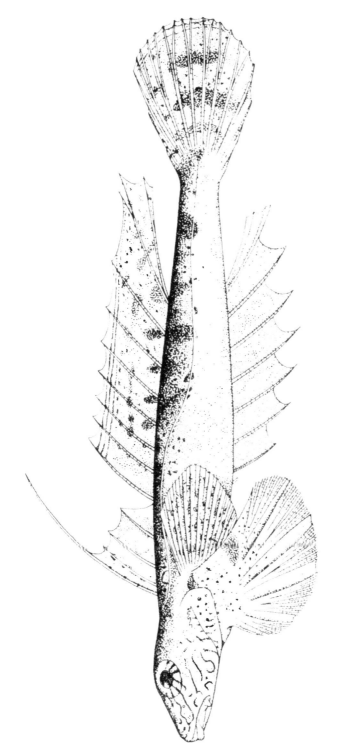

Figure 162. *Calymmichthys xenicus*. Length 135 mm (Jordan and Thompson, 1914).

Province (Mori, 1952: 133) and for Beidaihe in Liodun Bay (Zhang et al., 1955: 179). In Japan known from the Pacific coast of Hokkaido, where it was caught on the sandy coast near Tomakomai, east of Muroran.

[Genus *Calymmichthys* Jordan and Thompson, 1914][6]

Calymmichthys Jordan and Thompson, Mem. Carnegie Mus., 6, 4, 1914: 296 (type: *C. xenicus* Jordan and Thompson).

Differs from other genera of the family in thin dermal fold on side of body (not shown in drawing given by authors), which extends from apex of pectoral fin up to base of caudal fin. Japanese authors mistook this fold for a second lateral line. Conical process of upper jaw quite typical, which projects out near corner of mouth and is directed downward (Figure 162). Gill opening equal in size to pupil and located near upper corner of operculum.

1 species, *C. xenicus* Jordan and Thompson, 1914, known from the Pacific coast of Japan.

2. Genus *Callionymus* Linné, 1758—Dragonets

Callionymus Linné, Syst. Nat., ed. X, pt. I, 1758: 249 (type: *C. lyra* Linné).

Calliurichthys Jordan and Fowler, Proc. U.S. Nat. Mus., 25, 1903: 941 (type: *C. japonicus* Jordan and Fowler).

Two dorsal fins. First dorsal with 3-4 weak spines, second with 9-11 rays. Body and head extremely flat. Opercular valve, conical process of upper jaw, and longitudinal dermal fold below lateral line not present. Gill opening located on dorsal side of head. Spine in form of cusp directed forward present on lower side of bony process of preopercle; upper side of process minutely serrate or with large bent spines tapering to a point, or terminates in a hook. Soft rays of dorsal fin simple, only last ray branched. Last ray of pelvic fin connected through membrane with base of pectoral fin.

Many species. In waters of Japan 16 species known, of which 10 reported from the Sea of Japan.

Key to Species of Genus Callionymus[7]

1 (4). Preopercular bony process saber-shaped, straight, with pointed tip; upper margin minutely dentate or serrate (Figure 159, A).

209 2 (3). 2 bony, radially tubercular plates or processes behind eyes on

---

[6]This genus is included in the synonym of *Diplogrammus* Gill, 1865 (Schultz et al., 1960: 399).

[7]From Ochiai et al., 1955, with additions.

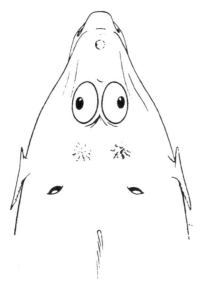

Figure 163. *Callionymus japonicus.* Head, dorsal view. No. 22851. Japan.

upper surface of head (Figure 163); space between plates covered with smooth skin.[8] Second dorsal fin with 9 and anal fin with 8 soft rays (last ray in both fins bifurcate). First dorsal fin with black spot between 3rd and 4th rays; first 2 rays of this fin in male elongate, filamentous (Figure 166). Thorax of male with dark spot . . . . 1. **C. japonicus** Houttuyn.

3 (2). Upper surface of head smooth or two plates of tubercles barely defined, touch each other, and distinguishable in dry specimen under high magnification. Second dorsal fin and anal fin with 9 rays each. First dorsal fin without black spot; all rays in adult fish elongate (Figure 167). Lower side of head in male light-colored. . . . . 2. **C. doryssus** (Jordan and Fowler).

4 (1). Preopercular bony process with falcate tip; upper margin armed with three or more large curved spines (Figure 159, B).

5 (6). Second dorsal fin with 8 and anal fin with 7 soft rays (last ray in both fins bifurcate). Additional branch not present on dorsal side of caudal peduncle connecting lateral lines of opposite sides of body. Eyes relatively moderate in size, about 25% of head length.[9] Male, in distinction from female, with filamentous rays in first dorsal fin (Figure 168). . . . .
. . . . 3. **C. calliste** Jordan and Fowler.

[8]In *Callionymus variegatus* Temminck and Schlegel from southern Japan, plates fused into single structure.

[9]The closely related species, *C. phasis* Günther, known from off the coast of Ehime Prefecture, is distinguished by larger eyes—33% of head length.

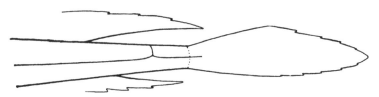

209   Figure 164. *Callionymus punctatus.* Connection of lateral lines on upper
surface of caudal peduncle. No. 23580. Japan.

6 (5). Second dorsal and anal fin with 9 soft rays each (last ray in both fins bifurcate). Additional branch present on dorsal side of caudal peduncle connecting lateral lines of opposite sides of body (Figure 164).[10]

210 7 (8). Preopercular bony process fairly stout, with 5-6 pointed spines along upper margin. Head flat, its width more than length. First dorsal fin of male with dark pattern of numerous dense broken lines and dots (Figure 170), and in female with minute dark spots (Figure 171, B)............... 4. **C. planus** Ochiai.

8 (7). Preopercular bony process rather thin, with 3-4 pointed spines along upper margin. Head moderately flat, width equal to less than length.

9 (10). Last spine of preopercular bony process serrate (Figure 173, a). Male with 2 middle rays of caudal fin distinctly elongate.
.... 5. **C. kaianus** Günther.

10 (9). Last spine of preopercular bony process not serrate. Rays of caudal fin not elongate, but if so almost all and not just middle 2 elongate.

11 (22). Genital papilla elongate (male) (Figure 165, A).

12 (13). Rays of caudal fin (middle 5) elongate-filamentous. Rays of first dorsal fin (except second ray) highly elongate-filamentous (Figure 174)................. 6. **C. flagris** Jordan and Fowler.

13 (12). Rays of caudal fin filamentous (not elongate or only slightly so).

14 (15). Rays of first dorsal fin not elongate. In male upper margin of membranes of first dorsal fin with characteristic coloration in form of narrow black stripe of crescent-shaped spots (Figure 176, A), and in female with oval black spots on membrane between 3rd and 4th rays (Figure 176, B)..............................
.... 7. **C. punctatus** Richardson.

15 (14). All or some rays of first dorsal fin elongate-filamentous.

16 (17). Only first ray of first dorsal fin elongate-filamentous; other rays without notable elongation (Figures 177 and 178, A);

[10]This connection is not shown in most illustrations.

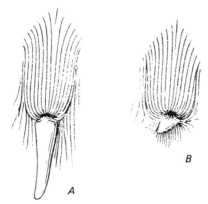

Figure 165. Difference between male and female of *Callionymus*. No. 23580. Tsuruga.

Urogenital papilla: A—male; B—female.

blackish spot present on membrane behind last ray. . . . .
. . . . 8. **C. lunatus** Temminck and Schlegel.

17 (16). All rays of first dorsal fin more or less elongate-filamentous.

18 (21). Anterior 2 rays of first dorsal fin longer than others. Membrane between rays of first dorsal fin same height or even lower than membrane between rays of second dorsal fin. Diameter of eye equal to or slightly less than snout length.

19 (20). Three middle rays of caudal fin longer than others (Figure 179). Posterior rays of dorsal and anal fins continue up to base of caudal fin and even slightly further. Large black spots not present on first dorsal fin; anal fin dull or black, with numerous oblique sinuous white stripes. . . . .
. . . . 9. **C. beniteguri** Jordan and Snyder.

211 20 (19). Caudal fin uniformly rounded (Figure 181). Posterior rays of dorsal and anal fins do not reach base of caudal fin. Several blackish spots present on first dorsal fin; anal fin light-colored but outer margin black. . . . .
. . . . 10. **C. valenciennesi** Temminck and Schlegel.

21 (18). All rays of first dorsal fin very highly elongate (Figure 182); membrane between rays almost twice higher than membrane between rays of second dorsal fin; membrane of last ray slightly attached to base of first ray of second dorsal fin. Diameter of eye more than snout length. . . . .
. . . . 11. **C. virgis** Jordan and Fowler.

22 (11). Genital papilla very short or rudimentary (female) (Figure 165, B).

23 (30). Last rays of dorsal and anal fins when latter folded against body, do not reach base of caudal fin.

24 (25). Rays of first dorsal fin more or less elongate-filamentous, first ray longest, and when fin folded against body, reaches base of second dorsal fin (Figure 175). Anal fin pale.............. .... 6. **C. flagris** Jordan and Fowler.

25 (24). Rays of first dorsal fin not elongate except first ray, which does not reach, or barely reaches base of first ray in second dorsal fin.

26 (29). First dorsal fin uniformly dark (Figure 178, B).

27 (28). Eyes moderate in size; their diameter almost equal to snout length. Weak dark stripe continues along midpoint of anal fin (Figure 178, F)........ 8. **C. lunatus** Temminck and Schlegel.

28 (27). Eyes fairly large; their diameter more than snout length. Anal fin with dark stripe. ... 11. **C. virgis** Jordan and Fowler.

29 (26). First dorsal fin with dark crescent-shaped spot on membrane between second and third ray; similar spot but larger on membrane between 3rd and 4th rays (Figure 184). .... 12. [**C. kitaharai** Jordan and Seale].

30 (23). Last rays of dorsal and anal fins in adults[11] when fins folded against body, reach base of caudal fin. Anal fin dull.

31 (32). First dorsal fin with numerous very large dark speckles and large white spots on membranes between 1st and 3rd rays (Figure 180, A); rest of fin uniformly dark................. .... 9. **C. beniteguri** Jordan and Snyder.

32 (31). First dorsal fin with large blackish spot on membrane between 3rd and 4th rays (Figure 176, B). .... 7. **C. punctatus** Richardson.

1. *Callionymus japonicus* Houttuyn, 1782—Japanese Dragonet
(Figure 166)

Callionymus japonicus Houttuyn, Verh. Maatsch., Wet. Harlem, 20, 1782: 311 (Nagasaki). Ochiai et al., Publ. Seto Mar. Biol. Lab., V, 1, 1955: 98 (synonyms).

Callionymus reevesi Richardson, Voy. Sulphur, Fishes, 1844: 60, pl. 36, figs. 1-3 (Canton).

212 *Callionymus longicaudatus* Temminck and Schlegel, Fauna Japonica, Poiss., 1845: 151, pl. 79A, fig. 1 (Nagasaki).

Calliurichthys japonicus, Jordan and Fowler, Proc. U.S. Nat. Mus., 25, 1903: 942, fig. 2. Abe, Enc. Zool., 2, Fishes, 1958: 154, fig. 456 (color diagram).

---

[11]In females 37 and 74 mm long examined by us, rays of the dorsal fin, and more so the anal, did not reach base of the caudal fin, although these specimens undoubtedly belong to *C. punctatus* on the basis of the typical spot on the first dorsal fin and other characters.

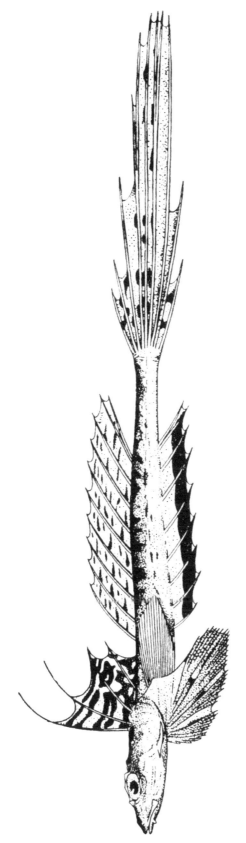

Figure 166. *Callionymus japonicus*—Japanese dragonet. Length 280 mm. Japan (Temminck and Schlegel, 1845).

1568. Japan. 1863. Maksimovich. 1 specimen.

22851. Nagasaki. February–March, 1901. P.Yu. Shmidt. 10 specimens.

22852. Kagosima. February 21, 1901. P.Yu. Shmidt. 2 specimens.

D IV, 9–10; A 8–9; P 18–21 (19.5) (Ochiai et al., 1955). In our 13 specimens with a standard length of 70–180 mm: D IV, 9; A 8.

Preopercular bony process strong, pointed, straight (Figure 159, A), equal in length to diameter of eye; upper margin of process serrate or minutely dentate. This species differs from other species with a similar process in presence of 2 well-separated radially tubercular bony processes on top of head. In male, first two rays of first dorsal fin filamentous, much longer than head, but distinctly shorter than twice its length. Bright black spot occurs between 3rd and 4th rays. Four middle rays of caudal fin also large, elongate, and almost equal to snout-to-vent length. In female, rays of first dorsal fin short and tips not filaments. Middle rays of caudal fin elongate, longest shorter than snout-to-vent length. Sexual dimorphism distinctly developed only in fish longer than 90 mm (standard length). Radially patterned bony tubercles on upper side of head in young fish poorly developed or absent.

Length, to 420 mm (Ochiai et al., 1955).

Distribution: In the Sea of Japan known from Pusan (Mori, 1952: 133); Aomori (Katoh et al., 1956: 22); Wakasa Bay (Takegawa and Morino, 1970: 382); and San'in region (Mori, 1956a: 22). Along the Pacific coast of Japan recovered from the central southern part of Honshu (Matsubara, 1955: 712); South China Sea (Zhu et al., 1962: 727); northeast coast of Australia, near Queensland and Indian coast to the Persian Gulf (Beaufort and Chapman, 1951: 54).

2. *Callionymus doryssus* (Jordan and Fowler, 1903) (Figure 167)

Calliurichthys doryssus Jordan and Fowler, Proc. U.S. Nat. Mus., 25, 1903: 945, fig. 4. Abe, Enc. Zool., 2, Fishes, 1958: 154, fig. 455.

Callionymus doryssus, Ochiai et al., Publ. Seto Mar. Biol. Lab., V, 1, 1955: 102 (synonyms).

22848. Tsuruga. September 3, 1917. V. Rozhkovskii. 1 specimen.

22849. Misaki. April, 1901. P.Yu. Shmidt. 5 specimens.

22850. Nagasaki. February 17, 1901. P.Yu. Shmidt. 1 specimen.

D IV, 9; A 9–10; P 18–20 (19.2) (Ochiai et al., 1955).

In our specimens with a standard length of 43–110 mm: D IV, 9; A 9.

Differs from *C. japonicus* in presence of 9 rays in anal fin and skin naked on head, without or with very poorly developed tubercles. In male, first ray of first dorsal fin very long, longer than twice head length, and when fin folded to body almost extends beyond end of base of second fin; second and third rays of first dorsal fin gradually reduce in length. First dorsal fin without black spot. Caudal fin distinctly longer than

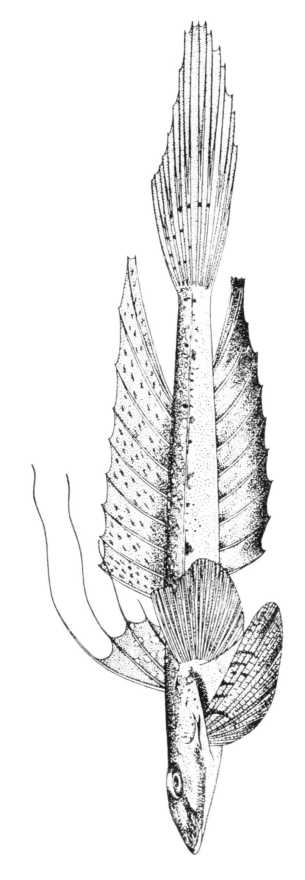

Figure 167. *Callionymus doryssus*. Length 180 mm. Japan (Jordan and Fowler, 1903).

head. In female, rays of first dorsal fin although elongate-filamentous much shorter than head length. Caudal fin moderate, slightly longer than head length. Eyes of young fish fairly large, their diameter longer than snout. Rays of first dorsal fin not elongate in female [sic], with filamentous tips in male.

Length, to 184 mm (Ochiai et al., 1955).

214   Distribution: In the Sea of Japan known from Hakodate (Snyder, 1912: 446); Aomori (Jordan and Fowler, 1903: 947); Toyama Bay (Katayama, 1940: 24); Tsuruga Bay (Shmidt, 1913b: 141); Wakasa Bay (Takewaga and Morino, 1970: 382); and San'in region (Mori, 1956a: 23). In the Yellow Sea found off Cheju-do Island (Uchida and Yabe, 1939: 13). Along the Pacific coast of Japan found from Sangaru Strait to Nagasaki (Matsubara, 1955: 712).

3. *Callionymus calliste* Jordan and Fowler, 1903 (Figures 168 and 169).

Callionymus calliste Jordan and Fowler, Proc. U.S. Nat. Mus., 25, 1903: 954, fig. 8 (Misaki). Shmidt, Tr. Tikhookeansk Kom. Akad. Nauk SSSR, 1931: 143, fig. 26. Ochiai et al., Publ. Seto Mar. Biol. Lab., V, 1, 1955: 104.

22859. Misaki. April 11, 1901. P.Yu. Shmidt. 7 specimens.

D IV, 8; A 7; P 17 (Ochiai et al., 1955). In our 7 specimens with a standard length of 36-62 mm: D IV, 8; A 7; P 17. Differs from closely related species with curved preopercular bony process in smaller number of rays in second dorsal (8) and anal (7) fins versus 9 and 9 respectively, and absence of additional branch in lateral line on dorsal side of caudal peduncle. Female differs significantly from male (Figures 168 and 169).

Length, to 100 mm (Jordan and Fowler, 1903).

Distribution: In the Sea of Japan reported from Otaru (Katoh et al., 1956: 321) and San'in region (Mori, 1956a: 22). Along the Pacific coast of Japan—Misaki (Matsubara, 1955: 712) and Nagasaki (Katoh et al., 1956: 321).

4. *Callionymus planus* Ochiai, 1955 (Figure 170)

Callionymus planus Ochiai et al., Publ. Seto Mar. Biol. Lab., V, 1, 1955: 106 (Aichi Prefecture).

D IV, 9; A 9; P 21; C 10.

Differs from other species with apically curved (not straight) preopercular process in shape and size of same. Process in this species very thick, broad, and with larger number of pointed spines (5-6 versus 3-4) along upper margin (Figure 171). Furthermore, pelvic fins with rather characteristic broad base, insert in front of posterior margin of preopercle. Color of first dorsal fin of male and female also serves as a distinguishing character.

Length, to 135 mm (Ochiai et al., 1955).

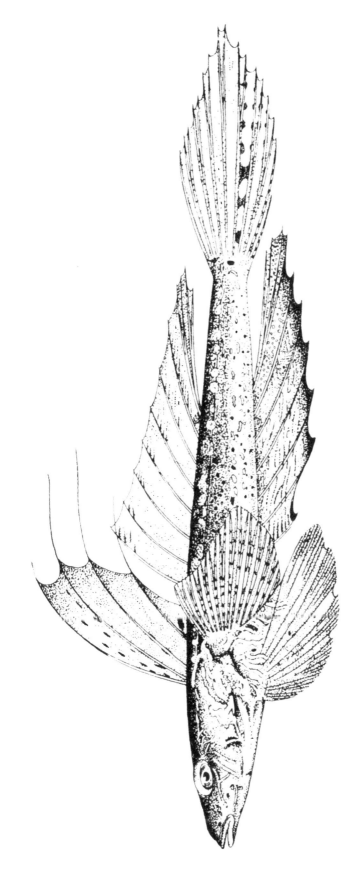

Figure 268. *Callionymus calliste.* Male. Length 85 mm. Japan (Jordan and Fowler, 1903).

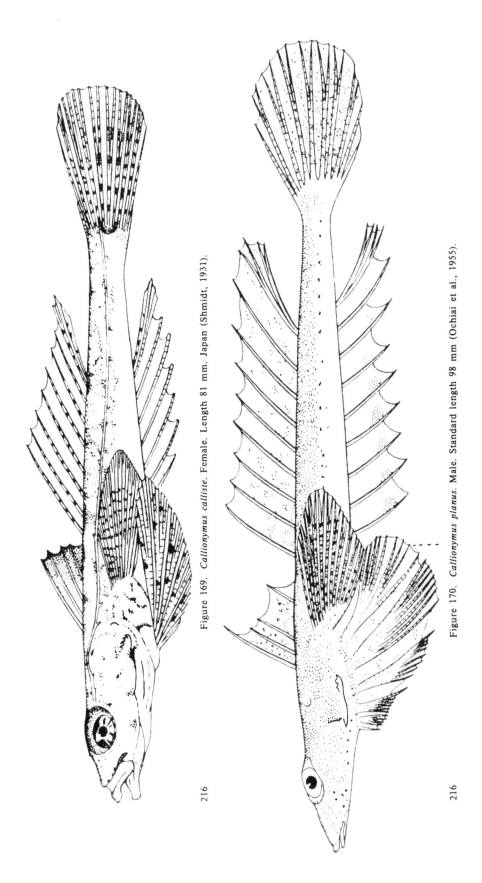

Figure 169. *Callionymus calliste.* Female. Length 81 mm. Japan (Shmidt, 1931).

Figure 170. *Callionymus planus.* Male. Standard length 98 mm (Ochiai et al., 1955).

216

216

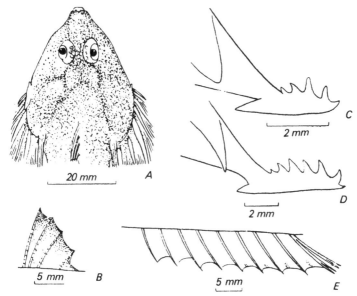

217

Figure 171. *Callionymus planus* (Ochiai et al., 1955).
A—head of male, dorsal view; B—first dorsal fin, female; C—preopercular
process, female; D—same, male; E—anal fin, female.

Distribution: In the Sea of Japan reported from Wakasa Bay (Takegawa and Morino, 1970: 382). Known from the Pacific coast of Honshu in Aichi Prefecture and in Kishu Province (Matsubara, 1955: 712).

5. *Callionymus kaianus* Günther, 1879 (Figure 172)

Callionymus kaianus Günther, Rep. Challenger Exp., Zool., 6 (1879) 1880: 44, pl. 19, fig. 6 (Japan). Weber and Beaufort, in: Beaufort and Chapman, Fish. Indo-Austr., Arch., IX, 1951: 66, fig. 12. Ochiai et al., Publ. Seto Mar. Biol. Lab., V, 1, 1955: 111, figs. 8, 9.

D IV, 9; A 9; P 18-21 (19.5); C 10.

Differs from other species with curved preopercular process in presence of small serration on posteriormost spine (Figure 173, a) and highly enlarged middle two rays of caudal fin.

Length, to 171.5 mm (Ochiai et al., 1955).

Distribution: In the Sea of Japan reported from Tsushima Islands (Arai and Abe, 1970: 91). Pacific coast of Japan in the south up to Indonesia (Matsubara, 1955: 713).

217 6. *Callionymus flagris* Jordan and Fowler, 1903
(Figures 174 and 175)

Callionymus flagris Jordan and Fowler, Proc. U.S. Nat. Mus., 25, 1903: 952, fig. 7 (Iokogama). Ochiai et al., Publ. Seto Mar. Biol. Lab., V,

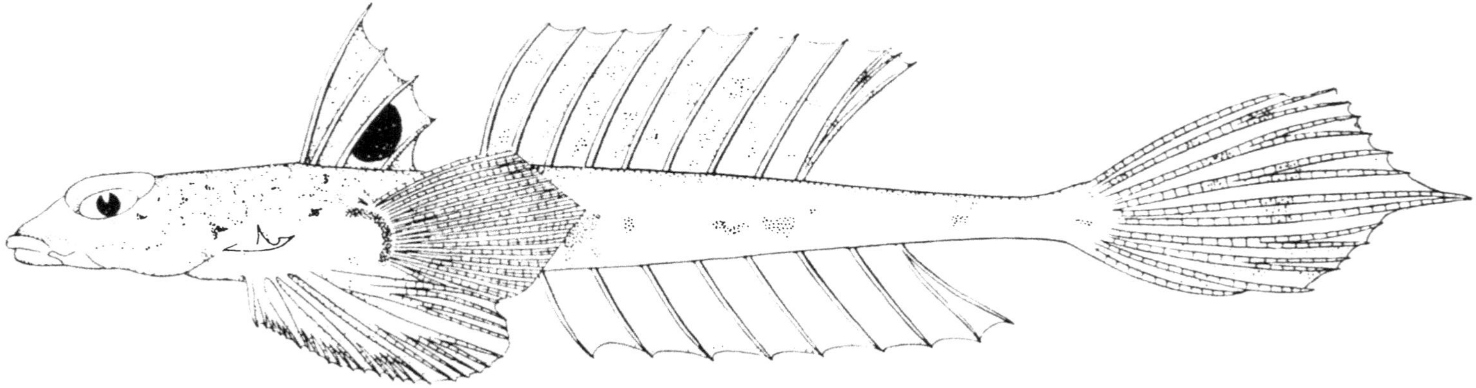

Figure 172. *Callionymus kaianus*. Male. Standard length 91 mm. Japan (Ochiai et al., 1955).

218

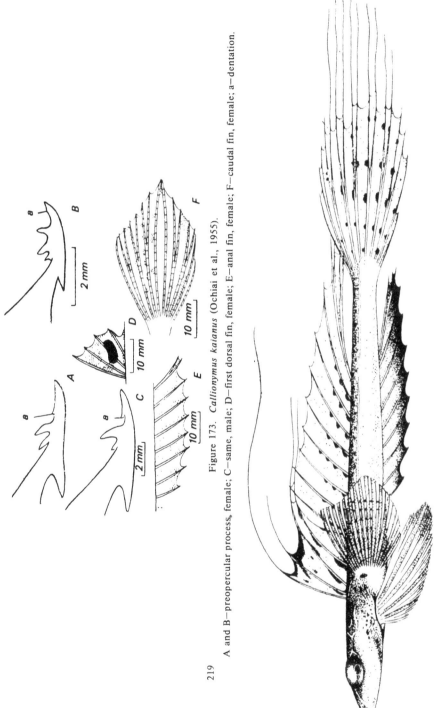

Figure 173. *Callionymus kaianus* (Ochiai et al., 1955).

A and B—preopercular process, female; C—same, male; D—first dorsal fin, female; E—anal fin, female; F—caudal fin, female; a—dentation.

Figure 174. *Callionymus flagris*. Standard length 110 mm. Japan (Jordan and Fowler, 1903).

219

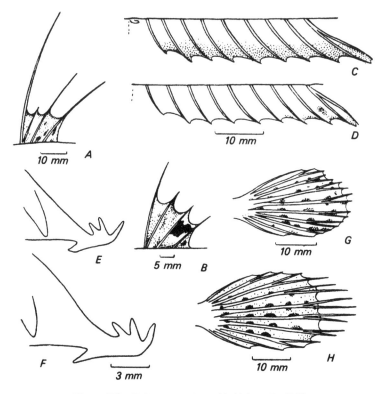

Figure 175. *Callionymus flagris* (Ochiai et al., 1955).
A—dorsal fin, male; B—same, female; C—anal fin, male; D—same, female;
E—preopercular process, female; F—same, male; G—caudal fin, female;
H—same, male.

1, 1955: 114, fig. 11 (synonyms). Abe, Enc. Zool., 2, Fishes, 1958: 153, fig. 453 (color figure). Zhu et al., Ryby Yazhno-Kitaiskogo Morya, 1962: 725, fig. 587.

23120. Obama. N. Grebenitskii. 1 specimen.

D IV, 7-9 (8.8); A 9-10 (9.1) (Ochiai et al., 1955).

In our specimen with a standard length of 102 mm: D IV, 9; A 9.

Males of this species, as in other closely related species, usually with 9 rays each in second dorsal and anal fins, and an additional branch in lateral line on dorsal side of caudal peduncle. This species differs from others, however, in presence of 6 greatly elongate-filamentous middle rays of caudal fin, as well as elongate rays of first dorsal fin (excluding short second ray). It differs from *C. calliste* in larger number of rays in second dorsal (9 versus 8) and anal fins (9 versus 7), as well as presence of an additional branch on dorsal side of caudal peduncle, connecting lateral lines of both sides of body. Females differ from males in elongation of second ray of first dorsal fin and first ray when fin folded to body reaching base of second dorsal fin.

Length, to 190 mm (Jordan and Fowler, 1903).

Distribution: In the Sea of Japan known from Hakodate (Snyder, 1912: 447); Aomori, Tsuruga (Jordan and Fowler, 1903: 952, fig. 7); Sado Island, Toyama Bay (Katoh et al., 1956: 321); and San'in region (Mori, 1956a: 22). South China Sea (Zhu et al., 1962: 725). Not found in the Yellow Sea, but possibly occurs there since listed in the fauna of the South China Sea (Zhu et al., 1962: 725) and in the fauna of the Korean peninsular coast (Chyung, 1961: 2). Found along the Pacific coast of Japan from Aomori to Nagasaki (Jordan and Fowler, 1903: 952).

218 7. *Callionymus punctatus* Richardson, 1846 (Figure 176)

Platycephalus punctatus (only name given).[12]

Callionymus japonicus (non Houttuyn) Cuvier and Valenciennes, Hist. Nat. Poiss., 12, 1837: 299 (specimen of Langsdorf, Japan).

Callionymus punctatus Richardson, Rep. British Assoc. Adv. Sci., for (1845) 1846: 210 (specimen of Langsdorf, according to Cuvier and Valenciennes, 1837). Boeseman, Revision..., 1947: 132, 133 (Nos. 2079b and 1011). Ochiai et al., Publ. Seto Mar. Biol. Lab., V, 1, 1955: 116, fig. 12 (synonyms, bibliography).

Callionymus richardsoni Bleeker, Nat. Tijdschr. Ned. Ind., 6, 1854: 414 (Nagasaki). Jordan and Hubbs, Mem. Carnegie Mus., 10, 2, 1925: 317. Shmidt and Lindberg, Izv. Akad. Nauk SSSR, 1930: 1150. Shmidt, Izv. Akad. Nauk SSSR, 1931: 122; Tr. Tikhookeansk, Kom. Akad. Nauk SSSR, 2, 1931: 143. Abe, Enc. Zool., 2, Fishes, 1958: 153, fig. 451 (color figure).

Callionymus valenciennesi (non Temminck and Schlegel) Jordan and Fowler, Proc. U.S. Nat. Mus., 25, 1903: 950, fig. 6 (= C. *punctatus,* see Boeseman, 1947: 132); and others.

1226. Japan. 1862. V. Schlegel. 1 specimen.

22855. Nagasaki. January–February, 1901. P.Yu. Shmidt. 6 specimens.

22856. Misaki. April, 1901. P.Yu. Shmidt. 5 specimens.

22991. Tsuruga. September, 1917. V. Rozhkovskii. 6 specimens.

22992. Tsuruga. September, 1917. V. Rozhkovskii. 1 specimen.

23118. Nagasaki. December, 1897. A. Bunge. 4 specimens.

23119. Obama. March 18, 1903. V. Rozhkovskii. 1 specimen.

23563. Tsuruga. August, 1917. V. Rozhkovskii. 2 specimens.

23580. Tsuruga. September, 1917. V. Rozhkovskii. 6 specimens.

28313. Nagasaki. March. 1901. V. Rozhkovskii. 6 specimens.

D IV, 8-10 (9.0); A 8-10 (9.0); P 17-22 (19.2); C 10 (Ochiai et al.,

[12]Name given on label of specimen from Japan presented by G.H. Langsdorf to the Berlin Museum.

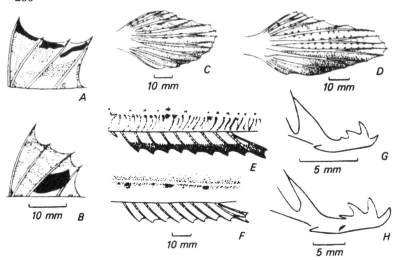

220

Figure 176. *Callionymus punctatus* (Ochiai et al., 1955).

A–dorsal fin, male; B–same, female; C–caudal fin, female; D–same, male;
E–anal fin, male; F–same, female; G–preopercular process, male; H–same,
female.

1955). In our 20 specimens with a standard length of 12 to 174 mm: D IV,
8 (4 specimens), 9 (13), 10 (3); A 8 (2)-9 (18); P 17 (16)-18 (4).

222   Included in group of species with terminally curved preopercular
bony process, additional branch of lateral line on dorsal side of caudal
fin, and higher number of rays in anal fin (not 7, but 8-10). Males of
this species characterized by absence of filamentous rays of caudal fin.
First dorsal fin also without elongate-filamentous rays.

In adult females last rays of dorsal and anal fins, when fin folded to
body, extend to base of caudal fin; in females 31 and 74 mm long these
rays shorter and an important character in their identification is the
characteristic large oval black spot on membrane of dorsal fin between
3rd and 4th rays. Nature of coloration of fins in male and female shown in
Figure 176, A-F.

Length, to 222 mm (Ochiai et al., 1955).

Distribution: In the Sea of Japan known from Pusan (Mori, 1952: 132);
Hakodate (Ochiai et al., 1955: 117); Sado Island (Honma, 1952: 132);
Niigata (Honma and Kitami, 1970: 71); Tsuruga (Shmidt and Lindberg,
1930: 1150); Wakasa Bay (Takegawa and Morino, 1970: 382); and San'in
region (Mori, 1956a: 22). In the Yellow Sea reported from the Gulf of
Chihli (Bohai). Along the Pacific coast of Japan distributed from the
central part of Honshu southward (Matsubara, 1955: 714). South China
Sea (Zhu et al., 1962: 724). Coasts of northern Australia (Johnson, 1971:
115).

8. *Callionymus lunatus* Temminck and Schlegel, 1845
   (Figures 177 and 178)

Callionymus lunatus Temminck and Schlegel, Fauna Japonica, Poiss., 1845: 155, pl. 78, fig. 4 (Nagasaki, male). Ochiai et al., Publ. Seto Mar. Biol. Lab., V, 1, 1955: 120, fig. 14 (synonyms). Abe, Enc. Zool., 2., Fishes, 1958: 152, fig. 450 (color figure).

22854. Kyushu Island, Pacific coast. February, 1901. P.Yu. Shmidt. 3 specimens.

22857. Nagasaki. February, 1901. P.Yu. Shmidt. 1 specimen.

D IV-V, 8-10 (9.0); A 8-9; P 18-21 (19.7); C 10 (Ochiai et al., 1955). In our 4 specimens with a standard length of 100-123 mm: D IV, 9; A 9; P 19; C 10.

223 In males of this species only first ray of first dorsal fin highly elongate and black spot located on membrane of last ray. In females first ray of first dorsal fin only slightly elongate and spot not present in fin (Figure 178, B); diameter of eye almost equal to snout length.

Length, to 210 mm (Abe, 1958: 152).

Distribution: In the Sea of Japan known from Pusan (Mori, 1952: 132); from Hakodate to Nagasaki (Jordan, Tanaka and Snyder, 1913: 374); Sado Island (Honma, 1952: 225); Toyama Bay (Katayama, 1940: 24); Wakasa Bay (Takegawa and Morino, 1970: 382); and San'in region (Mori, 1956a: 22). Along the Pacific coast of Japan found everywhere around Hokkaido and Honshu (Matsubara, 1955: 713).

9. *Callionymus beniteguri* Jordan and Snyder, 1900
   (Figures 179 and 180)

Callionymus beniteguri Jordan and Snyder, Proc. U.S. Nat. Mus., 23, 1900: 370, pl. XVII (Tokyo). Jordan and Fowler, Proc. U.S. Nat. Mus., 25, 1903: 956. Ochiai et al., Publ. Seto Mar. Biol. Lab., V, 1, 1955: 123, figs. 16, 17 (synonyms). Abe, Enc. Zool., 2, Fishes, 1958: 152, fig. 449 (color figure).

D IV, 8-9; A 9; P 18-22 (19.8); C 10 (Ochiai et al., 1955).

In males the first two rays of first dorsal fin elongate. Three middle rays of caudal fin equal in length and distinctly longer than others. Posterior rays of second dorsal and anal fins in adult males and females 224 reach base of caudal fin. Females with black spot in membrane between 3rd and 4th rays and membrane itself dark (Figure 180, A).

Length, to 250 mm (Abe, 1955).

Distribution: In the Sea of Japan known from Otaru, Hakodate, Aomori (Jordan and Fowler, 1903: 956); coasts of Akita Prefecture (Ochiai et al., 1955: 124); Sado Island (Honma, 1952: 226); Maidzuru (Ochiai et al., 1955: 124); Wakasa Bay (Morino, 1970: 382); and San'in region (Mori, 1956a: 22). In the Yellow Sea reported from the Gulf of Chihli (Bohai)

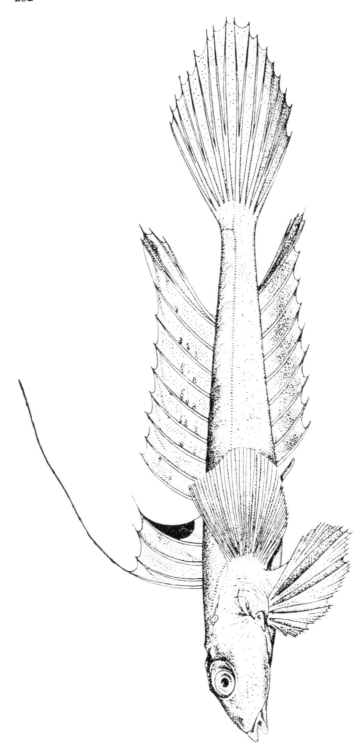

Figure 177. *Callionymus lunatus.* Standard length 155 mm. Japan (Temminck and Schlegel, 1845).

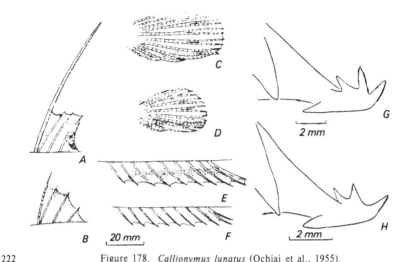

Figure 178. *Callionymus lunatus* (Ochiai et al., 1955).
A—dorsal fin, male; B—same, female; C—caudal fin, male; D—same, female;
E—anal fin, male; F—same, female; G—preopercular process, male; H—same,
female.

(Zhang et al., 1955: 182). Near the Pacific coast of Japan found from Hokkaido to Nagasaki (Matsubara, 1855: 714). Also noted for the coast of China (Zhu et al., 1963: 387).

10. *Callionymus valenciennesi* Temminck and Schlegel, 1845
(Figure 181)

Callionymus valenciennesi Temminck and Schlegel, Fauna Japonica, Poiss., 1845: 153, pl. 78, fig. 3 (Nagasaki). Ochiai et al., Publ. Seto Mar. Biol. Lab., V, 1955: 126.

D IV, 9; A 9 or 8; V I, 5; P 18; C 10 (Temminck and Schlegel, 1845).

Differs from *C. beniteguri* by rounded caudal fin, shorter posterior rays of second dorsal and anal fins, which do not reach base of caudal fin, and presence of blackish spots on first dorsal fin.

Length 155 mm, based on figure (Temminck and Schlegel, 1845).

Distribution: In the Sea of Japan probably absent, since all earlier records according to Ochiai et al. (1955: 114, 116) and our data, belong either to *C. flagris* or *C. punctatus*. However, 15 years after the work of Ochiai, this species was reported from Wakasa Bay (Takegawa and Morino, 1970: 382).

11. *Callionymus virgis* Jordan and Fowler, 1903 (Figures 182 and 183)

Callionymus virgis Jordan and Fowler, Proc. U.S. Nat. Mus., 25, 1903: 957, fig. 9. Ochiai et al., Publ. Seto Mar. Biol. Lab., V, 1, 1955: 126, figs. 18, 19. Abe, Enc. Zool., 2, Fishes, 1958: 153, fig. 452 (color figure).

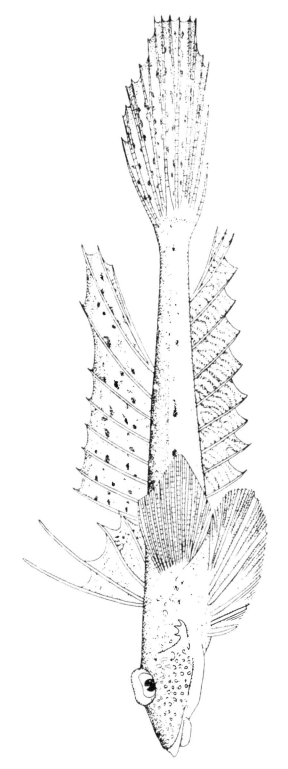

Figure 179. *Callionymus benitcguri.* Standard length 120 mm. Japan (Ochiai et al., 1955).

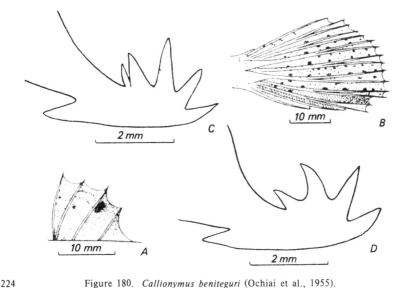

Figure 180. *Callionymus beniteguri* (Ochiai et al., 1955).
A—dorsal fin, female; B—caudal fin, female; C—preopercular process, female;
D—same, male.

36448. Yellow Sea, Tsingtao. June 7, 1957. E.F. Gur'yanova. 1 specimen.

227   D IV, 8-10 (9); A 8-10 (9); P 17-21 (19); C 10 (Ochiai et al., 1955). In our specimens: D IV, 10; A 10.

A fairly characteristic feature of the male in this species is a very high membrane between the elongate-filamentous rays of the first dorsal fin; its height almost twice exceeds height of membrane between rays of second dorsal fin. Diameter of eye less than snout length.

Length 110 mm (Abe, 1955).

Distribution: In the Sea of Japan reported from Wakasa Bay and the coastal prefectures Kyoto and Hyogo (Ochiai, 1955: 127) and San'in region (Mori, 1956a: 22). Along the Pacific coast of Japan known from Misaki and Koti (Matsubara, 1955: 714). Reported from the coast of China in the East China Sea (Zhu et al., 1963: 386). Our specimens from Tsingtao.

12. [*Callionymus kitaharai* Jordan and Seale, 1906] (Figure 184)

Callionymus kitaharai Jordan and Seale, Proc. U.S. Nat. Mus., 30, 1906: 148, fig. 6 (Nagasaki). Ochiai et al., Publ. Seto Mar. Biol. Lab., V, 1, 1955: 129. Zhang et al., Ryby Zaliva Bokhai..., 1955: 181.

36447. Yellow Sea, Tsingtao. June 23, 1957. P.V. Ushakov. 7 specimens.

D IV, 9; A 9 (Jordan and Seale, 1906).

Length of head 3.55 times and depth 9 times in standard length; eyes

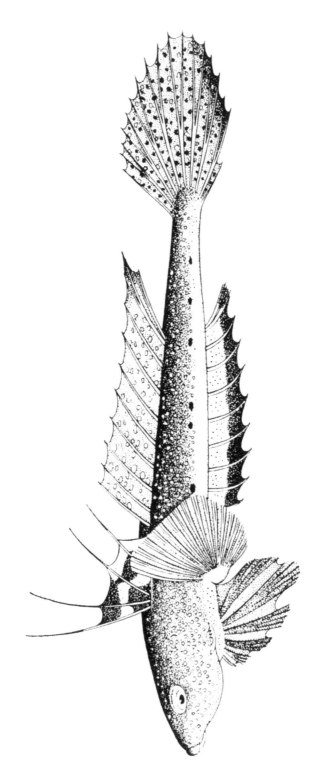

Figure 181. *Callionymus valenciennesi.* Standard length 120 mm. Japan (Temminck and Schlegel, 1845).

225

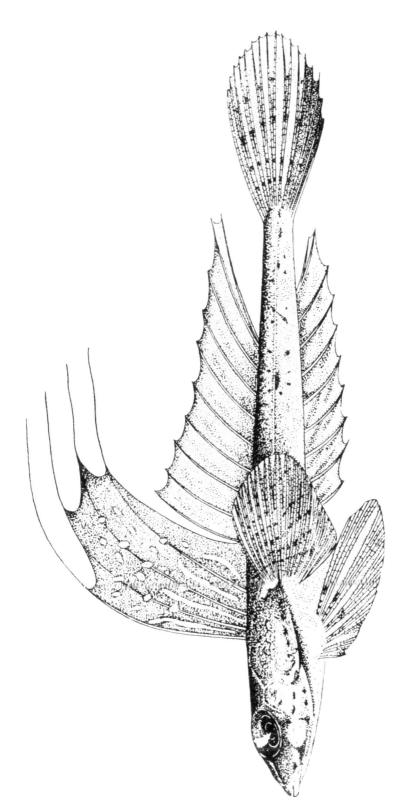

Figure 182. *Callionymus virgis*. Standard length 55 mm. Japan (Jordan and Fowler, 1903).

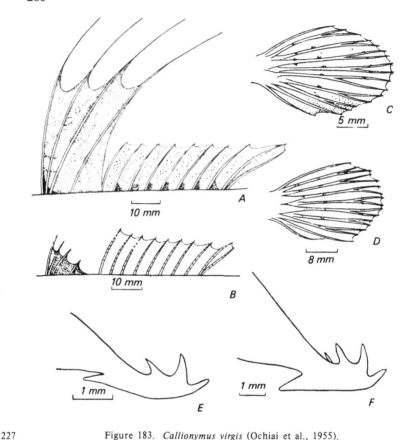

227

Figure 183. *Callionymus virgis* (Ochiai et al., 1955).
A–dorsal fin, male; B–same, female; C–caudal fin, male; D–same, female;
E–preopercular process, female; F–same, male.

2.5 times and snout 3 times in head length. Preopercular bony process with 4 bent spines on upper margin and one spine directed forward on lower margin. Distance between tips of preopercular processes of opposite 228 sides equal to head length. Depth of head equal to diameter of orbit. Head pointed toward front. Mouth small, with small teeth. Gill openings in form of small opening near upper margin of opercular bone. First dorsal fin low; first ray longest, barely exceeding diameter of eye; fin triangular; subsequent rays reduce gradually. Length of pectoral fin 1.4 times, pelvic fins 1.5 times, and caudal fin 1.1 times in head length (Jordan and Seale, 1906: 148).

Described on the basis of 1 specimen about 40 mm long, caught in Nagasaki Harbor.

Our collection from Tsingtao included 7 small specimens (L 20-25 mm), which almost completely correspond to the description of *C.*

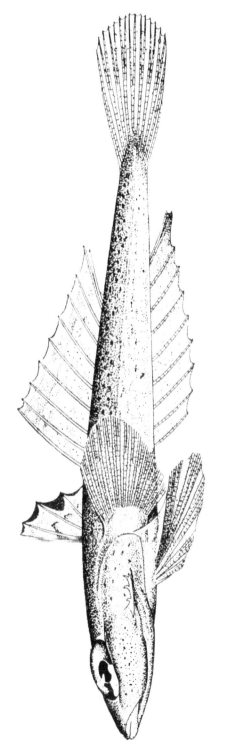

Figure 184. *Callionymus kitaharai*. Standard length 40 mm. Japan (Jordan and Seale, 1906).

kitaharai: D IV, 9 (5 specimens)–10 (2 specimens); A 9 (7 specimens). The teeth on the bony process of the preopercle in our specimens fully correspond to the characteristic shape and arrangement in the type specimen of this species.

Length, to 82 mm (Zhang et al., 1955).

Distribution: Not found in the Sea of Japan. In the Yellow Sea known from off the coast of Liaoning, Hebei, and Shantung Provinces (Zhang et al., 1955: 182). Reported from the western coast of the Korean Peninsula (Mori, 1952: 132). Our specimens from Tsingtao. For the Pacific coast of Japan reported from Nagasaki (Matsubara, 1955: 714), from where it was first described.

3. Genus *Synchiropus* Gill, 1859—Mandarin Dragonet

Synchiropus Gill, Proc. Acad. Nat. Sci., Philad., 11 (1859) 1860: 129 (type: *Callionymus lateralis* Richardson, design. by Jordan, 1919: 290). Fowler, Synopsis..., 12, 1959: 92. Beaufort and Chapman, Fish. Indo-Austr. Arch., IX, 1951: 70. Matsubara, Fish Morphol. and Hierar., 1955: 711.

Head and body cylindrical but slightly flat toward front and compressed in caudal part. Preopercular bony process without spine directed toward tip of snout on lower margin. Opercular dermal valve formed by overgrown branchiostegal membrane not present. First dorsal fin with 4 slender spines; second dorsal fin with 8–9 rays, all rays except first branched; anal fin with 7–8 rays, of which only last ray branched, others unbranched. Last ray of pelvic fin attached by membrane with base of 229 pectoral fin. Caudal fin cut truncate or barely rounded (Fowler, 1959: 92).

About ten species in the tropical and subtropical parts of the Indian and Pacific oceans. 5 species off the coast of Japan, of which 2 found in the Sea of Japan.

Key to Species of Genus Synchiropus

1 (2). Pectoral fins with 18 rays. Rays of spiny dorsal fin almost equal in length, and membrane between rays deeply notched. Second dorsal fin with 5 broad oblique dark stripes. Bases of dorsal fins touch each other............1. **S. ijimai** Jordan and Thompson.

2 (1). Pectoral fins with 19–21 rays. Rays of spiny fin not equal in length, first ray very high, and subsequent rays reduce gradually; membrane between rays without deep notches. Second dorsal fin very high with 6 narrow stripes brick-red in color. Bases of dorsal fins wide-set. .... 2. **S. altivelis** (Temminck and Schlegel).

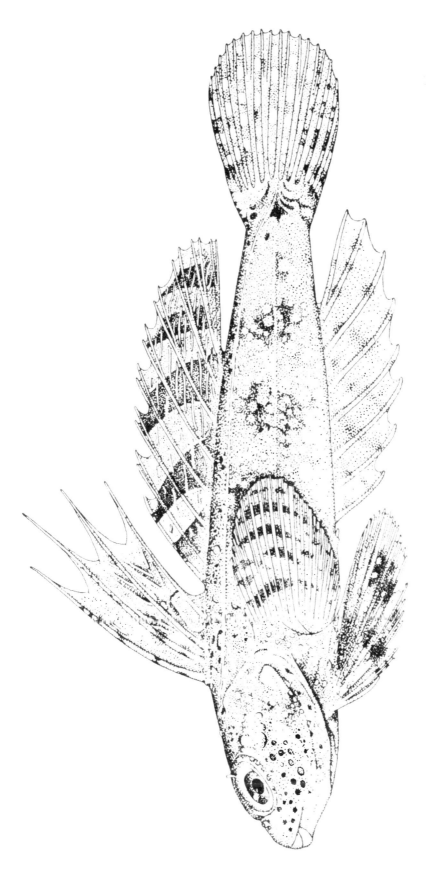

Figure 18&. *Synchiropus ijimai*. Standard length 65 mm. Japan (Jordan and Thompson, 1914).

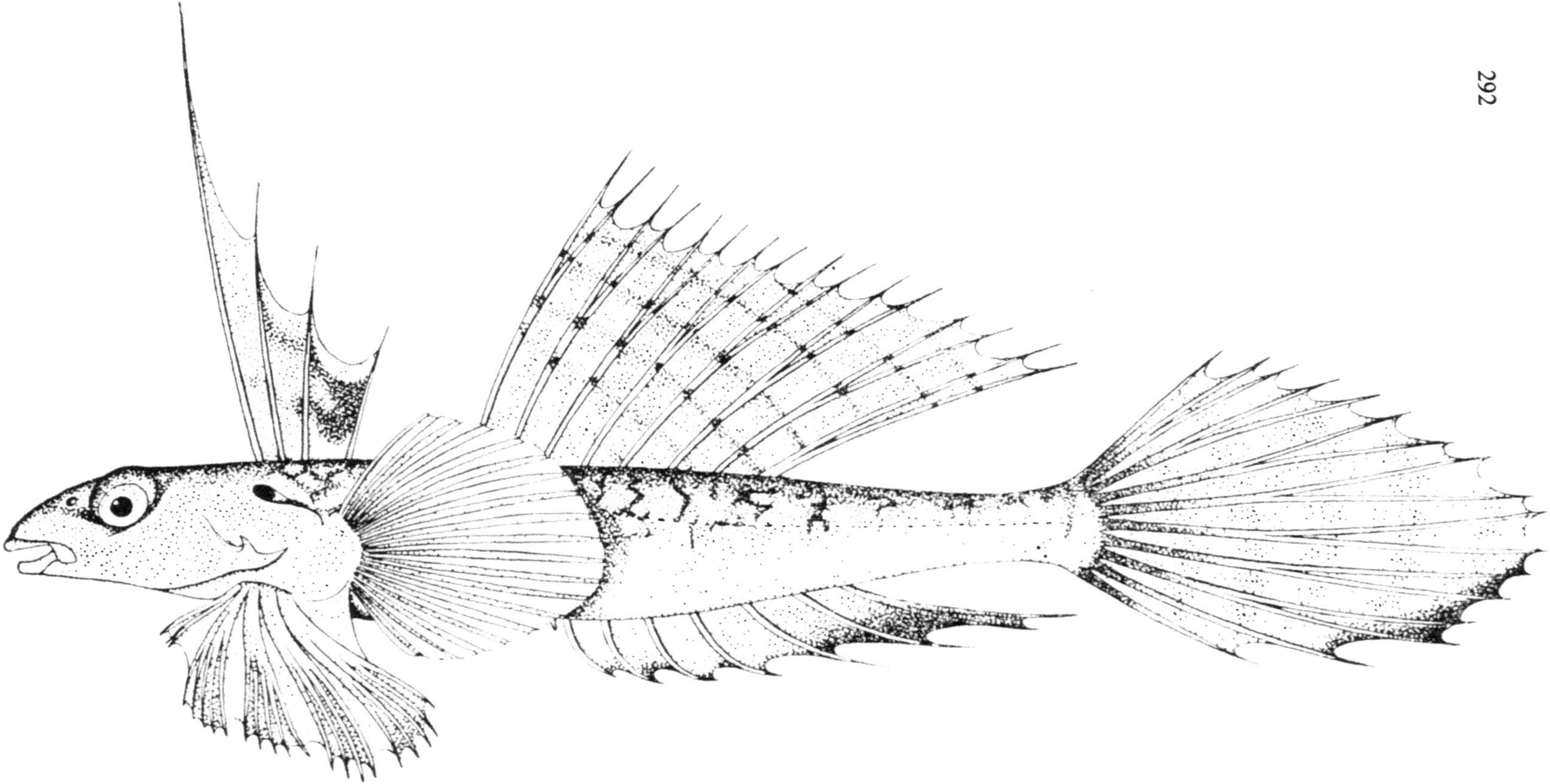

Figure 186. *Synchiropus altivelis*. Standard length 150 mm. Japan (Temminck and Schlegel, 1845).

1. **Synchiropus ijimai** Jordan and Thompson, 1914 (Figure 185)

Synchiropus ijimae Jordan and Thompson, Mem. Carnegie Mus., 6, 4, 1914: 295, pl. 36, fig. 1 (Misaki).

Synchiropus ijimai, Matsubara, Fish Morphol. and Hierar., 1955: 711. D IV, 8; A 7; P 18; V I, 5; *l. l.* 18-20.

Distinguishing characters of this species are: close-set bases of dorsal fins, deep notches in membrane between rays of first dorsal fin, length of rays in this fin almost equal, and broad oblique dark strips on second dorsal fin.

Length 65 mm (Jordan and Thompson, 1914).

Distribution: In the Sea of Japan reported from the coast of Hokkaido Island (Ueno, 1971: 83). Described from Misaki.

2. **Synchiropus altivelis** (Temminck and Schlegel, 1845) (Figure 186)

Callionymus altivelis Temminck and Schlegel, Fauna Japonica, Poiss., 1845: 155, pl. 79, fig. 1 (Nagasaki). Jordan and Fowler, Proc. U.S. Nat. Mus., 25, 1903: 948.

Synchiropus altivelis, Matsubara, Fish Morphol. and Hierar., 1955: 716. Abe, Enc. Zool., 2, Fishes, 1958: 152, fig. 448 (color figure).

22903. Nagasaki. January, 1901. P. Shmidt. 2 specimens.

D IV, 8; A 7; P 19; V I, 5 (Temminck and Schlegel, 1845).

D IV, 8; A 7; P 20-21; V I, 5 (Abe, 1958).

D IV, 8; A 7; P 20-21; V I, 5 (No. 22903).

The characters of this species distinguishing it from the other 4 Japanese species are: dorsal fins wide-set; very high first ray of first dorsal fin; height of subsequent rays reduces gradually; second dorsal fin high, all its rays almost equal in height to first ray of first dorsal fin. It differs from *S. lineolatus* (Cuvier and Valenciennes) in larger number of rays (19-21 versus 15) in the pectoral fin. Using the number of rays in the Japanese species as a character for differentiation, as done by Matsubara (1955: 715), can hardly be considered appropriate. In our specimens of *S. altivelis* (No. 22903) the pectoral fins had 20-21 rays; the same number has also been reported by Abe (1958: 152), although 19 rays
232 were reported for the type specimen and drawn in the figure. These differences in number of rays in the pectoral fins notwithstanding, it is clear that our specimens conform to the drawing given by Abe for this species. Another good distinguishing peculiarity is the red color of live fish. Biology of the species described by Akazaki (1957).

Length, to 240 mm (Abe, 1958).

Distribution: For the Sea of Japan reported from Pusan (Mori, 1952: 133) and San'in region (Mori, 1956a: 23). Along the Pacific coast from the central part of Honshu southward (Matsubara, 1955: 716). Hainan Island (Zhu et al., 1962: 729).

6. Suborder Siganoidei

232 Pelvic fins with two spines, inner and outer ones, with 3 soft rays between them. Prepalatine bone present, attached to maxilla in front of palatine. Nasals touch and firmly connect with mesethmoid. Anterior margin of mesethmoid ahead of vomer; mesethmoid entirely ahead of lateral ethmoids; middle plate originating from it extends backward to form internasal septum (similar to many members of Physostomi). Pelvic bones unique. Anal fin with 7–9 spines. Lower end of postcleithrum connected through strong fibrous ligament with anterior end of first radial of anal fin (Berg, 1940: 319).

1 family. Warm waters of the Indian and Pacific oceans.

CLIX. Family SIGANIDAE—Rabbitfishes

Body oblong, compressed laterally, covered with very minute, slightly elongate, thin cycloid scales. Sides of head more or less covered with scales. Lateral line simple and complete. Each side of snout with two nostrils. Suborbital stay absent. Mouth small, terminal, nonprotractile. Premaxilla without ascending process; jaws with one row of minute incisor-shaped teeth, compressed from sides, with one or more additional small cusps. Dorsal fin with 13 strong spines, slightly flat toward front; tips of rays when fin folded against body, directed alternately to both sides. A spine situated ahead of fin projects forward through the skin of the nape. Anal fin with 7 spines and 9 branched rays; soft part of fin equal in spread and shape to soft part of dorsal fin. Each pelvic fin with two spines, one on outer side, and the other on inner side; latter connected through membrane with skin of abdomen; 3 branched rays occur between spiny ones. Pelvic fins attached behind base of pectoral fins. Caudal fin truncate, with notch, or forked. Vertebrae 23. Pseudobranch well developed (Beaufort and Chapman, 1951: 95).

2 genera in Indian and Pacific oceans and eastern Mediterranean Sea. One genus recorded off the coast of Japan and also known from the Sea of Japan.

1. Genus *Siganus* Forskål, 1775—Rabbitfishes

Siganus Forskål, Descript. Animal., 10, 1775: 25 (type: *Scarus rivulatus* Forskål). Beaufort and Chapman, Fish. Indo-Austr. Arch., IX, 1951: 95 (synonyms).

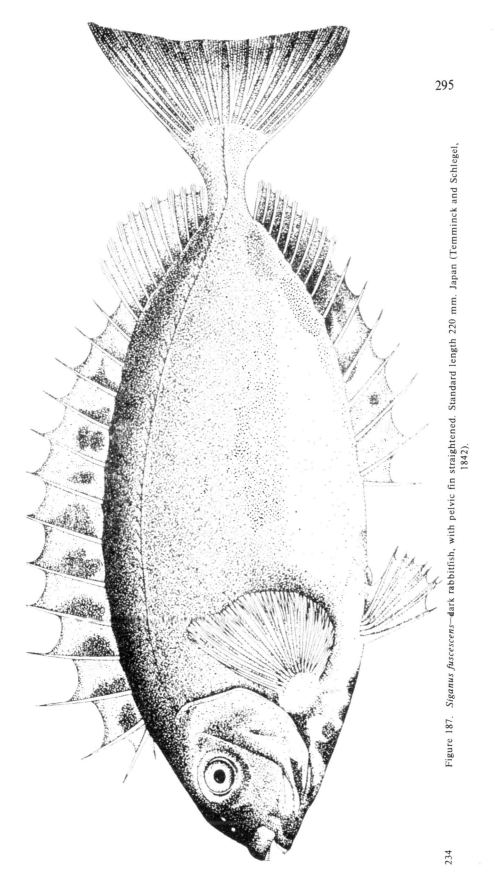

Figure 187. *Siganus fuscescens*—dark rabbitfish, with pelvic fin straightened. Standard length 220 mm. Japan (Temminck and Schlegel, 1842).

Characteristics of the genus given in description of the family. A closely related genus of this family, *Lo* Seale, 1906,[1] differs in an elongate tubular snout, which is normal in the genus *Siganus,* i.e., without distinct elongation.

233 Eastern part of the Mediterranean Sea (Norman, 1957: 378), Indian and Pacific oceans. Many species, of which 14 recorded off the coasts of Japan, of which 1 known in the Sea of Japan.

1. **Siganus fuscescens** (Houttuyn, 1782)—Dark Rabbitfish (Figure 187)

Centrogaster fuscescens Houttuyn, Verh. Holl. Maatsch. Weetensch., 20, 2, 1782: 333 (Japan).

Amphacanthum fuscescens, Cuvier and Valenciennes, Hist. Nat. Poiss., 10, 1835: 115 (156). Temminck and Schlegel, Fauna Japonica, Poiss., 1842: 127, pl. 68, fig. 1.

Teuthis fuscescens, Günther, Cat. Fishes British Mus., 3, 1861: 321.

Siganus fuscescens, Jordan and Fowler, Proc. U.S. Nat. Mus., 25, 1902: 560. Beaufort and Chapman, Fish. Indo-Austr. Arch., IX, 1951: 110. Abe, Enc. Zool., 2, Fishes, 1958: 122, fig. 358.

22950. Tsuruga. August 30–September 8, 1917. Rozhkovskii. 2 specimens, juv.

D XIII, 10; A VII, 9; P 17; V I, 3 I (Beaufort and Chapman, 1951).

In our specimens 44 and 86 mm long the pores of the lateral line are well developed, numbering 44, including the pores on the caudal peduncle. About 30 scales occur between the lateral line and the bases of the middle rays in the spiny part of the fin. Anterior nostril without valve.

S. fuscescens differs from other species with the last spine of the dorsal fin equal to or shorter than the first ray of this fin,[2] in fewer scales (25–30 versus 18–23 in transverse rows between lateral line and middle rays of dorsal fin), and absence of valve in anterior nostril. Color of our specimens uniformly dark chocolate-brown.

Length, to 350 mm.

Distribution: In the Sea of Japan known from off Sado Island (Honma, 1963: 22); Toyama Bay (Katayama, 1940: 15); near Tsuruga (Shmidt and Lindberg, 1930: 1144); off coasts of San'in region (Mori, 1956a: 29); and off Tsushima Islands (Arai and Abe, 1970: 95). In the Yellow Sea found in the Gulf of Chihli (Bohai) (Zhang et al., 1955: 185). Along the Pacific coast from Tokyo southward (Matsubara, 1955: 966) to Australia and the Indian Ocean (Beaufort and Chapman, 1951: 113).

[1]This genus is considered a synonym of *Siganus-Teuthis* (Tyler, 1970: 121).

[2]In Figure 358 in Tomiyama and Abe's work (1958) the last spine of the dorsal fin is distinctly longer than the first and its color highly variegated. In all probability this is a different species.

7. Suborder Acanthuroidei– Surgeonfishes

233 Posttemporal connected with skull through suture. Parasphenoid separates mesethmoid from vomer. Mesethmoid entirely in front of lateral ethmoids. Anal fin with 2–3 spines. Pelvic fins with 1 spine and 2–5 branched rays (Berg, 1940: 320).

Pelvic fins distinctly more than half length of pectoral fins and inserted partly under base of pectoral fins. Spine or lanceolate process present along each side of caudal peduncle, or scales covering body very small, almost not visible to naked eye, but impart notable roughness to skin and give the impression of short bristles.

2 families. Both from warm seas, especially among coral reefs.

Key to Families of Suborder Acanthuroidei

1 (2). Prickly spines, lanceolate processes, tubercles, or plates present
235 along sides of caudal peduncle. Teeth incisor-shaped and arranged in single row. Scales small, bristle-like. Spines of dorsal fin without filamentous apices.......................... CLX. **Acanthuridae**.

2 (1). Prickly spines, lanceolate processes, tubercles, or plates absent along sides of caudal peduncle. Teeth bristle-like, elongate, and protrude notably. Scales very small, bristle-like. Some spines of dorsal fin highly elongate with filamentous tips..... [**Zanclidae**].

CLX. Family ACANTHURIDAE–Surgeonfishes

Body oblong, compressed laterally, often fairly deep. Caudal peduncle with one or more spines or bony plates. Eyes moderate in size, set high. Mouth small, with small jaws, usually located below median line of body. Premaxilla slightly protractile, but not very movable. Jaws with one row of incisor-shaped teeth. Palatines without teeth. Each side of head with 2 nostrils. Gill rakers poorly developed. Pseudobranch large. Swim bladder large, posteriorly bifurcate. Intestine long. Vertebrae 21–23 (8–9 + 12–14). Scales very small, firmly attached to skin. Lateral line continuous and reaches base of caudal fin. One dorsal fin with strong spines; soft part of fin larger than spiny part. Anal fin similar to dorsal fin, but with fewer spines. Pectoral fins moderate in size (Fowler and Bean, 1929: 199).

298

Herbivorous tropical fishes, often brightly colored. Very dangerous, since spines on caudal fin inflict painful wounds. More than 10 genera[1] in tropical and subtropical waters. 8 genera off coast of Japan, of which 3 represented in the Sea of Japan.

Key to Genera of Family Acanthuridae[2]

1 (2). Caudal peduncle with very sharp lanceolate spine concealed in deep pit (Figure 188), but erectile and extrusile; when spine retracted in pit, detected with difficulty. Teeth fixed, broad, flat, lobate, each with 6 to 10 cusps.[3] Pelvic fins with 5 branched rays. Dorsal fin with 6-9 spines. Scales with spinules at posterior end directed not upward, but sideways....... 1. **Acanthurus** Forskål.

2 (1). Caudal peduncle with bony bucklers, with fixed tubercles with crestate process or pointed spine in center. Young fish without such bucklers.

3 (4). One or two bony bucklers on each side of caudal peduncle. Anal fin with 2 spines; pelvic fins with 3 soft rays. Adult fish with horn on forehead protruding forward.............. 2. **Naso** Lacépède.

4 (3). Three or more bony bucklers on each side of caudal peduncle. Anal fins with 3 spines; pelvic fins with 5 soft rays. Adult fish without horn on forehead protruding forward. ....
.... 3. **Prionurus** Lacépède.

1. Genus *Acanthurus* Forskål, 1775—Surgeonfishes

Acanthurus Forskål, Descript. Animal., 10, 1775: 59 (type: *Teuthis hepatus* Linné). Beaufort and Chapman, Fish. Indo-Austr. Arch., IX, 1951: 133. Tyler, Proc. Acad. Nat. Sci., 122, 2, 1970: 88.

Hepatus Gronow, Zoophylac., 1763: 113 (not binomial) (type: *Teuthis hepatus* Linné). Fowler and Bean, Bull. U.S. Nat. Mus., 100 (8), 1929: 207 (synonyms).

Body oblong, oval or moderately oval, never very deep. Teeth fixed, strong, lobate, each tooth with 6 to 10 cusps. Spines in dorsal fin 6-9. Soft dorsal and anal fins short. Caudal fin usual, with small or distinct notch. Pointed lanceolate spine on caudal peduncle concealed in pit but extrusile (Fowler and Bean, 1929).

Differs from closely related genera in extrusile lanceolate spine on

---

[1]Tyler (1970: 87) considers only six genera valid and presents their osteological characters in his work.

[2]From Fowler and Bean, 1929: 200, with additions.

[3]Teeth of genus *Ctenochaetus* movable, elongate, narrow, and serrate only on one side (Figure 189, B). Scales of the genera *Zebrasoma* and *Paracanthurus* with minute erect needle-like spines.

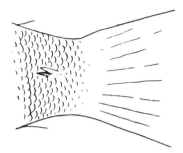

235   Figure 188. Shape of spines on caudal peduncle in the genus *Acanthurus*.
No. 23350. Japan.

each side of caudal peduncle, lobate teeth on jaws not movable and without petiolate base as in the genus *Ctenocheatus* Gill, 1885 (Figure 189, B); pelvic fins with 5 soft rays and not 3 as in the genus *Paracanthus* Bleeker, 1863; dorsal fin with 6–9 spines, and not 3–5; dorsal and anal fins low, and not high; caudal fin crescent-shaped, and not truncate as in the genus *Zebrasoma* Swainson, 1839.

Many species, mostly inhabiting coral reefs. About 15 species off the coast of Japan, of which only one species known from the Sea of Japan.

1. ***Acanthurus triostegus*** (Linné, 1758)–Banded Surgeon (Figure 190)
   Chaetodon triostegus Linné, Syst. Nat., ed. 10, 1758: 274 (India).
   Acanthurus triostegus, Cuvier and Valenciennes, Hist. Nat. Poiss., 10, 1835: 197. Beaufort and Chapman, Fish. Indo-Austr. Arch., IX, 1951: 144 (synonyms). Abe, Enc. Zool., 2, Fishes, 1958, fig. 373 (color figure).

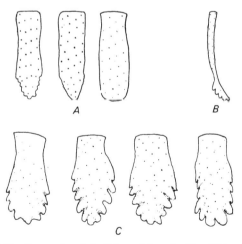

236   Figure 189. Shape of teeth in fishes of Acanthuridae (Tyler, 1970).
A–*Naso*; B–*Ctenochaetus*; C–other genera.

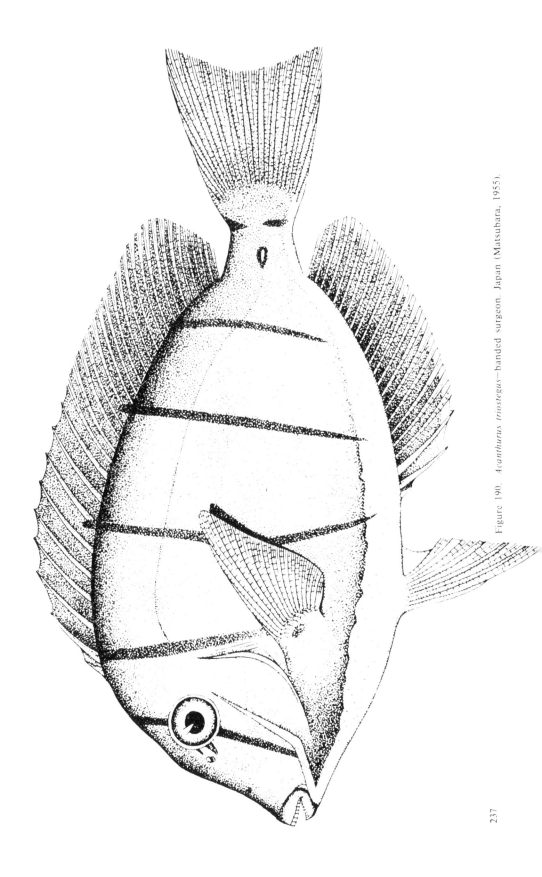

Figure 190. *Acanthurus triostegus*—banded surgeon. Japan (Matsubara, 1955).

237

Hepatus triostegus, Fowler and Bean, Bull. U.S. Nat. Mus., 100 (8), 1929: 249 (synonyms).

23350. Yaeyama, Ryukyu. February–March. P. Shmidt. 2 specimens. D IX, 22; A III, 19; P II, 13; *l. l.* 140–155 (Fowler and Bean, 1929: 251).

Differs from other species in pale margin around mouth of lower jaw; smaller number of soft rays in fins [D 22 (23) versus 24–29 and A 19–20 versus 23–27]; typical coloration in form of 5 narrow black vertical stripes, the first of which passes through the eye; and caudal peduncle with dark spot on upper and lower margins.

Length, to 210 mm (Abe, 1958).

Distribution: In the Sea of Japan reported from Sado Island (Honma, 1952: 221) and Toyama Bay and San'in region (Katoh et al., 1956: 325). Pacific coast of Japan from Tiba Prefecture and southward to Ryukyu, Hawaiian Islands, Taiwan (China), the Philippines, Melanesia, Polynesia, Australia, and South Africa (Matsubara, 1955: 953).

2. Genus *Naso* Lacépède, 1802—Unicorn Fishes

Naso Lacépède, Hist. Nat. Poiss., 3, 1802: 104 (type *N. fronticornis* Lacépède = *Chaetodon unicornis* Forskål). Fowler and Bean, Bull. U.S. Nat. Mus., 100 (8) 1929: 263 (synonyms). Tyler, Proc. Acad. Nat. Sci., 122, 2, 1970: 87.

Body oblong, compressed laterally. Caudal peduncle with one or two large rigid, bony, crestate plates in adult fish, none in young fish [in a fish 50 mm long (No. 9547) these plates were present]. Head of adult fish with horn protruding forward above eyes, which elongates with age and is absent in young fish. Dorsal fin with 6–7 spines and anal fin with 2; small anterior spiny ray absent in contrast to other genera of this family. Pelvic fins with 1 spine and only 3 soft rays (Fowler and Bean, 1929).

Two more genera are known from the Pacific coast of Japan, which have 2 fixed bony plates on each side of the caudal peduncle. In the genus *Cyphomycter* Fowler and Bean, 1929 adult fish have a swelling, compressed laterally, which resembles a crest on the upper side of the snout instead of a horn. In fishes of the genus *Callicanthus* Swainson, 1839 neither young nor adult fish have either a horn or crest. These genera are considered subgenera of the genus *Naso,* but Matsubara treats them as independent genera.

Indian and Pacific oceans. Many species. 2 species of the subgenus *Naso* known from the coasts of Japan, 1 of which is reported from the Sea of Japan.

1. *Naso unicornis* (Forskål, 1775)—Single-Horned Unicornfish
(Figure 191)

Chaetodon unicornis Forskål, Descript. Animal., 10, 1775: 63 (Jidda, Red Sea).

Naso unicornis, Fowler and Bean, Bull. U.S. Nat. Mus., 100 (8) 1929: 264: fig. 16 (synonyms). Beaufort and Chapman, Fish. Indo-Austr. Arch., IX, 1951: 173 (synonyms). Matsubara, Fish Morphol. and Hierar., 1955: 956. Abe. Enc. Zool., 2, Fishes, 1958: 122, fig. 360 (color figure).

1195. Japan. 1862. Schlegel. 1 specimen.

9547. Tokyo. 1891. Bunge. 1 juv.

D VI, 30; A II, 28; P II, 15; V I, 3.

Differs from the other Japanese species of unicorn fish, *N. brevirostris,* in a longer snout (Figure 192, A) that protrudes much beyond the horn in younger fish; only in adult fish does the horn protrude forward in line with the snout tip. The horn in *N. brevirostris* even during early stages protrudes forward the same distance as the snout, and subsequently protrudes considerably beyond the snout tip (Figure 192, B).

In adult specimens filamentous elongation of the marginal rays of the caudal fin has been reported. In our specimen with a standard length of 240 mm, such elongate rays were absent.

Length, to 700 mm (Abe, 1958: 122).

Distribution: Reported from the coast of Yamaguti Prefecture (Yoshida and Ito, 1957: 266), which on the northside is washed by the waters of 240 the Sea of Japan. Matsubara (1955: 956) indicated distribution of this species from the central part of Honshu southward, considering both coasts—Pacific and Sea of Japan. Reported from the Korean Strait and near Thonen in the Browton passage (Mori, 1952: 134). In the south found to Indonesia and the Indian Ocean to S. Africa (Matsubara, 1955: 956). Islands of the Pacific Ocean north to the Hawaiian Islands (Beaufort and Chapman, 1951: 175).

3. Genus *Prionurus* Lacépède, 1804—Sawtails

Prionurus Lacépède, Ann. Mus. Hist. Nat. Paris, 4, 1804: (205) 211 (type: *P. microlepidotus* Lacépède). Fowler and Bean, Bull. U.S. Nat. Mus., 100 (8) 1929: 287. Tyler, Proc. Acad. Nat. Sci., Philad., 122, 2, 1970: 87.

Xesurus Jordan and Evermann, Rep. U.S. Fish Comm., 21, 1895 (1896): 421 (type: *Prionurus punctatus* Gill).

This genus is characterized by a larger number of bony bucklers on the caudal peduncle, not 2 but 3–6 on each side. In other respects this genus is very close to *Naso* except that adult fish do not have a horn protruding forward on the forehead, anal fin with 3 spines versus 2, and pelvic fin with 5 soft rays versus 3.

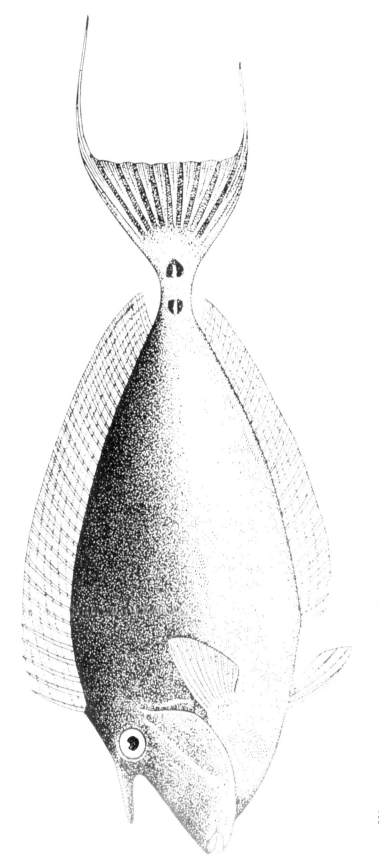

Figure 191. *Naso unicornis*—single-horned unicornfish. Japan (Abe, 1958).

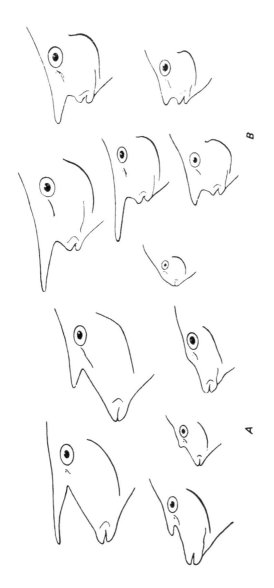

Figure 192. Change in shape of horn on snout (Fowler and Bean, 1929).
A—*Naso unicornis*; B—*Naso brevirostris*.

240

A few species known from the Pacific Ocean near the coast of America, and one species from the coast of Taiwan (China), Ryukyu, and Japan. Also found in the Sea of Japan.

1. *Prionurus microlepidotus* Lacépède, 1804—Small-Scaled Sawtail (Figure 193).

Prionurus microlepidotus Lacépède, Ann. Mus. Hist. Nat. Paris, 4, 1804: (205) 211 (without indication of locality). Fowler and Bean, Bull. U.S. Nat. Mus., 100 (8) 1929: 287. Abe. Enc. Zool., 2, Fishes, 1958: 122, fig. 539 (color figure).

Prionurus scalprum Cuvier and Valenciennes, Hist. Nat. Poiss., 10, 1835: 298 (Japan). Temminck and Schlegel, Fauna Japonica, Poiss., 1845: 129, pl. 70.

Xesurus scalprum Jordan and Fowler, Proc. U.S. Nat. Mus., 25, 1902: 556.

22628. Nagasaki. January 9–10, 1901. P. Shmidt. 3 specimens.

242   D IX, 22–23; A III, 22–23; P 17; V I, 5; bucklers on caudal peduncle 4; *l. l.* 33–35. In small specimens (147 mm) lateral line discernible but less distinct than in adults (265 and 285 mm), especially in posterior part. In large fish, course of lateral line distinct in form of dark spots the same color as the bucklers against a general body background of dark chocolate-brown.

Length 285 mm.

Distribution: In the Sea of Japan known from Primor'e (Novikov, 1957: 245); Pusan (Mori, 1952: 134); near Sado Island (Honma, 1963: 22); Toyama Bay (Katayama, 1940: 15); Echigo Province (Honma, 1957: 109); San'in region (Mori, 1956a: 23); and Tsushima Island (Arai and Abe, 1970: 95). Reported from Cheju-do Island (Mori, 1952: 134). Along the Pacific coast from Tiba Prefecture southward as well as from Taiwan (China) and Ryukyu (Matsubara, 1955: 957). Our collection comprised specimens (in addition to No. 22628) from Nagasaki (No. 7516), Iokogama (No. 8478), and Ohama (No. 23065).

[Family ZANCLIDAE—Moorish Idols]

Body deep, highly compressed laterally. Caudal peduncle without prickly spines. Mouth small. Teeth on jaws very long, thin, resemble bristles. Palatines without teeth. Hard thick bones on top of head form a median frontal process present in adults but absent in juveniles. Preopercle unarmed. Branchiostegal rays 4. Pyloric caeca 14. Intestine long. Vertebrae 22, of which 13 caudal. Scales small and rough. Dorsal fin one; spines in dorsal fin 7; third ray and subsequent ones with long filamentous tips. Anal fin similar in position and shape to dorsal fin, but

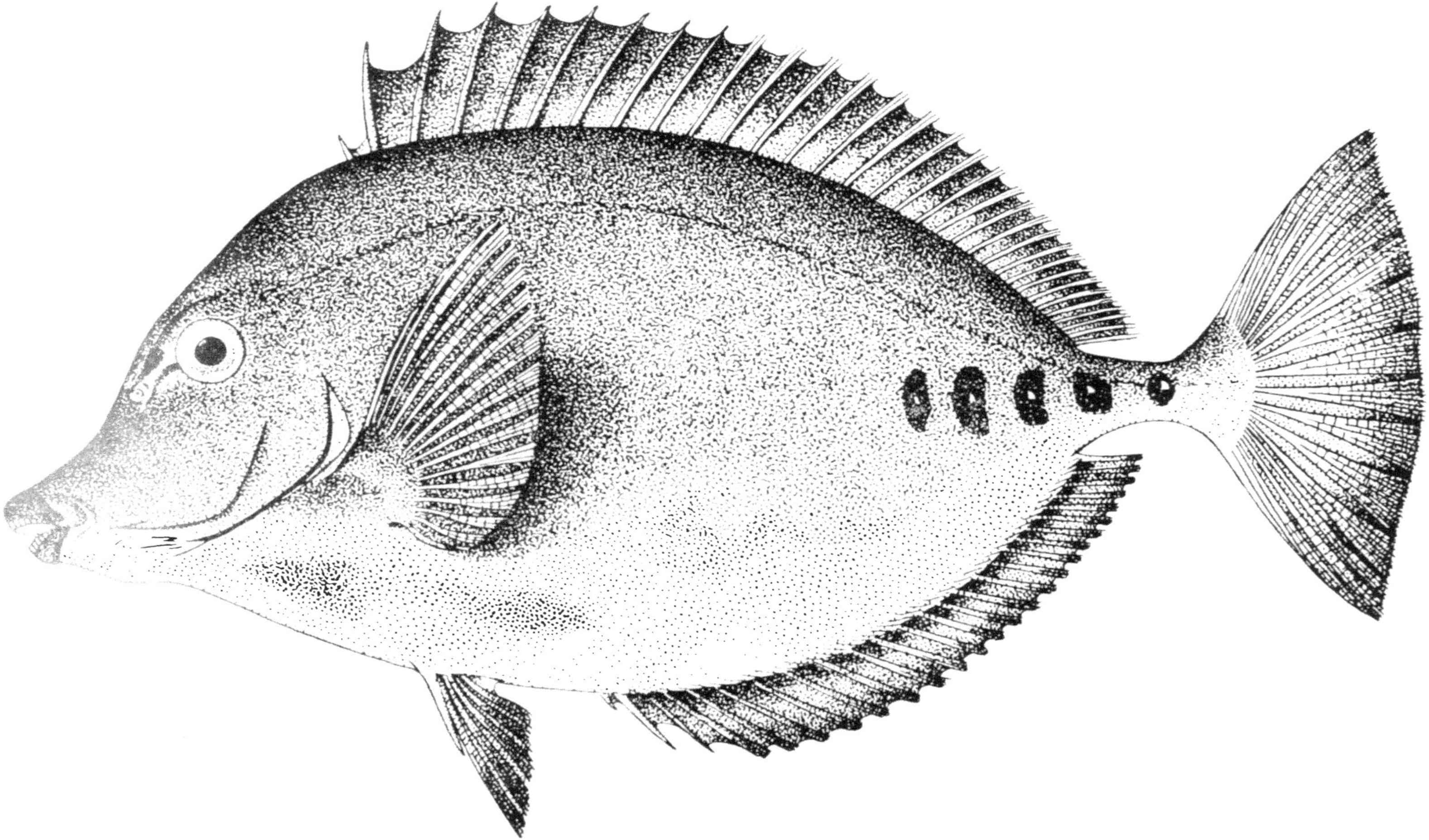

Figure 193. *Prionurus microlepidotus*—small-scaled sawtail. Length 155 mm. Japan (Temminck and Schlegel, 1845).

243   Figure 194. *Zanclus cornutus*—Moorish idol. Length 155 mm. No. 22330. Okinawa Islands.

without filamentous rays. Caudal fin broad, with very weak fork. Pectoral fins short; pelvic fins pointed (Fowler and Bean, 1929: 195).

1 genus in the Indian and Pacific oceans. Also distributed in Japan. Also possible in the Sea of Japan.

308

[Genus *Zanclus* Cuvier, 1831—Moorish Idol]

Zanclus Cuvier, Hist. Nat. Poiss., 7, 1831: 102 (type: *Chaetodon cornutus* Linné). Fowler and Bean, Bull. U.S. Nat. Mus., 100 (8) 1929: 196. Characteristics given in description of family.

Two species found off the coasts of Japan. *Z. cornutus* (Linné) and *Z. canescens* (Linné).

[*Zanclus cornutus* (Linné, 1758)—Moorish idol] (Figure 194)

Chaetodon cornutus Linné, Syst. Nat., ed. 10, 1758: 273 (eastern India).

Zanclus cornutus, Weber and Beaufort, Fish. Indo-Austr. Arch., 7. 1936: 170. Matsubara, Fish Morphol. and Hierar., 1955: 947. Abe, Enc. Zool., 2, Fishes, 1958: 127, fig. 375 (color figure).

22330. Okinawa. 1929. Awaya. 3 specimens.

39325. Hawaiian Islands. 1968. V. Fedorov. 2 specimens.

39483. Hawaiian Islands. 1968. V. Fedorov. 1 specimen.

D VII, 40-41; A III, 31-35.

Two species of the genus *Zanclus—canescens* and *cornutus—*described by Linnaeus are very similar to each other, especially the juveniles; in the adult of *Z. cornutus* the spiny process in front of the eyes was taken as a distinguishing character, which is absent in young fish of *Z. canescens*. The sharp differences between these species are described by Weber and Beaufort (1936: 173), who clarified problems in their identification. A distinguishing character, first mentioned by Bleeker, is

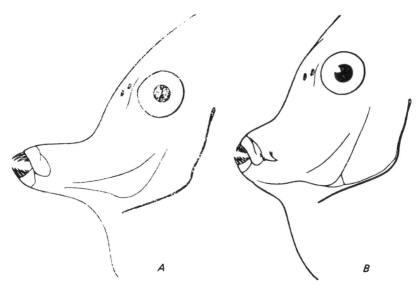

243

Figure 195. Shape of snout.

A—*Zanclus cornutus*; B—*Zanclus canescens*.

the presence of a bent prickly spine in *Z. canescens* in the anterior part of the preorbital, which is located above the posterior end of the maxilla (Figure 195, B). It should be noted that *Z. canescens,* in addition to this character, differs from *Z. cornutus* in shorter snout, absence of a chocolate-brown spot bordered by a dark stripe at the apex of the snout, and smaller dimensions—up to 80 mm. There is one specimen of *Z. canescens* in our collection (No. 3622, Caroline Islands, Martens) in which these peculiarities are well developed.

Length, more than 200 mm.

Distribution: Not found in the Sea of Japan. Known from the southern coast of Japan southward to Australia, and westward in the Indian Ocean up to the coast of eastern Africa (Weber and Beaufort, 1936: 172).

8. Suborder Trichiuroidei

Maxillae attached to non-protusile premaxillae. Bases of rays in caudal fin do not overlap hypural. Pectoral fins set low (Berg, 1940: 320).

Maxillae very firmly attached to non-protusile premaxillae, but jaws and teeth not fused together. Premaxillae form process protruding forward. Mouth not protractile. Caudal fin often absent and, if present, very small, although sometimes with deep notch or slightly forked, and its rays not hard. Body ribbonlike and, if moderately elongate, then head length significantly (twice) larger than maximum body depth, and snout without highly elongate process formed by maxillae. Caudal peduncle usually without keels.

Two families in pelagic waters of the Atlantic, Indian, and Pacific oceans, north to Japan and the Sea of Japan.

Key to Families of Suborder Trichiuroidei

1 (2). Body elongate but not ribbonlike. Caudal peduncle and fin well developed. Division between spines and soft rays of dorsal fin distinct. Pelvic fins sometimes absent, when present with 1 spine and 5 (or less) soft rays. Vertebrae less than 50. .... CLXI. **Gempylidae**.

2 (1). Body ribbonlike. Caudal peduncle poorly developed, low; caudal fin rudimentary, forked, or completely absent. In dorsal fin sometimes difficult to detect division between spines and soft rays. Pelvic fins rudimentary or absent. Vertebrae 100–160......... .... CLXII. **Trichiuridae.**[1]

CLXI. Family GEMPYLIDAE[2]–Snake Mackerels

Oblong or elongate body, compressed laterally. Dorsal fins two, usually with bases connected, and with deep notch; first fin spiny and base much longer than base of second fin. Anal fin similar to soft dorsal fin. Pelvic fins sometimes well developed, sometimes very small, reduced to a single spine, or even absent. Vertebrae from 31 (25 + 6) to 53 (28 + 25).

12 genera in tropical and subtropical seas at great depths. In Japan 9 genera, of which 1 genus is known from the Sea of Japan and another

[1]Lepidopidae, accepted by some authors (Tucker, 1956: 77) as a subfamily of Trichiuridae, differs in presence of pelvic fins, albeit they are very small.

[2]Anatomical description and analysis of the relationships within this family have been given by Matsubara and Iwai (1958: 23-54, 14 figs.).

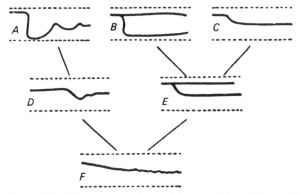

245

Figure 196. Different types of lateral lines in fishes of Gempylidae
(Matsubara and Iwai, 1958).

Genera in parentheses belong to Scombridae: A—*Lepidocybium*; B—
Epinnula, Neoepinnula (*Grammotorcynus*); C—*Promethichthys*; D—(*Scomberomorus*); E—*Rexea, Mimasea, Gempylus*; F—*Ruvettus* (*Scomber*).

for the southern coast of the Korean Peninsula. However, since fishes
of this family are found very rarely and have been poorly studied, only
genera found off the coasts of Japan are included in the key below.

Key to Genera of Family Gempylidae[3]

1 (2). Keel on each side of caudal peduncle. Lateral line single,
indistinctly expressed, highly curved, zigzag, almost reaching
back as well as belly (Figure 196, A)........................
.... [**Lepidocybium** Gill, 1862].[4]

2 (1). Keels absent on caudal peduncle. Lateral line less curved,
not zigzag.

3 (4). Belly with keel. Scales spinescent; *l. l.* poorly developed,
almost without curve (Figure 196, F). V I, 5.................
.... [**Ruvettus** Cocco, 1829].[5]

4 (3). Belly without keel. Scales smooth; *l. l.* well developed, with
curve or double.

5 (10). Pelvic fins developed, I 5. Finlets absent.

6 (9). Body fusiform, its depth about 1/4 standard length. Lower lateral
line passes near lower contour of body (Figure 196, B). Snout
does not protrude significantly ahead of anterior process of
premaxilla.

7 (8). Vomer with one to three teeth on each side. Two lateral lines

[3]From Matsubara and Iwai, 1952: 195.
[4]See Munro, 1950: 37, fig. 2. Widely distributed; in Japan found near the Pacific coast.
[5]Tropical and subtropical waters.

originate together near upper corner of gill opening (Figure 197). Dorsal fin originates behind vertical from upper end of gill opening. Spiny rays of dorsal fin weak or flexible. Pelvic fins reduced, 2.4 to 3.3 times in head length. Inner surface of gill rakers in angle of first gill arch armed with 2 rows of minute spines. Surfaces of oral and gill cavities black...............
.... [**Neoepinnula** Matsubara and Iwai, 1952].[6]

8 (7). Vomer without teeth. Two lateral lines originate at vertical from posterior margin of preopercle and branch at vertical between 5th and 6th spiny rays of dorsal fin (Figure 198). Spiny rays of dorsal fin fairly strong and prickly. Pelvic fins well developed, 1.3 times in head length. Inner surface of gill rakers in angle of first gill arch not armed with spines. Surface of oral and gill cavities pale................... [**Epinnula** Poey, 1854][7]

246 9 (6). Body elongate (depth 1/10 of standard length). Palatines without teeth. Lower lateral line passes along middle of body (Figure 196, E). Snout distinctly protrudes ahead of anterior process of premaxilla (Figure 199)......... [**Mimasea** Kamohara, 1936].[8]

10 (5). Pelvic fins reduced or absent. Finlets always present.

11 (12). Body highly elongate; its depth 12 times in standard length. Finlets 5–7, pelvic fins small, I 4–I 5; soft rays distinguishable with difficulty without using lens. Maxilla concealed for most part under preorbital. Gill rakers in angle of gill arch small, triangular. Snout large, protrudes ahead of anterior process of premaxilla. Two lateral lines; both originate at vertical from angle of gill opening (Figure 200). ....
.... [**Gempylus** Cuvier, 1829].[9]

12 (11). Body moderately elongate; depth less than 9 times in standard length. Finlets usually 2. Pelvic fins absent or in young fish represented by single spiny ray. Posterior margin of maxilla free, not concealed under preorbital. Gill rakers in angle of first gill arch I-shaped. Snout does not protrude or only slightly so ahead of anterior process of premaxilla.

13 (14). Lateral lines 2 (Figure 196, E). Pelvic fins absent in adult fish; in young fish represented by spine. .... 1. [**Rexea** Waite].

14 (13). Lateral line 1. Pelvic fins usually present.

249 15 (16). Dagger-shaped spine and small spine present behind vent. Lateral line almost straight (Figure 201); each pore of lateral line

---

[6]Pacific coast of Japan and Gulf of Mexico (Grey, 1960: 214).
[7]Carribean Sea. Pacific coast of Japan.
[8]Pacific coast of Japan.
[9]Tropical seas. Pacific coast of Japan.

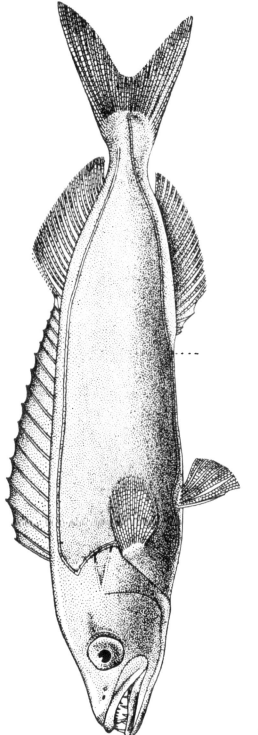

Figure 197. *Neoepinnula orientalis*. Standard length 143 mm (Matsubara and Iwai, 1952).

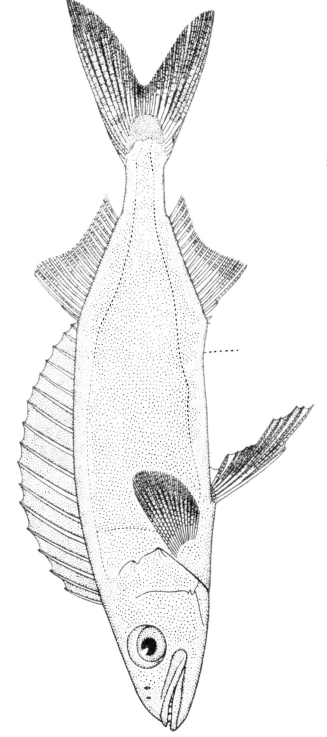

Figure 198. *Epinnula magistralis.* Standard length 188 mm (Matsubara and Iwai, 1952).

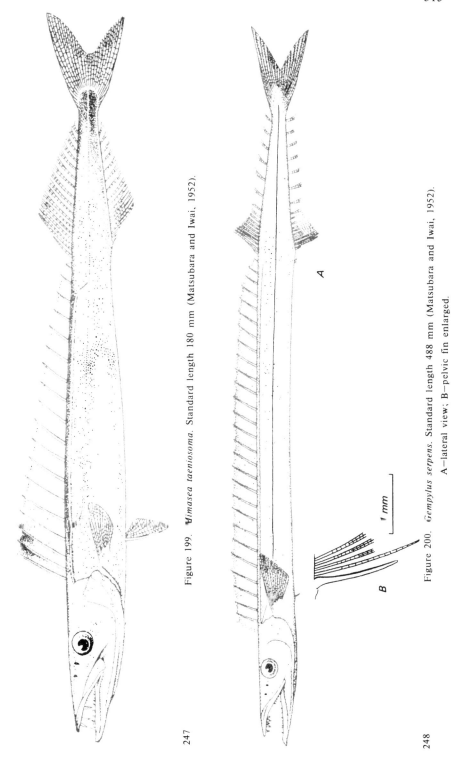

Figure 199. *Nimasea taeniosoma*. Standard length 180 mm (Matsubara and Iwai, 1952).

247

Figure 200. *Gempylus serpens*. Standard length 488 mm (Matsubara and Iwai, 1952).
A—lateral view; B—pelvic fin enlarged.

248

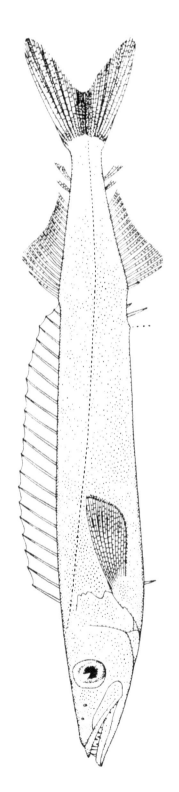

Figure 201. *Nealotus tripes.* Standard length 199 mm (Matsubara and Iwai, 1952).

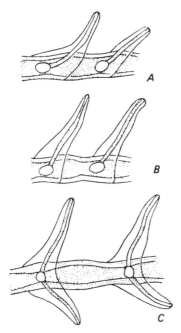

Figure 202. Branches of lateral line in fishes of Gempylidae (Matsubara and Iwai, 1952).

A—in anterior part of *Nealotus tripes*; B—in bent part of *Promethichthys prometheus*; C—on lower side behind bent part in *Promethichthys prometheus*.

with one oblique branch directed upward (Figure 202, A). Scales not embedded in skin .... [**Nealotus** Johnson, 1865].[10]

16 (15). Free spines not present behind vent. Lateral line distinctly curves toward front above pectoral fin (Figure 196, C); each pore of lateral line behind curve in anterior part with two oblique branches, upper and lower (Figure 202, C). Scales embedded in skin. .... ?. **Promethichthys** Gill.

1. [Genus *Rexea* Waite, 1911]

Rexea Waite, Proc. New Zealand Inst., 2 (1910), January 18, 1911: 49 (type: *R. furcifera* Waite = *Gempylus solandri* Cuvier). Beaufort and Chapman, Fish. Indo-Austr. Arch., IX, 1951: 201. Matsubara and Iwai, Pacific Sci., 6, 3, 1952: 205.

Jordanidia Snyder, Proc. U.S. Nat. Mus., 24, 1911: 527 (type: *J. raptoria* Snyder).

Body slightly oblong, fusiform. Scales small, cycloid, deciduous. Mouth large, with one row of dagger-shaped teeth on each jaw and a few large

---

[10]Eastern part of the Atlantic Ocean. Pacific coast of Japan.

canines in front. Teeth arranged in one row on palatines. Vomer without teeth. Dorsal fins two, their bases touching; first with 17-18 spines and second with two unbranched and 13-15 branched rays. Anal fin with two unbranched and 11-14 branched rays; anal fin similar to second dorsal fin in shape. Two finlets situated behind second dorsal and anal fin and separated from fins and from each other. Pectoral fins roundish. Pelvic fins rudimentary, distinguishable with difficulty, or completely absent in adult fish. Caudal fin bifurcate. Lateral line continues from upper end of gill opening along body side below bases of dorsal fins up to middle of base of soft dorsal fin or to its end; at vertical from membrane between 5th and 6th spines, branch proceeds from upper lateral line and extends downward and backward along middle of body, reaching almost to base of caudal fin. Pseudobranch present (Beaufort and Chapman, 1951: 201).

One species. Found in the southwestern part of the Pacific Ocean off the coasts of Japan, China, Australia, and New Zealand.

1. [*Rexea solandri* (Cuvier, 1831)] (Figure 203)

Gempylus solandri Cuvier. In: Cuvier and Valenciennes, Hist. Nat. Poiss., 8, 1831: 215 (New Zealand).

Thyrsites prometheoides Bleeker, Acta Soc. Sci. Indo-Néerl., 1, 1856 (1855): 42 (Amboina).

Jordanidia raptoria Snyder, Proc. U.S. Nat. Mus., 40, 1911 (May 26): 527; Proc. U.S. Nat. Mus., 42, 1912: 410, pl. 52, fig. 2 (Japan).

Jordanidia prometheoides, Shmidt, Tr. Tikhookeansk. Kom. Akad. Nauk SSSR, 2, 1931: 41, fig. 5.

Rexea solandri, Whitley, Austr. Mus. Rec., 17, 3, 1929: 120, pl. 33, fig. 2. Matsubara and Iwai, Pacific Sci., 6, 3, 1952: 204, fig. 8.

22455. Kagoshima. March, 1901. P.Yu. Shmidt. Eight specimens.

251 D XVII-XVIII, I, 15-16 + 2; A I, 14-16 + 2; P 13-14; V I (0); branchiostegal rays 7 (Matsubara and Iwai, 1952: 204). In our specimens: D XVIII, II, 14-15 + 2; A I, 14 + 2.

Characteristics of species given in description of genus.

Length, up to 400 mm (Abe, 1958).

Distribution: Not found in the Sea of Japan, but known from the Korean Strait, south of Pusan (Mori, 1952: 134). Pacific coasts of Japan and south up to Australia and New Zealand (Matsubara, 1955: 535).

2. Genus *Promethichthys* Gill, 1893

Prometheus (Quoy and Gaimard, Ms) Lowe, Trans. Zool. Soc. London, 2, 1841 (type: *P. atlanticus*); nom. praeocc.

Promethichthys Gill, Mem. Nat. Acad. Sci., 6, 1893: 115, 123 (type:

Prometheus atlanticus Lowe = *Gempylus prometheus* Cuvier). Jordan and Evermann, Fishes N. and M. Amer., 1, 1896: 882.

Body oblong, thin, fusiform. Mouth small, with two strong canines in anterior part of each jaw. Spiny dorsal fin long, connected with soft and fairly high fin. Two finlets behind dorsal and anal fins. Pectoral fins inserted fairly low. Caudal peduncle without keel. Pelvic fins represented by one pair of small spines. Dagger-shaped spine not present behind vent. Preopercle not armed, except in young fish. Lateral line single, forms a bend under the anterior part of dorsal fin. Scales very small, smooth. Predatory fishes of high seas, attaining moderate dimensions (Jordan and Evermann, 1896: 882).

One species. Also found in the Sea of Japan.

1. ***Promethichthys prometheus*** (Cuvier, 1831)—(Figure 204)

Gempylus prometheus Cuvier, Hist. Nat. Poiss., 8, 1831: 213, 222 (St. Helena Island).

Promethichthys prometheus, Shmidt, Tr. Tikhookeansk Kom. Akad Nauk SSSR, 2, 1931: 39. Matsubara and Iwai, Pacific Sci., 6, 3, 1952: 209. Grey, Copeia, 3, 1953: 139, fig. IC.

22489. Tokyo. March 15-26, 1901. P.Yu. Shmidt.

22489a. Misaki. April 1, 1901. P.Yu. Shmidt.

D XVIII-XIX, 18-20 + 2; A II, 16-18 + 2 (Abe, 1958).

In our specimens from Japan 230 to 440 mm long, D XVIII, II, 17-19 + 2; II, 14-15 + 2.

Characters given in descripition of genus.

Length, to 600 mm (Abe, 1958: 215).

Distribution: In the Sea of Japan known from the coast of San'in region (Mori, 1956a: 23). Tropical and subtropical waters of the Atlantic and Pacific oceans. Coast of Japan (Matsubara, 1955: 536).

CLXII. Family TRICHIURIDAE—Cutlassfishes
(Trichiuridae + Lepidopidae Tucker, 1956)

Spiny part of dorsal fin always distinctly pronounced[11]; in the genus *Diplospinus* longer than soft part, and in the genus *Aphanopus* only very slightly longer (individal specimens). Some rays of anal fin, if 252 not all, split, soft, support fin membrane (*Diplospinus, Aphanopus, Benthodesmus, Lepidopus, Evoxymetopon, Assurger*), but sometimes highly reduced or completely absent (*Trichirus, Lepturacanthus, Eupleurogrammus*). Two rays located before anal fin immediately behind vent; anterior ray very small and posterior enlarged to various degree and shaped like a small leaflet or keeled scute, sometimes like a strong spine.

---

[11]In some genera it is difficult or even impossible to detect the place of transition from spiny to soft part in the dorsal fin.

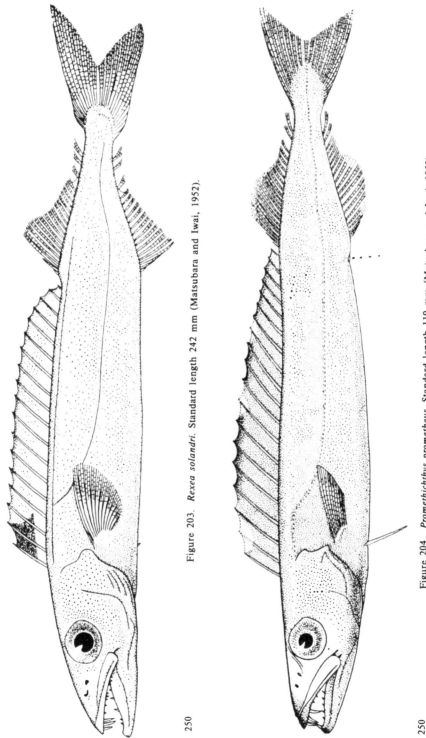

Figure 203. *Rexea solandri*. Standard length 242 mm (Matsubara and Iwai, 1952).

Figure 204. *Promethichthys prometheus*. Standard length 110 mm (Matsubara and Iwai, 1952).

Pelvic fins of some genera and possibly in all species with these fins, in form of scaly spine and rudimentary soft ray difficult to distinguish. Spines of dorsal fin and their basal and interneural bones always correspond to number of vertebrae; soft rays may be twice number (*Diplospinus*), slightly more (*Aphanopus, Benthodesmus*), or correspond to number of vertebrae (in remaining genera). Vertebrae vary from $34 + 24 = 58$ (*Diplospinus*) to $53 + 103 = 156$ (*Benthodesmus simonyi*) or $41 + 151 = 192$ (*Eupleurogrammus muticus*) (Tucker, 1956: 75).

3 subfamilies, 10 genera. Atlantic, Indian, and Pacific oceans.

Key to Genera of Family Trichiuridae[12]

1 (2). Occipital region of head almost at same level as preorbital region; Sagittal crest of frontals absent (Figure 205, A) (subfamily Aphanopodinae). Dorsal fin with more than 120 rays. Base of spiny dorsal fin constitutes half length of base of soft fin. Caudal fin present....... 1. **Benthodesmus** Goode and Bean.[13]

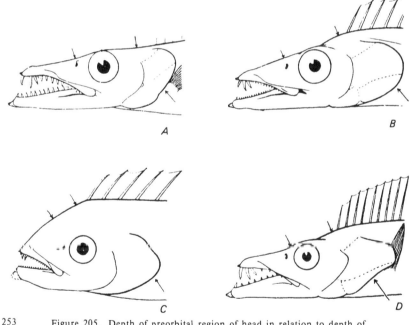

253   Figure 205. Depth of preorbital region of head in relation to depth of occipital region in various genera of Trichiuridae (Tucker, 1956).

A–*Aphanophus*; B–*Lepidopus*; C–*Evoxymetopon*; D–*Trichiurus*.

[12]From Tucker, 1956: 77.

[13]Rays fewer in dorsal fin of *Diplospinus* Maul, 1948 (72-73) and *Aphanopus* Lowe, 1839 (91-95), both of which are deep sea genera.

322

2 (1). Occipital region of head distinctly higher than preorbital; frontals with sagittal crest developed in occipital region, which sometimes continues to preorbital region (Figure 205, B-D).

3 (8). Pelvic fins, although very small, present. Lateral line with very weak curve anteriorly and extends almost along middie of body; distance from lateral line to ventral profile in region of vent much larger than half distance from lateral line to dorsal profile. Lower posterior margin of preopercle convex (subfamily Lepidopinae).

4 (7). Caudal fin present.

5 (6). Dorsal fin with 87-93 rays. Body depth 12-13 times in its length. . . . . 2. [**Evoxymetopon** (Poey) Gill].

6 (5). Dorsal fin with 120 rays. Body depth 20-28 times in its length. . . . . 3. [**Assurger** Whitley].

7 (4). Caudal fin absent. Body depth 14-18 times in its length. . . . . 4. [**Eupleurogrammus** Gill].

8 (3). Pelvic fins absent. Lateral line with fairly sharp curve and passes near ventral profile of body; distance from lateral line to ventral profile in region of vent less than half distance from lateral line to dorsal profile. Lower posterior margin of preopercle more or less concave. Caudal fin always absent (subfamily Trichiurinae).

9 (10). Spine of anal fin located behind vent very small, less than diameter of pupil; other rays of fin indistinguishable externally. Eyes fairly large, 5 to 7 times in head length. . . . . 5. **Trichiurus** Linne.

10 (9). Spine of anal fin hardly noticeable, equal to half diameter of eye; fin rays penetrate skin and protrude on surface in form of prickly spinules. Eyes small, 6.7 to 10.0 times in head length. . . . . [**Lepturacanthus** Fowler, 1905].[14]

1. Genus *Benthodesmus* Goode and Bean, 1882

Benthodesmus Goode and Bean, Proc. U.S. Nat. Mus., 4, 1882: 379 (type: *Lepidopus elongatus* Clarke). Tucker, Bull. Brit. Mus. Nat. Hist., 4, 3, 1956: 85.

Body greatly elongate; body depth 22 to 34 times in standard length. Vertebrae 47-53 + 75-103 = 123-156. Dorsal fin with 39-46 spines and 80-108 soft rays. Base of anterior part of dorsal fin equal to half length of base of posterior part of dorsal fin. Two spines behind vent; anterior spine very small and almost indistinguishable in adult fish, posterior

[14]*Lepturacanthus savala* (Cuvier, 1829). Indian and Pacific oceans, north to the East China Sea. Japan? (Beaufort and Chapman, 1951: 193).

spine in form of cordate scute with median keel. Anal fin with 70–76 unbranched rays or with only 25 rays. Pelvic fins in form of 1 scaly spine and one reduced soft ray. Caudal fin present, but very small and bifurcate (Tucker, 1956: 87).

Three species in the Atlantic, Indian, and Pacific oceans; one of them in the Sea of Japan.

1. *Benthodesmus tenuis* (Günther, 1877) (Figure 206)

Lepidopus tenuis Günther, Ann. Mag. Nat. Hist., 4, 20: 437; Challenger Reps. Zool., 22, 1887: 37, pl. 7, fig. B (35°11' N, 139°28' E, Sagami Bay, Japan).

Lepidopus aomori Jordan and Snyder, J. Coll. Sci. Univ. Tokyo, 15, 2, 1901: 303 (Aomori, Japan).

Benthodesmus tenuis, Tucker, Bull. Brit. Mus. Nat. Hist., 4, 3, 1956: 89, fig. 9.

255   D 120–133; A i + 1 + 70–76; vertebrae 123–131 (Tucker, 1956: 88).

Differs from other species of this genus in location of pelvic fins slightly ahead of vertical from anterior end of base of pectoral fin and lateral line; width of pectoral fins less than 1/15th length.

Length, to 2 m (Jordan and Snyder, 1901a).

Distribution: In the Sea of Japan known from the Japanese coast of Hokkaido (Ueno, 1971: 80); Hakodate and Aomori (Jordan and Snyder, 1901a: 303). Pacific coast of Japan, Hawaii, Tahiti (Matsubara, 1955: 536). Tropical part of the Atlantic Ocean, Gulf of Mexico (Tucker, 1956: 88).

2. [Genus *Evoxymetopon* (Poey) Gill, 1863]

Evoxymetopon Poey. In: Gill, Proc. Acad. Nat. Sci. Philad., 1863: 227 [type: *E. taeniatus* (Poey) Gill, 1863]. Tucker, Bull. Brit. Mus. Nat. Hist., 4, 3, 1956: 97.

Body elongate, maximum depth 12 to 13 times in total length. Upper profile of head convex from snout tip to origin of dorsal fin; interorbital space convex. Caudal fin present. Orbit large, 5 to 6 times in head length. Dorsal fin with 10 spines and 77 soft rays; first spine is sometimes equal to head length. Spine behind vent resembles keeled scute. In anal fin anterior rays, if present, barely protrude out of skin, but posterior rays, about 20, well developed and act as support for fin. Pelvic fins present, scale-like, located at a distance equal to diameter of eye behind posterior end of base of pectoral fin (Tucker, 1956: 99).

1 species. Known from the Atlantic, Pacific, and probably Indian oceans.

1. [*Evoxymetopon taeniatus* Poey, 1863] (Figure 207)

Evoxymetopon taeniatus Poey, in: Gill, Proc. Acad. Nat. Sci. Philad.,

324

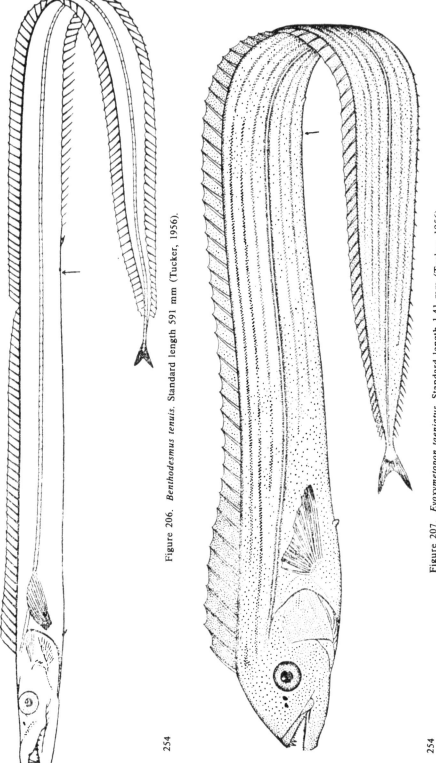

Figure 206. *Benthodesmus tenuis*. Standard length 591 mm (Tucker, 1956).

Figure 207. *Evoxymetopon taeniatus*. Standard length 1.41 mm (Tucker, 1956).

254

254

1863: 228 (Havana, Cuba). Tucker, Bull. Brit. Mus. Nat. Hist., 4, 3, 1956: 99, fig. 11.

D X, 77; A 19 (Tucker, 1956).

Characters given in description of genus.

Length, to 2 m (Tucker, 1956).

Distribution: Not found in the Sea of Japan, but known from Kojedo Island, which is slightly west of Pusan (Mori, 1952: 135). Described from the Carribean Sea and off Cuba (Havana). Report for Scotland erroneous since, according to Tucker, the Hoy specimen is a species of *Trachypterus* or *Regalecus*. From the Indian Ocean (Mascarene Islands, Mauritius), *Evoxymetopon poeyi* Günther, 1887, has been described, which was included by Tucker under the synonymy of *E. taeniatus* with a question mark.

3. [Genus *Assurger* Whitley, 1933]

Assurger Whitley, Rec. Austr. Mus., 19, 1933: 84 (type: *Evoxymetopon anzac* Alexander).

Body extremely elongate: head length 12 times and maximum depth 28 times in total length. Eyes small, 8 times in head length. Posterior margin of opercle rounded. Pelvic fins small, scale-like, located under posterior half of pectoral fins. Caudal fin present (Tucker, 1956: 106).

1 species, coast of Australia and known from the southern coast of the Korean Peninsula.

1. [*Assurger anzac* Alexander, 1916] (Figure 208)

Evoxymetopon anzac Alexander, J. Roy. Soc. W. Austr., 2, 1916: 104, pl. 7 (western Australia). Kamohara, Sci. Rept. Kochi Univ., 3, 1952: 31, fig. 26.

Assurger alexanderi (nom. emend.) Whitley, Rec. Austr. Mus., 19, 1933: 84 (western Australia).

Assurger anzac, Tucker, Bull. Brit. Nat. Hist., 4, 2, 1956: 107, fig. 16. D ca 120; A 14+; C 17; P 12; branchiostegal rays 7 (Tucker, 1956). Description given in characters of family and in key.

Length, to 1,415 mm (Tucker, 1956).

257 *Distribution*: Not reported from the Sea of Japan, but known off Komun'do Island near the southern coast of the Korean Peninsula (Mori, 1952: 135). Pacific coast of Japan off Shikoku Island (Tosa Bay, City of Koti), southwest Australia (Matsubara, 1955: 537).

4. [Genus *Eupleurogrammus* Gill, 1863]

Eupleurogrammus Gill, Proc. Acad. Nat. Sci. Philad., 1863: 226 (type: *Trichiurus muticus* Gray).

In external appearance and absence of caudal fin quite similar to the genus *Trichiurus,* but differs from it in presence of pelvic fins, albeit very small and rudimentary, and lateral line with curve and passing near ventral profile of body.

2 species, one known from the waters of the Yellow Sea.

1. [*Eupleurogrammus muticus* (Gray, 1831)] (Figure 209)

Trichiurus muticus Gray, Zool. Misc., 1, 1831: 10 (India).

Eupleurogrammus muticus, Gill, Proc. Acad. Nat. Sci. Philad., 1863: 226. Tucker, Bull. Brit. Mus. Nat. Hist., 4, 3, 1956: 105, fig. 15.

D III, 143–147; A 1 + 1 + 120–121; vertebrae 191–192 (Tucker, 1956).

Differs from *E. intermedius* (Gray, 1831) in larger number of rays in dorsal fin (not 111, but 123–131) and vertebrae (not 157–162, but 191).

Length, to 617 mm (Tucker, 1956).

Distribution: Not found in the Sea of Japan, but reported from Yamaguti Prefecture which partly borders the Sea of Japan. Reported from Napho in the Yellow Sea (Mori, 1952: 135). Pacific coast of Japan, China, Indonesia, and the Indian Ocean (Matsubara, 1955: 537).

5. Genus *Trichiurus* Linné, 1758—Cutlassfishes

Trichiurus Linné, Syst. Nat., ed. 10, 1758: 246 (type: *T. lepturus* Linné). Tucker, Bull. Brit. Mus. Nat. Hist., 4, 3, 1956: 113 (synonyms).

Body greatly elongate, ribbonlike, caudal part pointed. Caudal fin absent. Pelvic fins absent. Scales absent. Head elongate and pointed. Occipital region of head distinctly above preorbital region. Mouth large, with large dagger-like teeth. Lower posterior margin of subopercle more or less concave. Lateral line with sharp curve and passes along lower margin of body at some distance from it, near vent, which is less than half distance up to upper margin of body. Diameter of eye 5 to 7 times in the head length. Rudimentary spine of anal fin located behind vent very small, less than diameter of pupil; other rays of fin not discernible as concealed in skin.

One widely distributed species, also found in the Sea of Japan.

1. *Trichiurus lepturus* Linné, 1758—Cutlassfish (Figure 210)

Trichiurus lepturus (part) Linné (ex Artedi), Syst. Nat., ed. 10, 1758: 246 (South Carolina). Tucker, Bull. Brit. Mus. Nat. Hist., 4, 3, 1956: 114, fig. 18 (synonyms). Abe, Enc. Zool., 2, Fishes 1958: 214, fig. 636.

Clupea haumela Forskål, Descript. Animal., 1775: 72 (Red Sea).

Trichiurus lepturus japonicus Temminck and Schlegel, Fauna Japonica, Poiss., 1844: 102, Pl. 54 (Japan).

22457. Pusan. March 28, 1901. P.Yu. Shmidt, 3 specimens.

25183. Syauhu Inlet, Primor'e. 1934. G.U. Lindberg, 5 specimens.

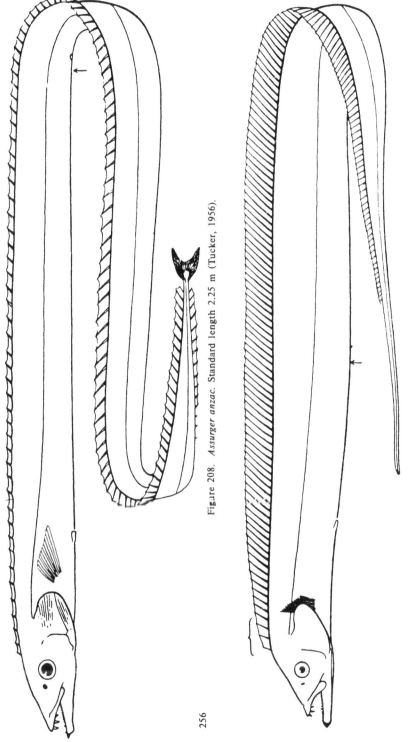

Figure 208. *Assurger anzac*. Standard length 2.25 m (Tucker, 1956).

Figure 209. *Eupleurogrammus muticus*. Standard length 617 mm (Tucker, 1956).

256

256

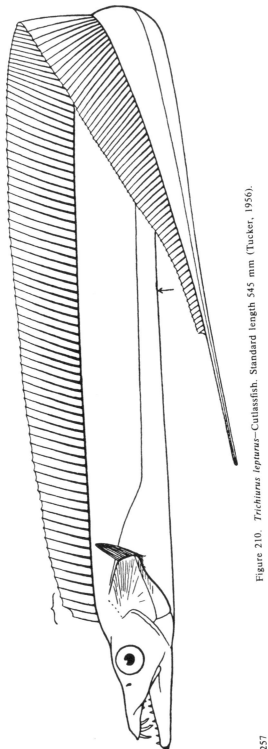

Figure 210. *Trichiurus lepturus*—Cutlassfish. Standard length 545 mm (Tucker, 1956).

31378. Yellow Sea, Dalyan City (Dal'nii). September 4–10, 1946. V.G. Gnezdplov, 2 specimens.

35597. Yellow Sea, Bikou. May 26, 1952. Academy of Sciences, China. 2 specimens.

35599. Yellow Sea, Chefoo. June, 1956. Academy of Sciences, China. 2 specimens.

D III, 137 (D 120–140); A i + I 105–108; vertebrae 39–40 + 123–128 = 162–168. Body depth 14.4–21.0 times and head length 7.0–9.4 times in body length; eye 5 to 7 times in head length (Tucker, 1956).

Length, to 1.5 m (Abe, 1958).

Distribution: In the Sea of Japan found along the continental coast beginning from Nelma (about 40° N) (Shmidt and Taranetz, 1934: 592) and southward: Olga Bay (Popov, 1933a: 141); Petrov Island (No. 20183); Peter the Great Bay (Soldatov and Lindberg, 1930: 114); Pusan (Shmidt, 1931b: 42); Tsushima Island (Arai and Abe, 1970: 88); along the coast of Japan from Hokkaido (Ueno, 1971: 80) and south to Sado Island (Honma, 1952: 143); San'in region (Mori, 1956a: 23); and Fukuoka Prefecture (Jordan and Hubbs, 1925: 222). In the Yellow Sea reported from the Gulf of Chihli (Bohai) (Zhang et al., 1955: 187) and Nampho and Inchon (Jordan and Metz, 1913: 27). This species, the only representative of the genus, is distributed in the tropical and subtropical waters of the Pacific, Indian, and Atlantic oceans.

9. Suborder Scombroidei[1]

258 Maxillae firmly attached to nonprotusile premaxillae,[2] which in some representatives form a more or less long xiphoid process produced forward. Mouth nonprotractile. Branchiostegal membranes not attached to isthmus. Caudal fin always present, well developed, crescent-shaped or highly notched, with hard rays (bases of rays completely overlap hypurals). Body fusiform, head length almost equal to maximum body
259 depth, if larger, not more than 1.5 times; if body elongate, then xiphoid process present and produced anteriorly. Caudal peduncle strong, with one to three keels on each side (Lindberg, 1971: 122).

In recent years considerable attention has been given by Soviet authors to the study of scombroids (Zharov et al., 1961, 1964; Svetovidov, 1964; Martinsen, 1965; Zharov, 1967; Parin, 1967) as well as researchers in other countries (Jordan and Evermann, 1926; Fraser-Brunner, 1949, 1950; Bullis and Mather, 1956; Jones and Silas, 1962, 1963; Collette and Gibbs, 1963; Iwai, Nakamura and Matsubara, 1965; Merrett and Thorp, 1965; Collette, 1966; Nakamura and Iwai, 1968).

Considering the present status of the composition of this suborder, we shall examine three families under it: Scombridae, Istiophoridae, and Xiphiidae. Members of all three families have been found in the Sea of Japan.

Members of genera of these families are either pelagic and mainly predators, living mostly in open parts of the ocean, or neritic and found above the slope of large depths. They are all capable of actively moving great distances in search of food and generally move very fast. To pounce on a prey they sometimes attain very high speeds, not possible for other fish species. Constant movement is essential to their life pattern; with cessation of movement respiration stops since the movement causing the gill operculum to open is associated with alternate bending of the body left and right in the process of oscillating the caudal fin. Water always enters the gill cavity through an open mouth in the course of forward body movement. Movement at high speed and over great distances consumes

---

[1]The family Thunnidae, assigned by Berg (1940: 333) to the order Thunniformes, is close to the family Cybiidae of Scombroidei. The Japanese ichthyologist Matsubara (1955) defined several gradual morphological transitions among members of Scombridae, Cybiidae, and Thunnidae and hence considered them merely subfamilies of Scombridae (Gasterochisminae, Scombrinae, Acanthocybiinae, and Sardinae), with which we do not agree.

[2]Maxillae also firmly attached to premaxillae in Scaridae and Oplegnathidae of the suborder Percoidei, but the jaw teeth in these families fuse to form a beak.

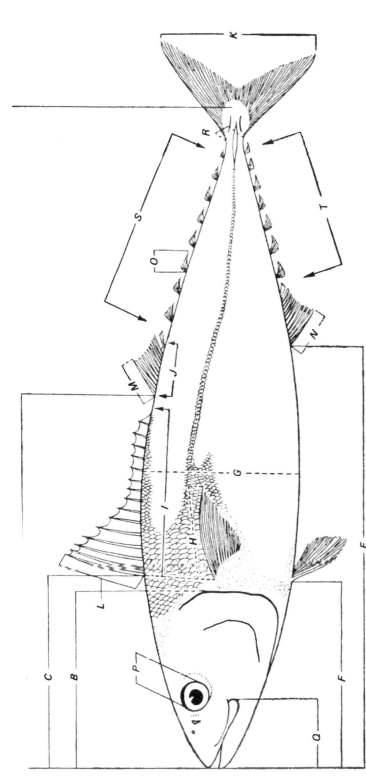

260   Figure 211.   Diagram depicting method of measurements of members of the suborder Scombroidei (Jones and Silas, 1960).

A – fork length to base of middle ray of caudal fin; B – head length; C – length to origin of base of second dorsal fin; D – length to origin of base of pectoral fin; E – preanal distance; F – length to origin of pelvic fins; G – maximum body depth; H – length of pectoral fin; I – length of base of first dorsal fin; J – length of base of second dorsal fin; K – distance between tips of lobes of caudal fin; L – height of dorsal fin; M – height of second dorsal fin; N – height of anal fin; O – height of finlet; P – diameter of eye; Q – length of upper jaw; R – least caudal peduncle depth; S – dorsal finlets; T – anal finlets.

considerable energy. Such a mode of life is closely linked with their unique morphological and physiological characteristics (Zharov, 1967: 215).

The biology of members of the suborder Scombroidei has been studied by Rass (1965a, 1965b), Osipov (1968a), and Gorbunova (1965a, 1965b, 1965c). Descriptions and figures of the larvae and juveniles have been given by Sun' Tszi-Zhen' (1960), Ehrenbaum (1924), Strasburg (1960), Matsumoto (1961), and Zhudova (1969). Parin (1967) has examined the dependence of scombroids on the abiotic factors of their environment and is convinced that the maximum variation among these fishes is seen in the Indian and Pacific Oceans. In his opinion, the present geographic center of distribution of the suborder is in the Indo-Pacific which, probably, also corresponds to the initial center of formation of the group.

Let it be noted hear that all the measurements of the suborder Scombroidei given in the text were taken by the method depicted in Figure 211 (Jones and Silas, 1961: 372), which is generally followed by ichthyologists.

Key to Families of Suborder Scombroidei [3]

1 (2). Snout not produced forward to form an elongate rostrum or bill. Pectoral fins usually inserted high. Pelvic fins with one spine and five soft rays each. .... CLXIII. **Scombridae.**

2 (1). Snout produced forward to form an elongate rostrum or bill formed by premaxillae and nasals. Pectoral fins inserted low. Pelvic fins with three rays each, or fins completely absent.

261 3 (4). Body distinctly elongate, flat on sides, covered with scales. Teeth small, but retained throughout life. Pelvic fins present. Two to three keels on each side of caudal peduncle. Base of first dorsal fin long, distinctly more than half body length and close to base of second dorsal fin. Snout in form of elongated process, almost round in cross section. .... CLXIV. **Istiophoridae.**

4 (3). Body less elongate, slightly fusiform, not covered with scales. Teeth not retained in adults. Pelvic fins absent. One keel on each side of caudal peduncle. Base of first dorsal fin short, distinctly less than half body length, and widely separated from base of second dorsal fin. Snout long, sword-shaped , with sharp margins, and not round but depressed in cross section. .... CLXV. **Xiphiidae.**

CLXIII. Family SCOMBRIDAE—Mackerels

Body elongate, fusiform, moderately compressed laterally. Caudal peduncle with two small keels on each side between lobes of caudal fin

[3]From Lindberg, 1971: 180.

and usually with large keel anterior to them also. Body covered with minute scales or posteriorly naked; a unique, more or less developed corselet of enlarged scales located in anterior part. Lateral line slightly curved or sinuous, often with transverse branches. Adipose eyelid present or absent. Two dorsal fins, more or less separated from each other or contiguous. Anal fin with one to three spines. The last rays of the second dorsal and anal fins are separate and appear like small fins. Pectoral fins inserted high. Pelvic fins thoracic, with one spine and five branched rays. Teeth minute or large, conical or more or less laterally compressed, sometimes knife-shaped on jaws. Sometimes teeth present on palatines and vomer. Gill rakers present or absent, few. Vertebrae 31–66 (Svetovidov, 1964: 384).

About 13 genera.[4] Tropical, subtropical, and partly temperate seas. Most members very important in world fisheries (Lindberg, 1927; Rass, 1948, 1960, 1965; Lindberg, 1971).[5] Eight genera known from the Sea of Japan.

Key to Genera of Family Scombridae[6]

1 (16). Teeth on jaws weak, conical.

2 (11). Interpelvic process separated from each other and from fins (Figure 212, A, B). Dorsal fins contiguous or slightly apart, with distance between them not more than diameter of eye.

3 (6). Body completely covered with scales. Scales of corselet and lateral line usually larger than other scales of body.

4 (5). Body rounded in cross section, not compressed laterally. Dark longitudinal stripes not present along upper half of body. Vomer and palatines with minute villiform teeth. Vertebrae 39–41.
.... 1. **Thunnus** South.

262 5 (4). Body slightly compressed laterally. From 5 to 10 narrow dark longitudinal stripes present in upper half of body. Vomer without teeth; palatines with one row of strong conical teeth. Vertebrae 44..................... .... 2 **Sarda** Cuvier.

6 (3). Body naked except for corselet and lateral line.

7 (8). Dorsal fins contiguous. Upper margin of first dorsal fin straight .... [**Gymnosarda** Gill, 1862][7]

8 (7). Dorsal fins not contiguous. Upper margin of first dorsal fin concave.

[4]Generic composition of this family analyzed by Collette (1962, 1966).

[5]Biological and technological characteristics of many scombroids given by Osipov and Myaksha (1961).

[6]From Matsubara (1955: 514), with additions.

[7]Found in the Kara Sea and near Australia, Indonesia, and Japan.

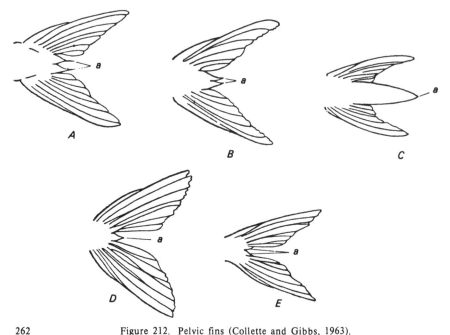

262   Figure 212. Pelvic fins (Collette and Gibbs, 1963).

A—*Thunnus*; B—*Euthynnus* and *Katsuwonus*; C—*Auxis*; D—*Scomber*;
E—*Scomberomorus*; a—interpelvic process.

9 (10). Teeth present not only on both jaws but also on palatines; sometimes also on vomer. Vertebrae 39. Longitudinal black stripes on lower part of body absent.........................
.... 3. **Euthynnus** Jordan and Gilbert.

10 (9). Teeth only on jaws. Vertebrae 41. Three to five distinct longitudinal black stripes in lower part of body..............
.... 4. **Katsuwonus** Linné.

11 (2). Interpelvic process single (Figure 212, C, D) or each merged with adjoining fin (Figure 212, E). Dorsal fins wide-set, distance between them more than diameter of eye.

12 (13). Body not completely covered with scales; corselet well developed. Posterior part of caudal peduncle with well-developed median keel and small keels at base of caudal fin. Vomer with teeth, palatines without teeth. Vertebrae 39.....
.... 5. **Auxis** Cuvier.

13 (12). Body completely covered with scales; corselet either poorly developed or absent. Posterior part of caudal peduncle without median keel, only with two small keels. Vertebrae 31.

14 (15). Vomer and palatines with teeth. Gill rakers moderately long

and slender, their number on lower arm of first gill arch less than
35. Body fusiform, depth less than head length.............
.... 6. **Scomber** Linné.

15 (14). Vomer and palatines without teeth. Gill rakers very long, broad, their number more than 35. Body compressed laterally, depth almost equal to head length..........................
.... **[Rastrelliger** Jordan and Dickerson, 1908].[8]

16 (1). Teeth on jaws strong, distinctly compressed laterally, almost triangular or knife-shaped.

17 (18). Lateral line bifurcates behind head (under first dorsal fin); one branch continues along top of back, the other along lower part of belly. Maxilla posteriorly does not extend beyond vertical from anterior margin of eye. Vertebrae 31...................
.... **[Grammatorcynus** Gill, 1862].[9]

18 (17). Lateral line single, does not bifurcate.

19 (20). Gill filaments do not form network. Gill rakers few. Spines in dorsal fin 14-20. Vertebrae 40-51. Snout much smaller than remaining part of head......... 7. **Scomberomorus** Lacépède.

20 (19). Gill filaments form network. Gill rakers absent. Spines in dorsal fin 25-26. Vertebrae 64. Snout same length as remaining part of head. ... 8. **Acanthocybium** (Cuvier and Valenciennes).

1. Genus *Thunnus* South, 1845

Thynnus Cuvier, Régne Anim., 2, 1817: 313 (type: *Scomber thynnus* L.) (preocc.).

Orcynus Cuvier, Régne Anim., 2, 1817: 314 (type: *Scomber germo* Lacépède) (preocc.).

Thunnus South. In: Smedley, Rose and Rose (eds.), Encyclopaedia Metropolitana, 25, 1845: 620 (type: *Scomber thynnus* L.). Kishinouye, J. Coll. Agr. Univ., Tokyo, 8, 3, 1923: 433 (description). Fraser-Brunner, Ann. Mag. Nat. Hist. (12), 3, 1950: 142 (synonyms, review of species). Svetovidov, Ryby Chernogo Morya, 1964: 386. Gibbs and Collette, Fish Bull. Fish Wildlife Serv., 66, 1, 1966: 97 (Synonymy, analysis of species and subspecies composition).

Germo Jordan, Proc. Acad. Nat. Sci. Philad., 40, 1888: 180 (type: *Scomber alalunga* Gmelin).

Parathunnus Kishinouye, J. Coll. Agr. Univ., Tokyo, 8, 3, 1923: 442

---

[8]Distributed in the Indian Ocean and western part of the Pacific Ocean (Zharov et al., 1961: 19).

[9]Known from the Kara Sea, off the coasts of Java, Celebes, Ryukyu, the Philippines, Marshall, and Hawaiian Islands, as well as the coast of Australia (Beaufort and Chapman, 1951: 216).

(type: *Thunnus mebachi* Kishinouye = *Thynnus obesus* Lowe). Rivas, Ann. Mus. Civico Storia Nat. Giacomo Doria, 72, 1961: 126.

Neothunnus Kishinouye, J. Coll. Agr. Univ., Tokyo, 8, 3, 1923: 445 (type: *Thynnus macropterus* Temminck and Schlegel = *Scomber albacares* Bonnaterre). Rivas, Ann. Mus. Civico Storia Nat. Giacomo Doria, 72, 1961: 126.

Kishinoella Jordan and Hubbs, Mem. Carnegie Mus., 10, 2, 1925: 219 (type: *Thunnus rarus* Kishinouye = *Thynnus tonggol* Bleeker).

Body moderately elongate, entirely covered with scales, usually enlarged along lateral line and in anterior part, and forms an armor (corselet). Lateral line usually slightly sinuous. Caudal peduncle with large median keel on each side and two small keels on upper and lower sides at posterior end of median keel (between caudal lobes). Dorsal fins contiguous or separated by space shorter than diameter of eye. Interpelvic process large, not fused, and forms two processes posteriorly pointed. Ring of infraorbital bones incomplete, without canal in system of lateral line. Teeth on jaws small, arranged in one row, conical; those on palatines and vomer villiform. Gill rakers short, not more than 30 on 264 lower half of first gill arch. Vertebrae 39-41. Swim bladder present or absent. This genus, like other closely related genera, characterized by strong development of cutaneous vascular system connected with vascular network in lateral muscles of body, in parts adjoining vertebral column on both sides. Vascular network also present on inner side of liver and in hemal canal (Svetovidov, 1964: 386).

Soviet literature has periodically been enriched with studies on the biology and fishery of tunas (Isipov, 1928; Rumyantsev and Kizevetter, 1949; Metelkin, 1957; Moiseev, 1957; Osipov, 1960, 1965a, 1965b, 1966, 1967, 1968a, 1968b, 1968c, 1968d, 1970; Zharov et al., 1961, 1964; Martinsen et al., 1965; Zharov, 1966; Shabotinets, 1968). In recent years a large volume of literature has appeared on the biology and migration of tunas by tagging (Mather, 1963a, 1963b; Brock, 1965; Fink, 1966; Kask, 1966; Mather and Bartlett, 1966; Yamanaka, 1966a, 1966b). Interesting work has also been undertaken on the larvae and juveniles (Yabe and Ueyanagi, 1962; Jones and Kumaran, 1963; Matsumoto, 1966; Scaccini, 1966); trophic relationships, feeding, and behavior (Talbot and Penrith, 1963; Thomas and Kumaran, 1963; Waldron and Ring, 1963; Magnuson, 1966; Sivasubramaniam, 1966; Ueyanagi, 1966a; H. Nakamura, 1969); and oceanic distribution of tunas depending on biotic and abiotic factors [Shepers, 1935; Godsil, 1945; Nakamura and Yamanaka, 1959; Hamre, 1963; Blackburn, 1965; Nakamura, 1966, 1969; Postel, 1969; Frey (ed.) 1971]. Analysis of the taxonomy of this genus, and anatomy, morphology, and biology of tunas are also available (Clothier, 1950; Watson, 1963; Mather, 1963a, 1963b; Iwai and Nakamura, 1964b). The bibliography compiled by the World Scientific Meeting on the Biology of Tunas deserves special mention (see: Proceedings..., 1964).

7 widely distributed species.[10] Tropical and subtropical seas in the Atlantic, Pacific, and Indian oceans. 3 species known from the Sea of Japan, and another 2 from adjacent waters. Japanese ichthyologists (Iwai et al., 1965), noting that the external characters of tunas vary with age, have developed three keys of species of this genus; one is based on external characters and the other two on anatomical features. These authors have given much importance to the species-dependent structural peculiarities of the olfactory organ and consider it possible to identify subspecies through this feature.

1. Key to Species of Genus Thunnus
(Based only on external morphological characters;
Iwai et al., 1965: 25)

1 (4). Length of pectoral fin less than 4/5 head length; tip does not reach vertical from origin of second dorsal fin.

2 (3). Keel of caudal peduncle yellow. Length of pectoral fin 4.4 to 4.6 times in body length ....
.... [**T. maccoyii** (Castelnau, 1872)].[11]

3 (2). Keel of caudal peduncle black (translucent in immature fish). Length of pectoral fin 4.6 to 6.0 times in the body length. .
.... 1. **T. thynnus** (L.).

265 4 (1). Length of pectoral fin more than 4/5 head length; tip either slightly shorter than vertical from origin of second dorsal or extends slightly beyond it.

5 (6). Posterior margin of caudal fin with white border. Pectoral fins very long, almost reach vertical from second dorsal finlet.
.... 2. **T. alalunga** (Bonnaterre).

6 (5). Posterior margin of caudal fin without white border. Pectoral fins in adult fish long but do not reach vertical from first dorsal finlet.

7 (10). Gill rakers on first gill arch 25–33.

8 (9). Body stout, its depth greater than longitudinal cleft of tail. Second dorsal and anal fins not very long. Anal fin originates behind midpoint between posterior margin of operculum and posterior end of caudal keel. Eyes and head large...........
.... 3. [**T. obesus** (Lowe)].

[10]Japanese (Iwai et al., 1965) and American ichthyologists (Gibbs and Collette, 1966a, 1966b), on the basis of analyses of proportions of body parts, skeleton, seismosensory system, and structure of internal organs, have convincingly demonstrated that there are only seven species in the genus *Thunnus*.

[11]Widely distributed in the southern part of the Pacific Ocean from New Zealand to Peru (Iwai et al., 1965: 34).

9 (8). Body fairly slender, its depth less than longitudinal cleft of tail. Second dorsal and anal fins distinctly elongate. Anal fin originates ahead of midpoint between posterior margin of operculum and posterior end of caudal keel. Eyes and head comparatively small............ 4. **T. albacares** (Bonnaterre).

10 (7). Gill rakers on first gill arch 19-25.

11 (12). Finlets of live fish not yellow (dorsal finlets of live fish bluish and anal finlets dark gray), turning yellow only after death. Origin of second dorsal fin behind midpoint between snout tip and end of caudal peduncle......................
.... [**T. atlanticus** Lesson, 1830].[12]

12 (11). Finlets of live fish yellow. Origin of second dorsal fin near midpoint between end of snout and posterior end of keel of caudal peduncle...................... 5. **T. tonggol** (Bleeker).

2. Key to Species of Genus Thunnus
(Based on internal and anatomical characters; Iwai et al., 1965: 25)

1 (6). Closed hemal arch appears on 10th vertebra. Blood vessels well developed on ventral side of liver.

2 (3). Parasphenoid narrow. First hemal spine distinctly broad and flat... 2. **T. alalunga.**

3 (2). Parasphenoid wide. First hemal spine not compressed.

4 (5). Alisphenoid bone extends below center or orbit and distinctly attached to parasphenoid in adult fish........ 1. **Th. thynnus.**

5 (4). Alisphenoid bone externds in center of orbit but distinctly not attached to parasphenoid in adult fish..... [**Th. maccoyii**].

6 (1). Closed hemal arch appears on 11th vertebra. Blood vessels absent or if present on ventral surface of liver, only at edges.

7 (8). Blood vessels on ventral side of liver present only at edges. Inferior foramina very small. Posthemal zygapophysis short...
....3. **Th. obesus.**

8 (7). Blood vessels on lower surface of liver absent. Inferior foramina large. Posthemal zygapophysis long, spine-shaped.

9 (10). Depression present in center of lower surface of parasphenoid. .... [**Th. atlanticus**].

10 (9). Depression absent in center of lower surface of parasphenoid, surface flat.

11 (12). Swim bladder present....................... 4. **Th. albacares.**

12 (11). Swim bladder absent........................ 5. **Th. tonggol.**

---

[12]Distributed in the Atlantic basin within the limits of the subtropical, partly temperate, and tropical regions (Zharov et al., 1964: 9).

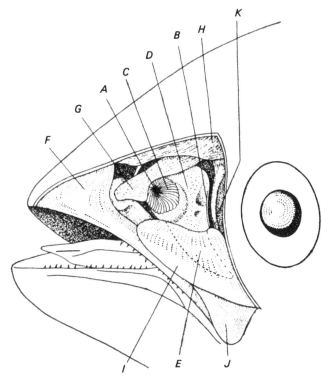

Figure 213. Longitudinal section of the nasal cavity of *Thunnus* (Iwai et al., 1965).

A—anterior nostril; B—posterior nostril; C—olfactory rosette; D—nasal cavity; E—additional nasal sac; F—premaxilla bone; G—nasal bone; H—frontal bone; I—lachrymal bone; J—posterior end of maxilla bone; K—sclera.

3. Key to Species of Genus Thunnus
[Based on structure of olfactory organ
(Figure 213); Iwai et al., 1965. 28]

1 (4). Olfactory rosette with outer fleshy process (Figure 214, A, B).

2 (3). Mucosal folds not reaching outer part of olfactory rosette (Figure 214, A)............................ 2. **Th. alalunga.**

3 (2). Mucosal folds in outer part of olfactory rosette noticeable although not significant (Figure 214, B)......... [**Th. obesus**].

4 (1). Olfactory rosette without outer fleshy process (Figure 214, C-G).

5 (8). Margins of mucosal folds usually entire, not serrate (Figure 214, C, D).

6 (7). Margins of mucosal folds in nasal cavity smooth (Figure 214, C). .... 1. **Th. thynnus**.

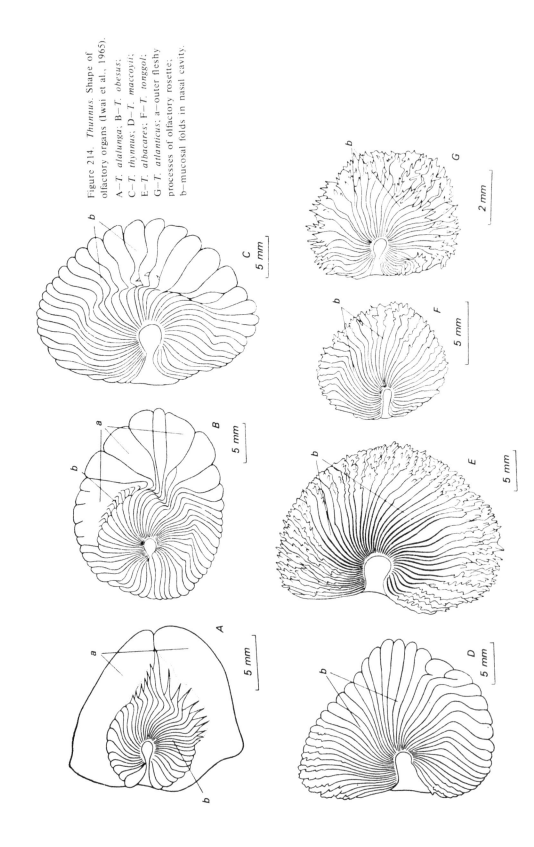

Figure 214. *Thunnus*. Shape of olfactory organs (Iwai et al., 1965).

A—*T. alalunga*; B—*T. obesus*; C—*T. thynnus*; D—*T. maccoyii*; E—*T. albacares*; F—*T. tongol*; G—*T. atlanticus*; a—outer fleshy processes of olfactory rosette; b—mucosal folds in nasal cavity.

268   7 (6). Margins of mucosal folds in adult fish with fine serrations (Figure 214, D)............................... **[Th. maccoyii]**.

   8 (5). Margins of mucosal folds not even and serrate to different degrees (Figure 214, E-G).

   9 (10). Margins of all mucosal folds in nasal cavity highly serrate (Figure 214, E). First gill arch with 27—34 gill rakers. ......
.... 4. **Th. albacares.**

   10 (9). Margins of mucosal folds with poorly defined serrations. Gill rakers 19-25.

   11 (12). Margins of mucosal folds slightly sinuate (Figure 214, F). ..
.... 5. **Th. tonggol.**

   12 (11). Margins of mucosal folds highly sinuate (Figure 214, G)....
.... **[Th. atlanticus]**.

1. ***Thunnus thynnus orientalis*** (Temminck and Schlegel, 1844)—Oriental Bluefin Tuna (Figure 215)

Thynnus orientalis Temminck and Schlegel, Fauna Japonica, Poiss., 1844: 94 (Japan).

Orcynus schlegelii Steindachner. In: Steindachner and Döderlein, Beiträge..., 3, 1884: 10, pl. 3, fig. 1 (Tokyo).

Thunnus thynnus, Jordan and Evermann, Fish. N. and M. Amer., 1896: 870. Walford, Univ. Calif. Press, Berkeley, 1937: 7, pl. 34 (color figure). Godsil and Byers, Calif. Fish and Game, Fish Bull., 60, 1944: 88, fig. 48 (anatomy). Fraser-Brunner, Ann. Mag. Nat. Hist., 12, 3, 1950: 143, fig. 4. Matsubara, Fish Morphol. and Hierar., 1955: 515. Zharov et al., Tuntsy..., 1961: 25, fig. 8 (description, synonyms, common names). Collette and Gibbs, Edit. U.S. Nat. Mus., 5, 1963: 38, pl. 10. Osipov et al., Tuntsy..., 1963: 19, fig. 9. Osipov et al., Tuntsy i Mecheobraznye..., 1964: 10. Martinsen et al., VNIRO, 1965: 1-122, figs.

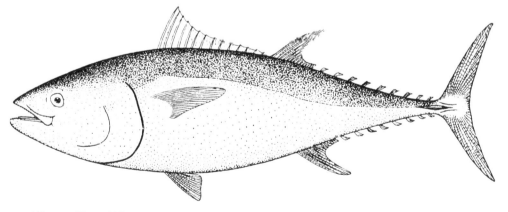

268   Figure 215. *Thunnus thynnus orientalis*—Oriental bluefin tuna (Iwai et al., 1965).

9, 10 (synonyms, description, map of distribution). Iwai et al., Misaki Mar. Biol. Inst., Spec. Rep., 2, 1965: 31, fig. 16 (detailed list of synonyms).

Thunnus orientalis, Kishinouye, Proc. Sci. Fisher, Assoc. Tokyo, I, 1915: 17, pl. 1, fig. 99. Kishinouye, J. Coll. Agric. Univ., Tokyo, 8, 3, 1923: 437, figs. 3, 21, 43, 44, 50. Metelkin, Promysel Tuntsvov, 1957: 7.

Thunnus thynnus orientalis, Serventy, Austr. J. Mar. and Freshwat. Res., 7, 1956: 11. Abe, Enc. Zool., 2, Fishes, 1958: 221, fig. 665 (color figures of adult and young specimens). Jones and Silas, Indian J. Fish., 7, 2, 1960: 381, fig. 8. Gibbs and Collette, Fish. Bull. U.S. Fish and Wildlife Serv., 66, 1, 1966: 117 (description, synonyms). Parin Scumbrievidnye Ryby, 1967: 91 (description, maps of distribution).

269   D XIII-XV, 14; dorsal finlets 8-9; A 13-15; anal finlets 7-8; P 31-38; *l. l.* about 230; gill rakers 9-16 + 21-28, total 32-43.[13]

Body depth 3.2 to 4.3 times in its length. Head length 3.2 to 3.5 times in body length. Caudal part of body fairly long, caudal peduncle slender. Body covered with minute cycloid scales, only in region of pectoral fins are scales in form of large elongate plates. Lateral line well marked, with curve above pectoral fin. Pectoral fins short, 4.8 to 6 times in body length. Height of first and second dorsal fins about equal. Second dorsal and anal fins similar in size and falciform; in specimens more than 200 cm in length, these fins markedly elongate. Eyes of adult fish small. Mouth large; posterior margin of maxilla continues beyond vertical from anterior margin of eye. Both jaws with minute conical teeth. Fleshy processes in outer part of olfactory rosette not developed; mucosal folds well defined, without notch along margin (Figure 214, C). Membrane of first dorsal fin yellow; second dorsal and anal fins grayish-yellow. Finlets grayish-yellow with black margin. Back dark blue, ventral surface silvery-white. Keel on caudal peduncle black. Body sides of young fish with more than 10 pale transverse stripes. Vertebrae 18 + 21 = 39 (Iwai et al., 1965: 31).

Differs from Atlantic bluefin tuna, *T. thynnus thynnus* (L.), in larger maximum body dimensions and larger maximum weight (Zharov et al., 1961).

Fecundity more than one million eggs (Zharov et al., 1961). First spawning at three years of age when length about 100 cm and weight about 20 kg (Osipov et al., 1963). Spawning also known in waters of the Sea of Japan (Osipov, 1968c: 8).

The biology of bluefin tuna has been studied numerous times (Esipov, 1928; Shepers, 1935; Schaefer and Marr, 1948; Shimada, 1951a, 1951b; Metelkin, 1957; Mather and Schuck, 1960; Osipov, 1960, 1968c; Zharov et al., 1961; Genovese, 1962; Orange and Fink, 1963; Osipov et al., 1963;

---

[13]Zharov and associates (1961) report: D (X) XII-XIV (XV) (12) 13-14 (15), 8-10; A 12-14, 7-9; P 31-33; V 6; gill rakers 9-13 + 21-26.

Robins, 1963, 1966; Zharov, 1965; Martinsen et al., 1965; Clemens, 1966; Flittner, 1966; Nakamura and Matsumoto, 1966; Shingu, 1966; Ueyanagi, 1966c; Parin, 1967; Clemens and Flittner, 1969). Many researchers have studied the anatomy and morphology (Starks, 1910, 1911; Kishinouye, 1923; Godsil and Byers, 1944; Godsil and Holmberg, 1950; Morice, 1953a, 1953c; Honma, 1956; Iwai and Nakamura, 1964a; Iwai et al., 1956). Dependence of body weight and length on age have also been analyzed (Roedel, 1953; Bell, 1963, 1964; Phelan, 1966; Yukinawa and Yabuta, 1967). But the larvae of this fish have not received adequate attention (Matsumoto, 1961; Yabe, Ueyanagi and Watanabe, 1966). Interesting information on the initiation of tuna fishing in the Far East seas is given by Oral (1926).

Weight, up to 700 kg.

Length, up to 3.5 m (Osipov, 1968c: 8).

Distribution: In the Sea of Japan known from Vladimir Bay (Taranetz, 1935: 90); Peter the Great Bay (Soldatov and Lindberg, 1930: 207); Pos'et Bay (Rumyantsev, 1950: 160); southern coast of Sakhalin (Parin, 1967: 91); west coast of Hokkaido (Ueno, 1971: 79); Sado Island (Honma, 1952: 143; 1963:18); Toyama Bay (Katayama, 1940: 8); and San'in region (Mori, 1956a; 23). Reported from the Sea of Okhotsk (Ueno, 1965c: 3); southern Kuril Islands (Parin, 1967: 91). Found off southern coast of Korean Peninsula (Mori, 1952: 173). Widely distributed in waters of the Pacific and Indian oceans (Nakamura and Warashina, 1965: 9).

270 **2. *Thunnus alalunga*** (Bonnaterre, 1788)–Albacore (Figure 216)

Scomber pinnis pectoralibus longissimus Cetti, Amfibi et pesci di Sardegna. Sassari, 1777: 191-193.

Scomber alalunga Bonnaterre, Tableau encyclopédique et méthodique des trois régnes de la nature. Ichthologie. Paris, 1788: 139 (original description, based on Cetti) (Mediterranean Sea).

Germo alalunga, Jordan and Evermann, Fish. N. and M. Amer., 1896: 871 (description, synonyms). Osipov et al., Tuntsy i Mecheobraznye..., 1964: 67. Zharov et al., Tuntsy..., 1961: 28, fig. 9 (description, synonyms). Martisen et al., Tuntsy..., 1965: 10, fig. 3 (description, synonyms).

Thunnus germo, Kishonouye, J. Coll. Agric. Univ., Tokyo, 8, 3, 1923: 434, figs. 20, 46, 52. Godsil and Byers, Calif. Fish and Game, Fish Bull., 60, 1944: 70, figs. 36-47.

Thunnus alalunga, Fraser-Brunner, Ann. Mag. Nat. Hist., 12, 3, 1950: 143, fig. 5. Matsubara, Fish Morphol. and Hierar., 1955: 516. Abe, Enc. Zool., 2, Fishes, 1958: 222, fig. 658. Jones and Silas, Indian J. Fish., 7, 2, 1960: 382, fig. 9. Iwai et al., Misaki Mar. Biol. Inst., Spec. Rep., 2, 1965: 28, fig. 13. Collette and Gibbs, Edit. U.S. Nat. Mus., 5, 1963: 37.

344

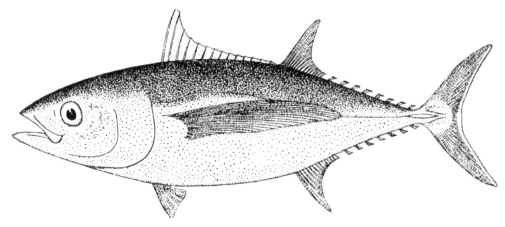

270   Figure 216. *Thunnus alalunga*–albacore (Iwai et al., 1965).

D XIII-XIV, 14-16[14]; dorsal finlets 7-8; A 14-15; anal finlets 7-8; P 31-34; *squ.* 120; gill rakers 7-10 + 18-22 = 25-31.

Body fusiform, its depth 3.6 to 4 times in its length. Head large (3.1 to 3.6 times in the body length). Caudal peduncle comparatively short, narrows sharply toward caudal fin. Body covered with minute cycloid scales. Scales on thorax in form of large plates but pectoral armor or corselet indistinct. Lateral line above pectoral forms smooth arch. Pectoral fins usually very long (2.2 to 3 times in body length). Height of first and second dorsal fins about equal. Anal fin almost same in size as second dorsal fin. Mouth large; posterior margin of oral slit extends beyond vertical from anterior margin of eye. Teeth on jaws arranged in 1 row, minute and conical. Fleshy processes in outer part of nasal rosette present; margins of olfactory folds with serrations (Figure 214, A). Membrane of first dorsal fin yellow. Finlets pale yellow with black border.
271 Posterior margin of caudal fin with white border. Back deep blue, belly silvery-white[15] (Iwai et al., 1965: 28).

Liver trilobate, middle lobe largest; large number of blood vessels on ventral side of liver (Morice, 1953b; Iwai et al., 1965).

Albacore attain sexual maturity at the age of five or six years, when the body length reaches 90 to 95 cm. Fecundity, about three million eggs.

Many authors have studied the anatomical characters of the albacore (Kishinouye, 1923; Godsil and Byers, 1944; Iwai et al., 1965; I. Nakamura, 1965), and have determined its age on the basis of skeletal structure

[14]Zharov and associates (1961: 29) report: D XIII-XV, 11-15; dorsal finlets 7-9; A 13-15; anal finlets 7-8; gill rakers about 27 (9 +18).

[15]Comparative characteristics of albacore from the Indian and Pacific oceans have been given by Kurogane (1959).

and scales (Otsu and Uchida, 1959a; Bell, 1962), spawning period, sexual maturity, and nature of spawning (Otsu and Uchida, 1959b; Otsu and Hansen, 1962; Yoshida, 1965; Kikawa and Ferraro, 1967). The oceanic distribution of larvae and juveniles has also been analyzed (Marchal, 1963; Nakamura and Matsumoto, 1966; Higgins, 1967).

Several researchers have studied the biology and behavior (Thompson, 1917; Otsu, 1960; Alverson, 1961; Clemens, 1961, 1962, 1966; Zharov et al., 1961; Radovich, 1962; Otsu and Uchida, 1963; Yoshida and Otsu, 1963; Walters and Fierstine, 1964; Clemens and Craig, 1965; Fink, 1966; Otsu and Yoshida, 1967; Parin, 1967; Osipov, 1968c; Rothschild and Yang, 1970). Albacore feed on small fishes, including Pacific saury, which constitutes about 50% of its diet (McHugh, 1952), squids, and planktonic crustaceans. The stocks of albacore in the central and especially the northern parts of the Pacific Ocean have not been studied sufficiently (Graham and McGary, 1961).

The food quality of the flesh of albacore ranks higher than that of all other tunas. Maximum weight, 45 kg.

Length, to 1.5 m[16] (Martinsen et al., 1965: 12).

Distribution: Rarely found in the Sea of Japan (Matsubara, 1955: 516). Known from the southern coast of the Korean Peninsula (Mori, 1952: 173). In the Sea of Okhotsk found near the southern Kuril Islands (Ueno, 1971: 79). In the northern Pacific Ocean found up to 45° N. Widely distributed in warm seas throughout the world (Iwai et al., 1965: 29).[17]

3. [*Thunnus obesus* (Lowe, 1839)−Bigeye Tuna] (Figure 217)

Thynnus obesus Lowe, Proc. Zool. Soc. Lond., 7, 1839: 78 (Madeira).

Thynnus sibi Temminck and Schlegel, Fauna Japonica, Poiss., 1844: 97, pl. 50 (Japan). Beaufort and Chapman, Fish. Indo-Austr. Arch., IX, 1951: 222 (synonyms, description). Rivas, Ann. Mus. Storia Nat. Genova, 72, 1961: 135 (synonyms).

Parathunnus mebachi Kishinouye, J. Coll. Agric. Univ., Tokyo, 8, 3, 1923: 442, figs. 4, 22, 47, 49. Godsil and Byers, Calif. Fish and Game, Fish Bull., 60, 1944: 105, fig. 59 (description, anatomy).

Parathunnus sibi, Abe, Enc. Zool., 2, Fishes, 1958: 221, fig. 657.

Parathunnus obesus, Matsubara, Fish Morphol. and Hierar., 1955: 516. Zharov et al., Tuntsy..., 1961: 30, fig. 10 (description, synonyms).

Thunnus obesus, Fraser-Brunner, Ann. Mag. Nat. Hist., 12, 3, 1950: 144, fig. 6 (description, synonyms). Rivas, Ann. Mus. Storia Nat. Genova, 72,

[16]Sometimes size more than 1.5 m in length and weight greater than 45 kg (Zharov et al., 1961).

[17]Maps of distribution of albacore have been published by Martinsen and associates (1965: 13, Figure 4) and Parin (1967), but the Sea of Japan not included in the area of distribution of this species.

1961: 133 (description, synonyms). Collette and Gibbs, Edit. U.S. Nat. Mus., 5, 1963: 40 (description, comparison with closely related species). Iwai et al., Misaki Mar. Biol. Inst., Spec. Rep., 2, 1965: 34, fig. 19 (synonyms, description). Gibbs and Collette, Fish. Bull. U.S. Fish and Wildlife Serv., 66, I, 1966: 109. Nakamura, Bull. Misaki Mar. Biol. Inst., Kyoto Univ., 8, 1965: 18 (anatomy). Parin, Skumbrievidnye Ryby..., 1967: 99. Osipov, Okeanskie Pelagicheskie Ryby, 1968: 10, fig. 3.

272   *Parathunnus obesus mebachi,* Jones and Silas, Indian J. Fish., 7, 2, 1960: 383, fig. 10, B (description).

D XIV-XV, 13-15; dorsal finlets 8-9; A 13-15; anal finlets 8-9; P 32-35; *squ.* 190; gill rakers $7-10 + 18-19 = 26-28$.

Body fusiform, thick. Body depth 3.3 to 3.5 times in length. Caudal part rather short, narrows sharply toward base of caudal fin. Scales small, cycloid, cover entire body. Pectoral corselet quite distinct in adult fish, not seen in young fish. Lateral line above pectoral fin curves sharply upward. Pectoral fins relatively long, 3.9 to 4.2 times in the body length. In young fish these fins reach a vertical from first dorsal finlet, and in adult fish almost up the vertical with the origin of second dorsal fin. Second dorsal fin slightly higher than first, falciform, and similar in shape and size to anal fin. Eyes relatively large. Mouth large. Teeth on jaws minute, conical. Fleshy processes developed in outer half of olfactory rosette; olfactory folds also developed, their margins elongate and serrate (Figure 214, B).

Membrane of first dorsal fin grayish-yellow. Second dorsal and anal fins pale yellow with black border.[18] Back deep yellow, body sides violet with yellowish tinge, ventral surface silvery-white. Young fish with grayish spots on ventral surface.

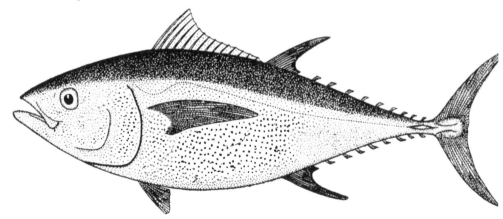

272   Figure 217. *Thunnus obesus*—bigeye tuna (Iwai et al., 1965).

[18]Zharov and associates (1951: 32) state that the finlets are not yellow, but dark, while Martinsen and associates (1965) add that the caudal and pectoral fins are reddish-black.

Liver trilobate, middle lobe largest. Blood vessels located along margins of liver on ventral surface. Vertebrae $18 + 21 = 39$ (Iwai et al., 1965: 34).

The anatomy of this species has been studied by scientists abroad (Godsil and Byers, 1944; Nakamura, 1965) and the biology studied in the Soviet Union (Metelkin, 1957; Zharov, 1961; Martinsen et al., 1965; Parin, 1967; Osipov, 1968).

Bigeye tuna move in large schools.[19] It feeds on small fishes and squids. The largest congregations of bigeye tuna form at a temperature of 21 to 22°C (Martinsen et al., 1965). Metelkin (1957) has stated that this is the only tuna species which dwells permanently at a depth greater than 20 m, but Martinsen has recorded it at depths greater than 200 m. Sexual maturity is achieved at the age of about three years and a body length of 90 to 100 cm.[20] Fecundity, 2,900 to 6,300 thousand eggs. Eggs pelagic, small (1.0 to 1.38 mm). Embryonic development takes place at 28 to 29°C and continues for about 21 hours (Parin, 1967). A few workers in other countries have provided information on the larvae and juveniles (Nakamura and Matsumoto, 1966; Higgins, 1967), and on growth and sexual dimorphism (Shomura and Keala, 1962).

Maximum weight, 197 kg; information has been published on catches of bigeye tuna weighing 272 kg (Parin, 1967).

Length, up to 236 cm (Parin, 1967).

273 *Distribution*:[21] Not found in the Sea of Japan. Known from southern Kuril Islands (Ueno, 1971: 79). Indicated off the coast of the Korean Peninsula (Mori, 1952: 173). Abe (1958: 221) has reported wide distribution from the middle part of Honshu, but Matsubara (1955: 516) has indicated only the Pacific coast for Japan. In the Atlantic Ocean found slightly more often north of 40° N. Common in tropical and subtropical waters of the world oceans (Mather and Gibbs, 1958: 237).

4. ***Thunnus albacares*** (Bonnaterre, 1788)—Yellowfin Tuna (Figure 218)
   Scomber albacares Bonnaterre, Tableau Encyclopédique et Méthodique..., Ichthyologie, Paris. 1788: 140 (Madeira).
   Thynnus macropterus Temminck and Schlegel, Fauna Japonica, Poiss., 1844: 98, pl. 52 (young specimen) (Japan).
   Neothunnus macropterus, Kishinouye, J. Coll. Agric. Univ., Tokyo, 8, 3, 1923: 445, fig. 45. Soldatov and Lindberg, Obzor..., 1930: 109.

[19]Kume (1966) reports the migration of this species, and Torin (1969) its vertical distribution.

[20]Sexual maturity and spawning of this species have been described by Japanese scientists (Kikawa, 1966; Kikawa and Ferraro, 1967), and information on its feeding habits is available in the work of Maksimov (1969).

[21]Maps of distribution and places of commercial fisheries published by Parin (1967) and Martinsen (1965).

Godsil and Byers, Calif. Fish and Game, Fish. Bull., 60, 1944: 47, fig. 20. Jones and Silas, Indian J. Fish., 7, 2, 1960: 385, fig. 12.

Semathunnus guildi Fowler, Proc. Acad. Nat. Sci. Philad., 85, 1933: 163-164, pl. 12 (Tahiti).

Thunnus macropterus, Beaufort and Chapman, Fish. Indo-Austr. Arch., IX, 1951: 223, fig. 39 (description, synonyms).

Thunnus albacora, Fraser-Brunner, Ann. Mag. Nat. Hist., 12, 3, 1950: 144, fig. 7 (description, synonyms).

Thunnus albacares, Rivas, Ann. Mus. Storia Nat., Genova, 72, 1961: 136 (synonyms). Collette and Gibbs, Edit. U.S. Nat. Mus., 5, 1963: 41 (comparison with other species). Iwai et al., Misaki Mar. Biol. Inst., Spec. Rep., 2, 1965: 36, figs. 20, 21 (detailed list of synonyms, description). Merrett and Thorp, Ann. Mag. Natur. History, London, 1965: 375. Parin, Skumbrievidnye Ryby..., 1967: 102 (synonyms, biology). Zharov, Zheltoperyi Tunets..., 1970: 1-121 (description of species, synonyms).

Thunnus argentivittatus, Rivas, Ann. Mus. Storia Nat., Genova, 72, 1961: 131.

Neothunnus albacora, Matsubara, Fish Morphol. and Hierar., 1965: 516. Abe, Encl. Zool., 2, Fishes, 1958: 221, fig. 655 (color figure). Martinsen et al., Tuntsy..., 1965: 5, fig. 1 (synonyms).

Neothunnus albacore, Zharov et al., Tuntsy..., 1961: 32 (synonyms, description).

D XII-XIV, 14-15; dorsal finlets 8-9; A 14-15; anal finlets 8-9; P 32-35; *l. l.* 220-270; gill rakers 8-11 + 19-24, total 27-34.

274 Body more oblong than in other tunas, body depth 3.6 to 4.1 times in its length. Anal fin shifted slightly forward, caudal part slender and long. Head not very large (3.1 to 4.0 times in body length). Scales small, cycloid, cover entire body. Corselet fairly large, better developed in young fish than in adults (barely distinguishable). Lateral line well developed, curves twice above origin of pectoral fin. Pectoral fins comparatively long (3.1 to 4.2 times in the body length), shorter in adults, and in older fish does not reach vertical from origin of second dorsal fin. Second dorsal and anal fins of larger fish distinctly elongate. First rays of adult fish in both fins equal to head length or longer. Eyes comparatively large. Mouth large. Jaws with minute conical teeth. Fleshy processes absent on margins of nasal rosettes; olfactory folds distinctly developed, their margins serrate (Figure 214, D). Membranes of dorsal and anal fins golden-yellow, finlets with black border. Back deep blue, sides golden, abdomen silvery. In fish up to 130 cm in length, body sides with sinuous pattern. Ventral surface of liver without blood vessels. Vertebrae $18 + 21 = 39$ (Iwai et al., 1965: 36).

The anatomy of the yellowfin tuna has been studied rather well (Kishinouye, 1923; Godsil and Byers, 1944; Iwai et al., 1965; Nakamura,

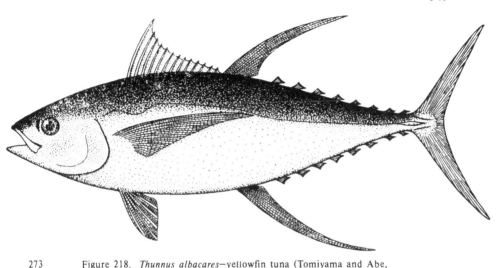

273   Figure 218. *Thunnus albacares*—yellowfin tuna (Tomiyama and Abe,
1958).

1965). Its fecundity and the dependence of fecundity on body size have also been well studied (Joseph, 1963). Other aspects covered in detail include maturation and spawning (Moore, 1951; Orange, 1961; Kikawa and Ferraro, 1967; Yuen, 1967), larvae and juveniles (Nakamura and Matsumoto, 1966; Higgins, 1967), population density and composition (Godsil and Greenhood, 1951; Hennemuth, 1961; Joseph et al., 1964), feeding (Nakamura, 1950; Reintjes and King, 1953; King and Ikehara, 1956; Watanabe, 1958; Alverson, 1963), and other biological characters (Nakamura et al., 1951; Iversen, 1956; Metelkin, 1957; Yabuta and Yukinawa, 1958; Zharov et al., 1961; Kolesnikov et al., 1961; Mimura and staff, 1962; Radovich, 1962; Davidoff, 1963; Ronquillo, 1963; Schaefer et al., 1963; Mimura, 1964; Osipov et al., 1964; Royce, 1964; Martinsen et al., 1965; Joseph, 1966; Kikawa, 1966; Zharov, 1967, 1970a, 1970b, 1970c, 1970d; Parin, 1967; Osipov, 1968b; Geft, 1970).

Average weight about 50 kg. Rare specimens have weighed up to 200 kg (Zharov et al., 1964: 16).

Length, up to 170 cm (Iwai et al., 1965: 38).

Distribution:[22] In the Sea of Japan known from west coast of Hokkaido (Ueno, 1971: 79); southward (Matsubara, 1955: 516); indicated for Sado Island (Honma, 1963: 18); Toyama Bay (Katayama, 1940: 8); and San'in

---

[22]Distribution of yellowfin tuna reported by Martinsen and associates (1965: 8, Figure 2). Parin (1967: 103) in Figure 21 depicts not only the boundaries of the area of distribution, but also the areas of commercial catches and the boundaries of spawning in the area of distribution. Neither author includes the Sea of Japan in the area of distribution of this species.

350

region (Mori, 1956a: 23). Found off the south coast of the Korean Peninsula (Mori, 1952: 174) and farther south up to Ryukyu Islands and Taiwan (China). Hawaiian Islands (Okada and Matsubara, 1938: 149). Widely distributed in the Pacific, Indian, and Atlantic oceans (Iwai et al., 1965: 36).

5. *Thunnus tonggol* (Bleeker, 1852)—Longtail Tuna (Figure 219)

Thynnus tonggol Bleeker, Nat. Tijdschr. Ned. Ind., I, 1851: 356; Verh. Bat. Gen., XXIV, 1852, Bijdr. Makreelacht. Visschen: 89 (description, synonyms) (Indonesia).

Neothunnus rarus Kishinouye, J. Coll. Agric. Univ., Tokyo, 8, 3, 1923: 448, figs. 24, 48, 64 (Japan).

Thunnus tonggol, Fraser-Brunner, Ann. Mag. Nat. Hist., 12, 3, 1950: 145, fig. 8 (description, synonyms). Beaufort and Chapman, Fish. Indo-Austr. Arch., IX, 1951: 224 (synonyms, description). Iwai et al., Misaki Mar. Biol. Inst., Spec. Rep., 2, 1965: 39, fig. 23 (description, synonyms). Zhu et al., Ryby Yuzhno-Kitaiskogo Morya, 1962: 766, fig. 620. Collette and Gibbs, Edit., U.S. Nat. Mus., 5, 1963: 43 (description, comparison with other species). Gibbs and Collette, Fish. Bull. U.S. Fish and Wildlife Serv., 66, I, 1966: 65 (systematics, anatomy).

Neothunnus tonggol, Matsubara, Fish Morphol. and Hierar., 1955: 516.

Kishinoella tonggol, Jones and Silas, Indian J. Fish., 7, 2, 1960: 384, fig. 11. Zharov et al., Tuntsy..., 1961: 36, fig. 14 (synonyms, description). Martinsen et al., Tuntsy..., 1965: 27, fig. 11 (synonyms, biology).

D XIII, 14-15; dorsal finlets 8-9; A 13-14; anal finlets 8-9; P 30-35; *l. l.* 210-220; gill rakers 5-8 + 14-17 = 20-25.

Body fusiform, distinctly elongate, body depth 4.0 to 4.6 times in its length. Caudal part comparatively long. Head relatively small, 3.5 to 4.0 times in the body length. Scales small, cycloid, cover entire body. Corselet indistinguishable. Lateral line forms double curve above

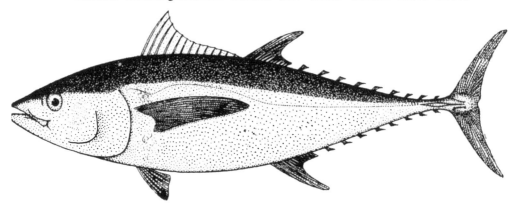

Figure 219. *Thunnus tonggol*—longtail tuna (Iwai et al., 1965).

pectoral fins. Pectoral fins rather long, 4.8 to 6.4 times in body length. Mouth relatively large. Both jaws with minute conical teeth. Fleshy processes absent on margins of olfactory rosette; olfactory folds developed and serrate along margin (Figure 214, F). Unpaired fins yellow. Finlets with black border. Back deep blue, abdomen silvery-white, body sides with minute pale spots. Swim bladder absent. Liver trilobate, right lobe large and highly elongate. Blood vessels absent on ventral surface of liver. Vertebrae $18 + 21 = 39$ (Iwai et al., 1965; 39).

Anatomical details of this species have been reported by Soviet and other researchers (Kishinouye, 1923: 448; Iwai et al., 1965: 39; Martinsen et al., 1965: 27; Nakamura, 1965: 24).

Detailed descriptions and biology have been furnished by Zharov and associates (1961), Martinsen et al. (1965) and Jones (1963). Distribution of larvae and juveniles has been studied by Matsumoto (1966).

Length, up to 100 cm (Iwai et al., 1965: 39).

Distribution: In the Sea of Japan known from Wakasa Bay (I. Nakamura, 1969: 160). Found off the south coast of the Korean Peninsula (Mori, 1952: 174). Pacific coast of Japan south to Kyushu Island (Matsubara, 1955: 516). South China Sea (Zhu et al., 1962: 766). Tropical and subtropical waters of the Indian and Pacific oceans, and south to 37° S (Martinsen et al., 1965: 27).

276

2. Genus *Sarda* Cuvier, 1829—Bonitos

Sarda Cuvier, Régne Animal., ed, 2, 2, 1829: 199 (type: *Scomber sarda* Bloch). Jordan and Evermann, Fish N. and M. Amer., 1896: 871. Kishinouye, J. Coll. Agric. Univ., Tokyo, 8, 3, 1923: 424. Soldatov and Lindberg, Obzor..., 1930: 112. Godsil, Calif. Fish and Game, Fish Bull., 99, 1955: 42. Svetovidov, Ryby Chernogo Morya, 1964: 389.

Body rather elongate, not obese, covered with minute scales, which form a more or less distinct corselet in the region of the pectoral fins. Caudal peduncle slender, with strong keel. Head large, pointed, compressed laterally. Mouth large. Teeth on jaws fairly strong, conical, noncutting, slightly compressed on sides; such teeth also present on palatines; vomer and tongue without teeth. Maxilla not concealed under preorbital. Gill rakers long and strong. First dorsal fin long and fairly low, with 18-22 spines, that gradually reduce in size posteriorly. Space between fins short. Second dorsal fin small, followed by 8-9 finlets. Anal fin similar in shape and size to second dorsal fin. Paired fins small. Lateral line simple (Soldatov and Lindberg, 1930: 112).

2[23] or 3 species found in the Atlantic and Pacific oceans. 1 species known in the Sea of Japan.

[23]Godsil (1955) recognizes only 2 species—*S. sarda* and *S. orientalis*; in his opinion the other 5 species proposed by various authors are no more than geographic variations.

352

1. *Sarda orientalis* (Schlegel, 1844)—Oriental Bonito (Figure 220).

Pelamys orientalis Schlegel. In: Temminck and Schlegel, Fauna Japonica, Poiss., 1844: 99, pl. 52 (Japan).

Sarda orientalis, Kishinouye, J. Coll. Agric. Univ., Tokyo, 8, 3, 1923: 424, fig. 33 (synonyms, description). Soldatov and Lindberg, Obzor..., 1930: 113 (description, synonyms). Fraser-Brunner, Ann. Mag. Nat. Hist., 12, 3, 1950: 147, fig. 12. Zharov et al., Tuntsy..., 1961: 58 (description, synonyms).

D XIX, 15 + 7-8; A 15 + 5-6; gill rakers 3-4 + 7-9; vertebrae 25 + 20.

Body elongate, fusiform in adults, but fairly short and compressed laterally in young fish. Head large, its length 3.25 to 3.50 times in body length (from tip of snout up to end of keel on caudal peduncle). Mouth broad; maxilla extends beyond eye. Teeth large, strong, curved; about 16 on upper jaw and 10-13 on lower jaw (Soldatov and Lindberg, 1930).

Not only the body shape but also color changes with age.

The anatomy of this species has been studied in detail by scientists abroad (Starks, 1910; Kishinouye, 1923; Godsil, 1955).

The biology of adults, larvae, and juveniles of oriental bonito has only been studied in the last decade (Klawe, 1961; Silas, 1962, 1963; Kikawa, 1963; Magnuson and Prescott, 1966).

The oriental bonito is of commercial value. Weight about 3 kg (Soldatov and Lindberg, 1930).

Length, up to 1 m (Abe, 1958).

Distribution: In the Sea of Japan known from Pos'et Bay (Tokarev, 1948: 43); west coast of Hokkaido (Ueno, 1971: 79); near Aomori (Soldatov and Lindberg, 1930: 113); Sado Island (Honma, 1952: 142); Toyama Bay (Katayama, 1940: 8); San'in region (Katoh et al., 1956: 316); and farther south of central Honshu. Found in large numbers at Kyushu (Matsubara, 1955: 516). South China Sea (Zhu et al., 1962: 768). Tropical and subtropical waters of the Pacific, Indian, and Atlantic oceans (Fraser-Brunner, 1950: 148).

3. Genus *Euthynnus* Lütken, 1882—Little Tunas

Euthynnus Lütken. In: Jordan and Gilbert, Bull. U.S. Nat. Mus., 16, 1882: 429 (type: *Thynnus thunnina* Cuvier and Valenciennes). Kishinouye, J. Coll. Agric. Univ., Tokyo, 8, 1923: 456.

Body thick, roundish, with corselet in anterior part. Mouth usually large; maxilla reaches vertical from center of eye. Teeth better developed and more numerous than in the genus *Katsuwonus*; present not only on jaws but also on palatines and sometimes on vomer; teeth on palatines arranged in single row. Dark spots on part of back and usually a few gray spots under pectoral fin (Kishinouye, 1923: 456).

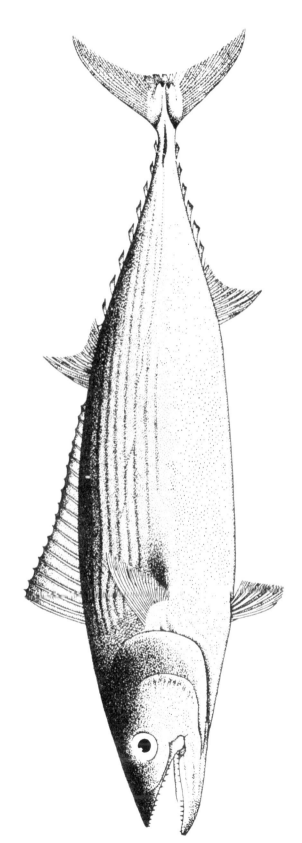

Figure 220. *Sarda orientalis*–Oriental bonito. Length 450 mm. Japan (Temminck and Schlegel, 1844).

354

Several species. In the western part of the Pacific Ocean 1 species with 2 subspecies known, 1 of which is found in the Sea of Japan.

Key to Subspecies of Euthynnus affinis

1 (2). Snout length equal to or less than half length of postorbital part of head. Head length about 3 3/4 to 3 5/6 times in the body length (up to end of caudal keel). Second dorsal fin originates anterior to vertical from midpoint between margin of operculum and end of caudal keel............ **[E. a. affinis** (Cantor, 1850)].[24]

2 (1). Snout length more than half postorbital part of head. Head length 3 1/4 to 3 1/3 in the body length (up to end of caudal keel). Second dorsal fin originates at or behind vertical from midpoint between margin of operculum and end of caudal keel. ....
.... 1. **E. a. yaito** Kishinouye.

1. *Euthynnus affinis yaito* (Kishinouye, 1923)−Yaito Tuna, (Figure 221)
Thynnus thunnina Schlegel, Fauna Japonica, Poiss., 1844: 95, pl. 48 (Japan).

Euthynnus yaito Kishinouye, Proc. Sci. Fisher. Assoc., Tokyo, 1, 1915: 22, pl. 1, fig. 15 (Japan). Kishinouye, J. Coll. Agric. Univ., Tokyo, 8, 3, 1923: 457, pl. 30, fig. 54 (synonyms, description).

Euthynnus affinis yaito, Fraser-Brunner, Ann. Mag. Nat. Hist., 12, 2, 1949: 624, fig. 1 (b) (synonyms). Matsubara, Fish Morphol. and Hierar., 1955: 517. Schultz et al., Bull. U.S. Nat. Mus., 202, 2, 1960: 415, pl. 123, C (description).

D XV–XVI, 12–13; 8 dorsal finlets; A 13; 7 anal finlets; gill rakers 8–10 + 22–24. Upper jaw with 27–30 teeth; lower jaw with 24–27 (Kishinouye, 1923: 458).

Differs from the typical subspecies in longer upper jaw, larger size of head, and shorter caudal part (Fraser-Brunner, 1949: 625).

Back bluish-black with numerous dark oblique stripes. Belly silvery, with three or more grayish spots under pectoral fin. Fins black or gray; pelvic fins partly black, but along margin chalk-white. Black spot under each eye.

Nongregarious; voracious, feeds on small fishes and plankton. Spawns in May off Taiwan (China). Flesh with pleasing taste (Kishinouye, 1923). Biology reviewed by Kikawa (1963) and Williams (1963).

Length, to 1 m (Osipov, 1968c: 13).

Distribution: In the Sea of Japan known from San'in region (Mori, 1956a: 24); Tsushima Islands (Arai and Abe, 1970: 87); and reported from the central part of Honshu southward (Matsubara, 1955: 517). Cheju-do

[24]Distribution from Taiwan (China) southward: Indonesia, Australia, Indian Ocean (Matsubara, 1955: 517).

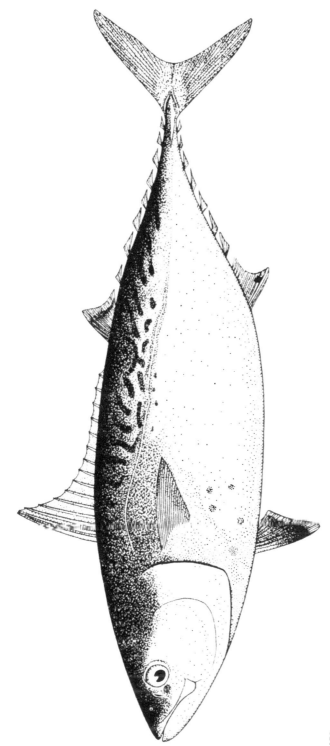

Figure 221. *Euthynnus affinis yaito*—Yaito tuna. Length 600 mm. Sea of Japan (Kishinouye, 1923).

Island (Mori, 1952: 174). South China Sea (Zhu et al., 1962: 772); Taiwan (China); and southwestern part of the Pacific Ocean (Matsubara, 1955: 517).

4. Genus *Katsuwonus* Kishinouye, 1915—Skipjack Tuna

Katsuwonus Kishinouye, Proc. Sci. Fisher. Assoc., Tokyo, 1, 1915: 21 (type: *Scomber pelamis* L.). Kishinouye, J. Coll. Agric. Univ., Tokyo, 8, 3, 1923: 452. Soldatov and Lindberg, Obzor..., 1930: 104. Matsubara, Fish Morphol. and Hierar., 1955: 517.

Body thick, roundish in cross section. Teeth present only on jaws, about 40 teeth on each jaw. In this respect *Katsuwonus* differs from the genus *Euthynnus*. Bases of dorsal fins almost contiguous, which differentiates this genus from *Auxis* (Soldatov and Lindberg, 1930). Studies on the blood composition of Atlantic and Pacific skipjack tunas (Grinols, 1969) deserve attention, as they establish interpopulation differences within the species.

One species. Widely distributed in the subtropical and tropical waters of the Atlantic, Indonesia, and Pacific oceans. Also known from the Sea of Japan.

1. *Katsuwonus pelamis* (Linné, 1758)—Skipjack Tuna (Figure 222)

Katsuwonus pelamis Linné, Syst. Nat., ed. 10, 1758: 297 (tropical seas). Kishinouye, J. Coll. Agric. Univ., Tokyo, 8, 3, 1923: 453, figs. 5, 14, 19, 25, 52, 57 (synonyms, description). Soldatov and Lindberg, Obzor..., 1930: 104 (synonyms). Schultz et al., Bull. U.S. Nat. Mus., 202, 2, 1960: 413 (synonyms, description). Zharov et al., Tuntsy..., 1961: 37, fig. 15 (synonyms, description). Martinsen et al., Tuntsy..., 1965: 15, fig. 5 (description, synonyms). Parin, Skumbrievidnye Ryby..., 1967: 106.

Euthynnus (*Katsuwonus*) *pelamis,* Fraser-Brunner, Ann. Mag. Nat. Hist., 12, 3, 1950: 152, fig. 19 (synonyms).

Euthynnus pelamis, Beaufort and Chapman, Fish. Indo-Auster. Arch., IX, 1951: 217 (synonyms, description).

D XII–XVII, 11–14; 8 dorsal finlets; A 11–15; 7 anal finlets; gill rakers on first gill arch 15–20 + 36–39; extremely slender (Soldatov and Lindberg, 1930). Vertebrae 20 + 21 (Abe, 1958). Scales on body only in region of corselet and along lateral line. Back with light blue hue, sides and belly white, and lower part of sides with dark bluish-chocolate-brown longitudinal stripes that extend from pectoral fin up to tail.[25] Vivid coloration dulls quickly after death (Roedel, 1953). First dorsal fin falciform. Swim bladder absent.

---

[25]Matsumoto et al. (1969) report that in some specimens of this species such stripes are absent.

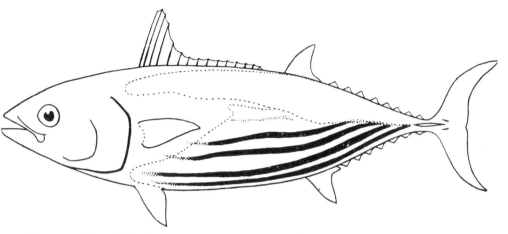

280   Figure 222. *Katsuwonus pelamis*–skipjack tuna (Jones and Silas, 1960).

280   The anatomy of this species has been studied fairly well (Kishinouye, 1923; Godsil and Byers, 1944; Godsil, 1954). Detailed studies have been conducted by Japanese ichthyologists (Fujino, 1966, 1967, 1969a, 1969b; Fujino and Kang, 1968a, 1968b; Fujino and Kazama, 1968) on the blood of skipjack tuna from various regions of the world oceans to determine their genetic relations.

The skipjack tuna is one of the most warmth-loving and smallest of tunas. It is generally found in the surface water layers and never submerges more than 100 m. It forms large schools of up to 50,000 fish. Spawning takes place throughout the year in batches. The skipjack feeds on sardines, young fish, squids, mollusks, and small crustaceans (Martinsen et al., 1965: 16). The biology of this species has been studied rather well by Soviet specialists (Metelkin, 1957; Zharov et al., 1961; Osipov et al., 1963, 1964; Martinsen, 1965; Parin, 1967; Osipov, 1968a, 1968c), as well as scientists abroad (Brock, 1954; Rivero and Fernandez, 1954; Matsumoto, 1959; Orange, 1961; Ch. Roux, 1961; Radovich, 1962; Jones and Silas, 1963; Waldron, 1963; E. Nakamura, 1965; Yao, 1966; Rothschild, 1967; Inoue et al., 1968; Yuen, 1970).[26]

The flesh of skipjack tuna is commercially canned. Weight up to 25 kg (Osipov, 1968c: 15).

Length, to 1 m (Metelkin, 1957: 5).

Distribution: In the Sea of Japan known from the coast of Hokkaido (Ueno, 1971: 79); Sado Island (Honma, 1963: 18); Toyama Bay (Katayama, 1940: 8); San'in region (Mori, 1956a: 23); and southern Kuril Islands (Ueno, 1971: 79). Cheju-do Islands (Mori, 1952; 174). Along the

[26]Bibliography on the biology of skipjack tuna published by Klawe and Miyake (1967).

Pacific coast of Japan south of Hokkaido (Matsubara, 1955: 517); specifically found near Koti Prefecture (Kamohara, 1959: 6). South China Sea (Zhu et al., 1962: 771) and Indonesia (Weber and Chapman, 1951: 152). Tropical and subtropical seas throughout the world (Matsubara, 1955: 517).

5. Genus *Auxis* Cuvier, 1829—Frigate Mackerels

Auxis Cuvier, Régne Animal, 2, 2, 1829: 119 (type: *Scomber rochei* Risso, 1810 = *Scomber thazard* Lacépède, 1802). Jordan and Evermann, 281 Fish. N. and M. Amer., 1, 1896: 867. Kishinouye, J. Coll. Agric. Univ., Tokyo, 8, 3, 1923: 460. Soldatov and Lindberg, Obzor..., 1930: 104. Fraser-Brunner, Ann. Mag. Nat. Hist., 12, 3, 1950: 152. Beaufort and Chapman, Fish. Indo-Austr. Arch., IX, 1951: 226. Collette and Gibbs, Edit. U.S. Nat. Mus., 117, 5, 1963: 32.

Body elongate, stout, naked on back, front covered with minute scales. Scales in pectoral region larger, forming corselet. Snout rather short, conical, slightly compressed laterally. Mouth relatively small. Jaws equal in length; teeth very small and mostly arranged in single row. Caudal portion of body slender, flat, with fairly large keels on each side. First dorsal fin short, separated from second fin by large space. Second dorsal and anal fins small, with 7–8 finlets behind each. Pectoral fins and pelvic fins small. Swim bladder absent. Branchiostegal rays 7. Pyloric caeca branched. Gill rakers numerous, very long, and slender. Vertebrae 39 (Soldatov and Lindberg, 1930: 104).

Found everywhere in tropical and subtropical waters; also known from the Sea of Japan. The species composition of the genus *Auxis* needs to be confirmed by special studies. At present some ichthyologists (Fraser-Brunner, 1950; Smith, 1950; Beaufort and Chapman, 1951; Jones and Silas, 1960; Zharov et al., 1961; Martinsen et al., 1965; Osipov, 1968a) believe that only one species exists—*A. thazard* Lacépède. Others (Kishinouye, 1923; Matsubara, 1955; Collette and Gibbs, 1963; Parin, 1967) recognize, in addition to *A. thazard,* the species *A. rochei* (Risso), which is distributed in the Indian Ocean. A study by Gorbunova (1961b) confirmed the difference between these species at all stages of postembryonic development. Matsumoto (1960) believes that in the waters of the Hawaiian Islands another independent species, *A. thynnoides* Bleeker, coexists with *A. thazard.*

1. *Auxis thazard* (Lacépède, 1802)—Frigate Mackerel (Figure 223)
   Scomber thazard Lacépède, Hist. Nat., Poiss., 3, 1802: 9 (New Guinea).
   Auxis hira Kishinouye, J. Coll. Agric. Univ., Tokyo, 8, 3, 1923: 462, figs. 55, 59 (Sea of Japan). Soldatov and Lindberg, Obzor..., 1930: 104.

Auxis maru Kishinouye, J. Coll. Agric. Univ., Tokyo, 8, 3, 1923: 463, figs. 2, 15, 27, 56, 60 (Yellow Sea). Soldatov and Lindberg, Obzor..., 1930: 104.

Auxis thazard, Fraser-Brunner, Ann. Mag. Nat. Hist., 12, 3, 1950: 152, fig. 20. Beaufort and Chapman, Fish. Indo-Austr. Arch., IX, 1951: 226 (synonyms, description). Smith, Sea Fish. S. Africa, 1950: 298, pl., 65, fig. 828 (color figure). Zharov et al., Tuntsy..., 1961: 41, fig. 18 (synonyms, description). Martinsen et al., Tuntsy..., 1965: 32, fig. 15 (synonyms, description).

D IX-XI, 11-13 + 6-9 finlets; A 12-15 + 6-8 finlets; P 23; gill rakers 9-10 + 30-36. Sinuous yellow to light blue spots present on back behind corselet. 3 keels on each side of caudal peduncle similar to other tunas. First and second dorsal fins short and separated from each other by space greater than length of base of first fin (Zharov et al., 1961: 42).

The biology of this species has not been studied well; only a few publications are available, and even these provide incomplete information (Hotta, 1955; Fitch and Roedel, 1963; Uchida, 1963; Williams, 1963; Yoshida and Nakamura, 1965).

The flesh of the frigate mackerel is edible but not in great demand (Zharov et al., 1964: 26).

Length, to 400 mm (Martinsen et al., 1965: 33).

282 *Distribution*: In the Sea of Japan known from Peter the Great Bay (Soldatov and Lindberg, 1930: 105); Pusan (Mori, 1952: 175); off Tsushima Islands (Arai and Abe, 1970: 87); west coast of Hokkaido (Ueno, 1971: 79); Sado Island (Honma and Kitami, 1967: 8); Toyama Bay (Katayama, 1940: 8); Wakasa Bay (Takegawa and Morino, 1970: 379); and San'in region (Mori, 1956a: 24). Along both coasts of Japan from Hokkaido southward (Matsubara, 1955: 518). Known from the Yellow Sea (Wang, 1935: 399), East China and South China seas (Zhu et al., 1962:

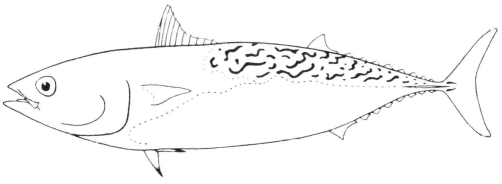

282   Figure 223. *Auxis thazard*—frigate mackerel (Fraser-Brunner, 1950).

770; 1963: 406). Tropical and subtropical waters of the Atlantic, Indian, and Pacific oceans (Martinsen et al., 1965: 33).

6. Genus *Scomber* Linné, 1758—Mackerels

Scomber Linné, Syst. Nat., ed. 10, 1, 1758: 297 (type: *S. scombrus* L.). Fraser-Brunner, Ann. Mag. Nat. Hist., 3, 26, 1950: 153 (review of species). Matsubara, Fish Morphol. and Hierar., 1955: 518. Svetovidov, Ryby Chernogo Morya, 1964: 397 (synonyms, description).

Pneumatophorus Jordan and Gilbert, Proc. U.S. Nat. Mus., 5, 1882: 593 (as subgenus, type: *S. colias* Gmel. = *S. japonicus* Houttuyn). Soldatov and Lindberg, Obzor..., 1930: 102 (synonyms, description). Manacop, Philipp. J. Fish., 4, 2, 1958: 80 (review of species).

Body fusiform, only slightly compressed laterally, completely covered with minute scales; corselet consists of enlarged scales in anterior part of body, slightly developed, or absent. Lateral line almost straight, with short sinuous curve. Caudal peduncle with two small lateral keels on each side between caudal lobes; oblong middle keel absent. Dorsal fins separated by wide space, greater than snout length. Interpelvic process small, fused, forms unpaired pointed appendage (Figure 212, D). Teeth small and conical on jaws; also present on palatines and vomer. Gill rakers medium long, thick, not fimbriate, and not more than 35 in the lower half of first gill arch. Vertebrae (30) 31 (32). Swim bladder present or absent (Svetovidov, 1964).

Few species. 2 species[27] known from the Sea of Japan.

283   *Key to Species of Genus Scomber*[28]

1 (2). First doral fin with 9-10 spines.[29] Body compressed laterally. Ventral surface silvery-white, without spots. ....
.... 1. **S. japonicus** Houttuyn.
2 (1). First dorsal fin with 11-12 spines. Body not compressed laterally. Ventral surface with many minute black spots.[30]..............
.... 2. **S. tapeinocephalus** Bleeker.

1. ***Scomber japonicus*** Houttuyn, 1782—Chub Mackerel (Figure 224)
Scomber japonicus Houttuyn, Verh. Holl. Maatsch. Wet., Haarlem., 20,

---

[27]Possibly Kishinouye (1923: 403) and Zharov (1961: 13) are right in recognizing only one species in Japan. Special studies are required to confirm this assumption, however. At present, we concur in the opinion of the Japanese ichthyologists (Matsubara, 1955: Abe, 1958).

[28]From Matsubara (1955).

[29]Often 9 according to studies conducted by Abe and Takashima (1958).

[30]Judging from the photograph published by Kadzawara and Ito (1953), a series of dark spots is well developed in *S. tapeinocephalus* along the median line of the body.

2, 1782: 331 (Japan). Kishinouye, J. Coll. Agric. Univ. Tokyo, 8, 3, 1923: 403, figs. 1, 7, 16, 28–30. Fraser-Brunner, Ann. Mag. Nat. Hist., 12, 3, 1950: 153, fig. 21 (synonyms, description). Smith, Sea Fish. S. Africa, 1950: 300, text-fig. 839, pl. 68, fig. 389 (color figure). Okada, Fishes of Japan, 1955: 134, fig. 124 (description). Matsubara, Fish Morphol. and Hierar., 1955: 518. Abe, Enc. Zool., 2, Fishes, 1958: 218, fig. 648 (color figure).

Pneumatophorus australasicus, Manacop, Philipp. J. Fish., 4, 2, 1958: 80, text-figs. 2, 3.

Pneumatophorus japonicus, Soldatov and Lindberg, Obzor..., 1930: 103, pl. 15. Abe, Enc. Zool., 2, Fishes, 1958: 218, fig. 647 (color figure). Zharov et al., Tuntsy..., 1961: 13, fig. 4 (synonyms and description).

1563. Japan. 1863. Maksimovich. 1 specimen.

7478. Honshu Island, Pacific coast. 1884. Polyakov. 1 specimen.

9541. Tokyo. 1891. Bunge. 1 specimen.

38467. Vladivostok. October 23, 1929. E.P. Rutenberg. 3 specimens.

In our specimens 120 to 340 mm long, first dorsal fin has 9 spines, gill rakers 13 + 1 + 26.

D (VIII) IX–X (XI), (I) II 10–11 + (4) 5; A I–III 9–11 + (4) 5; P II (16) 17–19; V I, 5; *l. l.* 200–233; vertebrae 31. Scales 40–60 anterior to first dorsal fin. Length of head 27.5 to 29.0% of standard length, of snout 9.0 to 9.5%, upper jaw 10.0 to 11.5%, predorsal distance 35.5 to 37.5%, and prepelvic distance 32.5 to 35.0%. Scales between second dorsal fin and lateral line 19–26, usually less than 23 (Soldatov and Lindberg, 1930).

The anatomy of chub mackerel has been described in foreign literature (Starks, 1910; Hotta, Abe and Takashima, 1958).

284 In live fish the back is greenish-blue with a metallic hue, with numerous narrow, sinuous, dark blue transverse stripes; sides white with yellowish tinge; and belly silvery-white. Warm water, schooling fish; remains in waters of Primor'e from the beginning of summer up to autumn. This mackerel lays eggs in water in June up to mid-July and remains close to the coast at this time (1 to 10 nautical miles out). Feeds on small crustaceans, squids, and small fishes (Kaganovskii et al., 1947). Its biology has been studied by many Soviet researchers (Pushkov, 1913; Pavlenko, 1919; Okhryamkin, 1931; Kaganovskii et al., 1947; Tokarev, 1948; Vedenskii, 1951, 1953, 1954a, 1954b, 1962; Kaganovskii, 1951; Probatov, 1951; Tyan Ir Khan, 1957; Dekhnik, 1959; Pushkareva, 1960; Zharov et al., 1961; Zvyagina, 1961; Kundius, 1964; Gorbunova, 1965a, 1965b, 1965c; Berenbeim, 1968; Fedorova, 1968; Vyskrebentsev, 1969; Sokolovskii, 1970, 1971, 1972; Chigirinskii, 1970; Latysh and Sokolovskii, 1972), and some scientists abroad [Kadzawara and Ito, 1953; Kimuro, 1953; Roedel, 1953; Okada, 1955; Frey (ed.), 1971].

Fecundity, about 40,000 eggs (Siro Isii, 1947).

The flesh of this mackerel is very tasty.

Length, to 600 mm (Taranetz, 1938).[31]

Distribution: In the Sea of Japan known from Olga Bay, Peter the Great Bay (Soldatov and Lindberg, 1930: 102); Pusan (Mori, 1952: 135); Tsushima Islands (Arai and Abe, 1970: 87), southwestern coast of Sakhalin (Probatov, 1951: 146); Sea of Japan coast of Hokkaido (Ueno, 1971: 79); Sado Island (Honma, 1952: 143); Toyama Bay (Katayama, 1940: 8); and San'in region (Mori, 1956a: 23). In the Sea of Okhotsk found off the coasts of Hokkaido and the southern Kuril Islands (Kaganovskii et al., 1947: 3). Along the Pacific coast of Japan from Hokkaido to Nagasaki (Jordan and Hubbs, 1925: 212). Off Cheju-do Island. Yellow Sea near the coast of the Korean Peninsula (Mori, 1952: 135); Gulf of Chihli (Bohai) (Zhang et al., 1955: 190); throughout the Yellow Sea (Wang, 1935: 394); Taiwan (China) (Matsubara, 1955: 518). East China Sea (Zhu et al., 1963: 399). North to the northeast coast of Kamchatka (Andriyashev, 1939a: 189) and east to California (Roedel, 1953: 180).

2. *Scomber tapeinocephalus* Bleeker, 1854—Spotted Mackerel
(Figure 225)

Scomber tapeinocephalus Bleeker, Nat. Tijdschr. Ned. Ind., 6, 1854: 407 (Nagasaki). Bleeker, Ichthyol. Japan, 1854–1857: 97, pl. 7, fig. 2.

Pneumatophorus tapeinocephalus Jordan and Hubbs, Mem. Carnegie Mus., 10, 2, 1925: 212. Soldatov and Lindberg, Obzor..., 1930: 103.

Pneumatophorus japonicus, Manacop, Philipp. J. Fish., 1958: 84, text-fig. 4.

Pneumatophorus japonicus tapeinocephalus, Abe, Enc. Zool., 2, Fishes, 1958: 218, fig. 647 (color figure).

7508. Nagasaki. 1893. Polyakov. One specimen.

D XI, I, 11 + 5; A I, 10 + 5; P 19; V I, 5.

Our specimen with a standard length of 378 mm has 11 spines in the first dorsal fin, 13 + 1 + 22 gill rakers on the first gill arch, and less than 200 perforated scales in the lateral line. Scales 24–32 anterior to first dorsal fin; scales 15–19 between the second dorsal fin and lateral line.

285 Body sides below lateral line in adult fish with many black spots; spots absent in young fish (Abe, 1958).[32] In very young fish, as pointed out by Okada (1955: 136), considerable yellow pigment seen along middle part of tail. Okada reports that spawning of this species takes place

[31]Mackerel weighing more than 3 kg with a length of 625 mm have been reported from the coast of California (Roedel, 1953).

[32]Such spots are prominently depicted in the work of Manacop (1958: 85, text-fig. 4, A) in fish up to 215 mm in length, and also in the work of Katzawara and Ito (1953).

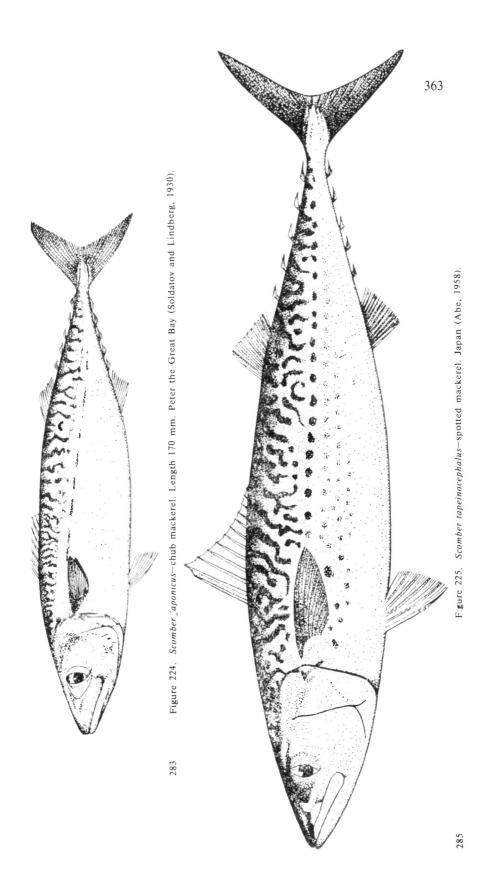

363

Figure 224. *Scomber japonicus*—chub mackerel. Length 170 mm. Peter the Great Bay (Soldatov and Lindberg, 1930).

283

Figure 225. *Scomber tapeinocephalus*—spotted mackerel. Japan (Abe, 1958).

285

in May–July and eggs are similar to those of *S. japonicus*. Spawning and larvae have been studied by authors abroad (Tanoue et al., 1960; Tanoue and Tamari, 1960; Tanoue, 1961).

Length, up to 535 mm (Bleeker, 1854).

Distribution: In the Sea of Japan known from near Pohang (Mori, 1952: 136); west coast of Hokkaido (Ueno, 1971: 79); Hakodate (Soldatov and Lindberg, 1930: 103); Sado Island (Honma, 1952: 143); Toyama Bay (Katayama, 1940: 8); and San'in region (Mori, 1956a: 23). Along the Pacific coast of Japan in Volcano Bay (Hikita, 1950: 7) and farther, from the central part of Honshu south to Nagasaki. Taiwan (China) (Matsubara, 1955: 518).

7. Genus *Scomberomorus* Lacépède, 1802—Spanish Mackerels

Scomberomorus Lacépède, Hist. Nat. Poiss., 3, 1802: 292 (type: *S. plumieri* Lacépède). Fraser-Brunner, Ann. Mag. Nat. Hist., 12, 3, 1950: 157 (synonyms). Soldatov and Lindberg, Obzor..., 1930: 111 (synonyms, description). Collette and Gibbs, Edit. U.S. Nat. Mus., 1963: 21, pl. 6 (characteristics).

286   Body elongate, entirely covered with rudimentary scales that do not form corselet. Head attenuate and small. Mouth large. Strong cutting teeth on jaws. Vomer and palatines with small teeth. Maxilla not concealed under preorbital bone. Caudal peduncle with simple keel. Soft dorsal and anal fins short, similar, sometimes fairly high, falciform. Pelvic fins small. Fish of warm seas, attractive in shape and color, with the best taste qualities among food fishes (Soldatov and Lindberg, 1930).

Many species. 4 known from the Sea of Japan and 1 from adjacent waters.

Key to Species of Genus Scomberomorus[33]

1 (4). Lateral line not sinuous, with one sharp curve. Swim bladder present.

2 (3). Pectoral fins relatively small, about equal to snout length, with pointed tip. Many transverse stripes on body sides............
.... 1. **S. commersoni** Lacépède.

3 (2). Pectoral fins large, about 1.5 times snout length, with roundish tip. One or two rows of indistinct round spots on body sides.......
.... 2. **S. sinensis** (Lacépède).

4 (1). Lateral line sinuous, with several slight curves. Swim bladder absent.

5 (8). Tongue with teeth. Maximum body depth more than head length.

[33]From Kishinouye, 1923: 416, with modifications.

Length of base of first dorsal fin much shorter than length of bases of dorsal finlets.

6 (7). Body depth only slightly more than head length. ....
.... 3. [**S. guttatus** Block and Schneider].

7 (6). Body depth considerably more than head length. ....
.... 4. **S. koreanus** (Kishinouye).

8 (5). Tongue without teeth. Maximum body depth less than head length. Length of base of first dorsal fin about equal to length of bases of dorsal finlets................. 5. **S. niphonius** (Cuvier).

1. *Scomberomorus commersoni* (Lacépède, 1800)–Barred Spanish Mackerel (Figure 226)

Scomber commerson Lacépède, Hist. Nat. Poiss., 2, 1800: 598, 600 (Madagascar).

Cybium commerson, Kishinouye, J. Coll. Agric. Univ., Tokyo, 8, 3, 1923: 416, fig. 36 (synonyms, description).

Scomberomorus commersoni, Fraser–Brunner, Ann. Mag. Nat. Hist., 12, 3, 1950: 161, fig. 34 (synonyms). Smith, Sea Fish. S. Africa, 1950: 301, pl. 64, fig. 840 (color plate). Beaufort and Chapman, Fish. Indo-Austr. Arch., IX, 1951: 230 (synonyms, description). Matsubara, Fish Morphol. and Hierar., 1955: 520. Abe. Enc. Zool., 2, Fishes, 1958: 217, fig. 644 (color figure). Jones and Silas, Indian J. Fish., 8, 1, 1962: 194, fig. 2 (description). Zharov et al., Tuntsy..., 1961: 48, fig. 22 (synonyms, description). Collette and Gibbs, Edit. U.S. Nat. Mus., 5, 1963, pl. 6. Osipov, Okeanskie Ryby..., 1968: 56, fig. 37 (description).

288   D XVI-XVII, 15-17 + 9-10; A 14-17 + 9-10; P 22-23; gill rakers 1 + 2-3. Back dark blue with silvery hue; body sides silvery with many dark transverse bars (in young fish elongated transverse spots occur in place of bars). Sexual maturity is attained in the third year of life at a body length of about 90 cm. Spawning observed off the northeast coast of Australia from July to September. Inhabits surface water layer (Zharov et al., 1961: 48).

Eggs and larval development of Pacific mackerel, their distribution, and population density of the species have been described by Kramer (1960) and Jones (1961). Information on the biology of this species has been published by Smith (1950), Zharov (1961), and Osipov (1968c). Anatomy of the Spanish mackerel described much earlier by Kishinouye (1923).

This fish predominantly feeds on other fishes. Its flesh is greatly valued for human consumption. It attains a weight of more than 45 kg (Osipov, 1968c: 56).

Length, to 1.8 m (Osipov, 1968c).

Distribution: In the Sea of Japan known from Pusan (Mori, 1952: 136);

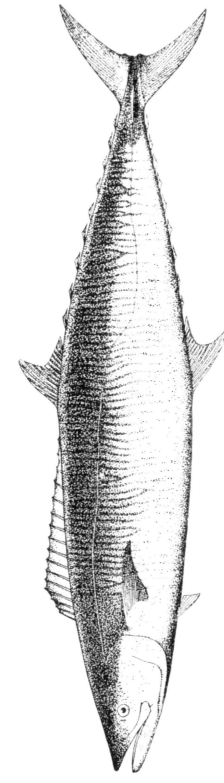

Figure 226. *Scomberomorus commersoni*—Barred Spanish mackerel. Length 1.25 m. Sea of Japan (Kishinouye, 1923).

Sado Island (Katoh et at., 1956: 316); and south of Yamaguti Prefecture (Abe, 1958: 217). Entire west coast of the Korean Peninsula (Mori, 1952: 136). South China Sea (Zhu et al., 1962: 755); Taiwan (China), Australia, Indonesia, and the Indian Ocean (Matsubara, 1955: 520); east to southwest Africa (Smith, 1950: 300).

2. *Scomberomorus sinensis* (Lacépède, 1802)—Chinese Mackerel
   (Figure 227)

Scomber sinensis Lacépède, Hist. Nat. Poiss., 3, 1802: 23 (locality not indicated).

Cybium chinense Schlegel. In: Temminck and Schlegel, Fauna Japonica, Poiss., 1944: 100, pl. 53, fig. 1 (Japan). Kishinouye, J. Coll. Agric. Univ., Tokyo, 8, 3, 1923: 418, figs. 34, 40 (synonyms, description, and details of anatomy).

Scomberomorus sinensis, Soldatov and Lindberg, Obzor..., 1930: 111 (synonyms, description). Abe, Enc. Zool., 2, Fishes, 1958: 217, fig. 643 (color figure). D'Aubenton and Blanc, Bull. Mus. Nat. Hist. Nat., Paris, 2, 37, 2, 1965: 233, figs. 1, 2 (synonyms, classification, and biology).

Scomberomorus cavalla, Fraser-Brunner, Ann. Mag. Nat. Hist., 12, 3, 1950: 160, fig. 33 (synonyms). Zharov et al., Tuntsy..., 1961: 471 (synonyms, description).

D XVI, 15 + 18; A 16 + 7[34]; gill rakers 2 + 9.[35] Vertebrae 18 + 22.

Upper profile of head concave, head large, pointed. Snout elongate. Pectoral fins large, rounded at tip. Indistinct round spots along body sides arranged in one or two rows (Soldatov and Lindberg, 1930). Back gray, sides silvery, spots on sides sometimes absent. Lateral line under end of first dorsal or second dorsal fin forms sharp, deep, downward curve (Zharov et al., 1961). Weight up to 80 kg. Anatomy of the species detailed by Starks (1910) and biology by Zharov (1961) and D'Aubenton and Blanc (1965).

Length, to 2 m (Abe, 1958).

289   *Distribution:* In the Sea of Japan known from near l'usan (Mori, 1952: 137); Akita (Soldatov and Lindberg, 1930: 111)[36]; and San'in region (Mori, 1956a: 23). Along the Pacific coast of Japan known from Tiba Prefecture (Matsubara, 1955: 520). South coast of the Korean Peninsula, in China, south coast of Taiwan (Abe, 1958: 217).

3. [*Scomberomorus guttatus* Bloch and Schneider, 1801—Spotted Spanish
   Mackerel (Figure 228)]

Scomber guttatus Bloch and Schneider, Syst. Ichthyol., 1801: 23, pl. 5 (Malabar coast).

---

[34]D XV-XVI, 14-15 + 7-8; A 16-17 + 7; P 1 + 20; V I, 5 (D'Aubenton and Blanc, 1965).
[35]2-3 + 5-9 (Zharov et al., 1961).
[36]Erroneously given as Akuma-ken.

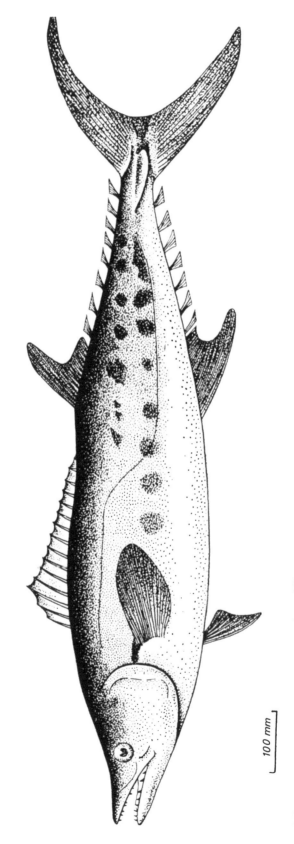

100 mm

Figure 227. *Scomberomorus sinensis*—Chinese mackerel (D'Aubenton and Blanc, 1965).

287

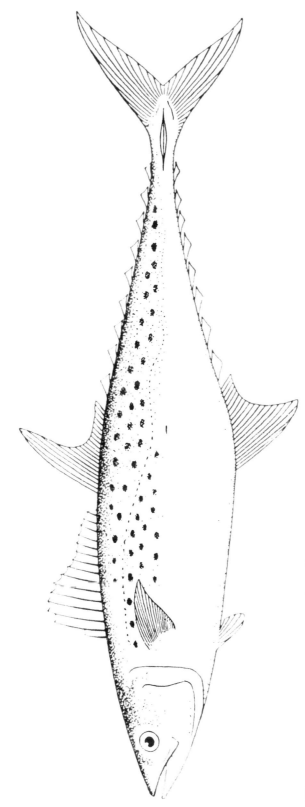

Figure 228. *Scomberomorus guttatus*—spotted Spanish mackerel (Zharov et al., 1961).

Cybium guttatum, Kishinouye, J. Coll. Agric. Univ., Tokyo, 8, 3, 1923: 419, fig. 61 (synonyms, description).

Scomberomorus guttatus, Fraser-Brunner, Ann. Mag. Nat., Hist., 12, 3, 1950: 160, fig. 31 (synonyms). Beaufort and Chapman, Fish. Indo-Austr. Arch., IX, 1951: 232 (synonyms, description). Zharov et al., Tuntsy..., 1961: 53, fig. 29 (synonyms).

D XV-XVI, 18-20 + 8-9; A 20 + 8-9; P 20-23; gill rakers 2-3 + 8-12. Lateral line without sharp curve, slightly sinuous. Length of upper jaw equal to half head length. Dark spots scattered on body sides. Meat tasty (Zharov et al., 1961).

Length, to 600 mm (Zharov et al., 1961).

Distribution: Not known in the Sea of Japan. Found in the Yellow Sea (Wang, 1953: 398). East China and South China seas (Zhu et al., 1962: 756; 1963: 402). Taiwan (China). Tropical and subtropical seas and coastal waters of the Pacific and Indian oceans in the region of Indonesia 290 (excluding Australian waters), Sri Lanka (Zharov et al., 1961; 53). Southeast coast of Africa (Smith 1950; 301).

4. *Scomberomorus koreanus* (Kishinouye, 1915)—Korean Mackerel
(Figure 229)

Cybium koreanum Kishinouye, J. Coll. Agric. Univ., Tokyo, 8, 3, 1923: 420, fig. 35 (Yellow Sea).

Sawara koreanum, Soldatov and Lindberg, Obzor..., 1930: 112.

Scomberomorus semifasciatus, Fraser-Brunner, Ann. Mag. Nat. Hist., 12, 3, 1950: 159, fig. 30. Zharov et al., Tuntsy..., 1961: 53, fig. 28.

Scomberomorus koreanus, Matsubara, Fish Morphol. and Hierar., 1955: 520. Zhu et al., Ryby Vostochno-Kitaiskogo Morya, 1963: 403, fig. 302.

D XIV, 19-21 + 9; A 18-21 + 7[37]; gill rakers 3 + 10; vertebrae 20 + 26. Teeth on jaws long and sharp; 16-19 teeth on upper jaw and 13-15 on lower jaw. Teeth on vomer, palatines, and tongue small, villiform (Kishinouye, 1923).

Off the southwest coast of the Korean Peninsula this species spawns in July. Feeds on sardines, anchovies, and small crustaceans, and reaches 15 kg in weight. Korean mackerel is a very tasty fish (Okada 1955: 150).

Length, to 1.5 m (Okada, 1955: 150).

Distribution: In the Sea of Japan found near Chongjin and Pusan (Mori, 291 1952: 136). Gulf of Chihli (Bohai) (Zhang et al., 1955: 194). West coast of the Korean Peninsula and north coast of Taiwan (China) (Matsubara, 1955: 520). East China Sea (Zhu et al., 1963: 403).

---

[37]Apparently 8 anal finlets depicted by mistake in the figure given by Okada (1955: 150).

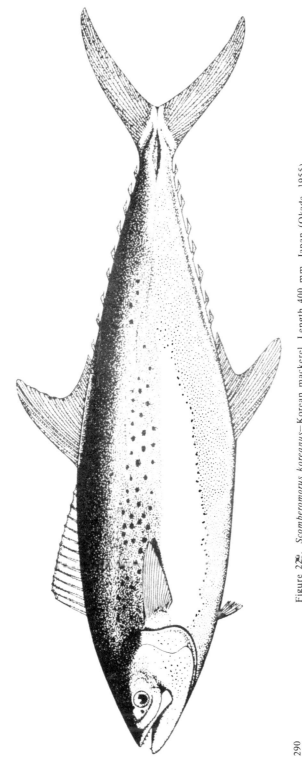

Figure 224. *Scomberomorus koreanus*— Korean mackerel. Length 400 mm. Japan (Okada, 1955).

5. *Scomberomorus niphonius* (Cuvier, 1831)—Sawara, Japanese Spanish Mackerel (Figure 230).

Cybium niphonium Cuvier. In: Cuvier and Valenciennes, Hist. Nat. Poiss., 8, 1831: 180 (Japan). Kishinouye, J. Coll. Agric. Univ., Tokyo, 8, 3, 1923: 421, figs. 6, 9, 32, 41 (synonyms, description).

Scomberomorus niphonius Fraser-Brunner, Ann. Mag. Nat. Hist., 12, 3, 1950: 158. Matsubara, Fish Morphol. and Hierar., 1955: 520. Zharov et al., Tuntsy..., 1961: 51, fig. 26 (synonyms, description).

Sawara niphonia, Soldatov and Lindberg, Obzor..., 1930: 112 (synonyms, description)..

18462. Ussurii Bay. September 17, 1913. Far East Expedition. 1 specimen.

38810. Yellow Sea. June 4, 1956. Institute of Zoology, Academy of Sciences, China. 2 specimens.

D XIX, 16 + 9; A 16 + 8; gill rakers 3 + 9.

Body elongate, compressed laterally, covered with minute scales, corselet indistinguishable. First dorsal fin very long. Pectoral fins with notch on posteroventral margin. Lateral line forms distinctly long gentle arch. Teeth lanceolate; upper jaw with 25 and lower with 19 teeth. Minute thin teeth on vomer and palatines poorly discernible but nonetheless present. Tongue without teeth (Soldatov and Lindberg, 1930: 112).

Many minute spots on body sides. Lateral line without sharp curves but sinuous, with numerous poorly discernible short branches that originate at a right angle.

This species dwells in water with a temperature ranging from 10 to 20°C; in summer found in surface layers and in winter moves deeper. Spawns in April–May in inlets. Flesh very tasty (Zharov et al., 1961). Reaches 4.5 kg in weight.

Length, to 1 m (Soldatov and Lindberg, 1930: 112).

Distribution: In the Sea of Japan known from Peter the Great Bay (Soldatov and Lindberg, 1930: 112); Pusan (Mori, 1952: 136); Sado Island (Honma, 1952: 143); Toyama Bay (Katayama, 1940: 8); and San'in region (Mori, 1956a: 23). Along the Pacific coast of Japan from Hokkaido southward (Matsubara, 1955: 520). In the Yellow Sea found in the Gulf of Chihli (Bohai) (Zhang et al., 1955: 192), Cheju-do Island (Uchida and Yabe, 1939: 8). East China Sea (Zhu et al., 1963: 401). Taiwan (China). Australia (Abe, 1958: 217).

8. Genus *Acanthocybium* Gill, 1862—Wahoo

Acanthocybium Gill, Proc. Acad. Nat. Sci. Philad., 14, 1862: 125 (type: *Cybium sara* Bennett). Kishinouye, J. Coll. Agric. Univ., Tokyo, 8, 3, 1923: 410 (description). Beaufort and Chapman, Fish. Indo-Austr. Arch.,

IX, 1951: 227 (description). Matsubara, Fish Morphol. and Hierar., 1955: 520.

Body elongate, more or less compressed laterally, covered with minute narrow scales. Snout long, beaklike. Mouth large, teeth on jaws three-faceted, compressed, immovable, slightly serrate, arranged in one row. Minute teeth form villiform strip on vomer and palatines. Two dorsal fins: first long, with 26-27 spines; second short, with 10-11 rays. Anal fin similar to second dorsal fin, but originates slightly behind vertical from 293 origin of second dorsal fin. Dorsal finlets 9-10 and anal finlets 8-9. Vertebrae 23 + 31-33 (Beaufort and Chapman, 1951).

1 species. Known from the Sea of Japan.

1. *Acanthocybium solandri* (Cuvier, 1831)—Wahoo (Figure 231)

Cybium solandri Cuvier. In: Cuvier and Valenciennes, Hist. Nat. Poiss., 8, 1831: 192 (open sea, exact locality not indicated).

Acanthocybium solandri, Kishinouye, J. Coll. Agric. Univ., Tokyo, 8, 3, 1923: 411, figs. 10, 31, 39 (synonyms, description). Fraser-Brunner, Ann. Mag. Nat. Hist., 12, 3, 1950: 161, fig. 35. Smith, Sea Fish. S. Africa, 1950: 301, pl. 64 (color figure). Beaufort and Chapman, Fish Indo-Austr. Arch., IX, 1951: 227 (synonyms, description). Matsubara, Fish Morphol. and Hierar., 1955: 520. Abe, Enc. Zool., 2, Fishes, 1958: 216, fig. 642 (color figure). Zharov et al., Tuntsy..., 1961: 46, fig. 20 (synonyms, description). Jones and Silas, Indian J. Fish., 8, 1, 1962: 192, fig. 1 (description).

D XXV-XXVII, 7-13 + 9-10; A 12-13 + 8-9; gill rakers on first gill arch absent. Teeth on jaws very strong, cutting[38]; vomer with teeth. Spawning recorded in the Pacific Ocean off Ogasawara Islands (Japan), and in the Atlantic Ocean in the region of Cape Green and Cape Dakar. Confined near surface of water. Generally in pairs, rarely in groups, does not form schools (Zharov et al., 1961).

Color of back, upper part of head, and fins dark; sides and belly light. Narrow, uneven dark stripes extend along body from head to caudal fin. Color of dead fish dulls and stripes disappear. Unlike Spanish mackerel (*Scomberomorus commersoni*), wuhoo is found at some distance from the shore. It is an active predatory fish. Feeds on squids and fishes. In Japan consumed raw or fried (Osipov, 1968c).

Length, to 2 m (Abe, 1958).

Distribution: In the Sea of Japan recorded off Pusan (Mori, 1952: 137); Sado Island (Honma, 1963: 18); Toyama Bay (Katayama, 1940: 8); and San'in region (Mori, 1956a: 23). Along both coasts of Japan from central part of Honshu south. Kyushu. South coast of the Korean Peninsula. Taiwan (China). Subtropical and tropical seas of the Pacific and Atlantic oceans (Matsubara, 1955: 520).

---

[38]Abe (1958) has reported 50 to 55 triangular teeth on both jaws.

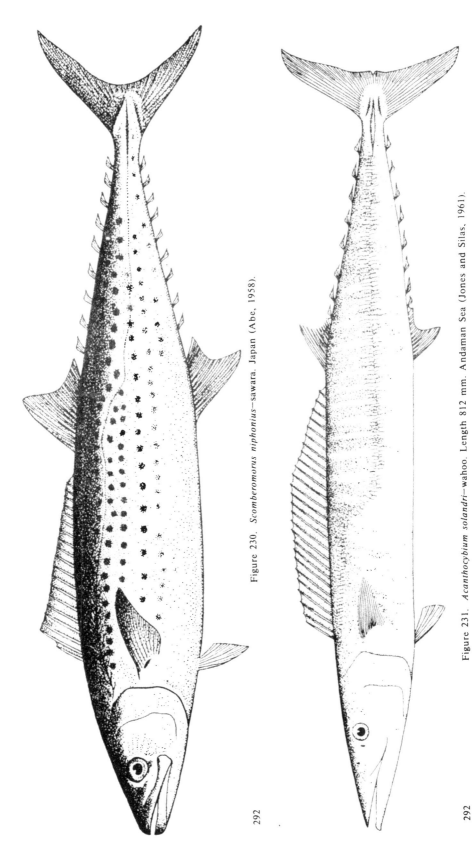

Figure 230. *Scomberomorus niphonius*—sawara. Japan (Abe, 1958).

Figure 231. *Acanthocybium solandri*—wahoo. Length 812 mm. Andaman Sea (Jones and Silas, 1961).

CLXIV. Family ISTIOPHORIDAE—Sailfishes, Spearfishes

Body distinctly elongate, compressed laterally, and covered with minute oblong scales embedded in skin. Snout produced in form of elongate process almost circular in cross section, formed by premaxillary and nasal bones. Pectoral fins inserted low. Base of first dorsal fin long, distinctly more than half body length, and close to base of short second dorsal fin; latter not separate from first dorsal in young fish. First anal fin with deep notch. Pelvic fins present, with one to three elongate rays. Caudal fin crescent-shaped with thin but strong rays. Two fleshy keels along each side of caudal peduncle (Lindberg, 1971: 180).

3 genera, widely represented in all oceans,[39] all known from the Sea of Japan.

294

Key to Genera of Family Istiophoridae[40]

1 (2). First dorsal fin distinctly sail-shaped, its height greater than body depth. Rays of pelvic fins long, falling slightly short of vent....
.... 1. **Istiophorus** Lacépède.

2 (1). First dorsal fin not sail-shaped, its height not greater or slightly more than body depth, equal to, or slightly less. Pelvic fins of adult fish relatively short, far from reaching vent.

3 (4). Anterior part of first dorsal fin almost equal to body depth. Nape not highly elevated between vertical of anterior margin of eye and origin of dorsal fin. .... 2. **Tetrapturus** Rafinesque.

4 (3). Anterior part of first dorsal fin less than body depth (about 1.5 to 2.0 times in depth). Nape highly elevated between vertical margin of eye and origin of dorsal fin. .... 3. **Makaira** Lacépède.

1. Genus *Istiophorus* (Lacépède, 1802)—Sailfishes

Istiophorus Lacépède, Hist. Nat. Poiss., 3, 1802: 374 (type: *Scomber gladius* Broussonet). Morrow and Harbo, Copeia, 1, 1969: 34 (revision of genus).

Histiophorus Cuvier. In: Cuvier and Valenciennes, Hist. Nat. Poiss., 8, 1831: 291 (correction of name).

One distinguishing feature of this genus is the very high first dorsal fin in the form of a sail. Rays of pelvic fins very long, almost reaching vent. Body sides with more than 10 transverse stripes of pale blue spots, and highly compressed. Keel present on head from vertical line of anterior margin of eye to origin of first dorsal fin. Skull narrow and thin. Vertebrae $12 + 12 = 24$ (Nakamura et al., 1968: 49).

---

[39]Comprehensive information on the biology of the family is given by Strasburg (1969) and on the skeletal structure of sailfishes by Gregory and Conrad (1937).

[40]From Nakamura et al. (1968), with modifications.

American ichthyologists (Morrow and Harbo, 1969), after analyzing the morphological and meristic characters of the species of this genus, think that *Istiophorus* comprises a single species, namely, *Istiophorus platypterus*. This species is found in tropical, subtropical, and temperate waters of the Indian, Pacific, and Atlantic oceans, and exhibits only insignificant local variations in morphological characters. Also found in the Sea of Japan.

1. ***Istiophorus platypterus*** (Shaw and Nodder, 1792)—Sailfish (Figure 232).

Xiphias platypterus Shaw and Nodder, Natural. Misc. ..., 1972: 28, pl. 88 (Indian Ocean).

Histiophorus orientalis, Temminck and Schlegel, Fauna Japonica, Poiss., 1844: 103, pl. 55 (Japan).

Istiophorus orientalis Fowler, Occ. Pap. Bishop Mus., 8, 7 1923: 375. Soldatov and Lindberg, Obzor..., 1930: 115 (description).

Istiophorus albicans Nakamura, Iwai and Matsubara, Review of Sailfish..., 4, 1968: 57, fig. 13.

Istiophorus platypterus Nakamura, Iwai and Matsubara, Review of Sailfish..., 4, 1968: 55, fig. 12. Morrow and Harbo, Copeia, 1, 1969: 34.

I D XLII-XLIII, II D 6-7; in first dorsal fin, first three rays spiny, next 9 soft, and all subsequent rays spiny. I A 12-15, II A 6-7; anterior two rays of first anal fin spiny, others soft. P 17-20; V I, 2.

296   Body considerably elongate (body depth 6.4-7.2 times in its length), distinctly compressed laterally. Snout long. Scales in form of triangular plates. Both jaws and palatines with minute rasp-like teeth. Lateral line quite distinct, with an upward curve above pectoral fin, and thereafter straight to caudal fin. Head large (length from tip of lower jaw about 4.3 to 4.8 times in standard length).[41] Keel continues along upper part of head from vertical line with anterior margin of eye to base of first dorsal fin. Caudal fin strong, with deep notch. Posterior part of caudal peduncle with two keels on each side. Pectoral fins long (1.2 to 1.4 times in head length from tip of lower jaw), with pointed tip. First dorsal fin in form of sail; second dorsal small, similar to second anal fin in shape and size. Pelvic fins very long, almost reaching vent. First dorsal fin deep bluish with minute scattered black spots.[42] Other fins black to chocolate-brown, sometimes with bluish stripes; silvery-white stripe passes at base of second dorsal and second anal fins. Large number of pale bluish spots seen on body sides, forming more than 10 vertical rows. Back black with

[41]Length, from tip of upper jaw 3 times in standard length (Soldatov and Lindberg, 1930: 115).

[42]Ovchinnikov (1963) reports considerable variation in color of the dorsal fin of sailfish, which may even be violet.

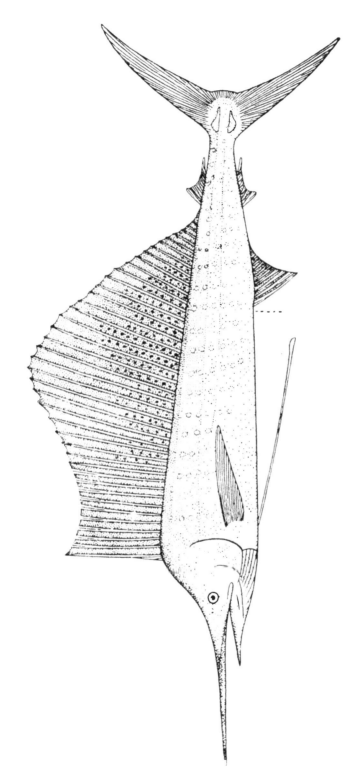

Figure 232. *Istiophorus platypterus*—sailfish (Nakamura et al., 1968).

378

bluish tinge. Sides chocolate-brown with bluish hue. Belly silvery-white (Nakamura et al., 1968).

Biological characteristics of sailfishes have been described by Soviet researchers (Zharov et al., 1961; Ovchinnikov, 1963, 1964; Osipov et al., 1964; Osipov, 1968c) and scientists abroad (Deraniyagala, 1933; Okada, 1955; Ueyanagi, 1963; Merrett and Thorp, 1965; Merrett, 1970). The cycle of development of sailfishes has been studied by Gehringer (1956) and Jones (1962). Information on the thyroid gland of these fishes is given by Honma (1956). Spawning takes place in August–September. Flesh particularly tasty in summer and autumn (Okada, 1955).

Length, to 3 m (Soldatov and Lindberg, 1930: 116).

Distribution: In the Sea of Japan known from Peter the Great Bay (Soldatov and Lindberg, 1930: 115); Sado Island (Honma, 1963: 18); Toyama Bay (Katayama, 1940: 9); and San'in region (Mori, 1956a: 24). In autumn of every year it passes along Tsushima Strait (Nakamura et al., 1968: 57). Found off south coast of the Korean Peninsula (Mori, 1952: 137) and Cheju-do Island (Uchida and Yabe, 1939: 8). Along the Pacific coast of Japan found from Tohoku region to Kyushu; Taiwan (China) (Matsubara, 1955: 529). East China and South China seas (Zhu et al., 1962: 758; 1963: 404). The Philippines, Solomon Islands, and Pacific coast of Mexico (Nakamura et al., 1968: 57). Widely represented in the Indian and Atlantic oceans, and the central and western parts of the Pacific Ocean (Morrow and Harbo, 1969: 35).

2. Genus *Tetrapturus* Rafinesque, 1810–Spearfishes

Tetrapturus Rafinesque, Caratteri..., 1810: 54 (type: *T. belone* Rafinesque).

Body oblong, compressed laterally. Minute teeth in bands on jaws and palatines. First dorsal fin low, but its anterior part in form of high lobe, height being almost equal to body depth. Second dorsal fin with 7 rays. First anal fin with 12 to 15 spiny rays. Second anal fin with 6 to 7 soft rays, located under second dorsal fin and similar to it in shape. Pelvic fins consist of single spine and 1 or 2 soft rays, which are relatively shorter in adult fish than in young (Beaufort and Chapman, 1951: 237).

Five species. Two species known from the Sea of Japan.

Key to Species of Genus *Tetrapturus*[43]

1 (6). Rays in posterior part of first dorsal fin almost equal in height. Vent quite anterior to anal fin. Origin of second anal fin anterior to vertical with origin of second dorsal fin (Figure 233, A).

[43]From Nakamura et al. (1968).

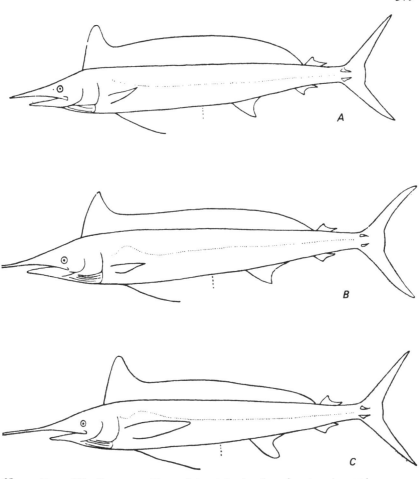

97   Figure 233. *Tetrapturus.* Shape of dorsal fin, location of vent, and second
anal fin (Nakamura et al., 1968).
A—*T. angustirostris*; B—*T. belone*; C—*T. pfluegeri.*

2 (5). Pectoral fins narrow, short, with acute tip. Less than 2 times in
head length from tip of lower jaw.

3 (4). Snout of adult fish relatively short; space between tips of jaws more
than 4 times in the head length. .... 1. **T. angustirostris** Tanaka.

98 4 (3). Snout in adult fish relatively long; space between tips of jaws about
3 times in the head length (Figure 233, B). . . . .
. . . . [**T. belone** Rafinesque, 1810][44]

5 (2). Pectoral fins broad and fairly long, almost equal to length of head
from tip of lower jaw (Figure 233, C). . . . .
. . . . [**T. pfluegeri** Robins and Sylva, 1963].[45]

[44]Found in the Mediterranean Sea.
[45]Found in the northwestern part of the Atlantic Ocean.

380

6 (1). Rays in posterior part of first dorsal fin decrease in height posteriorly. Vent situated immediately before origin of anal fin. Origin of second anal fin at vertical with origin of second dorsal fin or behind it.

7 (8). Pectoral fins lobate, broad, with roundish tip. Tips of first dorsal and anal fins roundish.......... [**T. albidus** Poey, 1860].[46]

8 (7). Pectoral fins keel-shaped, with sharp tip. Tips of first dorsal and anal fin acute....................... 2. **T. audax** (Philippi).

1. *Tetrapturus angustirostris* Tanaka, 1915—Shortbill Spearfish
(Figure 234)
Tetrapturus angustirostris Tanaka, Fig. and Descrip., XIX, 1915: 324, pl. 88, fig. 285 (Sagami Bay). Zharov et al., Tuntsy..., 1961: 64 (synonyms). Osipov, Okeanskie Ryby, 1968: 19, fig. 9 (description). Nakamura et al., Misaki Mar. Biol. Inst. Kyoto Univ., Spec. Rept., 4, 1968: 59, fig. 15 (synonyms, description).

D XLVII-L, 6-7; A 12-15, 6-7; P 18-19; V I, 2.

Body oblong, highly compressed laterally, relatively low (depth 8.3 to 10.4 times in body length). Snout relatively short. Scales in form of sharply serrated plates at margin. Jaws and palatines with small teeth. Head large (4.2 to 4.7 times in body length). Caudal fin large, with deep notch, its lobes fairly narrow. Caudal peduncle with two keels. Pectoral fins inserted low, short (1.6 to 2.3 times in head length from tip of lower jaw). First dorsal fin originates above posterior margin of preopercle. Second dorsal and anal fin similar in shape and size. Second anal fin distinctly anterior to vertical from origin of second dorsal fin. Pelvic fins longer than pectoral fins. Body shape and ratio of its parts change notably with age. Membrane of first dorsal fin deep blue, other fins black to chocolate-brown. Silvery-white stripes at bases of anal fins. Back deep blue. Sides of body chocolate-brown and light blue, without stripes. Belly silvery-white (Nakamura et al., 1968).

Length, to 160 cm (Osipov, 1968c: 20).

Distribution: In the Sea of Japan known from off Sado Island (Honma, 1952: 144). Found along Pacific coast of Japan (Matsubara, 1955: 527); confined to Kuro-Shio Current, Taiwan (China), Hawaiian Islands; northwestern, central and southern parts of the Pacific Ocean; migration toward coast of California observed (Nakamura et al., 1968: 60).

2. *Tetrapturus audax* (Phillipi, 1887)—Striped Spearfish (Figure 235)
Histiophorus audax Phillipi, Anal. Univ. Chile, 71, 1887: 34-39 (Chile).
Tetrapturus mitsukurii Jordan and Snyder, J. Coll. Sci. Univ., Tokyo, 15, 1901: 303, pl. XVI, fig. 5.

[46]Found in the northern part of the Atlantic Ocean and in the Mediterranean Sea.

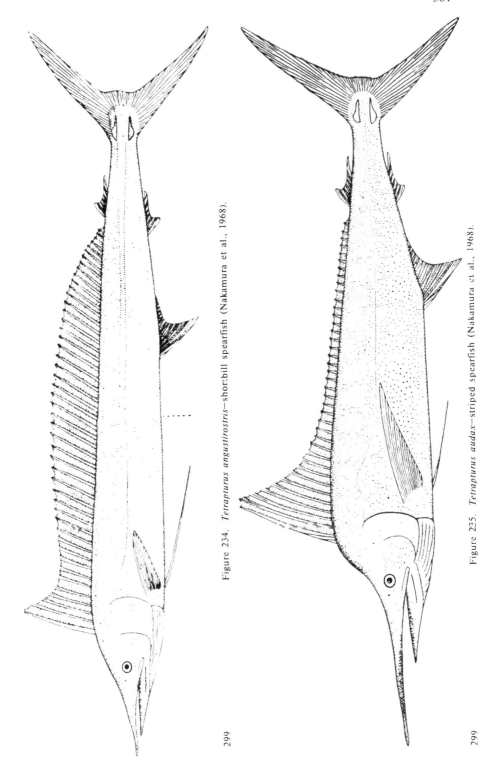

Figure 234. *Tetrapturus angustirostris*—shortbill spearfish (Nakamura et al., 1968).

Figure 235. *Tetrapturus audax*—striped spearfish (Nakamura et al., 1968).

299

299

300   *Makaira mitsukurii,* Matsubara, Fish Morphol. and Hierar., 1955:
528. Zharov et al., Tuntsy..., 1961: 66, fig. 41 (synonyms, description).

Tetrapturus audax, Nakamura et al., Misaki Mar. Biol. Inst. Kyoto
Univ., Spec. Rep., 4, 1968: 25, 67, fig. 21 (synonyms, description). Parin,
Scumbrievidnye Ryby..., 1967: 110 (synonyms, description).

I D XXXVII–XLII, II D 6; I A 13–18, II A 5–6; P 18–22; V I, 2.

Body depth 5.9 to 7.3 times in its length; body highly compressed
laterally. Snout relatively long (2/3 head length from tip of rostrum). Head
large (head length from tip of lower jaw 3.6 to 3.8 times in body length). In
young fish, pelvic fins longer than pectoral fins, they are shorter in adult
fish.

Membrane of first dorsal fin deep blue, back blackish-blue, belly
silvery-white. More than 10 cobalt-colored stripes on sides of body. All
fins black to chocolate-brown, sometimes with deep light blue stripes.
Bases of first and second anal fins with silvery-white stripes (Nakamura et
al., 1968).

Biology of this species given in publications by both Soviet and other
authors (Hubbs and Wisner, 1953; Okada, 1955; Zharov et al., 1961; Parin,
1967; Nakamura, 1968; Osipov, 1968c).

Length, to 3 m (Nakamura et al., 1968).

Distribution: In the Sea of Japan known from off Sado Island (Honma,
1952: 144); Toyama Bay (Katayma, 1940: 9); San'in region (Mori, 1956:
24); south coast of the Korean Peninsula (Mori, 1952: 138); and Cheju-
do Island (Uchida and Yabe, 1939: 8). Along the Pacific coast of Japan
found from Hokkaido to south of Taiwan (China) (Matsubara, 1955: 528).
Almost everywhere in temperate and tropical waters of the Indian and
Pacific oceans (Nakamura et al., 1968: 68).

3. Genus *Makaira* Lacépède, 1803—Marlins

Makaira Lacépède, Hist. Nat. Poiss., 4, 1803: 688 (type: *M. nigricans*
Lacépède).

Genus distinguished by relatively low anterior part of dorsal fin (1.5
to 2 times in body depth) and short rays in the pelvic fins, which do
not reach vertical with tip of pectoral fin. Nape highly elevated.
Vertebrae $11 + 13 = 24$ (Nakamura et al., 1968: 49).

Several species. Two known from the Sea of Japan.

Key to Species of Genus Makaira[47]

1 (4). Pectoral fins not rigid, readily folded against sides of body.

2 (3). Lateral line forms loops (Figure 236).........................
.... 1. **M. mazara** (Jordan and Snyder).

---

[47]From Nakamura et al. (1968), with modifications.

3 (2). Lateral line forms cells. .... [**M. nigricans** Lacépède, 1803][48]
4 (1). Pectoral fins rigid, cannot be folded against body; located
　　　perpendicular to body and immovable. Lateral line single......
　　　......................................2. **M. indica** (Cuvier).

1. *Makaira mazara* (Jordan and Snyder, 1901)—Striped Marlin
　　(Figure 236)
　　Tetrapturus mazara Jordan and Snyder, J. Coll. Sci. Univ., Tokyo, 15,
　　2, 1901: 305 (Mexico, Japan).
　　Makaira mazara, Jordan and Evermann, Calif. Acad. Sci., 12, 1926: 53,
301　pl. 11, fig. 2 (description). La Monte, Bull. Amer. Mus. Nat. Hist., 107,
　　1955: 336 (synonyms and description). Zharov et al., Tuntsy..., 1961:
　　67, fig. 42 (synonyms and description). Nakamura et al., Misaki Mar. Biol.
　　Inst. Kyoto Univ., Spec. Rep., 4, 1968: 68, fig. 22 (synonyms).
　　Makaira nigricans La Monte, Marine Game Fishes of the World, 1952:
　　190. Rass, Tr. Inst. Okeanol. Akad. Nauk SSSR, 80, 1965: 1. Parin,
　　Scumbrievidnye Ryby..., 1967: 113.
　　I D 40–44, II D 6; I A 12–15, II A 6–7; P 21–23; V I, 2.
　　Body depth 4.0 to 4.4 times in its length (from tip of lower jaw up to
keel). Snout long. Lateral line complex, with looping branches; in adult
fish almost indistinct. Pectoral fins inserted low; in fish 1 m long,
pectoral fins relatively short (2 times in head length); in fish 1.9 m long,
almost equal to head length. First anal fin comparatively large, triangular,
with acute tip. Pelvic fins in large specimens shorter than pectoral fins.
Body shape and proportion of its parts change with age (Figure 237). First
dorsal fin blackish with deep bluish stripes; other fins blackish to
chocolate-brown with blue stripes. Silvery-white stripes at bases of first
and second anal fins, back blackish-blue, belly silvery-white (Nakamura et
al., 1968).
　　Biology of this species given by Zharov et al. (1961), Osipov et al.
(1964), and Parin (1967).
　　Length, to 4.35 m (La Monte, 1955).
　　Distribution: In the Sea of Japan known from Toyama Bay (Katoh et al.,
1956: 317) and central part of Honshu southward (Matsubara, 1955: 528).
Off the south coast of the Korean Peninsula (Mori, 1952: 138) and near
303　Cheju-do Island (Uchida and Yabe, 1939: 8). Subtropical and tropical
waters of the Indian and Pacific oceans (Nakamura et al., 1968: 69).

2. *Makaira indica* (Cuvier, 1831)—Black Marlin (Figure 238)
　　Tetrapturus indicus Cuvier. In: Cuvier and Valenciennes, Hist. Nat.
Poiss., 8, 1831: 286 (Sumatra Island).
　　Makaira marlina Jordan and Hill. In: Jordan and Evermann, Occ. Pap.

[48]Distributed in the Atlantic Ocean.

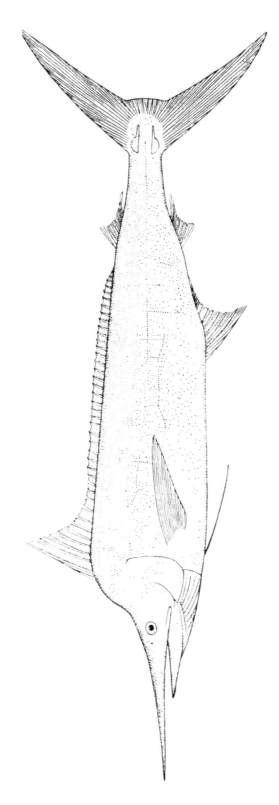

Figure 236. *Makaira mazara*—striped marlin (Nakamura et al., 1968).

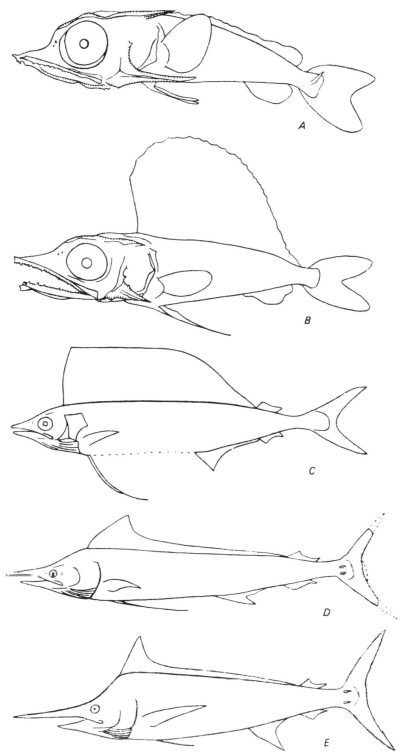

302 Figure 237. Age-related changes in shape of body and fins in *Makaira mazara* (Nakamura et al., 1968).

Body length: A–11.6 mm; B–23.2 mm; C–276 mm; D–792 mm; E–1.8 m.

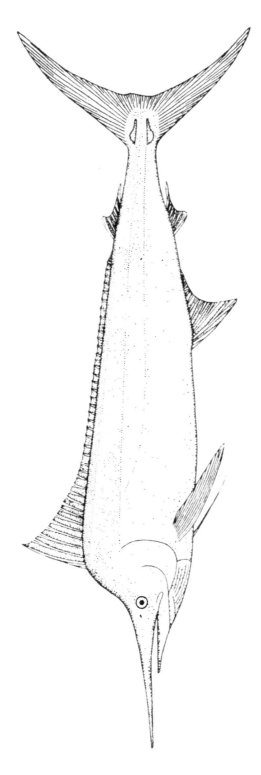

Figure 238. *Makaira indica*—black marlin (Nakamura et al., 1968).

Calif. Acad. Sci., 12, 1926: 59 (California). Matsubara, Fish Morphol. and Hierar., 1955: 528.

Makaira indica, Nakamura et al., Misaki Mar. Biol. Inst. Kyoto Univ., Spec. Rep., 4, 1968: 72, fig. 26 (detailed synonymy).

I D 38-42, II D 6-7; I A 13-14, II A 6-7; P 19-20; V I, 2.

Body depth about 5 times in its length from tip of lower jaw. Snout long. Lateral line forms neither loops nor cells. Head large (about 4 times in body length from tip of lower jaw). Pectoral fins almost perpendicular to sides of body and do not fold against it, contrary to other marlins. Pelvic fins of adult fish smaller than pectoral fins. Dorsal fin deep blue, other fins black to chocolate-brown. Sides of body without vertical stripes. Belly silvery-white. After death, ash-white stripes appear on body (Nakamura et al., 1968).

Biology described by Marrett and Thorp (1965) and Osipov (1968). Length, to 4.6 m (Nakamura et al., 1968).

Distribution: In the Sea of Japan known from Wakasa Bay (Takegawa and Morino, 1970: 379) and reported from central Honshu southward (Matsubara, 1955: 528). Known near the south coast of the Korean Peninsula (Mori, 1952: 138) and off Cheju-do Island (Uchida and Yabe, 1939: 8). East China Sea. Indonesia. Pacific coast of Central America. Indian Ocean (Nakamura et al., 1968: 73).

CLXV. Family XIPHIIDAE—Swordfishes

Body elongate, naked in adult fish. Upper jaw highly elongate, sword-shaped, formed by very long premaxillae and nasals, as well as maxillae, which are closely connected with them near base of rostrum and mesethmoid. Lower jaw much shorter than upper. Teeth absent in mouth of adults. Gill structure unique. Young fish with one long dorsal fin and one anal fin; with age, each fin divides into 2 fins (Figure 239). First dorsal fin high, originates on occiput, its height not greater than maximum body depth; second dorsal fin small, inserted near tail. Longitudinal lateral keel present on caudal peduncle. Pectoral fins inserted low. Pelvic fins absent and pelvic bone also not developed. Swim bladder large. Vertebrae 26 (Andriyashev, 1954: 326).

One genus. Known from the Sea of Japan.

1. Genus *Xiphias* Linné, 1758—Swordfishes

Xiphias Linné, Syst. Nat., ed. 10, 1758: 248 (type: *X. gladius* L.). Nakamura et al., Misaki Mar. Biol. Inst. Kyoto Univ., Spec. Rep., 4, 1968: 5 (synonyms).

Adult fish without scales and teeth. Pelvic fins and pelvic girdle absent. One keel each side of caudal peduncle. Base of first dorsal fin in adult

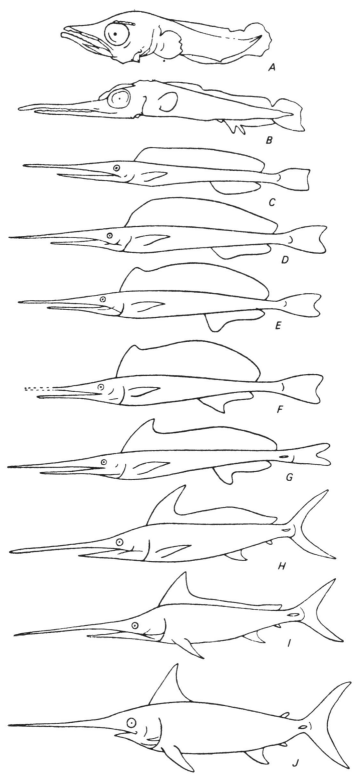

305 Figure 239. Age-related changes in shape of body and fins of *Xiphias gladius* (Nakamura et al., 1968).

Body length; A–6.4 mm; B–11.0 mm; C–160 mm; D–240 mm; E–359 mm; F–380 mm; G–554 mm; H–827 mm; I–1.2 m; J–3.0 m.

fish short; second dorsal fin considerably separated from first. Snout long, broad, and flat. Body rounded in cross section and almost not compressed laterally (Nakamura et al., 1968).

One species. Widely distributed; known from the Sea of Japan.

1. *Xiphias gladius* Linné, 1758—Swordfish (Figure 240).

Xiphias gladius Linné, Syst. Nat., ed. 10, 1758: 248 ("Habitat in Oceano Europae"). Zharov et al., Tuntsy..., 1961: 60 (synonyms, description). Merrett and Thorp, Ann. Mag. Nat. Hist., 13, 8, 1965: 377 (synonyms, remarks). Nakamura et al., Misaki Mar. Biol. Inst. Kyoto Univ., Spec. Rep., 4. 1968: 52, figs. 10, 11 (synonyms, description).

I D 38-45 (first three rays well developed, spiny), II D 4-5; I A 12-16 (first two rays well developed, spiny), II A 3-4; P 17-19. Body depth 4.5 to 5.3 times in its length from tip of lower jaw. Snout very long. Lower jaw distinctly shorter than upper; posterior margin of maxilla extends beyond vertical from posterior margin of eye. Head large (3.7 to 4.3 times in body length from tip of lower jaw). Eyes relatively large. Lateral line distinct in fish about 1 m long, curves slightly anteriorly and sinuous throughout length. Pectoral fins inserted very low, relatively long (1.2 to 1.4 times in head length), falciform, directed backward and downward. Height of anterior part of first dorsal fin in fish less than 1 m long greater than body depth; posteriorly relatively reduced. Second dorsal fin of adult fish small, similar in shape and size to second anal fin, and located slightly behind vertical from origin of second anal fin. First 306 anal fin falciform. Caudal fin deeply forked. Body shape and proportions of body parts change considerably with age (Figure 239). First dorsal fin deep black. Other fins chocolate-brown with chocolate-brown to black stripes. Sides of body black to chocolate-brown. Belly light chocolate-brown. No distinct boundaries between color of other body parts (Nakamura et al., 1968).

In their key to tunas, Nakamura and associates used the structure of the nasal rosette to characterize swordfish. In this particular species the rosette is roundish and consists of 37 to 39 radially divergent olfactory lamellae, on the surface of which blood capillaries and fleshy processes are distinctly visible.

Biological characteristics of this species have been described by both Soviet (Barsukov, 1960; Zharov et al., 1961; Osipov et al., 1964; Parin, 1967, 1968; Osipov, 1968c; Gorbunova, 1969a; Ochinnikov, 1969) and foreign ichthyologists (Cheeseman, 1876; Deraniyagala, 1933; Copley, 1936; Nichols and La Monte, 1937; Gudger, 1938; Nakamura, 1955; Royce, 1957; Jones, 1958, 1962; Yabe et al., 1959; Fitch, 1960; Scheer, 1961; Cavaliere, 1963; Eschmeyer, 1963; de Sylva, 1963; Scott and Tibbo, 1968; Strasburg, 1969). Information on larvae and young fish is available

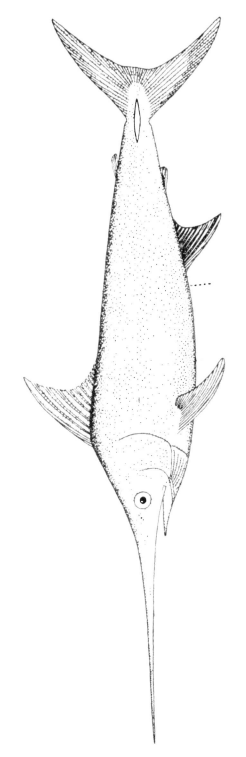

Figure 240. *Xiphias gladius*—swordfish (Nakamura et al., 1968).

306

in works from other countries (Nakamura et al., 1951; Yabe, 1951; Jones, 1962; Scott and Tibbo, 1968). The food value of the flesh of this fish has been assessed by Myaksha (1964).

Length, to 6 m (Golenchenko, 1960).

Distribution: In the Sea of Japan known from Peter the Great Bay[49]; Toyama Bay (Katayama, 1940: 9); and San'in region (Mori, 1956a: 24).[50] Found off the south coast of the Korean Peninsula (Mori, 1952: 138) and near Cheju-do Island (Uchida and Yabe, 1939: 8). Taiwan (China); the Philippines; Australia; New Zealand; Hawaiian Islands; from California to Peru; Chile; and the Atlantic, Indian, and Pacific oceans (La Monte, 1955: 253).

Detailed information on the distribution of this species is reported by Andriyashev (1954) and La Monte (1955).

307

[49]In the Museum of TINRO there is a photograph of a swordfish (taken by V. Aleksandrov) 402 cm long, which died in the mouth of the Shmidtovka on De'Fries Peninsula on August 20, 1954.

[50]Parin (1967: 116, fig. 25) in reporting the area of distribution of swordfish in the Pacific Ocean did not include the Sea of Japan.

10. Suborder Luvaroidei

Similar to Scombroidei, but premaxillae are not produced in form of a rostrum. Epiotics contiguous above supra-occipital. Bases of radials of dorsal and anal fins fused. Posttemporal very large, fused with supra-cleithrum. Pelvic bones fused. Vertebrae 23 (Berg, 1940: 323).

Maxillae very firmly attached to rigid premaxillae. Mouth nonprotractile, small, terminal. Premaxillae not transformed into xiphoid process. Snout reduced, blunt. Branchiostegal membranes broadly joined to isthmus. Dorsal fin originates above midpoint of dorsum.

1 family with 1 genus, distributed in pelagic zone of tropical and subtropical waters, north to Japan and the Sea of Japan.

CLXVI. Family LUVARIDAE—Louvars

Body oblong, laterally compressed. Head moderate in size. Eyes small. Mouth small. Teeth soft, arranged in rows on jaws. Branchiostegal rays 5. Pseudobranchs present. Body covered with very small granular scales. One dorsal and one anal fin with unbranched and wide flexible rays. In adult fish keel present at base of caudal fin. Pelvic fins absent or reduced; in the latter case consist of 1 spine and 4 soft rays, or one spine and 2 soft rays, and inserted near vent. Caudal fin deeply forked (Fowler, 1936: 642).

One genus.

1. Genus *Luvarus* Rafinesque, 1810—Louvars

Luvarus Rafinesque, Caratteri..., 1810: 22 (type: *L. imperialis* Rafinesque). Fowler, Bull. Amer. Mus. Nat. Hist., 70, 2, 1936: 643 (synonyms). Whitley, Rec. Austr. Mus., 20, 5, 1940: 325 (synonyms).

Body elongate, broad anteriorly, laterally compressed, and attenuate posteriorly. Mouth terminal and small. Teeth soft, arranged in single row on jaws; also present on palatines and tongue in young fish. Swim bladder large. Pyloric caeca few. Bones soft and fragile. Scales deciduous. Longitudinal keel on each side of caudal peduncle in adult fish. Dorsal fin consists of soft, wide-set rays, which lengthen with age; in adult fish fin base located only in posterior half of body. Anal fin similar to dorsal. Pelvic fins tend to change with age, inserted on breast, but sometimes absent (Fowler, 1936: 643).

One species.

1. **Luvarus imperialis** Rafinesque, 1810—Louvar (Figure 241)

Luvarus imperialis Rafinesque, Caratteri..., 1810: 22 (Italy). Fowler, Bull. Amer. Mus. Nat. Hist., 70, 2, 1936: 643 (synonyms). Ueno, Japan J. Ichthyol., 12, 3/6, 1965: 99-101, fig. 1.[1]

308 D 22; A 17; P 18; V II; C 25; gill rakers on first gill arch 5 + 13. Branchiostegal rays 5.

Head 4.04, depth 3.10, length from tip of snout to annus 3.60, length of base of dorsal fin 1.44, and length of anal fin 2.65 times in standard length. Diameter of eye 6.56, snout 2.56, interorbital space 2.80, oral slit 4.72, length of caudal peduncle 2.56, its depth 9.45, length of pectoral fin 1.08, length of pelvic 10.10, height of longest ray in dorsal fin 1.78, length of keel on caudal peduncle 2.90, length of postorbital part of head 2.25, and height of occipital crest above eyes 1.75 times in head length.

Body elongate, oval, highly laterally compressed, deepest at level of base of pectoral fin, and gradually attenuating posteriorly. Caudal peduncle very narrow, its minimum depth less than 1/3 length and notably less than 1/10 body depth. Anterior part of head very high, resembles crest with highly pointed median keel. Eyes small, rounded, set below midpoint of depth of head. Snout compressed laterally, rising almost vertically upward in profile. Mouth small, horizontal. Maxilla broad and short; lower jaw protrudes slightly forward when mouth closed. Teeth on jaws small, slightly crestate, arranged in 1 row. Palatines with narrow band of minute teeth. Tongue and vomer without teeth. Gill rakers short, wide, bluntly pointed at tips, and with a few minute spines along inner margin. Head and body covered with minute, very thin, granular, deciduous scales. Base of dorsal fin occupies posterior part of trunk and consists of unsegmented rays; however, anterior 10 rays spiny, wide-set, and situated in narrow groove. First ray of anal fin not elongate. First five rays of anal fin similar to rays of dorsal, and also situated in groove (anterior rays D and A not shown in Figure 241). Pectoral fin rather weak,

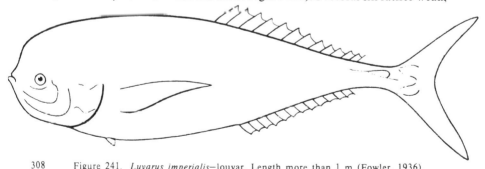

308 Figure 241. *Luvarus imperialis*—louvar. Length more than 1 m (Fowler, 1936).

[1]This publication presents a detailed description of a specimen from Japan which is reproduced as such here.

its tip reaching vertical from base of 9th ray of dorsal and 5th ray of anal fin. Pelvic fins very small. Vent immediately behind pelvic fins. Caudal fin relatively small, but deeply forked. Each side of caudal peduncle with horizontal keel. Body color in formalin pale chocolate-brown, with silvery tinge on sides of belly but not lower surface of anal region and caudal part of body. Dorsal and anal fins dark chocolate-brown, other fins dull (Ueno, 1965: 99).

Length, to 1,830 mm (Fowler, 1936).

Distribution: In the Sea of Japan a specimen (standard length 610 mm) 309 was found on the coast of Hokkaido, which had been washed ashore during stormy weather near the city of Ioiti (Ueno, 1965); reported from Pusan (Matsubara, 1955: 540). This rare fish has been described from the Mediterranean Sea, and is known from the Atlantic coast of Europe, the Indian Ocean—Mozambique Strait, and the Pacific Ocean—north to the Sea of Japan and California (Matsubara, 1955: 540), and south to Australia, and New Zealand (Philipps, 1941: 231).

11. [Suborder Tetragonuroidei]–
Squaretails

309 Pelvic bones not connected with pectoral girdle. Pelvic fins slightly behind pectoral fins. Esophagus with lateral pharyngeal sacs (Figure 242), with small papillae. Unique rhombic scales with keels (Figure 243) arranged in oblique transverse rows; scales of each row closely contiguous (Figure 244). Dorsal fin long, its anterior part spiny. Swim bladder absent (Berg, 1940: 323).

Some authors, including Haedrich (1967: 52, 94), include the family Tetragonuridae, the only family of the suborder isolated by L.S. Berg, in the suborder Stromateoidei. However, we consider the separation of this family into an independent suborder sufficiently well founded. In addition to the characters given by L.S. Berg, it may be pointed out that even Haedrich (1967: 95, 96) listed several characters which differentiate *Tetragonurus* from other genera of the suborder Stromateoidei and recognized that it had possibly branched off long ago during the process of evolution from the common stem of this order. Nevertheless it cannot be isolated from the presently extant members of the order. *Tetragonurus* shares some similarities with members of the family Nomeidae, but differs distinctly in structure of pharyngeal sacs and scale cover. The pharyngeal sacs are highly elongate (Figure 242) and the papillae on the inner surface poorly ossified and much reduced in size (Figure 245). The upper pharyngeal bones of the fourth pair are fused with the bones of the third pair and highly elongate. These elongate bones are

310 covered with teeth and protrude far inside the pharyngeal sacs (Figure 246), where they no doubt play an active part in macerating food and in supporting the muscles. Teeth on jaws highly specialized (Figure 247), with curved tips, close-set, and forming a continuous cutting margin. Similar teeth found in some species of *Psenes,* but members of the latter genus have no teeth on the tongue. In our specimen (No. 39256) of *Tetragonurus cuvieri* (standard length 260 mm) teeth were detected on the tongue after removing the mucus. These teeth were small and moderately spaced along the margin of the tongue and the sides of its base. An important distinguishing feature of Tetragonuridae noted by Haedrich is the absence of tubules in the scales of the lateral line; however, we observed them in our specimen, even though reduced and partly concealed by imbricate scales.

1 family, Tetragonuridae, with 1 genus, *Tetragonurus.*

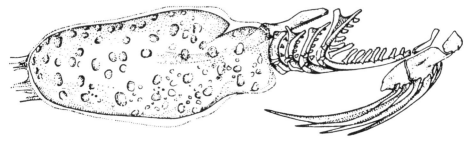

309   Figure 242. Esophagus with lateral pharyngeal sacs in *Tetragonurus cuvieri*
(Haedrich, 1967).

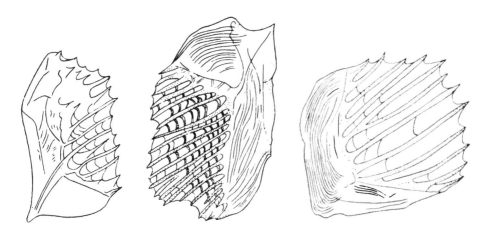

309   Figure 243. Scales in *Tetragonurus cuvieri* (Grey, 1955).

310   Figure 244. Arrangement of scales in *Tetragonurus cuvieri*. No. 39256.

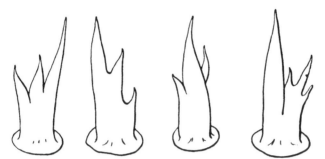

310

Figure 245. Papillae of pharyngeal sac in *Tetragonurus cuvieri*. No. 39256.

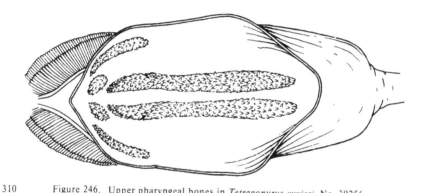

310

Figure 246. Upper pharyngeal bones in *Tetragonurus cuvieri*. No. 39256.

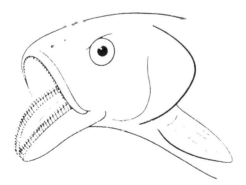

310

Figure 247. Teeth on jaws in *Tetragonurus cuvieri*. No. 39256.

398

[Family TETRAGONURIDAE—Squaretails]

Body oblong, thin, rounded on sides. Caudal peduncle thickened, triangular in cross section, with two keels on each side in posterior part formed by special scales. Two dorsal fins with 10 to 20 short spines situated in groove in which rays lie when fin folded. Base of first dorsal fin almost equal to or longer than base of second dorsal fin. Anal fin with 1 spine connected with soft rays by membrane. Second dorsal and anal fins opposite each other, almost equal in length at base, and each with 10 to 17 rays. Last ray of pelvic fins attached to abdomen at a distance almost equal to length of ray; when folded, fin lies in abdominal groove. Scales moderate in size, ctenoid, with highly developed longitudinal crest, compactly embedded in skin, and arranged in rings around body. Lateral line with very gentle arch in anterior part of body, then continues along middle of body sides and caudal peduncle. Skin thick. Margins of opercle and preopercle with or without minute serration; operculum thick, with barely distinguishable spines. Branchiostegal membrane with 5-6 rays. Mouth fairly large; upper jaw continues beyond vertical from anterior margin of eye. Vomer, palatines, and usually tongue with teeth. Supramaxilla absent. Eyes large but without adipose tissue. Vertebrae 43-58 (Haedrich, 1967: 94).

Length of adult fish 300 to 600 mm. Color uniformly dark chocolate-brown. Oceanic fishes of tropical, subtropical, and temperate seas.

1 genus and 3 species, 2 species known form the Pacific coast of Japan.

[Genus *Tetragonurus* Risso, 1810—Squaretails]

Tetragonurus Risso, Ichthyologie de Nice, 1810: 345 (type: *Tetragonurus cuvieri* Risso). Haedrich, Bull. Mus. Comp. Zool., 135, 2, 1967: 96.

Ctenodax Macleay, Proc. Linn. Soc. New South Wales, 10, 1885: 718 (from: *Ctenodax wilkinsoni* Macleay = *Tetragonurus altanticus* Lowe, 1839).

In characterizing *Tetragonurus* and distinguishing it from genera of the suborder Stromateoidei, Haedrich mentions a combination of such characters as: elongate body and caudal peduncle; special scales forming two keels on each side of caudal peduncle; origin of first dorsal fin barely or definitely behind base of pectoral fin, with base longer than that of second fin; thick keel-shaped scales; and unique lower jaw, with strong ensiform teeth.

3 species, 2 known from the Pacific coast of Japan.

[*Tetragonurus cuvieri* Risso, 1810—Squaretail] (Figure 248)

Tetragonurus cuvieri Risso, Ichthyologie de Nice, 1810: 345, pl. 10, fig. 37 (Nice, Mediterranean Sea). Grey, Dana Rep., 41, 1955: 24, figs.

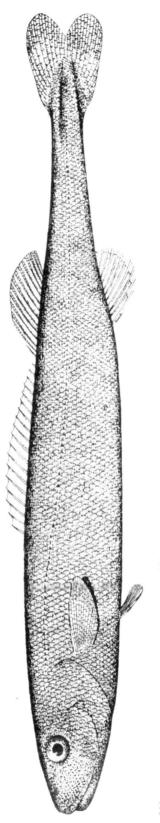

Figure 248. *Tetragonurus cuvieri*—squaretail. Standard length 230 mm. No. 39256.

1C, 10, 11, 14. Abe, Japan J. Ichthyol., 3, 1, 1953: 42–45; 4, 1–3, 1955: 115, 116. Matsubara, Fish Morphol. and Hierar., 1955: 576. Tomiyama and Abe, Enc. Zool., 2, Fishes, 1958: 203, fig. 601 (color figure). Haedrich, Stromateoid Fishes, 1967: 98.

39256. Atlantic Ocean, 33°34′ N, 02°40′ E. December 5–6, 1968. V.L. Yukhov. 15 specimens.

312   D XV–XXI, 10–17; A I, 10–15; P 14–21 (?); *l. l.* up to beginning of caudal keels 97–114; vertebrae 52–58 (Grey, 1955).

Most characters of species given in description of family and genus. One may mention additionally that this species, like *T. atlanticus* Lowe, 1839, differs well from *T. pacificus* Abe (1953: 47) in larger number of rays in first dorsal fin (14–21 versus 10–11), larger number of pores in lateral line up to beginning of caudal keels (83–114 versus 73–78), and number of vertebrae (45–58 versus 40–43). This species differs from *T. atlanticus,* with which it is found along the Japanese coast (Abe, 1955: 116), in larger size of adult fish, greater number of pores in lateral line (97–114 versus 83–95), and greater number of vertebrae (52–58 versus 45–51) (Grey, 1955).

In our specimens (6) of *T. cuvieri* from the Atlantic Ocean, only 59 vertebrae were seen.

The young of *Tetragonurus* are generally found in association with medusae as well as salps (*Pyrosoma*). In this respect the behavior of *Tetragonurus* is similar to that of young fish of the suborder Stromateoidei. Presumably *Tetragonurus* feed almost exclusively on coelenterates and ctenophores; the large knife-shaped teeth on the lower jaw and the structure of the mouth are adapted to such food. Quite likely the meat of these fish is poisonous during the breeding season (Haedrich).

Tetragonurus are typical oceanic fish and, judging from their dark or even black color, mesopelagic or bathypelagic. They are relatively rare. Our specimens (15) were removed from the stomach of a whale.

Length, to 400 mm (Abe, 1958).

Distribution: Not found in the Sea of Japan, but known near the south coast of Hokkaido (Abe, 1955: 115); Volcano Bay (Ueno, 1965b: 1); Pacific coast of Honshu Island in the Fukushima, Tiba, and Kanagawa prefectures (Abe, 1955: 115; Matsubara, 1955: 576). In the Pacific Ocean found near the coast of California, Hawaiian Islands, New Zealand, and southeast Australia; northwest part of the Atlantic Ocean, Mediterranean Sea (Grey, 1955: 33, fig. 14), and southern part (our catch).

12. Suborder Stromateoidei

312 Body oblong, moderately deep or quite deep, and compressed laterally or rounded. Dorsal fin with one or two bases and spiny rays, although latter very weak in some species. Anal fin with 1 to 3 spines. Dorsal and anal fins terminate at same vertical line. Pelvic fins present or absent. Pectoral fins with 16 to 25 rays. Body covered with scales, but anterior end of head naked. Scales usually thin, cycloid, caducous, or weakly ctenoid in some nomeids and in *Schedophilus medusophagus*. Scales usually cover base of vertical fins. Lateral line present and represented by simple tubules. Bony shields or keels absent on caudal peduncle. Well-developed hypodermal system of mucous canals usually present, communicating with external media through small pores scattered on surface of head and body. Eyes very small to large, set on sides of head, and do not protrude in profile of head. Two nostrils on each side; anterior nostril roundish, posterior one in form of vertical slit. Teeth on jaws small, simple, or in form of small canines, more or less arranged in 1 row compactly or with gaps. Teeth on vomer and palatines present or absent. Teeth absent on entopterygoid and metapterygoid bones. Small teeth usually present on inner margins of gill rakers. Gills 4 pairs, with slit behind fourth gill. Gill rakers 10 to 20 in lower half of second gill arch. Well-developed pseudobranchs usually present, but absent in *Pampus*; rudimentary gill rakers usually present under pseudobranchs. Branchiostegal membranes not attached to isthmus except in species of *Pampus* (Haedrich, 1967: 45).

This suborder is characteristically distinguished from other suborders of Perciformes in the structure of the anterior part of the esophagus, which is located immediately behind the pharynx (behind the last gill 313 arch), equipped with processes in the form of lateral sacs, provided with papillae or longitudinal folds on the inner side, and bearing numerous simple denticles with stellate or irregularly shaped roots embedded in the muscular wall of the sacs (Figure 249). Similar esophageal sacs also present in species of the family Tetragonuridae, but their denticles lack roots. In addition, the unique nature of the scales and several other differences prompted us to agree with Berg (1940), who separated this family into an independent suborder.

Members of the suborder Stromateoidei are typical marine fishes dwelling in the continental shelf of the Atlantic, Indian, and Pacific oceans. Most are known from tropical and subtropical waters, but some

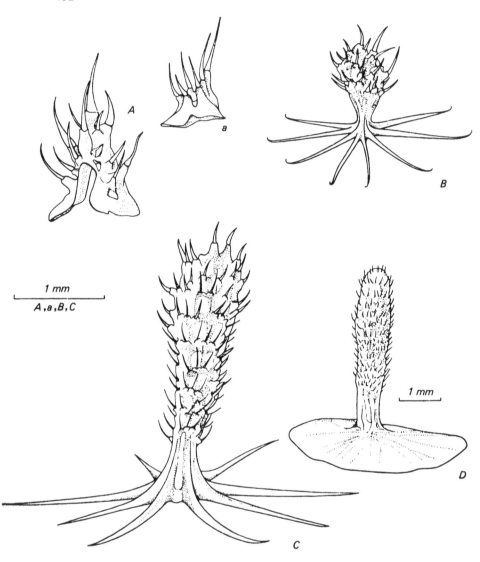

313   Figure 249. Papillae on inner surface of pharyngeal sacs of Stromateoidei
(Haedrich, 1967).

A—Centrolophidae, *Hyperoglyphe*; a—same, small papillae; B—Nomeidae, *Nomeus*;
C—Stromateidae, *Peprilus*; D—Ariommidae, *Ariomma*.

found in the Sea of Japan and north to Peter the Great Bay and Hokkaido
Island. 4 families, 3 of which are represented in the Sea of Japan and the
fourth family found off the Pacific coast of Japan.

Key to Families of Suborder Stromateoidei (Haedrich, 1967)[1]

1 (4). Two dorsal fins distinct although not entirely separate. First dorsal fin usually with 10 to 20 spines; if less than 10, longest spine almost equal in length to longest ray of second dorsal fin. Pelvic fins always present. Vomer and palatines with or without teeth.

2 (3). Vomer and palatines with minute, often almost imperceptible teeth. Caudal peduncle compressed laterally, in cross section not rectangular, its minimum depth more than 5% of standard length. Caudal keel absent. Second dorsal and anal fins usually with more than 16 rays each......................... CLXVII. **Nomeidae**.

3 (2). Vomer and palatines without teeth. Caudal peduncle rectangular in cross section, its minimum depth less than 5% of standard length. 2 low lateral keels located on each side of caudal peduncle near base of caudal fin. Second dorsal and anal fins with 14 to 16 rays each............................... CLXVIII. **Ariommidae.**

4 (1). Dorsal fin single, continuous; if two dorsal fins, not distinctly separate and first fin with less than 10 spiny rays. When spines present, longest less than half length of longest ray of second dorsal fin. Pelvic fins present or absent. Vomer and palatines without teeth.

5 (6). Pelvic fins always present. Soft rays either absent in anterior part of dorsal fin, or 1 to 5 weak or 5 to 9 strong spines present. Anal fin with 15 to 30 rays. Dorsal and anal fins never falciform; bases rarely equal in size. All teeth on jaws conical. Supplemental maxilla always present, but in some fish difficult to detect. Branchiostegal rays seven. Vertebrae 25 to 30 or 50 to 60.....
.... CLXIX. **Centrolophidae.**

6 (5). Pelvic fins always absent in adult fish, and rarely present in young ones. Strong spines not present in dorsal fin in front of soft rays, but some species with 5 to 10 discoid processes as tips of interneural processes which protrude in front of fin. Anal fin with 30 to 50 rays. Dorsal and anal fins often falciform; bases almost equal in length. Teeth along sides of jaws compressed laterally. Supplemental maxilla absent. Branchiostegal rays five to six. Vertebrae 30 to 48...................... CLXX. **Stromateidae.**

[1]Haedrich (1969: 5) described a new family, Amarsipidae, from equatorial waters of the Indian and Pacific oceans. The esophageal expansion typical of Stromateoidei is absent, but the pharyngeal bones armed with well-developed teeth, which are much larger than on the jaws. The pelvic fins are jugular and notably anterior to the base of the pectoral fins, whereas in other families of the suborder the pelvic fins are absent or inserted under the base of the pectoral fins. Small, almost transparent fishes up to 70 mm in length.

404

404

315   CLXVII. **Family NOMEIDAE—Man-of-War Fishes, Driftfishes**

Body oblong, sometimes relatively deep, compressed laterally. 2 distinct dorsal fins separated by deep notch. First dorsal fin with about 10 slender spines, which, when fin folded, lie in groove on back; longest spine almost equal in length to longest soft ray of second dorsal fin. One to three spines situated anterior to soft rays of anal fin, not separated from fin itself. Soft dorsal and anal fins located opposite each other and similar in shape and size; bases concealed in grooves formed by scales. Pelvic fins of young fish attached to belly by thin membrane and, when folded, concealed in depression on belly; young of *Nomeus* and some species of *Psenes* have highly enlarged pelvic fins. Scales on body very minute to very large, cycloid or slightly ctenoid, and caducous. Lateral line based high, follows profile of back, and often does not continue onto caudal peduncle. Skin thin. Margins of opercle and preopercle smooth or slightly serrate. Branchiostegal rays 6. Mouth small; maxilla rarely reaches under eye. Jaw teeth small, conical or canine-shaped in some species of *Psenes,* arranged in almost 1 row. Vomer and palatines with teeth. Supplemental maxilla absent. Adipose tissue moderately developed around eyes in most species. Vertebrae 30 to 38 or 41 to 42. Pharyngeal sacs with papillae in upper and lower halves. Papillae arranged in five to seven broad longitudinal bands. Bases of papillae stellate; denticles located on apices of central stalks (Figure 249, B). Adult fish about 300 mm long, but species of *Cubiceps* reach 900 mm. Color from silvery to bluish-chocolate-brown; some fishes with stripes and spots (Haedrich, 1967: 76).

Oceanic fishes of tropical and subtropical waters of the Atlantic, Indian, and Pacific oceans. Found in large numbers off the Philippines and near southern Japan.

Three genera. One represented in the Sea of Japan.

Key to Genera of Family Nomeidae[2]

1 (4). Body elongate, maximum depth usually less than 35% of standard length. Origin of dorsal fin behind insertion of pectoral fins, or in young fish above it.

2 (3). Anal fin with 1 to 3 spines and 14 to 25 soft rays. Pelvic fins inserted at vertical with posterior margin of base of pectoral fin or behind it. Oval plate on tongue with knob-like teeth. Vertebrae 30 to 33 . . . . [**Cubiceps** Lowe].

3 (2). Anal fin with 1 to 2 spines and 24 to 29 soft rays. Pelvic fins inserted anterior to vertical with posterior margin of base of pectoral fin or at vertical line; in very large specimens possibly even postoriorly. Oval plate with odontoid processes absent on tongue. . . . . [**Nomeus** Cuvier].

[2]From Haedrich, 1967: 78.

CLXVII. **Family NOMEIDAE—Man-of-War Fishes, Driftfishes**

Body oblong, sometimes relatively deep, compressed laterally. 2 distinct dorsal fins separated by deep notch. First dorsal fin with about 10 slender spines, which, when fin folded, lie in groove on back; longest spine almost equal in length to longest soft ray of second dorsal fin. One to three spines situated anterior to soft rays of anal fin, not separated from fin itself. Soft dorsal and anal fins located opposite each other and similar in shape and size; bases concealed in grooves formed by scales. Pelvic fins of young fish attached to belly by thin membrane and, when folded, concealed in depression on belly; young of *Nomeus* and some species of *Psenes* have highly enlarged pelvic fins. Scales on body very minute to very large, cycloid or slightly ctenoid, and caducous. Lateral line based high, follows profile of back, and often does not continue onto caudal peduncle. Skin thin. Margins of opercle and preopercle smooth or slightly serrate. Branchiostegal rays 6. Mouth small; maxilla rarely reaches under eye. Jaw teeth small, conical or canine-shaped in some species of *Psenes,* arranged in almost 1 row. Vomer and palatines with teeth. Supplemental maxilla absent. Adipose tissue moderately developed around eyes in most species. Vertebrae 30 to 38 or 41 to 42. Pharyngeal sacs with papillae in upper and lower halves. Papillae arranged in five to seven broad longitudinal bands. Bases of papillae stellate; denticles located on apices of central stalks (Figure 249, B). Adult fish about 300 mm long, but species of *Cubiceps* reach 900 mm. Color from silvery to bluish-chocolate-brown; some fishes with stripes and spots (Haedrich, 1967: 76).

Oceanic fishes of tropical and subtropical waters of the Atlantic, Indian, and Pacific oceans. Found in large numbers off the Philippines and near southern Japan.

Three genera. One represented in the Sea of Japan.

Key to Genera of Family Nomeidae[2]

1 (4). Body elongate, maximum depth usually less than 35% of standard length. Origin of dorsal fin behind insertion of pectoral fins, or in young fish above it.

2 (3). Anal fin with 1 to 3 spines and 14 to 25 soft rays. Pelvic fins inserted at vertical with posterior margin of base of pectoral fin or behind it. Oval plate on tongue with knob-like teeth. Vertebrae 30 to 33 . . . . [**Cubiceps** Lowe].

3 (2). Anal fin with 1 to 2 spines and 24 to 29 soft rays. Pelvic fins inserted anterior to vertical with posterior margin of base of pectoral fin or at vertical line; in very large specimens possibly even postoriorly. Oval plate with odontoid processes absent on tongue. . . . . [**Nomeus** Cuvier].

[2]From Haedrich, 1967: 78.

4 (1). Body rather deep, maximum depth usually more than 40% of standard length, but possibly less deep in very large fish. Origin of dorsal fin before insertion of pectoral fins, or above it in larger specimens............................ 1. **Psenes** Valenciennes.

316   [Genus *Cubiceps* Lowe, 1843—Cigarfishes]

Cubiceps Lowe, Proc. Zool. Soc., London, 11, 1843: 82 [type: *C. (Seriola) gracilis* Lowe]. Haedrich, Bull. Mus. Comp. Zool., 135, 2, 1967: 78 (description, synonyms).

Differs from other genera of the family Nomeidae in elongate body, long alate pectoral fins, insertion of pelvic fin on or behind vertical from posterior margin of base of pectoral fin, presence of scales on top of head, on cheeks and on operculum, and presence of oval plate with knoblike teeth on tongue. Found rarely; dwells in open parts of ocean. 8 species.

Two species found in the waters of Japan (Abe, 1959). Not known from the Sea of Japan.

[*Cubiceps gracilis* Lowe, 1843)[3]—Longfin Cigarfish] (Figure 250)

Seriola (Cubiceps) gracilis Lowe, Proc. Zool. Soc., London, 11, 1843: 82 (Madeira Island).

Cubiceps gracilis, Günther, Cat. Fish. Brit. Mus., 2, 1860: 389. Abe, J. Oceanogr. Soc., Japan, 11, 2, 1955b: 75, figs. 1-4 (description and synonyms); Enc. Zool., 2, Fishes, 1958: 202, fig. 599; Rec. Oceanogr. Works, Japan, 3, 1959: 225. Haedrich, Bull. Mus. Comp. Zool., 135, 2, 1967: 80, fig. 26.

Cubiceps squamiceps (non Lloyd) Matsubara, Fish Morphol. and Hierar., 1955: 579. Abe. Enc. Zool., 2, Fishes, 1958: 202, Fig. 600.

Cubiceps natalensis (non Gilchrist and Bonde) Kamohara, Rep. Kochi Univ., Nat. Sci., 3, 1952: 35.

1251. Madeira Island. 1850. Lakhtenberg. 1 specimen.

D IX-XI, I-II 20-22; A II-III, 20-23; P 20-24; gill rakers 8-9 + 1 + 14-17 (Haedrich, 1967. 80). Our specimens 190 mm long with similar formula: D X, 22; A 20; P 21.

Length to 1,140 mm (Abe, 1958).

Distribution: Not found in the Sea of Japan. Rather rare fish near the Pacific coast of Japan and caught away from coasts in the zone of Kuro-Shio Current. Until 1955 (publication of T. Abe), this species was reported from the Atlantic Ocean and the Mediterranean Sea.

---

[3]The other Japanese species, *C. pauciradiatus* Günther, found near the Pacific coast of Japan (Abe, 1959), differs in smaller number of rays in the second dorsal fin (16-18), anal fin (14-17), and pectoral fin (18-19), and probably also in size, rarely reaching 160 mm in standard length (Haedrich, 1967: 83).

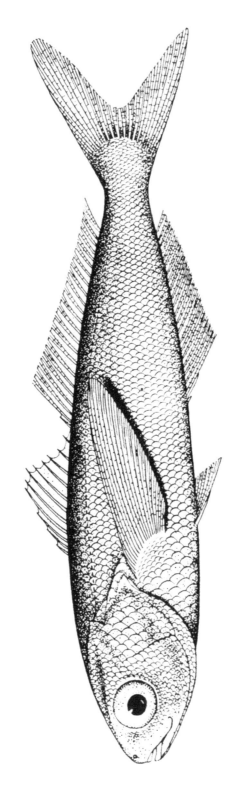

Figure 250. *Cubiceps gracilis*—longfin cigarfish. Standard length 164 mm (Haedrich, 1967).

317   **[Genus *Nomeus* Cuvier, 1817—Man-of-War Fishes]**

Nomeus Cuvier, Régne Animal., 2, 1817: 315 (type: *Gobius gronovii* Gmelin, 1788). Haedrich, Bull. Mus. Comp. Zool., 135, 2, 1967: 81.

This genus differs from other genera of the family Nomeidae in the following combination of characters: elongate body; black fanlike pelvic fins attached to belly throughout length of innermost ray by membrane and fold into groove on belly; abundance of dark spots on body and fins; and presence of 41 vertebrae. A biological peculiarity of the young of this monotypic genus is that *Nomeus gronovii* develops under the protection of medusa *Physalia*.

Temperate and tropical waters of the ocean. Only 1 species known.

[*Nomeus gronovii* (Gmelin, 1788)—Man-of-War Fish] (Figure 251)

Gobius albula Meuschen, Zoophylac. Gronov, 3, 1781, No. 278 (non *Gobius corpore*, etc. Gronov, 1763, Zoophylacium: 82, No. 278, America) (species non binomial). Fowler, Mar. Fishes West Africa, 2, 1936: 1279.

Gobius gronovii Gmelin, Syst. Nat. Linné, 1789: 1205 (Atlantic Ocean, tropical zone).

Nomeus gronovii, Cuvier, Régne Animal., 2, 1817: 315. Fowler, Mar. Fishes West Africa, 2, 1936: 660, fig. 296 (description, synonyms). Haedrich, Bull. Mus. Comp. Zool., 135, 2, 1967: 83.

Nomeus albula, Matsubara, Fish Morphol. and Hierar., 1955: 579, fig. 186. Abe, Enc. Zool., 2, Fishes, 1958: 201, fig. 597 (color figure).

D XII, 27; A 28; P 23–24; gill rakers $7 + 1 + 16$ (Abe, 1958); vertebrae $15 + 26 = 41$ (Haedrich, 1967: 83).

Characters of species given in generic diagnosis. Color of live fish: back bright blue and sides of body with light blue spots on lustrous silvery background. In specimens preserved in alcohol blue color changes to dark chocolate-brown. A specimen 225 mm long caught in bottom trawl was uniformly dark chocolate-brown, indicating that adult fish live near the bottom (Haedrich, 1967: 83).

Length to 250 mm (Abe, 1958: 201).

Distribution: Not found in the Sea of Japan; reported from waters of Japan but precise place not pinpointed.

1. Genus *Psenes* Valenciennes, 1833—Medusa Fishes

Psenes Valenciennes, Hist. Nat. Poiss., 9, 1833: 259 (type: *P. cyanophrys* Valenciennes, 1833). Haedrich, Bull. Mus. Comp. Zool., 135, 2, 1967: 84 (description, synonyms).

Icticus Jordan and Thompson, Mem. Carnegie Mus., 6, 1914: 242 (type: *I. ischanus* Jordan and Thompson, 1914).

Body deep, maximum depth usually more than 40% of standard length,

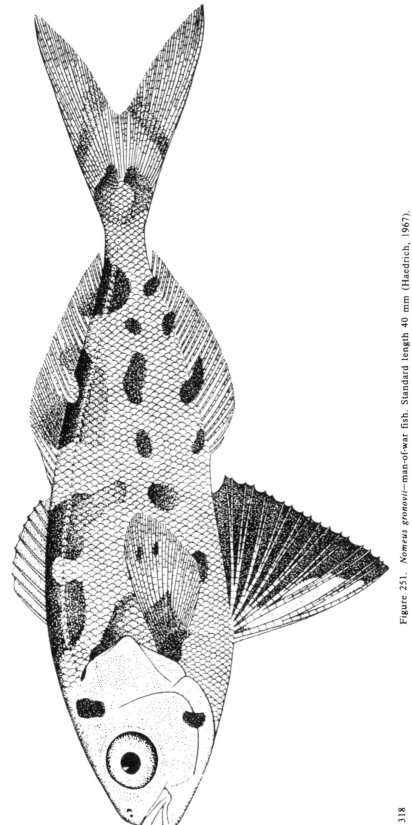

Figure 251. *Nomeus gronovii*—man-of-war fish. Standard length 40 mm (Haedrich, 1967).

but sometimes less in larger specimens; muscles compact or loose. Body at base of dorsal and anal fins highly compressed laterally, sometimes rather translucent. Caudal peduncle short, compressed laterally, and may be quite slender. Dorsal fins 2, but bases almost contiguous; origin of first dorsal fin anterior to insertion of pectoral fins, and with about 10 slender flexible spines; when folded, fin lies in deep groove on back. Rays of second dorsal fin 18 to 30, almost equal in length to longest ray of first dorsal fin. Vent slightly anterior to midpoint of standard length. Origin of anal fin almost at vertical from midpoint of body length; fin with 3 weak spines and 17 to 30 soft rays. Pectoral fins rounded or winglike; relative length of pectoral fins changes with age. Pelvic fins attached to belly under posterior part of base of pectoral by membrane[4] and fold
319 into groove on belly; they are relatively long in young fish and reduce notably with age. Caudal fin deeply forked. Scales small or minute,[5] very slightly ctenoid, very slender, deciduous, and cover base of dorsal and anal fins. Lateral line high, parallel to dorsal profile ending under base of last ray of second dorsal fin or sometimes extending onto caudal peduncle. Skin thin. Head length about 30% of standard length. Upper part of head naked; minute pores present in naked skin; scales on occiput continue forward almost to anterior margin of eyes. Eyes moderate or large; thin layer of adipose tissue surrounds them. Nostrils near tip of highly truncate snout; anterior nostril round, posterior one in form of slit. Upper jaw extends slightly beyond vertical from anterior margin of eye; premaxillae not protractile. Preorbital overlaps maxilla when mouth closed, leaving only its lower part outside. Supramaxilla absent. Teeth on both jaws arranged in single row, pointed, more minute or sparse in upper jaw, and larger and more compact on lower jaw. Teeth present on vomer head and arranged in one row on palatines. Opercle and preopercle thin, covered with scales, their margins barely serrate or smooth. Operculum with 2 weak flat spines difficult to discern; angle of preopercle projects backward slightly, rounded. Gill rakers slender, slightly shorter than gill filaments, serrate along inner margin, 14 to 19 on lower part of first gill arch. Branchiostegal rays 6. Vertebrae $13-15 + 18-23 = 31-38$ or $15 + 26-27 = 41-42$. Color of specimens preserved in alcohol from chocolate-brown to yellowish; some species with distinct traces of minute black spots or longitudinal stripes; unpaired and pelvic fins often darker than body. Body at base of dorsal and anal fins in *Psenes pellucidus* translucent (Haedrich, 1967: 84–85).

Young fish remain in surface layers of the ocean, often found close

[4]They are not connected in Figure 252.
[5]Scales of *Psenes arafurensis* Günther (1889) fairly large—*l. l.* 44 (Tomiyama, 1954: 1010, Fig. 543).

to or under flotsam, especially under *Sargassum* algae floating on surface; adult fish live in deeper layers.

This genus is widely represented in temperate and tropical parts of the Atlantic, Indian, and Pacific oceans. 6 species known. 3[6] species found in the Sea of Japan only near the Pacific coast, and 1 also found off the Sea of Japan.

1. **Psenes pellucidus** Lütken, 1880—Medusa Fish (Figure 252)

Psenes pellucidus Lütken, Kon. Denske Vid. Selsk. Skrift. Kjøbenhavn., 5, XII (Spolia Atlantica, 1880: 516, fig. 601) (Surabaja Strait, Java). Haedrich, Bull. Mus. Comp. Zool., 135, 2, 1967: 88, fig. 28, 30 (synonyms).

Icticus ischanus Jordan and Thompson, Mem. Carnegie Mus., 6, 4, 1914: 242, pl, 27, fig. 4 (Okinawa).

Icticus pellucidus, Tomiyama, Fig. and Descr..., 50, 1954: 1002, pls. 199, 200, figs. 539–542 (synonyms). Abe, Japan. J. Ichthyol., 3, 6, 1954: 246; Enc. Zool., 2, Fishes, 1958: 200, fig. 598.

D X–XI, I–II 27–32; A III, 26–31; P 18–20; gill rakers 8–9 + 1 + 14–16; vertebrae 15 + 26–27 (Haedrich, 1967: 88).

According to Haedrich, this meso- and bathypelagic species is distinguished from other species of this genus by soft muscles, long dagger-like teeth on lower jaw, dark color, and larger number of soft rays of dorsal fin (27 to 32 versus 19 to 28) and anal fin (26 to 31 versus 20 to 28), as well as vertebrae (41 to 42 versus 31 to 38).

Length, to 480 mm (Abe, 1954: 246).

Distribution: In the Sea of Japan reported from Toyama Bay (Katoh et al., 1956: 318) and San'in region (Mori, 1956a: 24). Reported from the south coast of the Korean Peninsula (Chyung, 1961). Off the Pacific coast of Japan from Kushiro (Abe, 1954: 246) and southward: Kesennuma, Gulf of Tokyo, Tosa Bay, Yamaguti and Miazaki prefectures (Matsubara, 1955: 578), and Pacific, Indian, and Atlantic oceans.

CLXVIII. Family ARIOMMIDAE

Body oblong, sometimes rather deep, rounded or compressed laterally. Caudal peduncle short, thin, with two low fleshy lateral keels on each side. Two dorsal fins: first dorsal fin with 10 to 11 slender spines, when folded concealed in groove on back; longest spine half length of soft rays of second dorsal fin. Anal fin with three spines not separated from fin. Second dorsal and anal fins located opposite each other and almost equal in basal length and height; each fin with 14 to 15 rays and bases of fins not covered with scales. Pelvic fins attached to belly by thin

[6]*Psenes cyanophrys* Cuvier and Valenciennes, 1833 = *P. kamoharai* Abe, Kojima and Kosakai, 1963 (Haedrich, 1967: 88); *Psenes maculatus* Lütken, 1880 (Kobayashi, 1961); *Psenes arafurensis* Günther, 1889 (Tomiyama, 1954: 1008).

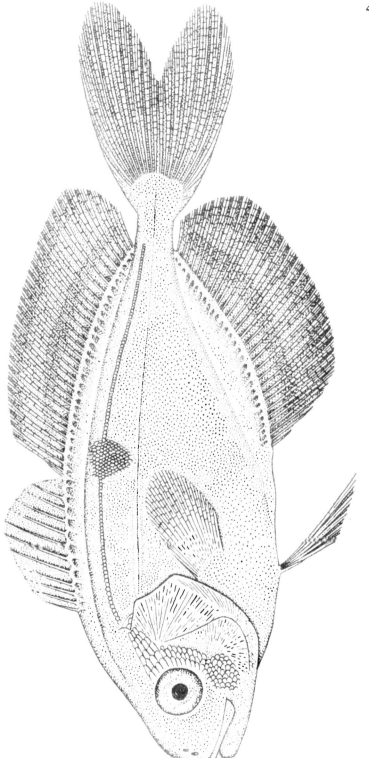

Figure 252 *Psenes pellucidus*—medusa fish. Standard length 130 mm (Haedrich, 1967).

membrane and when folded concealed in groove on belly. Scales large, cycloid, thin, and deciduous. Lateral line high, parallel to dorsal profile and does not extend onto caudal peduncle; tubules of scales in lateral line sometimes branched. Skin thin. Margins of opercle and preopercle smooth or minutely serrate. Operculum thin, brittle, with 2 weak ill-defined, flat spines. Branchiostegal rays 6. Mouth small; maxilla terminates slightly before eye. Teeth on jaws small, simple or tricuspid, and arranged in one row. Vomer and palatines without teeth. Supramaxilla absent. Eyes large; adipose tissue well developed and covers preorbital bone. Vertebrae 29 to 32. Papillae in pharyngeal sac present only in its upper part and arranged in same manner as in the family Stromateidae, and not in bands. In shape, papillae resemble stalks, laden throughout with teeth; bases of stalks in form of broad round plates (Figure 249, D). Color from silvery- to bluish-chocolate-brown. Some species covered with spots (Haedrich, 1967: 88).

Demersal fishes dwelling at great depths in tropical and subtropical waters of the Atlantic, Indian, and Pacific oceans.

One genus.

1. Genus *Ariomma* Jordan and Snyder, 1904

Ariomma Jordan and Snyder, Proc. U.S. Nat. Mus., 27, 1904: 942 (type: *A. lurida* Jordan and Snyder). Haedrich, Bull. Mus. Comp. Zool., 135, 2, 1967: 90.

Paracubiceps Belloc, Rev. Trav. l'Office Pêches Marit., 10, 3, 1937: 356 (type: *P. ledanoisi* Belloc).

Characters of the only genus, *Ariomma,* given in diagnosis of the family. This genus is distinguished from other genera of Stromateoidei as follows (Haedrich, 1967: 91): slender caudal peduncle with lateral keels; rigid, deeply forked caudal fin; about 15 rays with distinct bases on second dorsal and anal fins; well-developed adipose eyelid; presence of two dorsal fins separated by deep notch; and absence of teeth on palatines.

322

14 species described, but classification of genus needs thorough revision. One species known in the Sea of Japan and along the coast of Japan—*A. lurida,* although Haedrich (1967: 93) reported *A. indica* for southern Japan.

1. *Ariomma lurida* Jordan and Snyder, 1904 (Figure 253)

Ariomma lurida Jordan and Snyder, Proc. U.S. Nat. Mus., 27, 1904: 942 (Hawaiian Islands). Katayama, Japan J. Ichthyol., 2, 1, 1952: 31, 2 figs. Matsubara, Fish Morphol. and Hierar., 1955: 579. Abe, Enc. Zool., 2, Fishes, 1958: 201, fig. 596 (color figure). Ueno, Sci. Rep. Hokkaido Fish Exp. Sta., 4, 1965: 2, 8, fig. 5. Haedrich, Bull. Mus. Comp. Zool., 135, 2, 1967: 93.

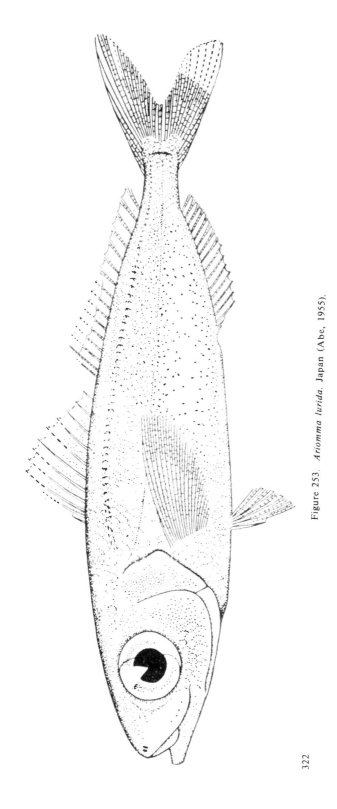

Figure 253. *Ariomma lurida*. Japan (Abe, 1955).

39326. Hawaiian Islands. March 3, 1968. V. Fedorov. 2 specimens. D X, I 15; A III, 14; P 22-23; gill rakers 10 + 1 + 20.

2 specimens in our collection with a standard length of 215 mm. *A. lurida* differs from the closely related species in the Hawaiian Islands *A. evermanni* in larger eyes, constituting 33% of the head length in our specimens versus less than 30% in *A. evermanni*. The difference in number of rays in the pectoral fins is probably not significant since the results reported by Abe (1958: 201)—P 20-24—reduce the divergence (20-21 and 25 rays).

323   *A. lurida* differs from *A. indica* (Day, 1870), possibly found off the south coast of Japan, in larger number of gill rakers (9 + 1 + 20 = 30 versus 7 + 1 + 15 = 23).

Length, to 430 mm (Abe, 1958).

Distribution: In the Sea of. Japan found in the coastal waters off Wakkani at the northwest tip of Hokkaido (Ueno, 1965b). Reported from the environs of the Hawaiian Islands. Along the Pacific coast of Japan found near Shikoku Island and Koti region (Katayama, 1952b: 31); Sagami Bay (Matsubara, 1955: 579); and Tsuruga Bay (Abe, 1954: 222).

CLXIX. Family CENTROLOPHIDAE—Ruffs

Body oblong, sometimes rather deep, usually slightly compressed laterally. Dorsal fin with one base, but in some species anterior part of fin represented by short prickly spines, and in others spiny rays weak and difficult to differentiate from soft rays. Anal fin with 3 spines. Pelvic fins usually attached to abdomen by thin membrane and when folded lie in groove on abdomen. Head conspicuously naked on top and usually covered with small pores. Body scales cycloid and usually deciduous. Lateral line continues onto caudal peduncle. Margin of preopercle slightly serrate but smooth in young fish. Operculum thin, with 2 flat weak spines. Branchiostegal rays 7. Mouth relatively large; maxilla extends beyond vertical from anterior margin of eye. Jaw teeth small, conical, almost arranged in one row; vomer and palatines without teeth. Supramaxilla present in most species, but absent in *Psenopsis*. Adipose tissue around eye indistinct. Vertebrae 25 to 30, except in *Icichthys,* 50 to 60. Pharyngeal sacs with irregularly shaped papillae arranged in 10 to 12 longitudinal bands; teeth situated directly on apex of bony base (Figure 249, a). Length of adult fish 300 to 1,200 mm. They are deeply pigmented and without spots (Haedrich, 1967: 53).

Pelagic fishes in open parts of seas near the continental shelf; however, *Psenopsis* and *Seriolella* found in shallows near the coast.

6 genera; 2 found in the Sea of Japan and 1 other in the Pacific waters of Japan.

Key to Genera of Family Centrolophidae[7]

1 (2). Spines of dorsal fin weak and gradually transform into soft rays. Vertebrae 50 to 60.[8]........ [**Icichthys** Jordan and Gilbert].

2 (1). Spines of dorsal fin (5 to 9) strong, short, and sharply delineated from successive soft rays. Vertebrae 25.

3 (4). Soft rays in dorsal fin 19 to 25, in anal fin 14 to 21. Margin of preopercle serrate. Scales not deciduous. Lateral line forms smooth arch in front; behind vertical from midpoint of anal fin extends in straight line to base of caudal fin. Adipose tissue around eyes negligible...................... 1. **Hyperoglyphe** Günther.

324 4 (3). Soft rays in dorsal fin 27 to 32, in anal fin 22 to 29. Margin of preopercle smooth or very finely serrate. Scales deciduous. Lateral line almost parallels dorsal profile, straightens behind vertical from base of fins. Adipose tissue around eyes well developed.................................. 2. **Psenopsis** Gill.

[Genus *Icichthys* Jordan and Gilbert, 1880]

Icichthys Jordan and Gilbert, Proc. U.S. Nat. Mus., 3, 1880: 305 (type: *I. lockingtoni* Jordan and Gilbert). Haedrich, Bull. Mus. Comp. Zool., 135, 2, 1967: 65.

Body moderately elongate, maximum depth less than 25% of standard length. Muscles soft. Caudal peduncle broad, compressed laterally, moderate in length. Dorsal fin continuous, originates far behind vertical from base of pectoral fin; weak spiny rays increase gradually in length and imperceptibly graduate into soft rays; total rays 39 to 43. Mid-dorsal ridge preceding dorsal fin. Vent almost midbody. Anal fin with three weak spines originating slightly behind vent; total number of rays in fin 27 to 32. Dorsal and anal fins with fleshy bases. Pelvic fins small, inserted under base of pectoral fins, not attached to abdomen by membrane; when folded, lie in groove on abdomen. Caudal fin broad, slightly rounded or emarginate. Scales moderate in size, cycloid, not deciduous, and extend onto fleshy base of vertical fins. Lateral line forms smooth arch in anterior part of body, then extends in straight line from vertical with origin of anal fin along side of body and partly continues onto caudal peduncle; judging from figure (Figure 254), lateral line does not reach base of rays of caudal fin.

Lateral line with about 120 scales. Skin rather thick. Head about 25% of standard length; top of head naked, occiput covered with scales. Eyes moderate in size, not surrounded by adipose tissue. Nostrils located near

[7]From Haedrich, 1967: 54.

[8]*Centrolophus* Lacépède has 25 vertebrae and thus differs notably from *Icichthys* Jordan and Gilbert.

truncate snout, both rounded. Angle of gape continues slightly behind vertical from anterior margin of eye. Premaxilla not protractile. Very small supramaxilla present. Jaw teeth small, pointed, arranged in single row, close-set; vomer and palatines without teeth. Opercle and preopercle thin, with weak serration along margins, covered with scales. Operculum with 2 weak flat spines; preopercle with rounded process. Cheeks covered with scales. Gill rakers thickened and slightly shorter than gill filaments, with denticles along inner margin, their number on lower branch of gill arch about 10. Pseudobranchs small. Branchiostegal rays 7. Vertebrae 50 to 60. Color of specimens preserved in alcohol from rusty to dark chocolate-brown, with darker fins. Spots absent (Haedrich, 1967: 65).

Parin (1958) listed this genus as a synonym of *Centrolophus*. However, according to Haedrich (1967: 67), this genus deserves independent status based on the following characters: cheeks covered with scales, differences in structure of caudal skeleton, and greater number of vertebrae (50 to 60 versus 25 in *Centrolophus*). In addition, *Icichthys* has a larger number of rays in the anal fin (27 to 32 versus 23 to 26) and a smaller number of scales in the lateral line (100 to 130 versus 160 to 230 in *Centrolophus*).

2 species. One reported from the northern part of the Pacific Ocean, and the other from the coast of New Zealand (Haedrich, 1967: 68).

[*Icichthys lockingtoni* Jordan and Gilbert, 1880] (Figure 254)

Icichthys lockingtoni Jordan and Gilbert, Proc. U.S. Nat. Mus., 3, 1880: 305-308 (Point Reyes, California). Abe, Bull. Tokai Reg. Fish. Res. Lab., 37, 1973: 27. Haedrich, Bull. Mus. Comp. Zool., 135, 2, 1967: 69, fig. 15. Ueno, Sci. Rep. Hokkaido Fish. Exp. Sta., 4, 1965: 4.

Schedophilus heathi Gilbert, Proc. Calif. Acad. Sci., Ser. 3, Zool., 1904: 255-271 (California).

Centrolophus californicus Hobbs, J. Wash. Acad. Sci., 19, 20, 1929: 460 (Monterey Bay, California).

Centrolophus niger (non Gmelin) Ueno, Bull. Fac. Fish. Hookaido Univ., 5, 3, 1954: 240. Matsubara, Fish Morphol. and Hierar., 1955: 577.

Centrolophus lockingtoni, Parin, Vopr. Ikhtiologii, 12, 11, 1958: 168, fig. 3.

D 39-43; A 27-32; P 18-21; gill rakers 4-6 + 1 + 11-13 (18-21); vertebrae 56-60 (Haedrich, 1967: 69).

325   Characters of species given in description of genus. According to Parin (1958: 170), this species belongs to bathypelagic fauna.

Length to 390 mm (Parin, 1958: 168).

Distribution: Not found in the Sea of Japan, but possibly in waters of the southern Kuril Islands. Known from the northern part of the Pacific Ocean near the coasts of northern California and British Columbia, and

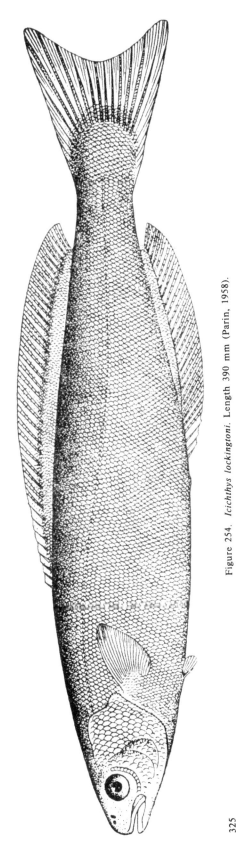

Figure 254. *Icichthys lockingtoni*. Length 390 mm (Parin, 1958).

in the Gulf of Alaska (Abe, 1963: 28); Pacific Ocean east of the Kuril Islands (44°27′ N, 157°50′ E (Parin, 1958: 168); southern coast of Hokkaido in Volcano Bay, near Cape Erimo and the City of Kushiro, as well as near the Pacific coast of Honshu Island in Tsuruga Bay (Ueno, 1965b: 1). Affinity of this species to fishes of amphi-Pacific distribution (Abe, 1963: 27) dubious, since as a bathypelagic fish it has been found near the eastern coast of Kamchatka as well as off the Aleutian Islands.

1. Genus *Hyperoglyphe* Günther, 1859—Barrelfishes

Hyperoglyphe Günther, Cat. Fish. Brit. Mus., 1, 1859: 337 (type: *Diagramma porosa* Richardson, 1845: 26 = *Perca antarctica* Carmichael, 1818: 501). Haedrich, Bull. Mus. Comp. Zool., 135, 2, 1967: 54.

Ocycrius Jordan and Hubbs, Mem. Carnegie Mus., 10, 2, 1925: 226 (type: *Centrolophus japonicus* Döderlein. In: Steindachner and Döderlein, 1885: 183).

Body moderately deep, maximum depth about 30 to 35% of the standard length. Musculature firm. Caudal peduncle broad, moderately long. Origin of dorsal fin above or slightly behind insertion of pectoral fin, continuous, with 6–8 short and almost equal-sized strong spines in anterior part; longest spines half length of longest soft ray—the first among 19 to 25 soft rays of fin. Vent in form of slit located in middle part of body. Anal fin originates slightly behind vent, with 3 spines and 15 to 20 soft rays. Pectoral fins in young fish rounded, pointed in adult fish. Pelvic fins inserted under lower end of base of pectoral fins, connected by membrane to abdomen, and when folded lie in groove on abdomen. Caudal fin broad, 326 emarginate or slightly forked in adult fish. Scales cycloid, moderate in size, somewhat deciduous, and cover bases of dorsal and anal fins. Lateral line forms smooth arch in anterior part of body, then extends in straight line along side of body, continuing onto caudal peduncle. Skin moderately thick. Anterior part of head naked up to occiput. Eyes moderate in size; adipose tissue almost absent around them. Nostrils large; near tip of blunt snout; anterior nostril round, posterior one slit-shaped. Angle of gape extends beyond anterior margin of eye. Premaxilla nonprotractile. Supramaxilla present. Jaw teeth small, sharp, arranged in one row; vomer and palatines without teeth. Opercle and preopercle thin; operculum with 2 weak flat spines, covered with scales, its margin slightly serrate or smooth; preopercle covered with scales, rugulose, its margin with numerous fine denticles, and angle slightly protruding and rounded. Gill rakers slightly longer than gill filaments and minutely serrated along inner margin. Lower part of gill arch with about 16 gill rakers. Branchiostegal rays 7. Vertebrae 10 + 15 = 25. Stomach in form of simple sac; intestine long. Color from greenish or bluish-gray to reddish-chocolate-brown.

Back dark, sides and belly lighter in color, sometimes silvery. Head dark, iris of eye in form of golden ring; operculum often silvery. Unpaired fins usually darker than body color. Irregular striped pattern and spots sometime distinctly visible on body. Oral cavity and gill chambers light-
327 colored. Peritoneum light-colored with minute dark spots. Young fish found near surface along edge of continental shelf; adults prefer deep waters, probably dwell near bottom (Haedrich, 1967: 54).

6 species in the Atlantic, Indian, and Pacific oceans; 1 species known from the Sea of Japan.

1. **Hyperoglyphe japonica** (Döderlein, 1885)—Japanese Barrelfish (Figure 255)

Centrolophus japonicus Döderlein. In: Steindachner and Döderlein, Denkschr. K. Akad. Wiss., Wien, 49, 1885: 183: Beiträge zur Kenntniss dar Fische Japan's, 3, 1884 (1885): 15 (Tokyo).

Ocycrius japonicus, Jordan and Hubbs, Mem. Carnegie Mus., 10, 2, 1925: 226, pl. IX, fig. 4. Matsubara, Fish Morphol. and Hierar., 1955: 577.

Mupus japonicus, Abe, Japan. J. Ichthyol., 4, 1-3, 1955: 113, figs. 1-2; Enc. Zool., 2, Fishes, 1958: 201, fig. 595 (color figure).

Palinurichthys japonicus, Parin, Vopr. Ikhtiologii, 12, 11, 1958: 166.

Hyperoglyphe japonica, Haedrich, Bull. Mus. Comp. Zool., 135, 2, 1967: 58.

22470. Kagoshima, Kyushu Island. February 27, 1901. P.Yu. Shmidt. 2 specimens.

D VII-VIII, 22-26; A III, 17-20; P 21-23; V I, 5; gill rakers 6-7 + 1 + 15-16; vertebrae 10 + 15 (Jordan and Hubbs, 1925; Parin, 1958; Haedrich, 1967).

There are 2 specimens in our collection with a standard length of 600 mm, one of which is in excellent condition, and the other slightly desiccated.

D VII, 23; A III, 18; *l. l.* 98; gill rakers 6 + 1 + 16. Our specimens conform well to the detailed description given by Döderlein. Body depth 83%* of the standard length. Diameter of eye slightly more than length of very blunt snout, but much smaller than wide interorbital space. Keel on head distinct in both specimens. Supramaxilla mostly concealed under preorbital bone, but posterior part visible. Opercles weakly serrate. Lateral line terminates slightly anterior to base of middle rays of caudal fin, straightening from vertical of first third of anal fin, and not from vertical of apex of pectoral fin as reported by Jordan and Hubbs (1925: 227), although in their figure it also originates from vertical of first third of anal fin. Spines of dorsal fin relatively short, equal in height,

---

*Obviously a misprint in the Russian text. In Figure 255 it appears to be about 38%— General Editor.

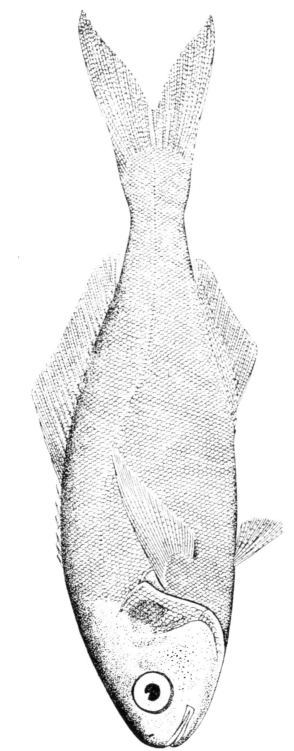

Figure 255. *Hyperoglyphe japonica*—Japanese barrelfish. Standard length 415 mm (Jordan and Hubbs, 1925).

326

and entirely accommodated in groove. In Japan this species constitutes a small deep-sea fishing enterprise.

Length to 900 mm and more (Okada, 1955: 160).

Distribution: In the Sea of Japan reported from the west coast of Hokkaido in Ioiti and Sutsu (Ueno, 1965b: 2); Sado Island (Honma, 1963: 19); San'in region (Yanai, 1950: 19; Mori, 1956a: 24); and Pusan (Mori, 1952: 139). Found along the Pacific coast of Japan in the central part of Honshu Island. Pacific Ocean east of southern Kuril Islands— 44°27' N, 157°50' E (Parin, 1958: 170).

2. Genus *Psenopsis* Gill, 1862

Psenopsis Gill, Proc. Acad. Nat. Sci. Philad., 14, 1862: 127 (type: *Trachinotus anomalus* Temminck and Schlegel, 1850: 107). Haedrich, Bull. Mus. Comp. Zool., 135, 2, 1967: 72.

Body oblong or slightly deep; maximum depth 30 to 45% of standard length, compressed laterally, but rather thick. Caudal peduncle short, deep, compressed laterally. Origin of dorsal fin above or slightly behind insertion of pectoral fin; continuous, with 5 to 7 short and gradually lengthening spiny rays, and 27 to 32 soft rays; last spine longest, but less than half length of longest soft ray. Vent slightly anterior to midpoint of body. Anal fin originates anterior to or slightly behind midpoint of body, with 3 gradually lengthening spines and 22 to 29 soft rays. Number of soft rays in dorsal fin never exceeds by more than 5 the number of soft rays in anal fin. Pectoral fins rounded in young fish and usually elongate in adults. Pelvic fins inserted right under origin of pectoral fins, connected to abdomen by membrane, and fold in groove extending up to vent.

Caudal fin slightly forked. Scales small, cycloid, deciduous, and cover fleshy bases of dorsal and anal fins. Lateral line located moderately high, parallel to dorsal contour, and continues onto caudal peduncle. Skin very thin. Head length about 30% of standard length. Upper part of head naked with distinct minute pores; naked skin does not continue or may continue beyond occiput. Eyes moderate to large; adipose tissue surrounds them and continues up to nostrils. Latter moderate in size and located near tip of truncate snout; anterior nostril rounded and posterior one slit-shaped. Maxilla extends slightly beyond vertical from anterior margin of eye. Premaxilla nonprotractile. Upper jaw, when mouth closed, mostly covered by preorbital. Supramaxilla absent. Jaw teeth minute, sharp, close-set, arranged in 1 row, and covered with membrane on both sides; vomer and palatines without teeth. Opercle and preopercle thin, naked, their margins smooth or very minutely serrate. Operculum with 2 weak flat spines; angle of preopercle in form of rounded process. Gill rakers about half length of gill filaments, serrate along inner margin, about 13 in lower half of first gill arch; small pseudobranch present. Branchiostegal rays seven. Vertebrae $10 + 15 = 25$. Color of specimens preserved in alcohol chocolate-brown[9] or bluish; specimens with deep body have silvery or whitish tinge. Black spot usually distinct in anterior part of lateral line. Fins slightly lighter in color than general body background. Adults confined closer to coast than other members of Centrolophidae. Caught in shallows near coasts of Japan. Sexually mature adult fish with standard length to 180 mm (Haedrich, 1967: 72).

This genus is distributed off the coasts of India and northwest Australia, and the sea frontiers of Southeast Asia north to Japan (Hokkaido). 3 species; 1 known from the Sea of Japan.

1. *Psenopsis anomala* (Temminck and Schlegel, 1850) (Figure 256)

Trachinotus anomalus Temminck and Schlegel, Fauna Japonica, Poiss., 1850: 107, pl. 57, fig. 2 (Tokyo).

Psenes anomalus, Günther, Cat. Fish. Brit. Mus., 2, 1860: 495.

Psenopsis anomala, Gill, Proc. Acad. Nat. Sci. Philad., 14, 1862: 127. Kamohara, Scombroidei, Fauna Nipponica, 15, 5, 1940: 186, fig. 96. Matsubara, Fish Morphol. and Hierar., 1955: 577. Abe. Enc. Zool., 2, Fishes, 1958: 200, fig. 593 (color figure). Zhu et al., Ryby Vostochno-Kitaiskogo Morya, 1963: 411, fig. 308. Ochiai and Mori, Bull. Misaki Mar. Biol. Inst., Kyoto Univ., 8, 1965: 2.

Psenopsis shojimai Ochiai and Mori, Bull. Misaki Mar. Biol. Inst., Kyoto Univ., 8, 1965: 4, fig. 2 (Sea of Japan). Haedrich, Bull. Mus. Comp. Zool., 135, 2, 1967: 75 (considered a synonym of *P. anomala*).

330   6506. Tokyo. 1882. A. Shneider. 1 specimen.

22649. Nagasaki. January 10, 1901. P.Yu. Shmidt. 2 specimens.

38133. Tonkinskii Bay. July 22, 1961. E.F. Gur'yanova. 5 specimens.

D V-VII, 27–32; A III, 25–29; P 20–23; gill rakers usually $6 + 1 + 13$, 12–15 in lower half of first gill arch, total 18–21; vertebrae $10 + 15$ (Haedrich, 1967: 75).

In our specimens from Nagasaki with a standard length of 114 to 135 mm: D VII, 29; A III, 27–28; P 21; gill rakers 6–7 + 13; length of pectoral fin 1.00–1.05 and length of caudal fin 0.90–0.95 times in the head length. These measurements fully conform to the characters given by Ochiai and Mori (1965: 2) for *P. anomala*. These authors have also described a new species, *P. shojimai,* whose distribution is restricted to the shoreline of the Sea of Japan from Henkai-Nada (northern tip of Kyushu Island) north to Hokkaido.

According to Ochiai and Mori (1965), *P. anomala* and *P. shojimai* differ as shown in the key below:

1 (2). Caudal fin relatively long, equal to or slightly shorter than

[9]Our specimens preserved in alcohol have a golden tinge.

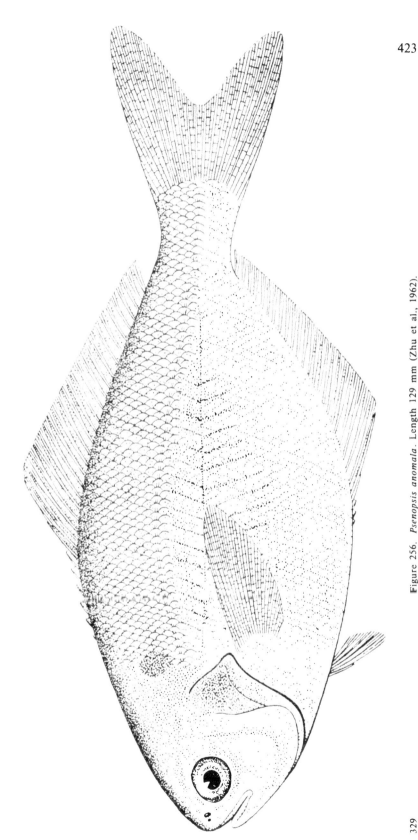

Figure 256. *Psenopsis anomala*. Length 129 mm (Zhu et al., 1962).

head length; pectoral fins elongate, 0.9 to 1.2 times in head length. Lateral line 55 to 63 (middle and southern Japan, South China and East China seas)................................... **P. anomala.**

2 (1). Caudal fin short, 1.3 to 1.5 times in the head length; pectoral fins posteriorly rounded, 1.1 to 1.3 times in head length. Lateral line 62 to 70 (Japan)..................................... **P. shojimai.**

The new species has also been found together with *P. anomala* in the coastal waters of Simane Prefecture and Wakasa Bay. The differences between the two are not significant, consisting only of a different ratio in length of caudal fin to head length. This single difference is insufficient for establishing an independent species, and the authors are inclined to agree with Haedrich, who considered this species a synonym of *P. anomala*.

This species is commercially important in Japan, and valued as a high quality fish, especially during summer.

Length, to 300 mm (Abe, 1958).

Distribution: In the Sea of Japan known from Pusan. Along the coast of Japan near Hokkaido (Ochiai and Mori, 1965: 45); Sado Island (Honma, 1963: 19); Toyama Bay (Katayama, 1940: 10); Wakasa Bay (Jordan and Hubbs, 1925: 226); San'in region (Yanai, 1950: 19; Mori, 1956a: 24); and Tsushima Strait (Tabeta and Tsukahara, 1967: 298). Along the Pacific coast from Matsushima Bay southward (Matsubara, 1955: 577). Cheju-do Island (Uchida and Yabe, 1939), west coast of the Korean Peninsula (Mori, 1956a: 24). East China and South China seas (Zhu et al., 1963: 308).

CLXX. Family STROMATEIDAE—Butterfishes

Body deep, compressed lateally. One dorsal fin. Sometimes 1–10 flat bladelike spines and 3–5 slender spines present anterior to dorsal and anal fins, which gradually lengthen in form of spiny rays situated in front of soft rays. Dorsal and anal fins almost similar in shape and size, usually falciform. Caudal fin deeply forked. Pectoral fins long and pointed. Pelvic fins absent, except in young fish of *Stromateus*. Scales small, cycloid, and extremely deciduous. Lateral line high, follows dorsal profile, and continues onto caudal peduncle. Margins of opercle and preopercle smooth. Operculum very thin, with 2 short, flat, weak spines. Branchiostegal membranes usually not attached to isthmus, but attached in *Pampus*. Branchiostegal rays 5–6. Mouth terminal to sub-terminal, small; angle of gape rarely extends beyond vertical from anterior margin of eye. Teeth very small, flat on sides, tricuspid, and arranged in 1 row on jaws. Vomer and palatines without teeth. Supramaxilla absent. Eyes comparatively small; adipose tissue around them usually poorly developed. Vertebrae 30 to 48. Papillae in pharyngeal sacs present in upper and lower halves and not arranged in bands; papillary bases stellate;

teeth arranged all along central stalk (Figure 249, C). Adults reach 300 mm in length. Color from silvery to blue; sometimes spots present (Haedrich, 1967).

Found in water column within the limits of the continental shelf and bays of tropical, subtropical, and temperate latitudes in the Atlantic, Indian, and Pacific oceans. 3 genera, one (*Pampus*) known from the coast of Japan and the Sea of Japan.

1. Genus *Pampus* Bonaparte, 1837—Pomfrets

Pampus Bonaparte, Iconographia..., 3, 2, 1837: 48 (type species of subgenus: *Stromateus candidus* Cuvier and Valenciennes, 1833: 391 = *Stromateus argenteus* Euphrasen, 1788: 53). Haedrich, Bull. Mus. Comp. Zool., 135, 2, 1967: 108.

Stromateoides Bleeker, Nat. Tijdschr. Nederl.-Indië, 1, 1851: 368 (type: *Stromateus cinereus* Bloch, 1793: 90 = *Stromateus argenteus* Euphrasen, 1788: 53).

Chondroplites Gill, Proc. Acad. Nat. Sci. Philad., 14, 1862: 126 (type: *Stromateus atous* Cuvier and Valenciennes, 1833: 389 = *Stromateus chinensis* Euphrasen, 1788: 54).

Body very deep, maximum depth more than 60% of standard length, highly compressed laterally, with firm musculature. Caudal peduncle very short, also compressed laterally. One continuous dorsal fin. Sometimes 5 to 10 flat, bladelike, pointed spines found in front of dorsal and anal fins, representing slightly outwardly protruding free tips of interneural processes.[10] In species with such spines, origin of dorsal fin slightly behind vertical through posterior end of base of pectoral fin; first bifurcate spine usually located above or slightly ahead of base of pectoral fin. In species devoid of bifurcate spines, dorsal fin originates above base of pectoral fin. Anal papilla situated well anterior to midpoint of body in form of slit. Anal fin originates at vertical from midpoint of body length or slightly ahead of it, and behind origin of soft rays of dorsal fin. Anteriormost rays of dorsal and anal fins elongate and often impart falciform shape to fin; subsequent rays shorter. In species with spines in front of dorsal fin, rays of posterior 2/3 of fin short and equal in height; anal fin with highly developed anterior lobe. In species without spines, rays of posterior 2/3 of fin gradually reduce in length and last ray shortest. Pectoral fins long, alate; base of fin at an angle of 45° to body axis. Pelvic fins absent, pelvic bones indistinguishable. Caudal fin consists of rather rigid rays, deeply forked in fish with spines; lower lobe of fin often distinctly elongate. Scales very small, cycloid, deciduous,

---

[10]In large specimens with a standard length of more than 150 mm, these spines are covered with skin.

426

332 and continue onto base of rays of all fins. Simple tubular scales of lateral line parallel to dorsal profile of body. Skin thin. Eyes small; adipose tissue around eyes continues up to nostrils. Nostrils large; anterior nostril round, posterior in form of long slit, located at tip of blunt snout above level of upper margin of eye. Mouth subterminal and small; maxilla rarely extends beyond vertical from anterior margin of eye. Premaxilla not protractile; maxilla fixed, covered with scales, and connected with cheek. Supramaxilla absent. Jaw teeth small, arranged in 1 row, flat, and many tricuspid; middle cusp largest, rounded. Vomer and palatines without teeth. Branchiostegal membranes broadly fused with isthmus; gill openings in form of straight slit covered with flap of skin. Gill rakers short, without denticles, rarely sessile. Pseudobranch absent. Branchiostegal rays 5. Vertebrae 33 to 41.

Color of live fish silvery with light bluish tinge on back. Color of specimens preserved in alcohol chocolate-brown or light bluish with silvery or whitish tinge. Dorsal, anal, and caudal fins yellowish with dark margins (Haedrich, 1967: 109).

The biology of species of this genus, in spite of their commercial value, has not been studied well. Young fish remain confined to shallow water and may be found in estuaries. Small mouth with cutting teeth and long pharyngeal sac suggest feeding on coelenterates. Most of the stomachs examined confirmed this conclusion, but remains of fish were also found in the stomach contents.

This genus is widely represented in tropical waters of the continental shelf from the Persian Gulf to Japan. Members are also found off the Hawaiian Islands as well as in the Adriatic Sea (Haedrich, 1967: 110).

3 species; 2 known within the limits of the Sea of Japan.

Key to Species of Genus Pampus

1 (4). Dorsal and anal fins falciform; anterior rays highly elongate and all posterior rays equal in length. Flat spines (5 to 10) occur in front of fins.

2 (3). Gill rakers 1-4 + 8-10. Dorsal fin with 5 to 10 spines and 38 to 43 rays; anal fin with 3 to 7 spines and 34 to 43 rays...
.... 1. **P. argenteus** (Euphrasen).

3 (2). Gill rakers 3-6 + 12-15. Dorsal fin with 8 spines and 42 to 49 rays; anal fin with 5 to 7 spines and 42 to 47 rays........
.... 2. **P. echinogaster** (Basilewsky).

4 (1). Dorsal and anal fins not falciform; anterior rays highly elongate but subsequent rays gradually reduce; last ray shortest. Flat spines not present in front of fins. Gill rakers 2-3 + 8-11............
.... [**P. chinensis** (Euphrasen)].

1. **Pampus argenteus** (Euphrasen, 1788)–Silvery Pomfret (Figure 257)

Stromateus argenteus Euphrasen, Vetensk. Acad. Nya Handl. Stockholm, 9, 1788: 49 (Castellum Chinense Bocca Tigris).

Stromateus cinereus Bloch, Naturgesch. Ausländisch. Fische, 9, 1795: 90, pl. 420 (India). Wang, Contr. Biol. Lab. Sci. Soc. China, 10, 9, 1935: 414.

Stromateus candidus Cuvier and Valenciennes, Hist. Nat. Poiss., 9, 1833: 361 (Pondicherry).

334   *Stromateus punctatissimus* Temminck and Schlegel, Fauna Japonica, Poiss., 1844: 121, pl. 65 (Nagasaki).

Stromateoides nozawae Ishikawa, Proc. Dept. Nat. Hist. Tokyo Imp. Mus., 6, 1, 1904: 8 (Kanagawa). Matsubara, Fish Morphol. and Hierar., 1955: 579. Zhu et al., Ryby Vostochno-Kitaiskogo Morya, 1963: 407, fig. 305.

Stromateoides punctatissimus, Soldatov and Lindberg, Obzor..., 1930: 126.

Stromateoides argenteus, Jordan and Metz, Mem. Carnegie Mus., 6, 1, 1913: 28, pl. 5. Zhu et al., Ryby Vostochno-Kitaiskogo Morya. 1963: 408, fig. 306.

Pampus argenteus, Beaufort and Chapman. In: Weber and Beaufort, Fish. Indo-Austr. Arch., IX, 1951: 92 (synonyms). Abe, Enc. Zool., 2, Fishes, 1958: 200, fig. 594 (color fig.). Abe and Kosakai, Japan. J. Ichthyol., 12 (1/2), 1964: 29 (*S. nozawae* Ishikawa = juvenile *P. argenteus*). Ueno, Sci. Rep. Hokkaido Fish. Exp. Sta., 4, 1965: 2. Haedrich, Bull. Mus. Comp. Zool., 135, 2, 1967: 112, figs. 45, 47.

31360. Yellow Sea, near City of Dal'nyi. September 4-10, 1947. D.G. Gnezdilov. 13 specimens.

39676. Sea of Japan, Tawaiza Inlet (45°10′ N, 136°50′ E). August 3, 1925. Dobrzhanskii. 1 specimen.

D V-X, 38-43; A V-VII, 34-43; P 24-27; vertebrae $14-16 + 20-25$ (Haedrich, 1967: 112); gill rakers from $2 + 8$ to $3 + 10$ (Abe and Kosakai, 1964. 30).

On the basis of external appearance, this species is difficult to distinguish from *P. echinogaster,* which raised doubts about the existence of two independent species. We have 33 specimens in our collection. *Pampus* sp.: standard length 63 to 130 mm (No. 31360). Externally, these fish do not differ from each other and only with an analysis of meristic characters could they be divided into 2 groups on the basis of number of gill rakers. Specimens of both groups are similar in size (small, medium, and large).

In the first group of 13 specimens the number of gill rakers [1-4 (average 2.7) + 8-10 (average 9.2)] distinctly differed from the second group of 20 specimens [3-5 (average 4.4) + 12-15 (average 13.8)]. The

428

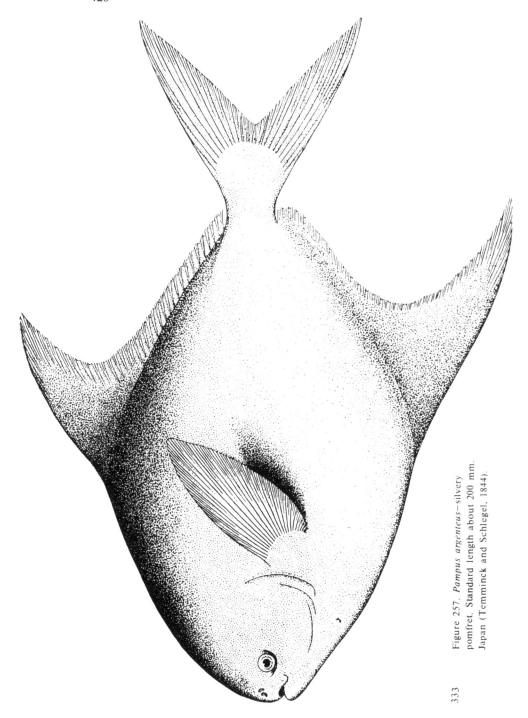

Figure 257. *Pampus argenteus*—silvery pomfret. Standard length about 200 mm. Japan (Temminck and Schlegel, 1844).

333

number of gill rakers differed most notably on the lower arch: 8-10 (average 9.2) versus 12-15 (average 13.8). This data accords well with that of Japanese ichthyologists. Hence fish with a smaller number of rakers could be included under *P. argenteus,* and those with a larger number of rakers under *P. echinogaster.* It is interesting that the number of flat bifurcated spines in front of the dorsal fin in fish with a smaller number of rakers was less [VII (11 specimens) and VIII (2 specimens)] compared to fish with a larger number of rakers [VI (1 specimen), VIII (1 specimen), IX (6 specimens), X (9 specimens), XI (2 specimens), and XII (1 specimen)]. On the average the former had 7.2 spines versus 9.6 in the latter. The difference in the number of soft rays in the vertical fins were less distinct: D 40-42 (45) (average 41.3); A (36) 39-44 (average 41.1) in *P. argenteus*; and D 44-50 (average 46.9); A 41-46 (average 44.3) in *P. echinogaster.* The number of flat spines in front of the anal fin was also less [IV (1 specimen), V (9 specimens), VI (3 specimens)−average 5.1] versus [V (5 specimens), VI (8 specimens), and VII (11 specimens)−average 6.5].

Length to 600 mm (Abe, 1958).

Distribution: According to our survey, in the Sea of Japan it is found rarely near the Primor'e coast north up to Tawaiza Inlet (45°10′ N); reported from Peter the Great Bay (Soldatov and Lindberg, 1930: 126); Pusan (Jordan and Metz, 1913: 28); along the coast of the Sea of Japan—Otaru on Hokkaido (Ueno, 1965b: 2); Sado Island (Honma, 1952: 144); Toyama Bay (Katayama, 1940: 10); San'in region (Mori, 1956a: 24). In the Yellow Sea found off the south and west coasts of the Korean Peninsula (Mori, 1952: 139); Gulf of Chihli (Bohai) (Zhang et al., 1957: 195) and Chefoo (Wang, 1935: 414-415). Along the Pacific coast of Japan from the central parts of Honshu Island southward (Matsubara, 1955: 579). Pacific and Indian oceans from Japan to the Persian Gulf (Haedrich, 1957: 112).

336 ## 2. *Pampus echinogaster* (Basilewsky, 1855) (Figure 258)

Stromateus echinogaster Basilewsky, Ichthyographia Chinae Borealis, 1855: 223 [Gulf of Chihli (Bohai)].

Stromateoides echinogaster, Jordan and Metz, Mem. Carnegie Mus., 6, 1, 1913: 28, pl. 5. Soldatov and Lindberg, Obzor..., 1930: 124.

Pampus echinogaster, Abe and Kosakai, Japan. J. Ichthyol., 12 (1/2), 1964: 29. Haedrich, Bull. Mus. Comp. Zool., 135, 2, 1967: 112.

22468. Pusan. March 28, 1901. P.Yu. Shmidt. 1 specimen.

35592. Yellow Sea. May, 1956. Academy of Sciences, China. 2 specimens.

39677. Yellow Sea, near City of Dal'nyi. September 4-10, 1946. V.G. Gnezdilov. 2 specimens.

39678. Yellow Sea, 34°00′ N, 123°00′ E. January 19, 1958.

39679. Nel'ma Inlet, 47°40′ N, 139°15′ E. July 17, 1958. Kamernitskaya. 3 specimens.

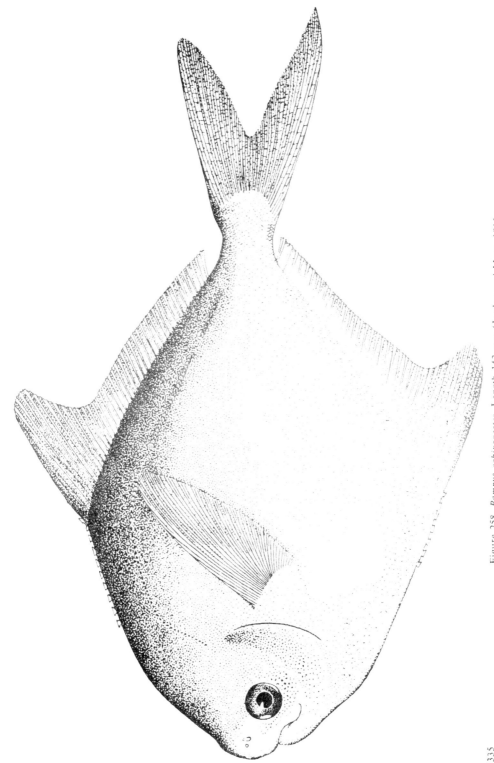

Figure 258. *Pampus echinogaster*. Length 143 mm (Jordan and Metz, 1913).

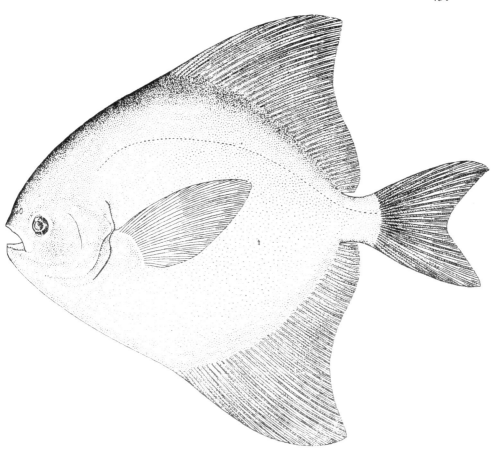

336   Figure 259. *Pampus chinensis*—Chinese pomfret. Length 126 mm (Zhu et al., 1962).

39763. East China Sea, 30°45′ N, 124°42′ E. March 1, 1958. 2 spcimens. 39764. Yellow Sea, 34°00′ N, 123°00′ E. January, 1958 1 specimen, D VIII-X, 42-49; A V-VII, 42-47; P 24-25; vertebrae 14-15 + 24-26 (Haedrich, 1967: 112); gill rakers from 3 + 12 to 5 + 15 (6 + 14) (Abe and Kasakai, 1964: 30).

As detailed in the characters of *P. argenteus,* 20 specimens of *P.*
337 *echinogaster* (with many rakers) were distinguished by the following meristic characters: D (VI) VIII-XII, 44-50; A V-VII, 41-46; gill rakers 3-5 + 12-15.

Length to 320 mm (Soldatov and Lindberg, 1930: 126).

Distribution: In the Sea of Japan, according to our survey, found rarely off the Primor'e coast north to Nelma Inlet (47°40′ N); reported from Peter the Great Bay (Soldatov and Lindberg, 1930: 124); and Pusan (No.

22468). In the Yellow Sea—Gulf of Chihli (Bohai) (Basilewsky, 1855: 223); Port Arthur, Namp'o (Jordan and Metz, 1913: 28). Along the Pacific coast reported from Tokyo (Abe and Kosakai, 1964: 29). China, the Korean Peninsula, and Japan (Haedrich, 1967: 112).

[*Pampus chinensis* (Euphrasen, 1788)—Chinese Pomfret] (Figure 259)

Stromateus chinensis Euphrasen, Vetensk. Acad. Nya Handl. Stockholm, 9, 1788: 53 (Gastellum Chinense Bocca Tigris).

Stromateoides sinensis, Zhu et al., Fishes of the South China Sea, 1962: 762, fig. 617; 1963: 409, fig. 307.

Pampus chinensis, Beaufort and Chapman, Fish. Indo-Austr. Arch., IX, 1951: 94. Matsubara, Fish Morphol. and Hierar., 1, 1955: 580. Haedrich, Bull. Mus. Comp. Zool., 135, 2, 1967: 111, fig. 44.

35855. Hainan Island, Haikou, mouth of river. May 26, 1958. D.V. Naumov. 6 specimens.

35856. Hainan Island, northern part. 1958. B.E. Bykhovskii and L.F. Nagibina. 1 specimen.

D 43–50; A 39–42; P 24–27; vertebrae 14 + 19 (Haedrich); gill rakers 2–3 + 8–11 (Zhu et al.).

In our collection 6 specimens had a standard length of 22 to 30 mm and 1 specimen of 73 mm. D 48–49; A 43 (3 specimens). This data completely conforms to the characters reported for this species by Haedrich. In our specimens neither the young nor the adult have flat bifurcate spines (a typical feature of other species of this genus) in front of the dorsal fin, nor do they exhibit even rudimentary structures of such spines. We paid special attention to this character because Zhu et al. (1963: 409–410) indicated the presence of 4–5 such spines in young fish, which become concealed in the skin in adult fish.

Length to 143 mm (Zhu et al., 1963: 410).

Distribution: Absent in the Sea of Japan; also not known from the coast of Japan. In the East China Sea reported from Taiwan (China) and farther along the southeast coast of Asia up to the India Ocean (Matsubara, 1955; Haedrich, 1967).

13. Suborder Gobioidei[1]

337 Spiny dorsal fin, if present, consists of one to eight flexible spiny rays. Pelvic fins below pectoral fins with one large flexible spiny ray and four to five branched rays, serving as a suctorial disk, and often fused and modified into a sucker. Occasionally, these rays absent as in *Expedio* (Figure 32). Parietals absent. Opisthotics large, reaching basioccipitals. Infraorbitals nonossified or absent. Gap present in-between preopercle, symplectic, and quadrate. Swim bladder usually absent. Coastal fishes of tropical, warm, and temperate seas; some species found in fresh waters (Berg, 1940: 325).

A large number of works have been published in recent years on the suborder Gobioidei. Particularly interesting studies with reference to fishes of the Sea of Japan and adjacent waters comprise those on the osteology of fishes of the family Gobiidae by Akihito (1963, 1967b, 1969, 1971), the sex characters of 25 species of gobies by Arai (1964), and an analysis of the scales, taxonomy, and ecology of Japanese gobies by Takagi (1963, 1966a, 1966b, 1966c).

Seven families. Five families represented in the Sea of Japan and adjacent waters.

443 *Addendum*: A detailed classification of the suborder Gobioidei has been published by Miller (1973, *J. Zool.*, London, 171, 3, 397–434, 11 figs.). We came to know of this work after the present book had gone to press.

338 *Key to Families of Suborder Gobioidei, Order Perciformes*

1 (8). Dorsal and anal fins not confluent with caudal fin, except in *Caragobius*.[2]

2 (7). Two dorsal fins, separate or with deep notch between them; rarely single short fin in posterior half of body.

3 (4). Pelvic fins separate, close-set, often approximate, but not united and do not form a sucker.

4 (3). Pelvic fins united, and usually form a sucker, except in *Expedio* (Figure 320).

5 (6). Eyes not movable, but if so, then bases of pectoral fins normal,

[1]Several authors have recently included the families Eleotridae and Periophthalmidae within the family Gobiidae. However, since the purpose of the present book is to provide keys for fishes in a limited water area, it appears more convenient to consider these families as independent taxonomic units.

[2]Recorded off the Philippines (Herre, 1927: 287).

not thickened, and not highly muscular; fins not used for crawling.................................. CLXXII. **Gobiidae.**

6 (5). Eyes movable. Bases of pectoral fins highly muscular and fins used for crawling. .... CLXXV. **Periophthalmidae.**

7 (2). Dorsal fin single, without notch. Lower jaw protrudes notably forward. Chin enlarged and constitutes part of upper profile of head. Dorsal fin with not more than 25 rays. ....
.... **[Kraemeriidae].**[3]

8 (1). Dorsal and anal fins confluent with caudal fin. Dorsal fins fused, without notch.[4]

9 (10). Pouchlike cavity in opercular region (Figure 324) situated above gill cavity. Body elongate... CLXXIII. **Trypauchenidae.**

10 (9). Opercular region without such pouchlike cavity. Body highly elongate........................ CLXXIV. **Gobioididae.**

CLXXI. Family ELEOTRIDAE—Sleepers

Pelvic fins separate, not fused into disk. Ascending branch of palatine directly joined with prefrontal behind origin of maxillary process. Mesopterygoid narrow, but well developed. Scapula and coracoid well developed; radial elements located on scapula and coracoid, as well as between them. Vertebrae 24–28 (Berg, 1949: 1055).

Marine, brackish, and fresh-water fishes, widely distributed in temperate and tropical regions.

More than 80 genera. Four genera known from the Sea of Japan and the occurrence of another quite possible.

443   *Addendum*: In the region of Niigata and Sado Island in the Sea of Japan, a fresh-water member of this family has been reported— *Odontobutis obscura* (Temminck and Schlegel)—by Honma and Tamura (1972) (Revised list of the gobioid fishes from waters adjacent to Niigata and Sado Island in the Sea of Japan, *Bull. Niigata Pref. Biol. Soc. Educ.*, 8, 33–38, 7 figs.).

338   *Key to Genera of Family Eleotridae*

1 (4). Preopercle with one or more spines.

2 (3). Angle of preopercle with single strong spine directed downward, sometimes concealed in skin. Head flat on dorsal side. Body anteriorly cylindrical and posteriorly slightly compressed laterally.
.... 1. **Eleotris** Gronow.

[3]Indian and Pacific oceans; Amami Islands (Ryukyu) (Matsubara and Iwai, 1959).

[4]Family Microdesmidae, known from the Central American and west African coasts, differs from the family described here in a reduced anal fin that extends forward to not far from midpoint of the body (Reid, 1936; Gosline, 1955).

3 (2). Angle of preopercle with one to six spines directed backward.
Head as well as body highly compressed laterally. . . . .
. . . . 2. [**Asterropteryx** Rüppell].

4 (1). Preopercle without spines.

5 (6). Chin with long flat barbel and a few (two to three) short barbels. Second dorsal fin with 24–26 branched rays, and anal fin with 24 rays. . . . . 3. **Vireosa** Jordan and Snyder.

6 (5). Chin without barbels. Second dorsal fin with less than 20 branched rays.

7 (8). Pelvic fins long and narrow, longer than head. Scales large; lateral line with not more than 30 scales. Second dorsal and anal fins with not more than 11 branched rays. . . . 4. **Eviota** Jenkins.

8 (7). Pelvic fins short, about half head length. Scales small; lateral line with about 100 scales. Second dorsal and anal fins with more than 15 branched rays. . . . . 5. **Parioglossus** Regan.

1. Genus *Eleotris* Gronow, 1763

Eleotris Gronow, Zoophylaceum, 1763: 83 (type: *Gobius pisonis* Gmelin). Koumans, Fish. Indo-Austr. Arch., 10, 1953: 292. Akihito, Japan. J. Ichthyol., 14, 4/6, 1967: 135–166.

Body oblong or elongate, cylindrical anteriorly, laterally compressed posteriorly, covered with 42 to 70 rows of scales. Scales in posterior part of body ctenoid, and in anterior part in front of first dorsal fin and on head cycloid. Head dorsally flat and covered with scales beginning at interorbital space or behind eyes; sides of head completely or partly flat. Eyes small. Mouth oblique; lower jaw protrudes forward. Jaws with several rows of teeth; teeth in outer row enlarged, canines absent. Tongue rounded. Angle of preopercle with curved spine sometimes concealed in skin. Ventrally, gill openings do not extend far anteriorly; interbranchial space broad. Dorsal fins separate; first dorsal fin with 6 rays; second with one unbranched and 8–9 branched rays. Caudal fin elongate (Koumans, 1953: 292).

Tropical seas; some species enter rivers.

Serveral species. Four species in Japan.[5] One species known from the Sea of Japan.

1. *Eleotris oxycephala* Temminck and Schlegel, 1845 (Figure 260)
Eleotris oxycephala Temminck and Schlegel, Fauna Japonica, Poiss., 1845: 150, pl. 77, figs. 4, 5 (Sea of Japan). Koumans, Fish. Indo-Austr. Arch., 10, 1953: 299 (description, synonyms).

Eleotris balia Jordan and Seale, Proc. U.S. Nat. Mus., 29, 1905: 526, fig. 6 (Siangan (Hong Kong?)).

[5]Akihito Prince, *Japan. J. Ichthyol.*, 14, 4/6, 1967a: 135–166, 31 figs.

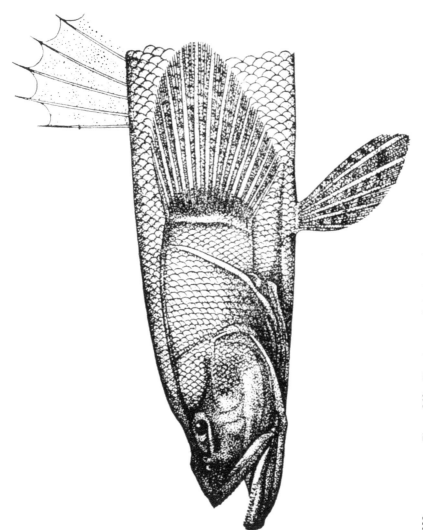

Figure 260. *Eleotris oxycephala.* Japan (Temminck and Schlegel, 1845).

7017. China, Fujian. 1884. Polyakov. 4 specimens.

D VI, I 8; A I, 8; P 16; *l. l.* 50–55 (Koumans, 1953).

Differs from other large-scaled species with less than 60 scales in lateral line, in presence of scales on snout and larger number of rows of scales from tip of snout to origin of dorsal fin (60 versus 45 to 40).

Mode of life described by Japanese researchers (Dotu and Fujita, 1959: 191).

Length to 270 mm (Dotu and Fujita, 1959).

Distribution: In the Sea of Japan known from marine and fresh waters (Tomiyama, 1936: 42); San'in region (Mori, 1956a: 24). In the Yellow Sea indicated for Cheju-do Island (Mori, 1952: 140). From the shores of Japan to southern China (Matsubara, 1955: 812).

2. [Genus *Asterropteryx* Rüppell, 1828]

Asterropterix Rüppell, Atlas, Reise Nord Africa Fische, 1828: 138 (type: *A. semipunctatus* Rüppell).

Asterropteryx Günther, 1861: 132 (type: *A. semipunctatus* Rüppell). Koumans, Fish. Indo-Austr. Arch., 10, 1953: 289 (synonyms).

Body slightly elongate, compressed laterally, and covered with large ctenoid scales (*l. l.* about 25). Head also compressed laterally, curved in profile. Dorsal surface covered with scales, as are areas behind eyes, cheeks, and gill covers. Mouth slightly oblique. Teeth arranged in several rows; outer teeth enlarged. Lower jaw with one canine on each side. Angle of preopercle with 1–6 strong spines directed backward. Ventrally, gill openings do not continue forward; interbranchial space broad. Base of first dorsal fin contiguous with base of second dorsal fin. First dorsal fin with 6 unbranched rays; second dorsal fin with 1 unbranched and 9–11 branched rays. Anal fin with 1 unbranched and 8–11 branched rays. Caudal fin often rounded (Koumans, 1953).

Two species. One possibly occurs in the Sea of Japan.

1. [*Asterropteryx semipunctatus* Rüppell, 1828] (Figure 261)

Asterropteryx semipunctatus Rüppell, Atlas, Reise Nord Africa Fische, 1828, pl. 34, fig. 4 (Red Sea). Tomiyama, Japan. J. Zool., 7, 1, 1936: 40, fig. 2. Koumans, Fish. Indo-Austr. Arch., 10, 1953: 290, fig. 73 (description, synonyms).

D VI, I 9–11; A I, 8–10; P 17; *l. l.* 24 (Koumans, 1953).

Differs from the closely related species, *A. ensiferus* (Bleeker, 1865), known from the Sulu Sea (Indonesia), in presence of 3–5 short spines instead of one at angle of preopercle. Body of live fish with large irregularly shaped transverse blackish to chocolate-brown spots. Almost every scale with small, rounded purple-blue brilliant spot.

341   Mode of life described by Japanese researchers (Dotu and Mito, 1963: 10).

438

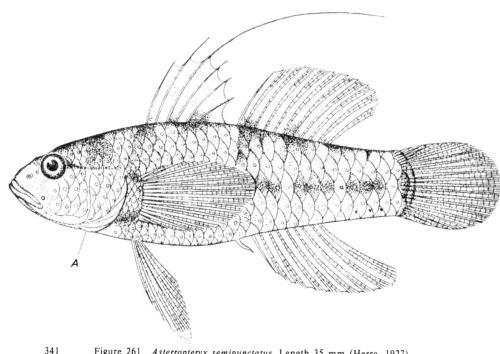

341   Figure 261. *Asterropteryx semipunctatus.* Length 35 mm (Herre, 1927).
A–preopercular spine.

Length to 60 mm (Tomiyama, 1936).

Distribution: This species has not been found in the Sea of Japan, but is known from Oita Prefecture and Bungo Strait (Tomiyama, 1936: 40), and hence its occurrence in the waters of the Sea of Japan is probable. In the Pacific Ocean known from Tokyo (Hatidze Island) south to Indonesia and farther west to the Red Sea (Matsubara, 1955: 812).

3. Genus *Vireosa* Jordan and Snyder, 1901

Vireosa Jordan and Snyder, Proc. U.S. Nat. Mus., 24, 1901 (1902): 38 (type: *V. hanae* Jordan and Snyder, 1901). Koumans, Fish. Indo-Austr. Arch., 10, 1953: 364.

Body slightly elongate, compressed laterally, and covered with minute separated cycloid scales partly embedded in skin. Head naked, relatively short, its anterior part blunt and roundish. Eyes fairly large. Chin with one relatively long flat barbel and three short barbels. Mouth large, almost vertical; some teeth elongate; minute canines present. Pelvic fins completely separate, each with one spine and four soft rays. Caudal fin with a few elongate-filamentous upper and lower rays. Rays of dorsal fin not elongate; first dorsal fin with 6 rays, and second with 25. Anal fin with long base. Gill openings broad, interbranchial space narrow; gill rakers long and slender; pseudobranchs present (Jordan and Snyder, 1901c).

Indonesia and Japan. One species. Also known from the Sea of Japan.

1. **Vireosa hanae** Jordan and Snyder, 1901 (Figure 262)

Vireosa hanae Jordan and Snyder, Proc. U.S. Nat. Mus., 24, 1901: 38, fig. 1 (Misaki, Japan). Koumans, Fish. Indo-Austr. Arch., 10, 1953: 365, fig. 90 (synonyms). Tomiyama, Fig. and Descr., Fish of Japan, 58, 1958: 1200, pl. 233, figs. 589, 590.

342 D VI, I 24–26; A I, 24: P 21; V I, 4 (Koumans, 1953).

Characters given in description of genus. Color of live fish characterized by specific red spot against bluish-green background at base of pectoral fin and longitudinal brick-red stripe above anal fin.

Length, excluding caudal filaments, up to 130 mm (Koumans, 1953).

Distribution: In the Sea of Japan known from Toyama Bay (Tomiyama, 1936: 50). Reported from the Korean Peninsula—Thonen (Mori, 1952: 146). Along the Pacific coast of Japan from Misaki southward. Reported from Djakarta, Java (Koumans, 1953: 366).

4. Genus *Eviota* Jenkins, 1902

Eviota Jenkins, Bull. U.S. Fish. Comm., 22, 1902 (1903): 501 (type: *E. epiphanes* Jenkins). Koumans, Fish. Indo-Austr. Arch., 10, 1953: 316.

Body slightly elongate, compressed laterally, covered with fairly large scales (*squ.* 22–28). Head also slightly compressed laterally and naked. Eyes large. Mouth oblique. Jaw teeth arranged in several rows, outer teeth large. Dorsal fins separate or contiguous only at bases. First dorsal fin with 6 rays, second with 8–10; anal fin with 1 unbranched and 7–9 branched rays. Pelvic fins narrow and long; rays fimbriate. Caudal fin rounded. Very small fishes (Koumans, 1953).

Indian and Pacific oceans, predominantly among coral reefs. Large number of species. About 10 species near the coasts of Japan. One species in the Sea of Japan.

1. **Eviota abax** (Jordan and Snyder, 1901) (Figure 263)

Asterropteryx abax Jordan and Snyder, Proc. U.S. Nat. Mus., 24, 1901 (1902): 40, fig. 2 (Misaki).

Eviota distigma Jordan and Seale, Bull. Bur. Fisher., 25, 1905 (1906): 389, fig. 79 (Pago Pago, Samoa). Koumans, Fish. Indo-Austr. Arch., 10, 1953: 319 (description, synonyms).

440

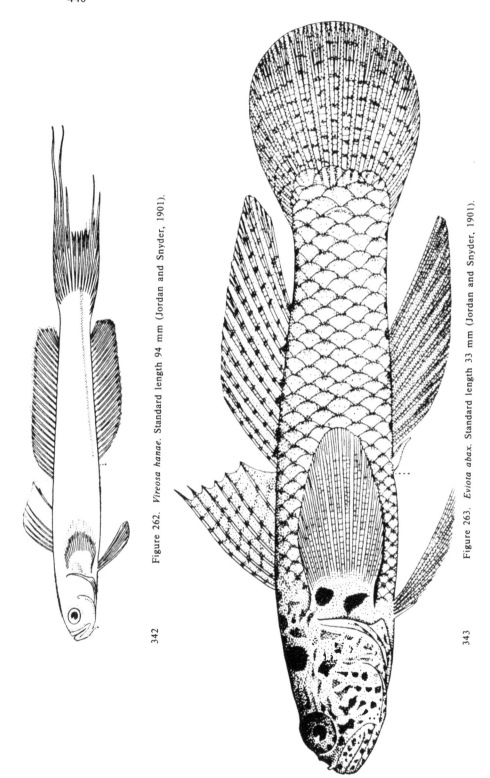

Figure 262. *Vireosa hanae*. Standard length 94 mm (Jordan and Snyder, 1901).

342

Figure 263. *Eviota abax*. Standard length 33 mm (Jordan and Snyder, 1901).

343

Eviota abax, Matsubara, Fish Morphol. and Hierar., 1955: 815, fig. 315.
Zhu et al., Ryby Yuzhno-Kitaiskogo Morya, 1962: 779, fig. 630.

D VI, I 9-10; A I, 8; *l. l.* 24 (Zhu et al., 1962).

Differs from other species in 1 spine and four soft rays in pelvic fins, color, and presence of two black spots at base of pectoral fin.

Mode of life, reproduction, and sexual dimorphism detailed in publications by Japanese researchers (Dotu, Arima and Mito, 1965: 41).

Length to 45 mm (Abe, 1958), usually smaller.

344 *Distribution*: In the Sea of Japan known from Sado Island (Honma, 1963: 21) and near Tsushima Islands (Arai and Abe, 1970: 94). Reported from Japan from the central part of Honshu southward (Matsubara, 1955: 815). Known near Cheju-do Island (Mori, 1956a: 146). Oceania, Indonesia, and Indian Ocean (Koumans, 1953: 320).

5. Genus *Parioglossus* Regan, 1912

Parioglossus Regan, Trans. Linn. Soc. London (2 ser.), 15, 2, 1912: 302 (type: *P. taeniatus* Regan). Koumans, Fish. Indo-Austr. Arch., 10, 1953: 363.

Body highly compressed laterally, covered with very small scales. Head also highly compressed laterally. Mouth protractile, very oblique. Each side of premaxilla with 3 canines in outer row, teeth in inner row small; each side of lower jaw with 3-5 canines; lateral teeth minute, arranged in one row. Gill openings do not continue forward ventrally. Dorsal fins separate. I D 6, II D I, 16-17; A I, 15-17; V I, 4 (Koumans, 1953).

Indian Ocean, north to Japan. One species in the Sea of Japan.

1. *Parioglossus dotui* Tomiyama, 1958 (Figure 264)

Parioglossus taeniatus Dotu (not Regan), Sci. Bull., Fac. Agric., Kyushu Univ., 15, 4, 1956: 489, 3 figs. (near Fukuoka).

Parioglossus dotui Tomiyama, Fig. and Descr. Fish of Japan, 57, 1958: 1179, pl. 230, fig. 582 (Nagasaki).

D VI, 17; A I, 18; P 18; V I, 4; C 13; *squ.* ca 110 (Tomiyama, 1958a).

This species differs from the similar *P. rainfordi* (from Australia) in number of rays in first dorsal fin (6 versus 5) and presence of a longitudinal fleshy fold in front of this fin.

345 Mode of life described by Dotu (1956: 489).

Length to 36 mm (Tomiyama, 1958a).

Distribution: In the Sea of Japan known from the environs of Fukuoka (Dotu, 1956: 489) and Tsushima Islands (Arai and Abe, 1970: 94). Pacific coast of Japan (Tomiyama, 1958a: 1179).

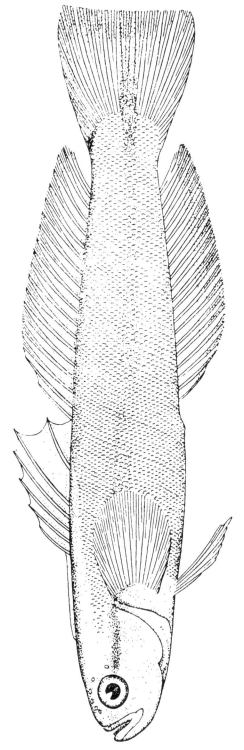

Figure 264. *Parioglossus dotui*. Standard length 36 mm (Zhu et al., 1962).

CLXXII. Family GOBIIDAE—Gobies

Pelvic fins, if present, fused and modified into suctorial disk. Eyes immovable. Palatines T-shaped with branch at end for articulation with frontals. Mesopterygoid rudimentary or absent. Scapula absent in adult and radial elements of skeleton of pectoral fins on clavicle, only lower part touching coracoid. Vertebrae 25–34 (Berg, 1949: 1060). Dorsal and anal fins not fused with caudal. Dorsal fins separate or with deep notch between; rarely single but short dorsal fin located in posterior half of body. Eyes immovable, but if movable, base of pectoral fins not thickened and not used for climbing.

Marine, brackish, and fresh-water fishes of temperate and tropical regions.

Over 200 genera. Five subfamilies and nearly 50 genera reported from the Sea of Japan.

443   *Addendum*: The photocopy sent by Takagi of his work in Japanese (*Studies of Gobioid Fishes in Japanese Waters: Comparative Morphology, Phylogeny, Taxonomy, Distribution, and Bionomics*, 1963: i–v + 273 pp.) could not be used unfortunately, since it was received after this volume had gone to press.

345   *Key to Subfamilies of Family Gobiidae*

1 (2). Body ovoid, compressed laterally. Teeth with single cusp, arranged in several rows on both jaws. Two dorsal fins. . . . . **[Gobiodontinae].**[6]

2 (1). Body oblong (sometimes very much so), and not compressed laterally.

3 (4). Teeth in outer row of both jaws tricuspid. . . . .
. . . . 1. **Tridentigerinae.**

4 (3). Teeth with single cusp, at least on lower jaw.

5 (8). Teeth on lower jaw in more than one row.

6 (7). First dorsal fin well developed, with 6 or more unbranched rays. . . . . 2. **Gobiinae.**

7 (6). First dorsal fin absent or reduced and with less than 6 unbranched rays. . . . . 3. **Luciogobiinae.**

8 (5). Teeth on lower jaw in one row.

9 (10). Second dorsal fin with long base. Teeth on lower jaw not vertical but slightly slanted. . . . . 4. **[Apocrypterinae].**

10 (9). Second dorsal fin with short base. Teeth on lower jaw vertical.
. . . . 5. **Sicydiaphinae.**

[6]South of Tokyo. Not found in the Sea of Japan.

444

1. Subfamily Tridentigerinae

Body oblong. Teeth in outer series of each row tricuspid, their middle cusp larger than lateral ones. Isthmus broad. Body covered with ctenoid scales. Dorsal fins separate. Second dorsal and anal fins with relatively short bases, not fused with caudal fin. Pelvic fins attached to belly only by bases, remaining partly free (Fowler, 1961: 234).

346

Key to Genera of Subfamily Tridentigerinae

1 (2). Barbels absent.............................. 1. **Tridentiger** Gill.
2 (1). Barbels present in large number on jaw or on lower side of head.
.... 2. **Triaenopogon** Bleeker.

1. Genus *Tridentiger* Gill, 1858

Tridentiger Gill, Ann. Lyceum Nat. Hist. New York, 7, 1858: 16 (type: *Sicydium obscurum* Temminck and Schlegel). Fowler, Quart. J. Taiwan Mus., 14, 3-4, 1961: 234.

Body slightly compressed laterally. Head broad, slightly flat dorsally. Cheeks .dilated. Eyes wide-set. Mouth slightly oblique; lower jaw protrudes forward slightly. Teeth well developed, arranged in 2 rows on each jaw; teeth in outer row tricuspid; middle cusp larger than lateral; teeth in second row small, pointed, with single cusp. Tongue roundish toward front. Barbels absent. Gill openings reduced. Body covered with relatively large ctenoid scales. First dorsal fin with 6 rays, second with 10 to 13 rays. Caudal fin rounded, similar to pectoral fin. Pelvic fins joined and well developed; funnel behind them not fused with belly (Fowler, 1961).

Three species along coasts of Japan as well as in the Sea of Japan and the Yellow, East China, and South China seas. Two species reported from the Sea of Japan.

Key to Species of Genus Tridentiger

1 (4). Transverse rows of scales 34-46. Two dark longitudinal stripes not present on sides of body.
2 (3). Nape and belly covered with scales.[7]..........................
.... 1. **T. obscurus** Temminck and Schlegel.
3 (2). Nape and belly naked. .... [**T. nudiventris** Tomiyama, 1934].[8]
4 (1). Transverse rows of scales 48-62. Two dark longitudinal stripes along sides of body................ 2. **T. trigonocephalus** (Gill).

---

[7]Mucus should be removed from the skin to detect scales in specimens preserved in alcohol.

[8]Ariake Bay, south of Kyushu Island.

1. **Tridentiger obscurus** (Temminck and Schlegel, 1845) (Figure 265)

Sicydium obscurum Temminck and Schlegel, Fauna Japonica, Poiss., 1845: 145, pl. 76, fig. 1 (rivers in the Gulf of Nagasaki).

Tridentiger obscurus, Berg, Ryby Presnykh Vod..., 3, 1949: 1102, fig. 833. Abe, Enc. Zool., 2, Fishes, 1958: 92, fig. 268 (color figure). Fowler, Synopsis..., 14, 3-4, 1961: 235 (description, synonyms). Zhu et al., Ryby Yuzhno-Kitaiskogo Morya, 1962: 817, fig. 663.

16972-16982. Mouth of Tumintsyan River. June-September, 1913. A.I. Cheiskii. 28 specimens.

D VI, I 10-12; A I, 9-11; *squ.* 34-37 (Berg, 1949).

Rays of first dorsal fin filamentous, especially in males. Color dark; head with a few minute light spots. First rays of each dorsal fin with 4 sharp dark spots. Second dorsal and anal fins white along margin. Caudal fin membrane black. White (orange in live specimens) patch near base of pectoral fin and dark spot near upper margin of base. In young fish 347 sometimes an indistinct narrow dark stripe present along sides, disappearing with age (Berg, 1949: 1103).

Mode of life described by Dotu (1958a: 343).

Length to 135 mm (Berg, 1949).

Distribution: In the Sea of Japan known from Peter the Great Bay and Tumintsyan River (Berg, 1949: 1103); near Wonsan and Pusan (Mori, 1952: 147); Hokkaido (Okada and Ikeda, 1938); Aomori, Niigata and Tsuruga (Jordan and Snyder, 1901c: 113); Sado Island (Honma, 1952: 225); San'in region (Mori, 1956a: 25); and Tsushima Islands (Shibata, 1968: 26; Arai and Abe, 1970: 72; Tomodo, 1970: 199). In the Yellow Sea 348 from Cheju-do Island, Inch'on (Mori, 1952: 147); Gulf of Chihli (Bohai) (Zhang et al., 1955: 216). Pacific coast of Japan, Ryukyu Islands (Matsubara, 1955: 821), South China Sea (Zhu et al., 1962: 817).

2. **Tridentiger trigonocephalus** (Gill, 1858) (Figure 266)

Triaenophorus trigonocephalus Gill, Proc. Acad. Nat. Sci. Philad., 10, 1858: 17 (China).

Tridentiger bifasciatus Steindachner, Sitzb. Akad. Wiss., 83, 1, 1881: 190 (Strelok Inlet near Vladivostok).

Tridentiger bucco Jordan and Snyder, Proc. U.S. Nat. Mus., 24, 1901: 113 (Misaki, Tokyo).

Tridentiger marmoratus Regan, Ann. Mag. Nat. Hist., (7), 15, 1905: 17, pl. 2, fig. 2 (Inner Sea of Japan).

Tridentiger trigonocephalus, Berg, Ryby Presnykh Vod..., 3, 1949: 1103, fig. 834. Fowler, Synopsis..., 14, 3-4, 1961: 236, fig. 66a. Zhu et al., Ryby Yuzhno-Kitaiskogo Morya, 1962: 817, fig. 664.

6609. Vladivostok. 1883. Polyakov. 2 specimens.

17812. Liman of Amur River near Chomi Island. August 13, 1910. V.K. Soldatov. 6 specimens.

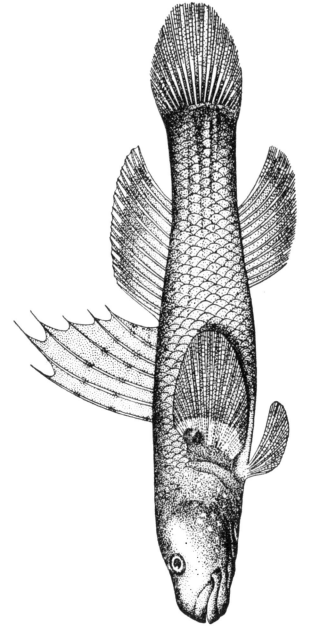

Figure 265. *Tridentiger obscurus*. Length 64 mm (Zhu et al., 1962).

347

21330. Vladivostov. 1910. V.K. Soldatov. 6 specimens.

22832. Nagasaki. March, 1901. P.Yu. Shmidt.

22834. Misaki. April 11, 1901. P.Yu. Shmidt.

35625. Yellow Sea. May 26, 1956. Academy of Sciences, China. 2 specimens.

36429. Yellow Sea, Tyantsin. June 11, 1957. E.F. Gur'yanova. 1 specimen.

D VI, I 12-14; A I, 10-11; *squ.* 52-53 (Berg, 1949).

Dark stripe in middle of body from tip of snout to base of caudal fin. Another similar but broader stripe above first continues along bases of dorsal fins. In older fish these stripes are sometimes indistinct. All fins, except pelvic fins, with rows of brown stripes. Sides of head and ventral surface with minute light-colored more or less rounded spots on brown background. Dark, curved stripe present near base of pectoral fin. Second dorsal and anal fins with white edging along margin. Suborbital series of cutaneous papillae does not reach forward to midpoint of eyes (Berg, 1949).

Mode of life described by Dotu (1958: 343).

Length to 110 mm (Berg, 1949).

Distribution: In the Sea of Japan known from Peter the Great Bay (Berg, 1949: 1103); Wonsan (Shmidt, 1931b: 136); Pusan (Mori and Uchida, 1934: 20; Yamagata Prefecture (Matsubara, 1955: 821); near Sado Island (Honma, 1963: 22); Toyama Bay (Katayama, 1940: 23); San'in region (Mori, 1956a: 25); near northern coast of Kyushu (Tabeta and Tsukahara, 1967: 299); and Tsushima Islands (Arai and Abe, 1970: 92). In the Yellow Sea—Inch'on (Jordan and Starks, 1905: 210); Gulf of Chihli (Bohai) (Zhang et al., 1955: 217); Chefoo (Wang and Wang, 1935: 197-198). The Philippines (Herre, 1927: 283, 285), East China Sea (Zhu et al., 1962: 817).

2. Genus *Triaenopogon* Bleeker, 1874

Triaenopogon Bleeker, Arch. Néerl. Sci. Nat., 9, 1874: 312 (type: *Triaenophorichthys barbatus* Günther). Fowler, Synopsis..., 14, 3-4, 1961: 239.

Close to *Tridentiger,* but differs in presence of several rows of short barbels on head, one row under eye above upper jaw, second row along posterior margin of upper jaw, third row along posterior margin of preopercle, and fourth row along margin of lower jaw. Opercle with a few isolated barbels. Teeth arranged in 2 rows on each jaw; teeth in outer row tricuspid, in inner row with one pointed cusp. First dorsal fin with 6 rays, second dorsal fin with 11 rays (Fowler, 1961).

One species in the waters of China, the Philippines, and Japan. Known from the Yellow Sea and the southern part of the Sea of Japan (Tsushima Strait).

350 1. *Triaenopogon barbatus* (Günther, 1861) (Figure 267)

Triaenophorichthys barbatus Günther, Cat. Fish. Brit. Mus., 3, 1861: 90 (China).

Triaenopogon japonicus Rendahl, Arkiv Zool., 16, 1924: 27 (Japan).

Triaenopogon barbatus Fowler, Synopsis..., 14, 3-4, 1961: 239, fig. 68 (description, synonyms). Zhu et al., Ryby Yuzhno-Kitaiskogo Morya, 1962: 819, fig. 665.

35622. Yellow Sea. May, 1956. Academy of Sciences, China. 2 specimens.

D VI, I 10-11; A I, 9-10; *squ.* 36/14 (Fowler, 1961).

Characters given in description of genus.

Mode of life described by Dotu (1957: 261).

Standard length, to 98 mm (Jordan and Snyder, 1910c).

Distribution: In the Sea of Japan known from Tsushima Islands (Tomiyama, 1936: 97). In the Yellow Sea near Ionamp'o (Mori, 1952: 147) and Gulf of Chihli (Bohai) (Zhang et al., 1955: 219). Along the Pacific coast of Japan from Tokyo southward (Matsubara, 1955: 821). The Philippines (Herre, 1927: 281), East China Sea (Zhu et al., 1963: 415) and South China Sea (Zhu et al., 1962: 819).

2. Subfamily Gobiinae

Body elongated to varying degrees. Head from small to moderate in size. Eyes well developed, normal in shape. Teeth simple and cusps without notch. Teeth of upper jaw arranged in 1 or more rows, teeth of lower jaw arranged in 2 or more rows. Vomer sometimes with a few teeth. Gill openings moderate or broad. Scales moderate or small in size. Snout and interorbital region naked. Body naked in some genera. Head with large number of cutaneous papillae, often arranged in rows or individual groups. Two dorsal fins well separated or divided by deep notch. Rays of first dorsal fin unbranched, flexible, and sometimes with elongate tips. Second dorsal fin larger than first. Sometimes upper rays of pectoral fin unattached. Pelvic fins completely or almost completely fused and attached only by bases (Fowler, 1961: 93).

Generally small fishes. Species numerous, widely distributed, and found along coasts.

Taxonomically the subfamily Gobiinae has been very poorly analyzed. The Japanese ichthyologist Takagi (1963) revised the gobies of Japan in terms of mucous canals, pores, and cutaneous papillae.[9] Unfortunately we

[9]The significance of these organs in the systematics of fishes was recognized even in the 1920's and 30's (Berg, 1949: 1060). Il'in (1927) provided a key to gobies of the Azov and Black seas, in which these organs were used for differentiation.

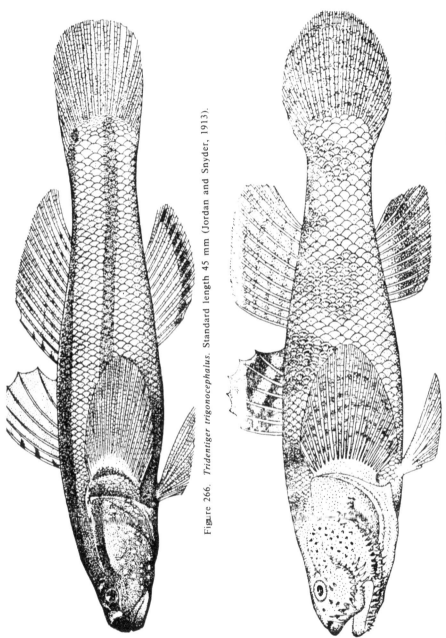

Figure 266. *Tridentiger trigonocephalus.* Standard length 45 mm (Jordan and Snyder, 1913).

Figure 267. *Triaenopogon barbatus.* Standard length 98 mm (Jordan and Snyder, 1901).

did not have this publication at our disposal. Interesting studies have been conducted by Akihito, who divided the Gobiidae into 4 groups on the basis of differences in the nature of arrangement of mesopterygium, postcleithrum, branchiostegal rays, pectoral fins, cleithrum, and infra-orbital bones. ·

Widely distributed in all warm seas, rarely found in rivers. About 200 genera with a very large number of species. About 40 genera with 100 species known in Japan, of which 13 genera with more than 30 species are found in the Sea of Japan.

Addendum: The genus *Suruga* Jordan and Snyder, 1901 has not been considered by us in the key to genera of the subfamily Gobiinae because the work of Honma and Tamura was received too late (see addendum to the family Eleotridae). This monotypic genus is close to *Acanthogobius* Gill in lacking free upper rays in the pectoral fins, but differs from it in a short snout and nonfimbriate border of the pelvic disc [not pectoral lobe as in Russian—Editor]. *Suruga fundicula* Jordan and Snyder is found in the waters of Niigata and Sado Island.

Key to Genera of Subfamily Gobiinae

1 (16). Barbels not present on head, not even a single pair; also absent on sides of head, under jaws, and symphysis of lower jaw.

351 2 (9). Tongue anteriorly rounded or straight; if with very weak notch, upper jaw does not extend beyond eyes.

3 (6). First dorsal fin with 6 rays.

4 (5). Length of postorbital part of head shorter than 2/3 length of head...1. **Gobius** Linné.

5 (4). Length of postorbital part of head equal to or more than 2/3 length of head. Transverse rows of scales 85 to 140.........
....2. **Cryptocentrus** Ehrenberg.

6 (3). First dorsal fin with (7) 8 rays or more.

7 (8). Pectoral fins without free upper rays. Border of pelvic disk fimbriate.............................. 3. **Acanthogobius** Gill.

8 (7). Pectoral fins with free upper rays. Border of pelvic disk smooth. .... 4. **Pterogobius** Gill.

9 (2). Tongue anteriorly with deep notch; if notch not very deep, upper jaw long and extends beyond eyes.

10 (11). Transverse rows of scales 25–40. Lower jaw protrudes forward notably. .... 5. **Glossogobius** Gill.

11 (10). Transverse rows of scales 50–100.

12 (15). Pectoral fins without free upper rays.

13 (14). Seismosensory canal of head consists of 3 parts: supraorbital, middle (near posterodorsal margin of orbit), and postorbital (Figure 268). .... 6. **Gymnogobius**

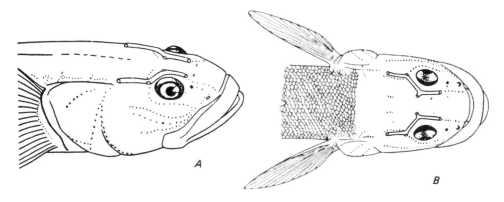

Figure 268. Head of *Gymnogobius macrognathus* (Berg, 1948).
A—lateral view; B—dorsal view.

14 (13). Seismosensory canal of head consists of only the middle part
(near posterodorsal margin of orbit) (Figure 269) ... 7. **Chloea.**

15 (12). Pectoral fin with free rays. Posterior part of interorbital
space without pores. .... 8. **Chasmichthys** Jordan.

16 (1). Barbels present on head; located on chin, or lower jaw, or
below head, or only one very short barbel on each side of
symphysis [in fishes with long low body (Figure 270)].

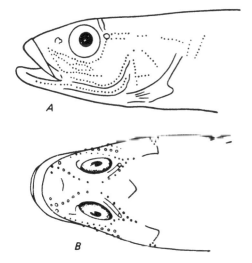

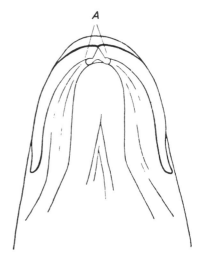

Figure 269. Head of *Chloea castanea*
(Berg, 1949).

A—lateral view; B—dorsal view.

Figure 270. Barbel (A) near symphysis of
lower jaw in *Synechogobius hasta.*

17 (18). First dorsal fin with 5–6 rays. Chin with many small barbels along margin of lower jaw. . . . 9. **Parachaeturichthys** Bleeker.[10]

352 18 (17). First dorsal fin with 7–9 rays.

19 (20). First dorsal fin with 7 rays. Head flat. Teeth arranged more or less obliquely. Lower surface of head with very large number of barbels. . . . . 10. **Lophiogobius** Günther.

20 (19). First dorsal fin with 8–9 rays. Head not depressed. Either a single pair or more than 10 to 12 barbels on each side of head. Teeth vertical.

21 (22). A single pair of barbels near symphysis (Figure 270). Body depth 2 times in head length . . . . 11. **Synechogobius** Gill.

22 (21). More than one pair of barbels. Body depth distinctly less than 2 times in head length.

23 (24). 3 pairs of barbels. Caudal fin rounded or pointed, but not truncate. . . . . 12. **Chaeturichthys** Richardson.

24 (23). 10 pairs of barbels. Caudal peduncle truncate, its length less than length of head. . . . . 13. **Sagamia** Jordan and Snyder.

1. Genus *Gobius* Linné, 1758—Gobies

Gobius Linné, Syst. Nat., 10th ed., 1758: 262 (type: *G. niger* Linné). Tomiyama, Japan. J. Zool., 7, 1, 1936: 59 (synonyms).

Ctenogobius Gill, Ann. Lyceum Nat. Hist. New York, 6, 1858: 374, 430 (type: *C. fasciatus* Gill). Fowler, Synopsis. . . , 13, 3–4, 1960: 104.

Rhinogobius Gill, Proc. Acad. Nat. Sci. Philad., II, 1859 (1860): 143 (type: *R. similis* Gill).

Coryphopterus Gill, Proc. Acad. Nat. Sci. Philad., 15, 1863: 263 (type: *C. glaucofraenum* Gill).

Acentrogobius Bleeker, Arch. Néerl. Sci. Nat., 9, 1874: 321 (type *Gobius chlorostigma* Bleeker). Fowler, Synopsis. . . , 13, 3–4, 1960: 141.

Zonogobius Bleeker, Arch. Néerl. Sci. Nat., 9, 1874: 323 (type: *Gobius semifasciatus* Kner = *Gobius semidoliatus* Val.). Fowler, Synopsis. . . , 13, 3–4, 1960: 150.

Bathygobius Bleeker, Arch. Néerl. Sci. Nat., 13, 1878: 54 (type: *Gobius nebulopunctatus* Valenciennes). Fowler, Synopsis. . . , 13, 3–4, 1960: 101.

Mugilogobius Smitt, Ofvers. K. Vet. Akad. Forh. Stockholm, 56, 1899: 543 (Jordan, The Genera of Fishes, 1963: 487). Fowler, Synopsis. . . , 13, 3–4, 1960: 155.

[10] A goby (*Paleatogobius uchidae* Takagi g.n. et sp. n. 1957: 118, fig. 6) has been described from the estuaries of rivers near the City of Fukuoka on Kyushu Island, which has barbels on the lower side of the head, but differs from members of the genus *Parachaeturichthys* in a notched tongue. Mode of life described by Dotu (1957: 97).

353 Species of this genus have often been separated as independent genera, then listed as synonyms. The presence of overlapping characters and the absence of detailed taxonomic analysis compels us to agree with Japanese ichthyologists (Tomiyama, 1936; Tomiyama and Abe, 1958) and accept the status of this genus as proposed by them.

Included in this genus are fishes lacking barbels on the lower side of the head, having a rounded tongue or a very weak notch at the end, possessing 6 rays in the first dorsal fin, 9 rays in the second, and having a reduced postorbital part of the head (shorter than 2/3 head length).

Many species, of which 8 and, possibly, one other are found in the Sea of Japan.[11]

Key to Species of Genus Gobius

1 (2). Anterior nostril located close to upper lip and resembling conical tubule directed ventrally toward lip . . . .
. . . . 1. **G. abei** (Jordan and Snyder).

2 (1). Anterior nostril located at some distance from upper lip, and if resembling tubule, then directed dorsally.

3 (4). Anterior half of body with about 7 white vertical stripes, clearly visible against dark background of body. . . . .
. . . . 2. **G. semidoliatus** Cuvier and Valenciennes.

4 (3). Anterior half of body without white vertical stripes.

5 (6). Upper lip definitely does not form anterior margin of snout. Dark vertical stripe located under eyes. . . . .
. . . . 3. **G. ornatus campbelli** (Jordan and Snyder).

6 (5). Upper lip definitely forms anterior margin of snout.

7 (8). Pectoral fins with free upper rays. . . . . . . 4. **G. fuscus** Rüppell.

8 (7). Pectoral fins without free upper rays.

9 (10). Sides of head not entirely naked: scales more or less cover upper part of operculum. Scales about 30. Dark spot above base of pectoral fin. . . . .
. . . . 5. [**G. caninus** Cuvier and Valenciennes].

10 (9). Sides of head naked.

11 (14). Snout short and more or less equal to diameter of eyes.

12 (13). Occipital region covered with scales. Dark longitudinal stripe located under eyes. Snout blunt. Upper jaw equal to diameter of eye. . . . . 6. **G. pflaumi** Bleeker.

[11]Honma et al. (1972: 53) have indicated another species in the Yamagata Prefecture, the Sea of Japan, *Ctenogobius dotui* Takagi, 1957, described for the first time from a rivermouth in Saga Prefecture.

443 *Addendum:* Rhinogobius brunneus (Temminck and Schlegal) has been reported for the region of Niigata and Sado Island. This species was identified as *Glossogobius giuris brunneus* (Temminck and Schlegel) before Takagi's work was published (see addendum to the family Gobiidae).

13 (12). Occipital region naked. Dark stripe under eyes absent. Snout
pointed. Upper jaw 2 times length of eye diameter..........
.................................7. **G. gymnauchen** Bleeker.

14 (11). Snout long, almost 2 times or even longer than eye diameter.

15 (16). Scales in longitudinal row 30. Scales of occiput continue
almost up to eye. Snout and sides of head marked with zigzag
stripes that vary in shape.............. 8. **G. giurinus** Rutter.

16 (15). Scales in longitudinal row 35 or more. Scales on occiput do
not continue behind upper margin of corner of preopercle. Snout
and sides of head without zigzag stripes.....................
.... 9. **G. similis** (Gill) Jordan and Snyder.

354 1. *Gobius abei* (Jordan and Snyder, 1901) (Figure 271)

Ctenogobius abei Jordan and Snyder, Proc. U.S. Nat. Mus., 24, 1901:
55, fig. 5 (Wakanoura, Japan).

Tamanka bivittata Herre, Gobies..., 1927: 224, pl. 17, fig. 4 (Hainan
Island).

Gobius abei, Tomiyama, Japan. J. Zool., 7, 1, 1936: 74. Abe, Enc. Zool.,
2, Fishes, 1958: 101, fig. 295.

Mugilogobius abei, Matsubara, Fish Morphol. and Hierar., 1955: 832,
fig. 323. Zhang et al., Ryby Zaliva Bokhai..., 1955: 201, fig. 127.
Fowler, Synopsis..., 13, 3-4, 1960: 160, fig. 31.

D VI, 9; A 9; P 16; *squ.* 36-41 (Jordan and Snyder, 1901c).

A characteristic feature of this species is the location of the anterior
nostril near the upper lip and the shape of its conical tubule directed
ventrally toward the lip. In the first dorsal fin rays 2-4 elongate. Occiput
and upper operculum covered with weak ctenoid scales, smaller than on
body. Light-colored median stripe in posterior part of body bordered
dorsally and ventrally by dark stripes. Caudal fin with flabelliform narrow
dark stripes.

Length to 60 mm (Abe, 1958)

Distribution: In the Sea of Japan known from Toyama Bay (Tomiyama,
1936: 74) and Tsushima Islands (Arai and Abe, 1970: 93). In the Yellow
Sea indicated for the Gulf of Chihli (Bohai) (Zhang et al., 1955: 201).
Along the Pacific coast of Japan from the central part of Honshu
southward; Ryukyu Islands and along the coast of China to Hainan
Island (Matsubara, 1955: 832).

2. *Gobius semidoliatus* Valenciennes, 1837 (Figure 272)

Gobius semidoliatus Valenciennes, Hist. Nat. Poiss., 12, 1837: (51)
67 (New Hebrides).

Zonogobius boreus Snyder, Proc. U.S. Nat. Mus., 36, 1909: 605; 42,
1912: 399, pl. 59, fig. 3.

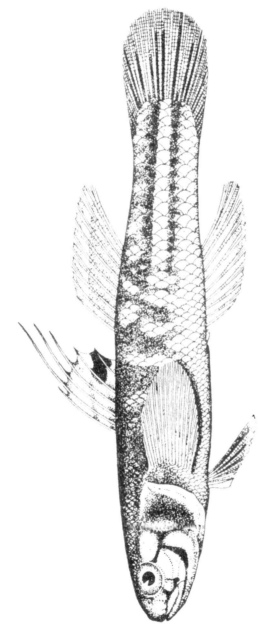

Figure 271. *Gobius abei.* Standard length 36 mm (Jordan and Snyder, 1901).

456

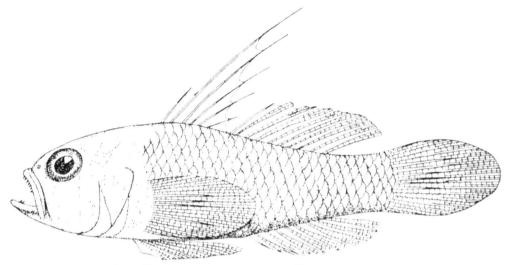

Figure 272. *Gobius semidoliatus.* Length 30 mm (Koumans, 1953).

Zonogobius semidoliatus, Koumans, Fish. Indo-Austr. Arch., 10, 1953: 149, pl. 36. Fowler, Synopsis..., 13, 3-4, 1960: 150, fig. 26.

2612. Red Sea. 1872. Klunzinger. One specimen.

443 *Addendum*: Honma and Tamura (see addendum to the family Eleotridae) have reported *Zonogobius boreus* Snyder, 1909 for the region of Niigata and Sado Island. This species differs from *Z. semidoliatus* in the red to chocolate-brown body color of live fish and the presence of more than 7 white stripes on the head, of which 2 are narrow; the last stripe continues from the occiput downward, cutting through the base of the pectoral fin.

354 D VI, I 9; A I, 9; *squ.* 22-25 + 3 (Fowler, 1961).

In this species and in closely related forms included in the genus *Zonogobius,* the border (membrane at base of pelvic fins) is poorly developed or almost absent. Color unique: 6-7 light-colored transverse stripes occur on head and anterior part of body.

Matsubara (1955: 826) separated two independent species, considered a single species by Koumans (1953). The difference between the two lies in coloration. *Z. boreus* is red to chocolate-brown in life. Head with about 7 white transverse stripes (2 narrow), and last stripe continues from occiput through base of pectoral fins (Tomiyama and Abe, 1958, fig. 304). *Z. semidoliatus* is dark red in anterior part of body, including head, with 7 or more distinct broad transverse white stripes in anterior part of body (Tomiyama and Abe, 1958, fig. 305). The characters reported by Matsubara are very similar and hardly suffice to distinguish two species. Judging from the drawings provided by Tomiyama and Abe, the rays in the first dorsal fin of *Z. semidoliatus* have long free tips and pigmentation occurs in the rays of all fins; perhaps, however, this is a sexual character of the male.

Length to 35 mm (Abe, 1958).

Distribution: In the Sea of Japan reported from the north at Hyoton and Mukoze Banks (Ouchi and Ogata, 1960: 183; Ouchi, 1963: 129); Pusan (Mori, 1952: 142); Sado Island (Honma, 1956: 21). Cheju-do Island; Pacific coast of Japan and Ryukyu Islands (Matsubara, 1955: 826). Indian Ocean and the Red Sea (Koumans, 1953: 150).

356 3. **Gobius ornatus campbelli** (Jordan and Synder, 1901) (Figure 273)

Ctenogobius campbelli Jordan and Snyder, Proc. U.S. Nat. Mus., 24, 1901: 62, fig. 8 (Wakanoura, Japan).

Gobius ornatus campbelli, Tomiyama, Japan. J. Zool., 7, 1, 1936: 72. Abe, Enc. Zool., 2, Fishes, 1958: 101, fig. 296.

D VI, 11; A 10; P 18; *squ.* 26/9 (Jordan and Snyder, 1901c).

Shape of snout characteristic for species; anterior margin protrudes slightly ahead of upper lip or located at same level. Anterior nostril distinctly separate from upper lip. Margin of pelvic fins entire. About 10 rows of scales in front of dorsal fin. Dark vertical stripe under eye.

Length to 90 mm (Abe, 1958).

Distribution: In the Sea of Japan known in the south near Tsushima Islands (Arai and Abe, 1970: 93). Matsubara (1955: 831) has indicated the central part of Honshu southward; usually this would include the coast of the Sea of Japan. Yellow Sea (Wang and Wang, 1935: 177) and south to Indonesia (Matsubara, 1955: 831).

4. **Gobius fuscus** Rüppell, 1828 (Figure 274)

Gobius fuscus Rüppell, Atlas Reise N. Afr. Fische, 1828: 137 (Red Sea). Tomiyama, Japan. J. Zool., 7, 1, 1936: 63. Abe, Enc. Zool., 2, Fishes, 1958: 103, fig. 302.

Gobius poecilichthys Jordan and Snyder, Proc. U.S. Nat. Mus., 24, 1901: 52, fig. 4 (Misaki).

Bathygobius fuscus, Koumans, Fish. Indo-Austr. Arch., 10, 1953: 187, fig. 45 (description, synonyms). Fowler, Synopsis..., 13, 3-4, 1960: 102, fig. 4. Zhu et al., Ryby Yuzhno-Kitaiskogo Morya, 1962: 788, fig. 638.

23393. Okinawa Island. January, 1927. P.Yu. Shmidt. 2 specimens.

D VI, 10-11; A 9-10; P 19-20; *squ.* 38-40; scales in transverse row 11-13; scales in front of first dorsal fin 24 or fewer (Koumans, 1953).

Differs from other species in free upper rays of pectoral fin, average number of scales in front of dorsal fin about 20, and not 30 or 10.

Length to 120 mm (Koumans, 1953).

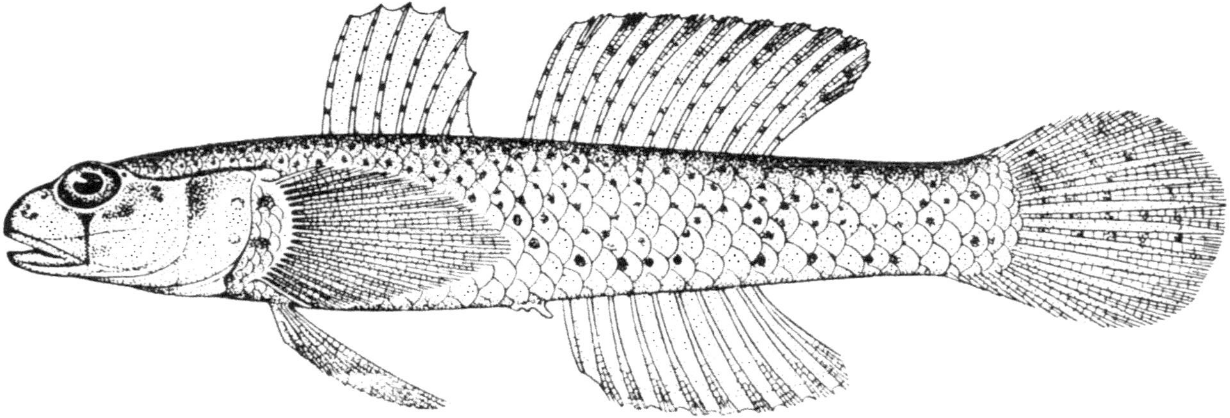

Figure 273. *Gobius ornatus campbelli*. Standard length 65 mm (Jordan and Snyder, 1901).

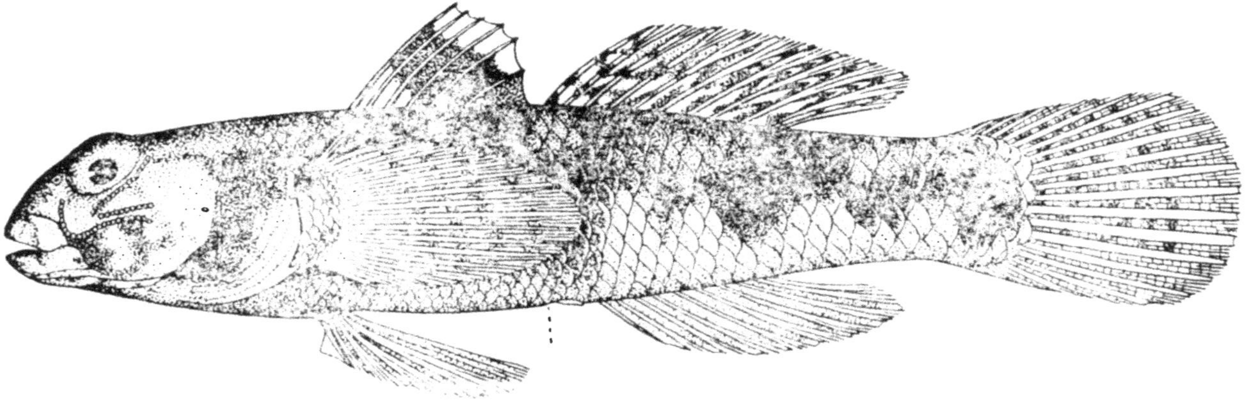

357

Figure 274. *Gobius fuscus*. Standard length 48 mm (Jordan and Snyder, 1901).

357

Distribution: In the Sea of Japan known from Wakasa Bay (Takegawa and Morino, 1970: 383); Tsushima Island (Arai and Abe, 1970: 92); and central part of Honshu southward to the Indian Ocean (Matsubara, 1955: 827).

5. [*Gobius caninus* Valenciennes, 1837] (Figure 275)

Gobius caninus Valenciennes, Hist. Nat. Poiss., 12, 1837: (65) 86 (Java). Tomiyama, Japan. J. Zool., 7, 1 1936: 70.

Coryphopterus bernadoui Jordan and Starks, Proc. U.S. Nat. Mus., 28, 1905: 207, fig. 9.

Rhinogobius caninus, Herre, Gobies..., 1927: 186, pl. 13, fig. 4.

Acentrogobius caninus, Koumans, Fish. Indo-Austr. Arch., 10, 1953: 61, fig. 16 (synonyms).

Vaimosa canina, Matsubara, Fish Morphol. and Hierar., 1955: 829.

Acentrogobius caninus, Fowler, Synopsis..., 13, 3-4, 1960: 145, figs. 23, 24 (synonyms).

D VI, 10; A 10; *squ.* 30/9 (Koumans, 1953).

Characters given in key. Scales on upper part of operculum. Dark spot located above base of pectoral fin.

Length to 132 mm (Koumans, 1953).

Distribution: Not found in the Sea of Japan. Reported from rivers of the Korean Peninsula (Mori, 1952: 141), but also known from marine waters (Tomiyama, 1936). Okinawa and the Korean Peninsula south to Indonesia (Matsubara, 1955: 829) and Sri Lanka (Koumans, 1953: 61).

359 6. *Gobius pflaumi* (Bleeker, 1853) (Figure 276)

Gobius pflaumi Bleeker, Verh. Bat. Gen., 25, 1853: 42, figs. 3, 18 (Nagasaki). Tomiyama, Japan. J. Zool., 7, 1, 1936: 66.

Ctenogobius virgatulus Jordan and Snyder, Proc. U.S. Nat. Mus., 24, 1901: 63, fig. 9 (Misaki).

Coryphopterus virgatulus, Jordan and Starks, Proc. U.S. Nat. Mus., 28, 1905: 206.

Ctenogobius chefuensis Wu and Wang, Contrib. Biol. Lab. Sci. Soc. China, Zool. ser., 8, 1, 1931: 6, fig. 4 (Chefoo).

Rhinogobius pflaumi, Matsubara, Fish Morphol. and Hierar., 1955: 830.

Ctenogobius pflaumi, Fowler, Synopsis..., 13, 3-4, 1960: 107 (synonyms).

13096. Wonsan. June, 1900. P.Yu. Shmidt. 6 specimens.

22984. Tsuruga. August 25, 1917. V. Rozhkovskii. 4 specimens.

30309. Peter the Great Bay. 1907. Brazhnikov. 1 specimen.

36414-36419. Yellow Sea. June, 1957. Expedition to China. 30 specimens.

D VI, 11; A 11; P 16; *squ.* 26; scales in transverse series 9 (Jordan and Snyder, 1901c).

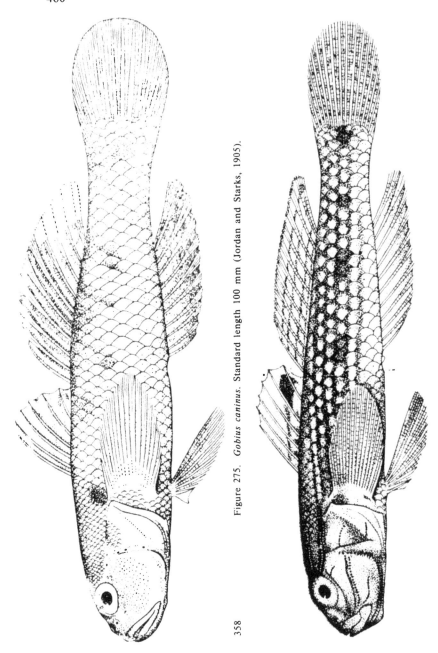

Figure 275. *Gobius caninus.* Standard length 100 mm (Jordan and Starks, 1905).

Figure 276. *Gobius pflaumi.* Standard length 50 mm (Jordan and Snyder, 1901).

358

358

Differs from other species in short snout, equal to diameter of eye; occiput covered with scales; stripe under eye; four dark spots along sides of body equal to eyes or slightly smaller; pelvic fins almost equal to length of pectoral fins, but terminate far from base of anal fin. Head naked except for small part of occiput.

Length to 80 mm (Tomiyama, 1936).

Distribution: In the Sea of Japan known from Peter the Great Bay (Soldatov and Lindberg, 1930: 420); near Wonsan, Pohang, and Pusan (Mori, 1952: 142); near Tsushima Islands (Arai and Abe, 1970: 92); coast of Japan from Aomori (Jordan and Snyder, 1901c: 65); Sado Island (Honma, 1952: 225); Toyama Bay (Katayama, 1940: 22); Tsuruga (Shmidt and Lindberg, 1930: 1149); San'in region (Mori, 1956a: 23); and even farther south. In the Yellow Sea reported from the Gulf of Chihli (Bohai) Zhang et al., 1955: 202) and Chefoo (Wang and Wang 1935: 180). Along the Pacific coast of Japan from Miyagi Prefecture south to the Philippines (Matsubara, 1955: 830).

7. *Gobius gymnauchen* (Bleeker, 1860) (Figure 227)

Gobius gymnauchen Bleeker, Act. Soc. Sci. Indo-Néerl. Japan, 6, 1860: 84, pl. 1, fig. 2 (Tokyo). Tomiyama, Japan. J. Zool., 7, 1, 1936: 67, fig. 21 (synonyms). Abe, Enc. Zool., 2, Fishes, 1958: 102, fig. 300.

Ctenogobius gymnauchen, Jordan and Snyder, Proc. U.S. Nat. Mus., 24, 1901: 58, fig. 6. Fowler, Synopsis..., 13, 3-4, 1960: 111, fig. 8.

36413. Yellow Sea. June 3, 1957. Expedition to China. 1 specimen. D VI, 10; A 10; *squ.* 23-27 (Abe, 1958).

This species differs from the closely related species *G. pflaumi* in a pointed snout versus blunt, length of upper jaw, absence of scales on occiput, and dark stripe under eye.

Length to 100 mm (Abe, 1958).

Distribution: In the Sea of Japan known from the coast of Hokkaido (Ueno, 1971: 87); Tohoku region (Tomiyama, 1936: 67); Sado Island (Katho et al., 1950: 324); Tsuruga (Jordan and Snyder, 1901c: 58); San'in region (Honma, 1956: 22); and Tsushima Islands (Arai and Abe, 1970: 92)

8. *Gobius giurinus* Rutter, 1897 (Figure 278)

Gobius giurinus Rutter, Proc. Acad. Nat. Sci. Philad., 1897: 86 (Shantou, China). Abe, Enc. Zool., 2, Fishes, 1958: 102, fig. 299.

Ctenogobius hadropterus Jordan and Snyder, Proc. U.S. Nat. Mus., 24, 1901: 60, fig. 7.

362 D VI, 9; A 9; P 19; *squ.* 28/9 (Jordan and Snyder, 1901c).

This long-snouted goby differs from the other long-snouted species, *G. similis,* in having large scales covering occipital region almost up to eyes. Snout and sides of head with zigzag pattern. Four to five dark spots

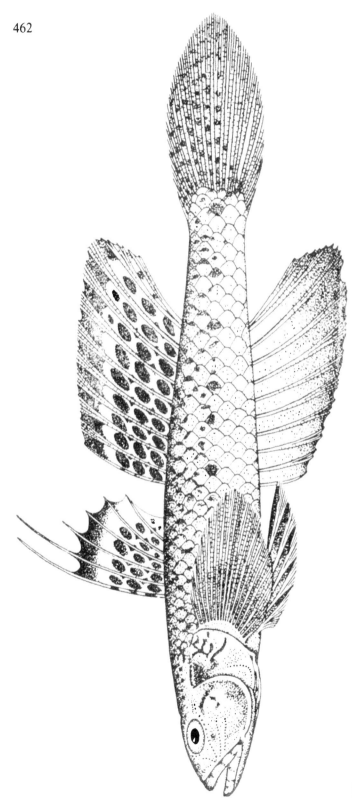

Figure 277. *Gobius gymnauchen.* Standard length 65 mm (Jordan and Snyder, 1901).

arranged along middle line of body sides that do not continue onto back.

Mode of life described by Dotu (1961: 120).

Length to 120 mm (Abe, 1958).

Distribution: Predominantly a fresh-water fish but in the Sea of Japan reported from Oshoro Bay (Kobayashi, 1962: 259). Known from near Sado Island (Honma, 1952: 225); San'in region (Mori, 1956a: 24); Tsushima Islands (Shibata, 1968: 26). Yellow Sea (Wang and Wang, 1935: 176). South China (Matsubara, 1955: 831).

9. *Gobius similis* (Gill, 1859) (Figure 279)

Rhinogobius similis Gill, Proc. Acad. Nat. Sci. Philad., 1859: 145 (near Shimoda, Japan). Berg, Ryby Presnykh Vod..., 3, 1949: 1077, fig. 1078.

Ctenogobius similis, Jordan and Snyder, Proc. U.S. Nat. Mus., 23, 1901: 759, fig. 35; 24, 1901: 56.

Gobius similis, Tomiyama, Japan. J. Zool., 7, 1, 1936: 68, fig. 22, A and B (synonyms). Abe, Enc. Zool., 2, Fishes, 1958: 102, fig. 298.

Rhinogobius bergi Lindberg, Tr. Zool. Inst. Akad. Nauk, SSSR, 3, 1936: 402, figs. 7, 8 (Maikhe River, Peter the Great Bay).

Rhinogobius similis lindbergi Berg, Ryby Presnykh Vod..., 3, 1949: 1078, figs. 810–812 (lower course of Amur River).

D VI, 9; P 19; *squ.* 31/11 (Jordan and Snyder, 1901c).

Long-snouted goby with small scales covering occipital region but not behind upper posterior corner of preopercle. Sides of head without zigzag dark stripes. Pelvic fins distinctly shorter than length of pectoral fins.

Length to 100 mm (Abe, 1958).

Distribution: Fresh-water fish of river basins of the Sea of Japan and the Sea of Okhotsk, Hokkaido Island (Ueno, 1971: 87); basin of the Pacific Ocean in Japan, Ryukyu Islands, and Taiwan (China); also found in estuaries of rivers.

2. Genus *Cryptocentrus* Ehrenberg, 1837

Cryptocentrus Ehrenberg. In: Cuvier and Valenciennes, Hist. Nat. Poiss., 12, 1837: 111 (type: *Gobius cryptocentrus* Cuvier and Valenciennes). Fowler, Synopsis..., 14, 1-2, 1961: 53.

Body moderately elongate, compressed laterally. Head compressed laterally, large, blunt, attenuates upward. Snout convex, equal in length to diameter of eye. Eyes close-set and high. Mouth broad, slightly oblique; lower jaw protrudes slightly. Upper jaw extends beyond eye. Teeth on jaws arranged in several rows, enlarged in outer rows; pair of canine-shaped teeth on lower jaw at end of series. Bony part of interorbital space narrow, about 1/4 diameter of eye. Gill openings broad, isthmus narrow. Inner margin of pectoral girdle without fleshy processes. Transverse rows

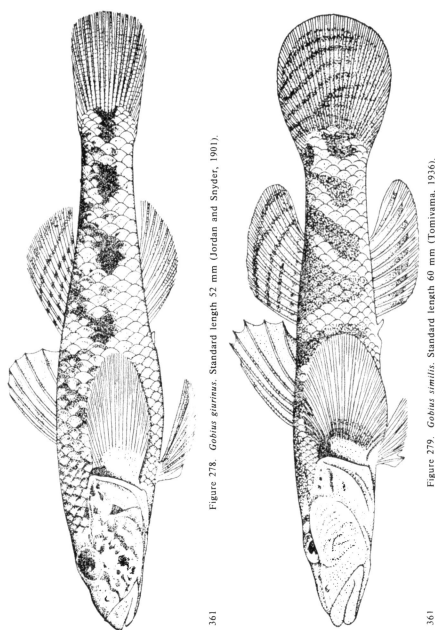

Figure 278. *Gobius giurinus.* Standard length 52 mm (Jordan and Snyder, 1901).

361

Figure 279. *Gobius similis.* Standard length 60 mm (Tomiyama, 1936).

361

of scales 85 to 140. Scales on anterior part of body cycloid and in caudal part sometimes ctenoid. Head behind eyes covered with scales or naked, as is occiput. Dorsal fins separate. First dorsal fin with 6 rays, second dorsal with 11 to 21 rays, and anal fin with 20 to 22 rays. Caudal fin pointed. Pectoral fins without free rays on upper side. Pelvic fins united and elongate (Fowler, 1961).

363   Indian and Pacific oceans. Several species. One from waters of the Sea of Japan.

1. *Cryptocentrus filifer* (Valenciennes, 1837) (Figure 280)

Gobius filifer Valenciennes, Hist. Nat. Poiss., 12, 1837: (80) 106 (Indian Ocean).

Cryptocentrus filifer, Jordan and Snyder, Proc. U.S. Nat. Mus., 24, 1901: 72, fig. 12. Wang and Wang, Contrib. Biol. Lab. Sci. Soc. China, Zool., ser. 11, 6, 1935: 182, fig. 14. Tomiyama, Japan. J. Zool., 7, 1, 1936: 82. Koumans, 1953: 86, fig. 17. Abe, Enc. Zool., 2, Fishes, 1958: 99, fig. 290. Fowler, Synopsis..., 14, 1-2, 1961: 57, fig. 35. Zhu et al., Ryby Vostochno-Kitaiskogo Morya, 1963: 423, fig. 319.

22817. Nagasaki. February 17, 1901. P.Yu. Shmidt. 1 specimen.

22818. Pusan. March, 1901. P.Yu. Shmidt. 1 specimen.

22985. Tsuruga. August 28-September 5, 1917. V. Rozhkovskii. 6+ specimens.

D VI, 11; A 10; *squ.* 100 (Abe, 1958).

Differs from *C. fontanesi,* recorded off Kagoshima, as well as other Japanese species of this genus in fewer number of rays in second dorsal fin (11-12 versus 14-16) and anal fin (10-12 versus 15-17). Differs from other species also in very small scales (more than 85 transverse rows) compressed body, and black spot in anterior part of first dorsal fin near its base.

Length to 150 mm (Abe, 1958).

Distribution: In the Sea of Japan repoted from Ulsan (Mori, 1952: 144); Pusan (Shmidt, 1931b: 131); Sado Island (Honma, 1952: 225); Toyama Bay (Katayama, 1940: 22); Tsuruga (Shmidt and Lindberg, 1930: 1149); coast of Yamaguti Prefecture (Yoshida and Ito, 1957: 268) and Tsushima Islands (Arai and Abe, 1970: 93). In the Yellow Sea—Gulf of Chihli (Bohai) (Zhang et al., 1955: 203), Chefoo (Wang and Wang, 1935: 182). Along the Pacific coast of Japan from the central part of Honshu southward. Coast of China south to Siangan (Hong Kong) (Zhu et al., 1963: 423); Indonesia (Koumans, 1953: 86).

3. Genus *Acanthogobius* Gill, 1859

Acanthogobius Gill, Proc. Acad. Nat. Sci. Philad., 11, 1859: 145 (type: *Gobius flavimanus* Temminck and Schlegel). Fowler, Synopsis..., 14, 3-4, 1961: 208.

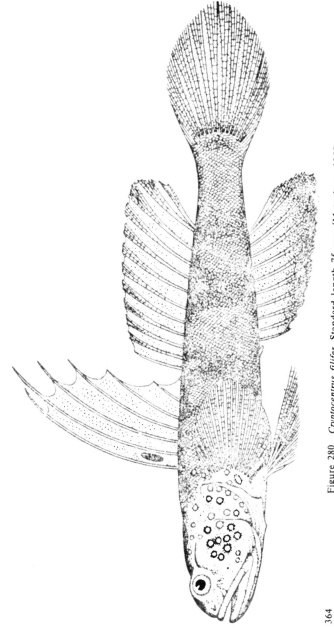

Figure 280. *Cryptocentrus filifer.* Standard length 75 mm (Matsubara, 1955).

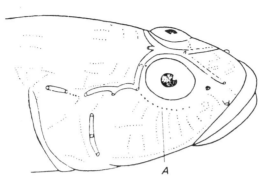

365 Figure 281. Figure of arrangement of cephalic sensory pores and cutaneous papilla system in fishes of the family Gobiidae (Berg, 1949).

A—suborbital series of cutaneous papillae.

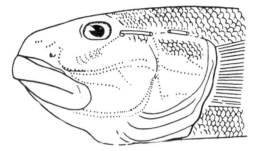

365 Figure 282. Arrangement of cephalic sensory pores and cutaneous papilla system in fishes of the genus *Acanthogobius* (Berg, 1949).

Aboma Jordan and Starks. In: Jordan, Proc. Calif. Acad. Sci., ser. 2, 5, 1895: 497 (type: *A. etheostoma* Jordan and Starks).

Sagamia Jordan and Snyder, Proc. U.S. Nat. Mus., 24, 1901: 100 (type: *S. russula* Jordan and Snyder).

Body oblong, posteriorly slightly compressed laterally. Head elongate, its profile rounded. Snout fairly long, longer than diameter of eye. Eyes close-set, almost in middle part of head. Mouth moderate in size, oblique; jaws equal. Teeth moderate in size, arranged in several rows on both jaws. Tongue blunt or slightly notched. Anterior nostril with very short tubule. Width of interorbital space less than 1/2 diameter of eye. Gill openings continue slightly forward ventrally. Isthmus rather broad. Branchiostegal rays 5. Fleshy processes absent on inner side of pectoral girdle. Mucous canals extend from corner of mouth up to preopercle and along lower jaw up to posterior margin of preopercle. Body scales medium in size, and smaller on cheeks. Dorsal fins separate; first with 7-9 rays, second with 14-15 rays. Anal fin with 12-14 rays.[12]. Pectoral fin without free rays in

[12] *Acanthogobius lactipes*: D VIII, 11-12; A 10-11.

365 upper part; base of fin covered with scales. Caudal fin blunt, slightly shorter than length of head (Fowler, 1961).

Berg (1949: 1062, 1075) recognized the independent status of the genus *Aboma,* listing the following differences from the genus *Acanthogobius*: *Aboma* with longitudinal, curved suborbital series of cutaneous papillae (Figure 281). Dorsal muscles do not reach eyes. Second dorsal fin with not more than 10-11 rays (including unbranched rays); anal fin with I 7-10 rays. Sides of head naked. *Acanthogobius,* instead of longitudinal, curved suborbital series of cutaneous papillae, with oblique series extending from posterior margin of eye forward and ventrally (Figure 282). Dorsal muscles continue upward almost to eyes. Second dorsal fin with 14-15 rays; anal fin with 12-13 (14) rays. Occiput and part of sides of head covered with scales.

East China and South China seas, Yellow Sea, and waters of Japan. Several species. Two known from the Sea of Japan.

Key of Species of Genus Acanthogobius

1 (2). Transverse rows of scales 35-40. Head naked..................
.... 1. **A. lactipes** (Hilgendorf).
2 (1). Transverse rows of scales 55-75. Upper part and part of sides of head covered with scales....................................
.... 2. **A. flavimanus** (Temminck and Schlegel).

1. *Acanthogobius lactipes* (Hilgendorf, 1878) (Figure 283)

Gobius lactipes Hilgendorf, Sitzber. Ges. Naturf. Freunde, Berlin, 1878: 109 (Tokyo).

Aboma lactipes, Jordan and Snyder, Proc. U.S. Nat. Mus., 24, 1901: 67, fig. 10. Wang and Wang, Contrib. Biol. Lab. Sci. Soc. China, 11, 6, 1935: 180, fig. 12. Berg, Ryby Presnykh Vod..., 3, 1949: 1080. Zhang et al.,Ryby Zaliva Bokhai..., 1955: 204, fig. 129. Zhu et al., Ryby Vostochno-Kitaiskogo Morya, 1963: 429, fig. 326.

Aboma lacticeps (error), Matsubara, Fish Morphol. and Hierar., 1955: 836.

Aboma tsushimae Jordan and Snyder, Proc. U.S. Nat. Mus., 23, 1901: 759 (Tsushima Islands). Jordan and Snyder, Proc. U.S. Nat. Mus., 24, 1901: 69, fig. 11.

Acanthogobius lactipes, Tomiyama, Japan. J. Zool., 7, 1, 1936: 84. Abe, Enc. Zool., 2, Fishes, 1958: 98, fig. 288. Fowler, Synopsis..., 14, 3-4, 1961: 210, figs. 54, 55.

22204. Vladivostok. October, 1927. E.P. Rutenberg. 1 specimen.

22816. Genzan. June, 1901. P.Yu. Shmidt. 1 specimen.

23106. Sonon Bay, Chosomman Bay. May 6, 1897. A. Bunge. 1 specimen.

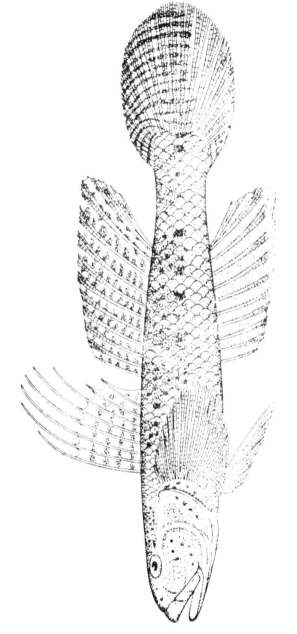

Figure 283. *Acanthogobius lactipes*. Standard length 70 mm (Jordan and Snyder, 1901).

25471. Mouth of the Sidemi River, Peter the Great Bay. July 11, 1929. A.Ya. Taranets. 2 specimens.

25481. Pos'et, Peter the Great Bay. August 10, 1928. A.Ya. Taranets. More than 6 specimens.

37287. Pos'et, Peter the Great Bay. July 20, 1962. O.A. Skarlato. 4 specimens.

38744. Pos'et, Peter the Great bay. 1967. A.N. Golikov. 2 specimens.

367 D VIII, 11-12; A 10-11; *squ.* 35-40 (Abe, 1958).

In addition to larger scales, this species differs from *A. flavimanus* in having fewer rays in the second dorsal fin (11-12 versus 13-15) and anal fin (10-11 versus 12).

Mode of life described by Dotu (1959: 196).

Length to 90 mm (Abe, 1958).

Distribution: In the Sea of Japan known from Peter the Great Bay (Soldatov and Lindberg, 1930: 421); near Wonsan (Shmidt, 1931b: 31); Shestakov Port—40° N (Shmidt, 1931a: 119); Pusan (Mori, 1952: 145); Tsushima Islands (Arai and Abe, 1970: 93); Aomori (Jordan and Snyder, 1901c: 67); Sado Island (Honma, 1963: 22); Toyama Bay (Katayama, 1940: 23); San'in region (Mori, 1956a: 25). Sea of Okhotsk—mouths of rivers and lakes of Hokkaido Island (Hikita, 1952: 15). Yellow Sea—Gulf of Chihli (Bohai) (Zhang et al., 1955: 204); Chefoo (Wang and Wang, 1935: 180). East China Sea—near coasts of China.

2. *Acanthogobius flavimanus* (Temminck and Schlegel, 1845) (Figure 284)

Gobius flavimanus Temminck and Schlegel, Fauna Japonica, Poiss., 1845: 141, pl. 74, fig. 1 (Nagasaki).

Acanthogobius flavimanus, Jordan and Snyder, Proc. U.S. Nat. Mus., 24, 1901: 98. Wang and Wang, Contrib. Biol. Lab. Sci. Soc. China, 11, 6, 1935: 191, fig. 20. Tomiyama, Japan. J. Zool., 7, 1, 1936: 85. Berg, Ryby Presnykh Vod..., 3, 1949: 1076, figs. 808, 809. Zhang et al., Ryby Zaliva Bokhai..., 1955: 206, fig. 130. Fowler, Synopsis..., 14, 3-4, 1961: 212, figs. 56, 57.

Gobius stigmathonus Richardson, Voy. Sulphur. Fishes, 1844: 147 (Canton).

Acanthogobius stigmathonus, Jordan and Metz, Mem. Carnegie Mus., 6, 1 (1914), 1915: 57.

Aboma snyderi Jordan and Fowler, Proc. U.S. Nat. Mus., 25, 1902: 575, fig. (Aomori, young specimen).

13094. Wonsan. June, 1900. P.Yu. Shmidt. 6 specimens.

22203. Peter the Great Bay. September, 1927. E.P. Rutenberg. 2 specimens.

22823. Pusan. March, 1901. P.Yu. Shmidt. 4 specimens.

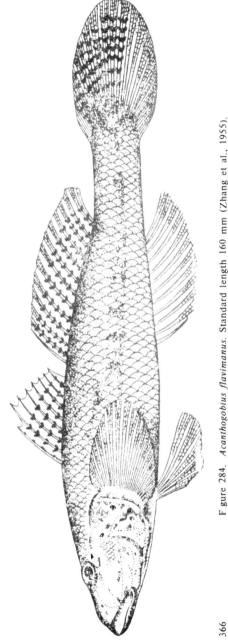

Figure 284. *Acanthogobius flavimanus*. Standard length 160 mm (Zhang et al., 1955).

22895. Wonsan. June, 1900. P.Yu. Shmidt. 3 specimens.

23839. Pos'et, Peter the Great Bay. October 1, 1925. Institute of Geography. 3 specimens.

32583. Wonsan. June 20, 1900. P.Yu. Shmidt. 3 specimens.

36406. Yellow Sea, Tsihgtao. June 18, 1957. EZIN. 2 specimens.

36407. Yellow Sea, Chefoo. June 28, 1957. EZIN. 1 specimen.

368   D VIII, 14-15; A 12-13 (14); *squ.* 47-54. Operculum covered with scales in upper part; cheeks with very few scales, which are very small, sometimes almost imperceptible, especially in young fish. Throat covered with small scales. Scales on dorsal surface of head continue to eyes. Base of pectoral fin covered with scales. Series of papillae "b" almost reach mouth. Eyes set high. Pelvic fins do not reach vent; pectoral fins reach vertical from posterior end of first dorsal fin. Collar (border) of pelvic disk without small lobe. Corner of mouth reaches vertical from anterior margin of eye. Lips not broadened toward corner of mouth. Dorsal fins separated by distance less than longitudinal diameter of eye. Head length 28.5 to 30.7% of standard length. Width of flat forehead less than longitudinal diameter of eye. Body yellowish, laterally with some dark spots, particularly sharp at base of caudal fin; series of minute dark spots on dorsal fin; zigzag stripes on caudal fin, almost imperceptible in lower third; pectoral fin dusky, with dark spots at base; operculum with dark spots; oblique dark spots on snout continue from eye to mouth (Berg, 1949).

Mode of life of this rather large species described in detail by Japanese researchers (Miyazaki, 1940: 159; Dotu and Mito, 1955: 153; Imamura and Hashitani, 1957: 45).

Length to 250 mm (Berg, 1949).

Distribution: In the Sea of Japan reported from Peter the Great Bay (Soldatov and Lindberg, 1930: 428); Wonsan and Pusan (Shmidt, 1913b: 134); Hakodate and Aomori (Jordan and Snyder, 1901: 98); Sado Island (Honma, 1952: 225); Toyama Bay (Katayama, 1940: 23); San'in region (Mori, 1956a: 25); Simonosaka (Jordan and Thompson, 1914: 289); and Tsushima Islands (Arai and Abe, 1970: 93). Yellow Sea—Gulf of Chihli (Bohai) (Zhang et al., 1955: 206); all coasts of the Korean Peninsula (Mori, 1952: 145); and Chefoo (Wang and Wang, 1935: 191). Pacific coast of Japan from Hokkaido southward (Matsubara, 1955: 836).

4. Genus *Pterogobius* Gill, 1863

Pterogobius Gill, Proc. Acad. Nat. Sci. Philad., 15, 1863: 266 (type: *Gobius virgo* Temminck and Schlegel). Fowler, Synopsis..., 14, 3-4, 1961: 203.

Body moderately elongate, slightly compressed laterally. Head not flat, rounded, and broad in region of eyes. Width of interorbital space and snout length equal to diameter of eye. Eyes set in anterior part of head. Mouth moderate in size, slightly oblique, terminal; lower jaw protrudes more or less. Tongue rounded or truncate. Teeth moderate in size, outer ones larger; teeth on lower jaw continue only up to half its length and last teeth larger. Barbels absent. Nostrils not tubular. Gill openings moderate in size, separated by wide isthmus or continue slightly forward ventrally. Body covered with very small cycloid or ctenoid scales, 65-135 transverse rows. Cheeks completely naked or with small areas covered with scales. Occiput and thorax covered with scales. First dorsal fin with 8 rays, sometimes elongate in males. Bases of second dorsal and anal fins long, each fin with 20-30 close-set rays. Caudal fin moderately long, bluntly rounded. Pectoral fin with free silky upper rays (Fowler, 1961: 203).

Seas of China and Japan. Four species from the Sea of Japan.

369

Key to Species of Genus Pterogobius

1 (4). Second dorsal fin with 20-22 rays. Transverse rows of scales 66-80. Inner margin of pectoral girdle smooth, without crest or collar [= fleshy ridge—Ed.].

2 (3). Color pale reddish; sides of body with 6-8 transverse yellow narrow stripes, width of stripe less than eye diameter.........
.... 1. **P. zonoleucus** Jordan and Snyder.

3 (2). Color violet; sides of body with 6-7 transverse dark-colored, fairly broad stripes, equal to diameter of eye, which continue behind dorsal and anal fins and are bordered by narrow yellow stripes............................... 2. **P. elapoides** (Günther).

4 (1). Second dorsal fin with 25-28 rays. Transverse rows of scales 100 or more. Inner margin of pectoral girdle with fleshy crest or collar [= fleshy ridge—Ed.].

5 (6). Five deep black transverse stripes located on pale, almost white background of body, their width almost equal to diameter of eye.
.... 3. **P. zacalles** Jordan and Snyder.

6 (5). Large number of longitudinal red, blue, green, and dove-blue stripes on light bluish background of body, as well as on head, sides of body, and dorsal and anal fins................................
.... 4. **P. virgo** (Temminck and Schlegel).

1. *Pterogobius zonoleucus* Jordan and Snyder, 1901 (Figure 285)

Pterogobius zonoleucus Jordan and Snyder, Proc. U.S. Nat. Mus., 24, 1901: 94, fig. 19 (Misaki). Tomiyama, Japan. J. Zool., 7, 1, 1936: 86. Abe, Enc. Zool., 2, Fishes, 1958: 97, fig. 285. Fowler, Synopsis..., 14, 3-4, 1961: 204, fig. 51.

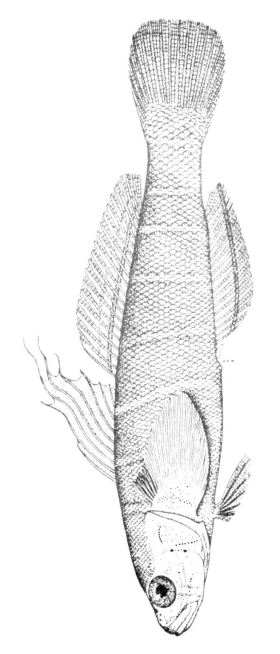

Figure 285. *Pterogobius zonoleucus.* Standard length 70 mm (Jordan and Snyder, 1901).

D VIII, 20; A 19; *squ.* 66 (Jordan and Snyder, 1901c).

This short-snouted species has a unique color pattern: 6-8 yellowish stripes, in width less than diameter of pupil, located on pale reddish or brick-red background.

Comments on reproduction given by Japanese researchers (Tsutsumi and Dotu, 1961: 149).

Length to 90 mm (Abe, 1958).

Distribution: In the Sea of Japan known from Pusan (Mori, 1952: 144); Sado Island (Honma, 1963: 22); Toyama Bay (Katayama, 1940: 24); San'in region (Mori, 1956a: 25). Along the Pacific coast of Japan from Misaki southward (Matsubara, 1955: 837).

2. *Pterogobius elapoides* (Günther, 1871) (Figure 286)

Gobius elapoides Günther, Proc. Zool. Soc., London, 1871: 665, pl. 63, fig. D (Japan ?).

Pterogobius elapoides, Jordan and Snyder, Proc. U.S. Nat. Mus., 24, 1901: 90. Abe, Enc. Zool., 2, Fishes, 1958: 97, fig. 284. Fowler, Synopsis..., 14, 3-4, 1961: 205, fig. 51a.

Pterogobius daimio Jordan and Snyder, Proc. U.S. Nat. Mus., 24, 1901: 91, fig. 17 (Misaki).

Pterogobius elapoides elapoides, Tomiyama, Japan. J. Zool., 7, 1, 1936: 86.

22821. Misaki. April 11, 1901. P.Yu. Shmidt. More than 6 specimens.
22822. Pusan. March 30, 1901. P.Yu. Shmidt. More than 6 specimens.

D VIII, 20-23; A 19-22; P 19-22; *squ.* 77-91 (Jordan and Snyder, 1901c—*P. elapoides + P. daimio*).

This species differs from the previous one in elongate snout and color. Sides of body with 6-7 transverse dark stripes located on violet
371 background, equal in width to diameter of eye. Stripes bordered by narrow yellow stripes; one such stripe located under eye.

Mode of life described by Japanese scientists (Dotu and Tsutsumi, 1959: 186).

Length to 110 mm (Abe, 1958).

Distribution: In the Sea of Japan reported from Pusan (Shmidt, 1931b: 133); Hakodate, Aomori (Jordan and Snyder, 1901c: 90); Sado Island (Honma, 1963: 22); Toyama Bay (Katayama, 1940: 23), San'in region (Mori, 1956a: 25); Yamaguti Prefecture (Yoshida and Ito, 1957: 268); Tsushima Islands (Arai and Abe, 1970: 93). Cheju-do Island (Mori, 1952: 144). Along the Pacific coast of Japan from Tohoku region (Honshu Island) southward (Matsubara, 1955: 837). Presence indicated for Siangan (Hong Kong). St. John Island (Tomiyama, 1936: 86; see Smitt, 1896: 196), but Chinese ichthyologists (Zhu et al.) do not mention this species.

3. *Pterogobius zacalles* Jordan and Snyder, 1901 (Figure 287)

Pterogobius zacalles Jordan and Snyder, Proc. U. S. Nat. Mus., 24, 1901: 93, fig. 18 (Misaki). Tomiyama, Japan. J. Zool., 7, 1, 1936: 86. Abe, Enc. Zool., 2, Fishes, 1958: 97, fig. 283.

D VIII, 25–27; A 25–27; *squ.* 100 (Abe, 1958).

Five deep black transverse stripes, almost equal in width to diameter of eye, located on very pale, almost white background of body. Anteriormost stripe located under origin of first dorsal fin and reduced; 3 stripes under base of second fin, and fifth stripe at base of caudal fin. Sometimes a semispherical dark stripe borders posterior margin of caudal fin. Absence of stripes on head a characteristic feature.

Length to 150 mm (Tomiyama, 1936).

Distribution: In the Sea of Japan recorded from Otaru and Ioiti on Hokkaido (Ueno and Abe, 1968: 37); Tohoku region (Okada and Matsubara, 1938: 371); Sado Island (Honma, 1952: 225); Toyama Bay (Katayama, 1940: 23). Along the Pacific coast of Japan from Tohoku region southward (Matsubara, 1955: 837); and Nagasaki (Tomiyama, 1936: 86).

4. *Pterogobius virgo* Temminck and Schlegel, 1845 (Figure 288)

Gobius virgo Temminck and Schlegel, Fauna Japonica, Poiss., 1845: 143, pl. 74, fig. 4 (Nagasaki).

Pterogobius virgo, Jordan and Snyder, Proc. U. S. Nat. Mus., 24, 1901: 88. Tomiyama, Japan. J. Zool., 7, 1, 1936: 86, fig. 34. Abe. Enc. Zool., 2, Fishes, 1958: 96, fig. 282. Fowler, Synopsis..., 14, 3–4, 1961: 206, fig. 52.

22820. Misaki. April 11, 1901. P.Yu. Shmidt. More than 6 specimens.

D VIII, 26–28; A 26–28; *squ.* 130 (Abe, 1958).

Differs from other species of this genus in having longitudinal rather than transverse stripes on head, sides of body, and on dorsal and anal fins, as well as very minute scales.

Length to 170 mm (Abe, 1958).

Distribution: In the Sea of Japan known from Sado Island (Honma, 1963: 22); Toyama Bay (Katayama, 1940: 23); San'in region (Mori, 1956a: 25); and Tsushima Islands (Arai and Abe, 1970: 93). In the Korean Strait near Thonen. Along the Pacific coast of Japan from Misaki south to Nagasaki (Matsubara, 1955: 837).

5. Genus *Glossogobius* Gill 1862

Glossogobius Gill, Ann. Lyceum Nat. Hist., New York, 7, 1858–1862 (1862): 46 (type: *Gobius platycephalus* Richardson, 1846 = *Gobius giuris* Buchanan-Hamilton, 1822). Fowler, Synopsis..., 13, 3–4, 1960: 127.

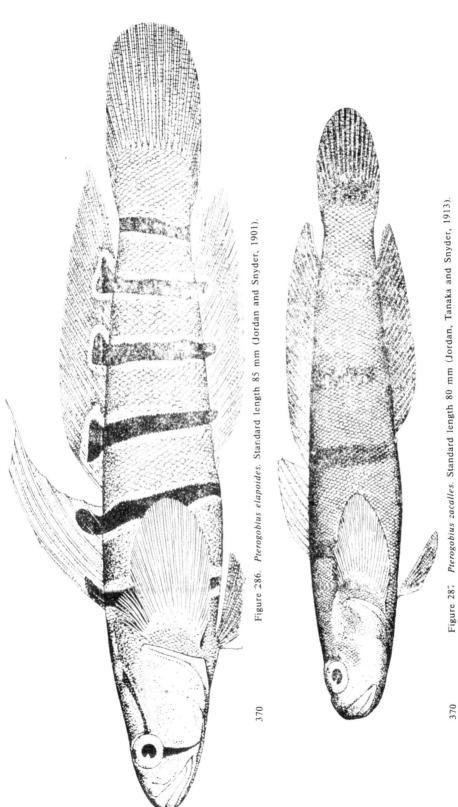

Figure 286. *Pterogobius elapoides.* Standard length 85 mm (Jordan and Snyder, 1901).

370

Figure 287. *Pterogobius zacalles.* Standard length 80 mm (Jordan, Tanaka and Snyder, 1913).

370

Body oblong, cylindrical in anterior part, and compressed laterally in caudal part. Head flat and pointed. Snout 1.5 to 2.0 times longer than eye. Eyes located in anterior half of head. Upper jaw extends backward but not beyond vertical from midpoint of eye. Mouth slightly oblique; lower jaw protrudes slightly. Teeth on both jaws arranged in several rows, some teeth enlarged, not equal in size, wide-set, and curved. Tongue with notch. Anterior nostril a short tube. Width of interorbital space from 1/3 to 3/4 diameter of eye. Gill openings broad, continue forward ventrally. Isthmus narrow. Inner margin of pectoral girdle without fleshy processes. Scales ctenoid, 25-40 transverse rows. Mucous canals continue from mouth up to posterior margin of operculum; one canal passes obliquely under opercle. Dorsal fins separate; first dorsal fin with 6 rays, second with 7-11 rays. Caudal fin oblong, sometimes pointed toward back. Pectoral fins without free rays on upper side, base covered with scales. Pelvic fins fused into long suctorial disk (Fowler, 1960).

Several species in the Indian Ocean and western part of the Pacific Ocean. One species reported from the Sea of Japan.

1. **Glossogobius olivaceus** (Temminck and Schlegel, 1845) (Figure 289)

Gobius oblivaceus Temminck and Schlegel, Fauna Japonica, Poiss., 1845: 143, pl. 74, fig. 3 (Japan).

Gobius fasciato-punctatus Richardson, Voy. Sulphur., Fishes, 1844: 145, pl. 62, figs. 13, 14 (Canton).

Glossogobius brunneus (non Temminck and Schlegel) Jordan and Snyder, Proc. U.S. Nat. Mus., 24, 1901: 74. Tanaka, Ann. Zool. Japan., 6, 4, 1908: 251.

Glossogobius giuris brunneus, Tomiyama, Japan. J. Zool., 7, 1, 1936: 88. Matsubara, Fish Morphol. and Hierar., 1955: 838. Abe, Enc. Zool., 2, Fishes, 1958: 96, fig. 281.

Glossogobius olivaceus, Akihito, Japan. J. Ichthyol., 13, 4/6, 1966: 73, figs 1-27.

1990. Japan. Salmin. 1 specimen.

D VI, (9) 10 (11); A (8) 9 (10); *squ.* 31-34 (Akihito, (1966).

Differs from *Glossogobius giuris* (Hamilton), according to Akihito, in dark spots on occiput, dorsal part of body,[13] and fin membranes. Differences between this species and *Glossogobius giuris* have been detailed by Akihito (1966).

Length to 200 mm (Abe, 1958).

Distribution: In the Sea of Japan known from Hakodate (Jordan and Snyder, 1901c: 74); Toyama Bay (Katayama, 1940: 23); San'in region (Mori, 1956a: 25); and Tsushima Islands (Shibota, 1968: 26). In the Sea of

[13]These spots are not depicted in Figure 289, which was drawn from figure of this species given by Temminck and Schlegel (1845, pl. 74, fig. 3).

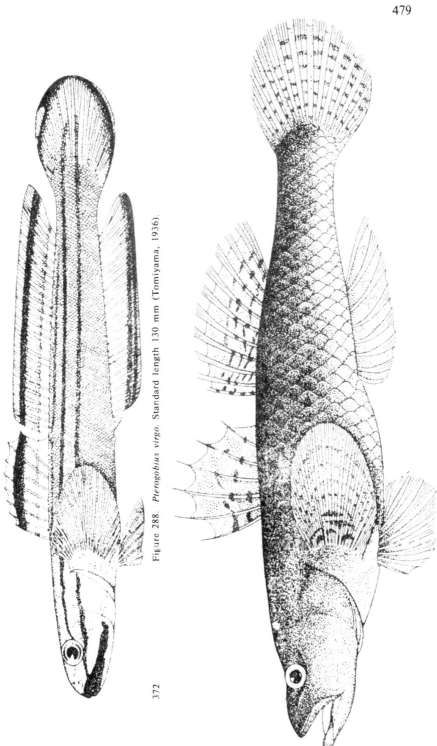

Figure 288. *Pterogobius virgo.* Standard length 130 mm (Tomiyama, 1936).

Figure 289. *Glossogobius olivaceus.* Standard length 135 mm (Temminck and Schlegel, 1845).

480

Okhotsk reported from Aniva Bay (Tanaka, 1908). Not known from the Yellow Sea, but reported from the East China Sea (Zhu et al., 1963: 420). Pacific coast of Japan (Akihito, 1966).

6. Genus *Gymnogobius* Gill, 1863

Gymnogobius Gill, Proc. Acad. Sci. Philad., 1863: 269 (type: *G. macrognathus*). Taranetz, Dokl. Akad. Nauk SSSR, 1934: 397. Berg, Ryby Presnykh Vod..., 1949: 1073.

374   Body covered with minute ctenoid or cycloid scales arranged in 65–95 transverse rows. First dorsal fin usually with 6–8 simple rays rarely 5; second dorsal fin with 11–13 and anal fin with 10–14 rays. Vertebrae 33–38. Sides of head naked. Barbels absent. Mouth large, its corners reaching vertical from posterior margin of eye or extending beyond it. Lower jaw usually protrudes forward. Tongue notched in front. Upper rays of pectoral fins not silky. Teeth on jaws arranged in several rows. Canines absent. Caudal fin moderately long. Isthmus narrow. Anterior nostril slightly elongate. Canals and pores of seismosensory system present (Figure 268). Suborbital series of cutaneous papillae present; vertical suborbital rows of cutaneous papillae absent; 3 horizontal rows present. Snout with two longitudinal rows of papillae. Supraorbital sensory canal consists of three parts: one above eye, second near its posterodorsal margin, and third behind eye above preopercle. Supraorbital canals on right and left sides not joined (Berg, 1949).

Some explanation is required with regard to the genus *Gymnogobius*. It has been listed among the synonyms of *Chaenogobius* Gill, 1859, the type species of which is *Chaenogobius annularis* Gill (Gill, *Ann. Lyceum Nat. Hist.,* 1858–1862, 7: 12). As indicated by Taranetz (1934: 398), not this species but a second species *Ch. megacephalus* Fowler was reported by contemporary ichthyologists. *Ch. megacephalus* most probably belongs to a different genus. Takagi (1966b: 29) has written that the fish included by Tomiyama under *Ch. annularis* should be included under *Ch. castaneus,* and not considered a synonym of *Ch. annularis.* In this context Takagi described a new genus, *Rhodonichthys,* with *Gobius laevis* Steindachner, 1879 as the type species.

Takagi (1966a, 1966b) attempted to clarify the chaos (his term!) enmeshing the classification of *Chaenogobius annularis* Gill. Unfortunately, his publication of 1963 was not available to us.

Seven species reported from the waters of Japan, the Yellow Sea, southern part of the Sea of Okhotsk, and the Pacific coast of Japan.

Key to Species of Genus Gymnogobius

1 (2). Lower jaw slightly shorter than upper. Body depth 6.6 to 7.4

times in standard length. Diameter of eye 5.1 to 7.9 times in head length. Head and body highly flattened.....................

.... 1. **G. raninus** Taranetz.

2 (1). Lower jaw protrudes beyond upper.

3 (4). Body low, 6.4 times in standard length. Diameter of eye 4 times in head length. D VII, 12, A 13; *l. l.* 80. Dark longitudinal band of spots extends along sides of body........................

.... 2. [**G. nigripinnis** (Wang and Wang)].

4 (3). Body deeper, 4.7 to 5.8 times in standard length.

5 (6). Lower jaw reaches vertical from middle or posterior margin of eye, but not beyond it. D (V) VI (VII), 11–12 (13); A 10–12; *squ.* 66–80...................... 3. **G. macrognathus** (Bleeker).

6 (5). Lower jaw extends distinctly beyond vertical from posterior margin of eye.

7 (8). Mouth superior; tip of lower jaw at level of upper profile of head. Fins and branchiostegal membranes black...........

.... 4. [**G. nigrimembranis** (Wu and Wang)].

8 (7). Mouth almost superior; tip of lower jaw on horizontal line with lower margin of eye or pupil.

9 (10). Length of pectoral fin, measured from upper end of fin base of its tip, equal to length of base of second dorsal fin. ......

.... 5. **G. bungei** (Schmidt).

375 10 (9). Length of pectoral fin, measured from upper end of fin base to its tip, distinctly shorter than length of base of second dorsal fin.

11 (12). Length of pectoral fin equal to length of base of anal fin. Transverse rows of scales 85–95. Vomerine processes absent or poorly developed. Eyes fairly small, about 7 times in head length. .... 6. **G. mororanus** (Jordan and Snyder).

12 (11). Length of pectoral fin distinctly shorter than anal fin base length. Transverse row of scales 65–78. Vomer with two distinct processes directed ventrally and located opposite lobes of tongue. Eyes relatively large, 4 times in head length.........

.... 7. **G. heptacanthus** (Hilgendorf).

1. ***Gymnogobius raninus*** Taranetz, 1934 (Figure 290)

Gymnogobius raninus Taranetz, Dokl. Akad. Nauk SSSR, 1934: 398 (Peter the Great Bay). Berg, Ryby Presnykh Vod..., 1949: 1075.

Chaenogobius cylindricus Tomiyama, Japan. J. Zool., 7, 1, 1936: 92, fig. 39 (Hiroshima).

25485. Mouth of Sidemi River, Peter the Great Bay. July 11, 1929. A.Ya. Taranetz. 1 specimen.

35325. Sea of Japan, Olga Bay. N.I. Tarasov. 1 young specimen.

D VI, 13-12; A 12-(10); *l. l.* 77-87; P 18.

Gill rakers short, 4 + 9-10 on first arch. Head depressed in adult and young fish. Body cylindrical in anterior part. Head 3.5 to 3.7 and body depth 6.6 to 7.4 times in standard length. Eyes 5.1 to 7.9 times in head length. Upper jaw long, distinctly continues beyond eye. Pelvic fins distinctly longer than snout (Taranetz, 1934).

Length to 67 mm (Taranetz, 1934).

Distribution: In the Sea of Japan known from Peter the Great Bay. *Ch. cylindricus* described from Hiroshima.

2. [*Gymnogobius nigripinnis* (Wang and Wang, 1935)] (Figure 291)

Chlosa nigripinnis Wang and Wang, Contrib. Biol. Lab. Sci. Soc. China, Zool. ser., 11, 6, 1935: 187, fig. 17 (Chefoo).

Chaenogobius nigripinnis, Fowler, Synopsis..., 14, 1-2, 1961: 68, fig. 42.

D VII, 12; A 13; P 19; *squ.* ca 80. Head length 3.6 and body depth 6.4 times in standard length. Depth of caudal peduncle 3.4, eye diameter 4.0, snout length 3.5, interorbital space 5.1, and maxilla 1.7 times in head length (Fowler, 1961).

Body oblong, fairly thin. Caudal peduncle compressed laterally, rather long, length twice its depth. Head long, slightly flattened anteriorly and compressed laterally in posterior part. Eyes large, located along sides of head, closer to tip of snout than to gill opening. Interorbital space broad, only slightly less than diameter of eye and slightly concave. Snout short, broad, its tip almost blunt. Nostrils well separated; anterior nostril behind upper lip, posterior one in front of eye. Mouth very broad, oblique; lower jaw protrudes forward; maxilla extends distinctly beyond vertical from posterior margin of eye. Teeth small, simple, arranged in narrow bands on both jaws. Tip of tongue with distinct notch. Gill openings continue slightly forward; isthmus narrow. Gill rakers long and thin, more than half diameter of eye, 13 on lower branch of gill arch. Body covered with very small and slightly ctenoid scales, except for breast and occiput which, like head, are naked.

Dorsal fins well separated; first dorsal fin quite short and when adpressed, does not reach origin of second dorsal fin; latter originates slightly in front of vertical from vent, its rays slightly shorter than rays of first dorsal fin. Anal fin originates at a vertical from third ray of dorsal fin and continues backward from it or slightly beyond. Caudal and pectoral fins rounded. Pelvic fins rather long and extend backward almost to vertical from end of pectoral fin.

377 Color of fish preserved in formalin dark gray in upper part and lighter on lower part. Sides with row of longitudinal or irregular spots arranged in form of more or less longitudinal stripe. Back with numerous

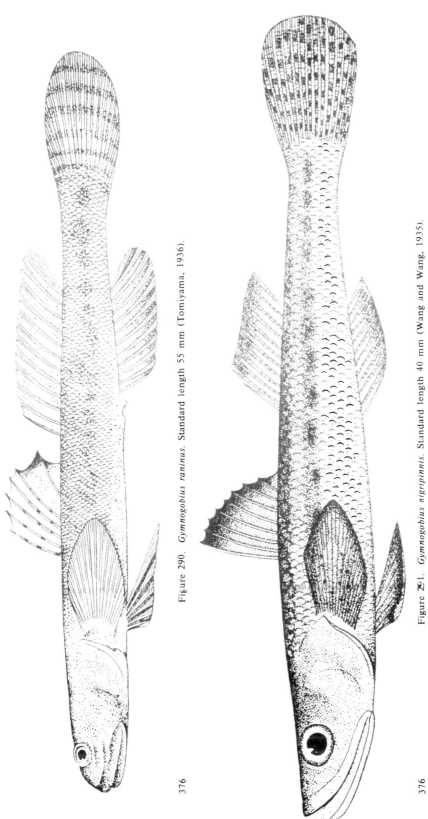

Figure 290. *Gymnogobius raninus.* Standard length 55 mm (Tomiyama, 1936).

Figure 291. *Gymnogobius nigripinnis.* Standard length 40 mm (Wang and Wang, 1935).

376

376

minute black spots forming reticulate pattern. Head dark; dark spot located in front of eye; lower surface of head blackish. First dorsal fin dark, almost blackish, especially in posterior part; second dorsal fin darkish, with dark border along margin. Caudal fin blackish underneath; anal fin dark with black stripe along margins; pelvic fins dark; and pectoral fins lighter in color, with a few darkish stripes. *G. raninus* is close to *Ch. mororana* and *Ch. heptacanthus,* but differs from them in larger size of mouth and darker fins (Wang and Wang, 1935: 187).

Standard length 40 mm (Wang and Wang, 1935).

Distribution: Not found in the Sea of Japan. Described from Chefoo in the Yellow Sea.

3. *Gymnogobius macrognathus* (Bleeker, 1860) (Figure 292)

Gobius macrognathos Bleeker, Acta Soc. Sci. Indo-Néerl., 8, 1860: 83, pl. I, fig. 1 (Tokyo).

Chaenogobius macrognathus, Tomiyama, Japan. J. Zool., 7, 1, 1936: 89, fig. 37.

Chaenogobius macrognathos, Jordan and Snyder, Proc. U.S. Nat. Mus., 24, 1901: 76, fig. 13.

Chloea aino Shmidt, Ryby Vostochnykh Morei..., 1905: 207 (Aniva Bay).

Chaenogobius urotaenia Hilgendorf, Sitz. Naturf. Freunde, Berlin, 1878: 108 (Tokyo). Takagi, Japan. J. Ichthyol., 2, 1, 1952: 14, 22, fig. 2.

Gymnogobius macrognathus, Berg, Ryby Presnykh Vod..., 1949: 1073, figs. 804–807.

16967. Peter the Great Bay, mouth of Tumyntsyan River, August 12, 1913. A.I. Cherskii. 4 specimens.

26102. Primor'e, mouth of Kvandagau River. August 27, 1934. G.U. Lindberg. 6 specimens.

26105. Primor'e, Prebrazheniya Inlet. September 26, 1934. G.U. Lindberg. 1 specimen.

26109. Primor'e, Kvandadan. September 27, 1934. G.U. Lindberg. More than 6 specimens.

38008. Primor'e, Petrov Island. 1961. Yu.I. Orlov. 5 specimens.

D (V) VI (VII), 11–12 (13); A 10–12; *squ.* 66–80. Gill rakers on first arch 9–12. Vertebrae 33–34. Head length 2.75 to 3.50 times in standard length. Head length in young fish 28 to 31.5% and in adult fish 29.5 to 34% of standard length. In juvenile fish with a standard length of 40 to 50 mm, head flat and cheeks inflated. Eye diameter 1.5 to 1.75 times in forehead width and 5 times in head length.

Head dorsally and laterally naked. Body covered with small cycloid or ctenoid scales (both types of scales often found on sides of body in the same specimen). Scales behind occiput and on breast very small; scales on belly minute and caducous. Distance between vertical from origin of anal fin and end of last vertebra constitutes 85 to 92% of predorsal distance. Dorsal fins separate. Anal fin originates at vertical from third or fourth ray of second dorsal fin. Second dorsal and anal fins flexible and do not reach caudal fin. Brownish spots on head, body, and fins. First dorsal fin with dark spot near tip of last unbranched ray. Ontogenetic variation and sexual dimorphism rather pronounced. Upper jaw of males much longer than in females, continues beyond vertical from posterior margin of eye. Head flat (Berg, 1949: 1074).

Tomiyama (1936: 90) reported that, in the opinion of Koumans, the description of this species (Figure 293) given by Jordan and Snyder (1901c: 76) is not identical to the description given by Bleeker, and therefore he included the fish described by these authors under this name as well as those mentioned by Berg (1949: 1073) under *Chaenogobius annularis urotaenia*, a fresh-water fish. According to our data, specimens of *G. macrognathus* from Peter the Great Bay are found in rivers as well as marine waters. Hence, like L.S. Berg, we have tentatively included these 379 fish under *Gymnogobius macrognathus*. However, the need for a detailed analysis of species of *Gymnogobius* is obvious. As mentioned earlier, such an attempt was made by the Japanese ichthyologist Takagi (1952, 1966a, 1966b).

Mode of life described by Dotu (1955: 367).

Length to 157 mm (Berg, 1949).

Distribution: In the Sea of Japan known from Peter the Great Bay (Soldatov and Lindberg, 1930: 423); Wonsan and Pusan (Mori, 1952: 143); Sea of Japan coast of Hokkaido (Ueno, 1971: 88); Oshoro Bay (Kobayashi, 1962: 259); Sado Island (Honma, 1952: 225); San'in region (Mori, 1956a: 25); and Tsushima Islands (Arai and Abe, 1970: 93). In the Yellow Sea reported from the Gulf of Chihli (Bohai) (Zhang et al., 1955: 210). In the Sea of Okhotsk reported from rivulets of Aniva Bay (Shmidt, 1950: 128).

4. [*Gymnogobius nigrimembranis* Wu and Wang, 1931] (Figure 294)

Gobius (*Chaenogobius*) *nigrimembranis* Wu and Wang, Contrib. Biol. Lab. Sci. Soc. China, Zool. ser., 8, 1, 1931: 4, fig. 3 (Chefoo). Fowler, Synopsis..., 14, 1-2, 1961: 70.

D VI, 13-14; A 12-13; P 23; *squ.* 82. Head length 3.8 to 4.0, body depth 5.6, and length of caudal peduncle 4.9 to 5.2 times in standard length. Snout length equal to diameter of eye, 4.1 to 4.3 times in the head length. Pectoral fins 1.4, pelvic fins 1.6, caudal fin 1.2 to 1.4, maximum depth of caudal peduncle 3.0 to 3.4, longest ray of first dorsal fin 2.3, and longest ray of second dorsal fin 2.0 to 2.1 times in head length. Width of interorbital space 1.3 to 1.5 times in the eye diameter.

Body compressed laterally; maximum body depth at vertical from

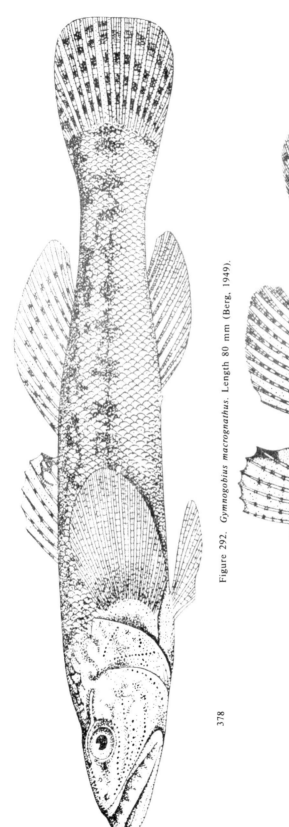

Figure 292. *Gymnogobius macrognathus.* Length 80 mm (Berg, 1949).

378

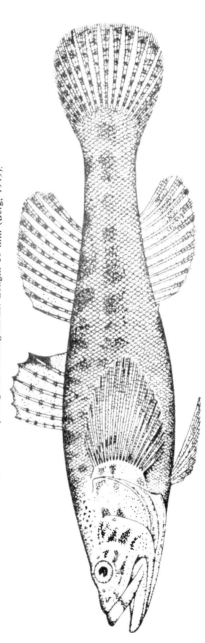

Figure 293. *Gymnogobius macrognathus.* Standard length 90 mm (Jordan and Snyder, 1901).

378

origin of dorsal fin. Head equal in height and width, 1.8 to 2.0 times in head length. Snout broad and blunt. Lower jaw long; profile of chin convex; mouth subterminal, very deep; end of upper jaw extends beyond vertical from anterior margin of eye. Teeth arranged in several rows, those in outer row not distinctly enlarged; canines absent. Interorbital space flat. Nostrils separate; anterior nostril in form of tubule. Branchiostegal membranes attached to isthmus at vertical from posterior margin of preopercle; gill rakers elongate, longer than gill filaments, 15 to 16 in the lower part of anterior arch.

Dorsal fins separate; distance between them 3 times in head length. Second dorsal fin, slightly apart from first dorsal, its origin closer to base of caudal fin than to eye; rays in middle part longest. Pectoral fins without free rays in upper part and do not reach vertical from end of base of first dorsal fin. Pelvic fins connected but not fused with belly, and continue almost three-fifths distance to anus. Anal fin originates slightly behind vertical from origin of second dorsal fin; tips of last rays of both fins reach same vertical line, but not half length of caudal peduncle. Caudal fin rounded. Scales very minute; occiput covered with scales; cheeks and throat naked.

Color of fish preserved in formalin greenish-gray; back and occiput with reticulate chocolate-brown pattern. Lateral body surface with 5 pairs of transverse chocolate-brown stripes—two pairs in trunk region and three pairs in caudal. Branchiostegal membranes, pelvic fins, and anal and first dorsal fins black; second dorsal and caudal fins dark; caudal fin with a few light-colored transverse stripes (Wu and Wang, 1931).

Length to 61 mm (Fowler, 1961).

Distribution: Not found in the Sea of Japan. Found in the Yellow Sea and described from Chefoo.

5. *Gymnogobius bungei* (Schmidt, 1931) (Figure 295)

Chloea bungei Schmidt, Izv. Akad. Nauk SSSR, ser. 7, 1, 1931a: 119, fig 5 (Hinnam, Sea of Japan).

23107. Hinnam, Korean Peninsula. May 5, 1897. A Bunge. 3 specimens.

D VII, 13; A 13; P 21; *squ.* 85. Head 3.4, body depth 5.3, and depth of caudal peduncle 12.0 times in standard length. Diameter of eye 5.1, interorbital space 5.5, snout 3.4, and length of upper jaw 1.8 times in head length. Body moderately compressed laterally. Head same width as body, but less deep, fairly long, and pointed. Eyes set high along sides of head; upper margin protrudes somewhat above head profile. Interorbital space slightly less than diameter of eye. Mouth large, oblique; maxilla long, extends beyond vertical from posterior margin of eye by distance equal to diameter of pupil. Lower jaw protrudes slightly. Three

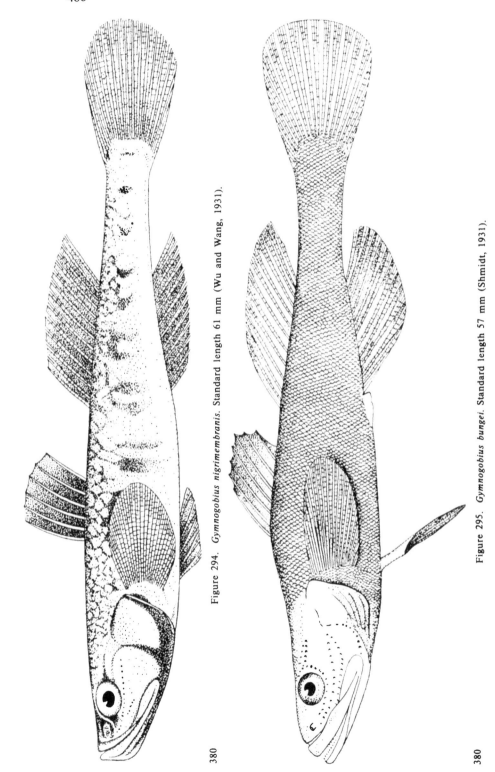

Figure 294. *Gymnogobius nigrimembranis*. Standard length 61 mm (Wu and Wang, 1931).

Figure 295. *Gymnogobius bungei*. Standard length 57 mm (Shmidt, 1931).

rows of pores under eye. Tongue with deep notch. Gill opening continues up to upper margin of base of pectoral fin. Gill rakers long and thin, 5 + 14 on first gill arch. Anterior nostril in form of short tubule. Head and occiput entirely naked. Body covered with moderately small scales. Dorsal fins well separated; rays of first dorsal fin, when folded, do not reach origin of base of second dorsal fin. First dorsal fin high in anterior part; height 2.5 times in head length. Pectoral fins rounded, 1.7 times in head length. Caudal fin bluntly rounded. Pelvic fins slightly longer than pectoral fins. Color of fish preserved in alcohol yellowish, with large number of chocolate-brown dotlike spots scattered on dorsal side of body and head. Branchiostegal membranes dark chocolate-brown. Pectoral fins yellowish; caudal fin chocolate-brown. Both dorsal fins with chocolate-brown dotlike spots, darker on first dorsal fin than on second. Pelvic and anal fins blackish (Schmidt, 1931a: 119-120).

In the opinion of P.Yu. Shmidt, this species is very close to *Chloea mororana* Jordan and Snyder found off Muroran (Hokkaido). It differs from this species in shorter head, larger scales, shorter maxilla, and darker color of branchiostegal membranes and dorsal and pectoral fins. This species was later included in the list of synonyms of *Ch. mororana,* (Tomiyama, 1936: 92; Fowler, 1961: 69), but differs from *Ch. mororana,* in addition to color, in greater body depth (as mentioned in the key). Taranetz (1934: 398) paid no attention to the greater body depth of the type specimens, considered color an ontogenetic character, and included this species as a synonym of *Ch. mororana.*

Length to 70 mm (Shmidt, 1931a).

Distribution: One specimen from Hinnam (Shestakov coast) in Chosonman Bay in the Sea of Japan.

6. **Gymnogobius mororanus** (Jordan and Snyder, 1901) (Figure 296)

Chaenogobius mororana Jordan and Snyder, Proc. U.S. Nat. Mus., 24, 1901: 80, fig. 14 (Muroran, Hokkaido). Abe, Enc. Zool., 2, Fishes, 1958: 95, fig. 277.

Chaenogobius heptacanthus mororana, Tomiyama, Japan. J. Zool., 7, 1, 1936: 92.

Gymnogobius mororanus, Taranetz, Dokl. Akad. Nauk SSSR, 1934: 398, ftnt.

22172. Vladivostok. October 1, 1927. E.P. Rutenberg. 5 specimens.

40951. Shikotan Islands. August 24, 1947. E.P. Rutenberg. 2 specimens.

Body low, its depth 6.25 times in standard length, moderately compressed laterally. Head slightly flat. Eyes set high; interorbital space flat. Snout pointed. Mouth very large; upper jaw extends beyond vertical from posterior margin of eye by a distance equal to diameter of eye; lower jaw protrudes beyond upper. Tongue with deep notch in front. Gill rakers

5 + 19 on first gill arch. Anterior nostril in form of tubule. Body covered with small scales, 26 scales in transverse series; head and occiput without scales (Jordan and Snyder, 1901c).

Differs from the closely related species *Ch. nigrimembranis* (Wu and Wang) in absence of black coloration of fins and branchiostegal membranes.

Length to 70 mm (Tomiyama, 1936).

382   *Distribution*: In the Sea of Japan reported from the northeast coast of the Korean Peninsula (Mori, 1952: 143) and the Sea of Japan coast of Hokkaido (Ueno, 1971: 88). Along the Pacific coast from Mororan to Matsushima Bay and Tokyo (Jordan and Snyder, 1901c: 82).

7. *Gymnogobius heptacanthus* (Hilgendorf, 1878) (Figure 297)

Gobius heptacanthus Hilgendorf, Sitzber. Ges. Naturf. Freunde, Berlin, 1878: 110 (Tokyo).

Aboma heptacanthus, Jordan and Snyder, Proc. U.S. Nat. Mus., 24, 1901: 70.

Chloea sarchynnis Jordan and Snyder, Proc. U.S. Nat. Mus., 24, 1901: 82, fig. 15 (Wakanoura).

Gymnogobius sarchynnis, Taranetz, Dokl. Akad. Nauk SSSR, 1934: 398.

383   *Chaenogobius heptacanthus heptacanthus,* Tomiyama, Japan. J. Ichthyol., 7, 1, 1936: 91.

Chaenogobius heptacanthus, Matsubara, Fish Morphol. and Hierar., 1955: 839. Abe, Enc. Zool., 2, Fishes, 1958: 95, fig. 278.

22173. Vladivostok. October 1, 1927. E.P. Rutenberg. 1 specimen.

25524. Tafuin, Peter the Great Bay. July 14, 1924. G.U. Lindberg. More than 6 specimens.

25457. Gaidamak, Peter the Great Bay. July 16, 1924. G.U. Lindberg. 3 specimens.

29618. Peter the Great Bay. May 18, 1914. A.I. Cherskii. 2 specimens.

35624. Yellow Sea. Chefoo. June 6, 1956. Academy of Sciences, China. 2 specimens.

36408. Yellow Sea, Tsingtao. June 24, 1957. E.F. Gur'yanova. 2 specimens.

D VIII, 13; A 13; P 20; *squ.* 70/20. Head length 3.6 and depth 5.5 times in standard length. Depth of caudal peduncle 3, eyes 4, snout 3.3, and maxilla 1.8 times in head length.

Body rather elongate, slightly compressed laterally. Head long and pointed. Eyes set along sides of head. Interorbital space almost equal to diameter of eye. Snout slightly longer than diameter of eye. Mouth large and oblique; lower jaw protrudes slightly beyond upper. Maxilla very long, extends beyond posterior margin of eye, much farther from corner of mouth, with posterior third free. Teeth simple, very small, thin, arranged

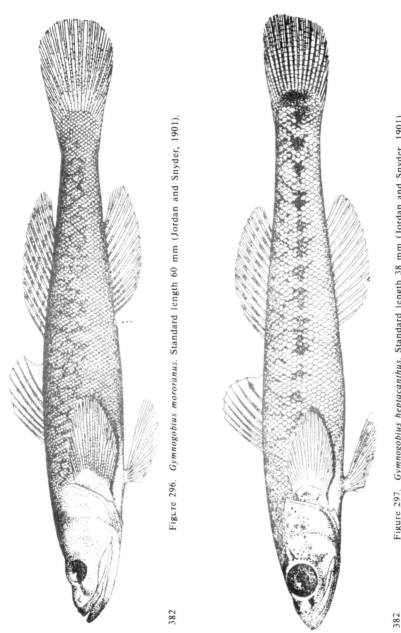

Figure 296. *Gymnogobius mororanus.* Standard length 60 mm (Jordan and Snyder, 1901).

382

Figure 297. *Gymnogobius heptacanthus.* Standard length 38 mm (Jordan and Snyder, 1901).

382

in form of narrow bands on both jaws. Tongue with very deep notch. Vomer with prominent lobes that protrude downward, each facing anterior lobes of tongue. Gill openings continue forward over moderate length; width of isthmus about equal to snout length. Papillae not present along inner margin of pectoral girdle. Gill rakers on first arch 6 + 14, long, and slender. Head without barbels, naked. Body covered with minute, weakly ctenoid scales, except for breast and occiput. Dorsal fins well separated; rays of first dorsal fin short and thin; rays of second dorsal and anal fins longer than in first dorsal fin, but when folded terminate far from base of caudal fin. Anal fin located under base of second dorsal fin; when folded, extends backward slightly farther than second dorsal fin. Caudal fin bluntly rounded or almost truncate. Pectoral fins pointed; upper rays not separated from fin. Pelvic fins not attached posteriorly. Sides of body with row of 15 or more minute dark spots, some of which fuse. Upper part of body with weakly expressed reticulate pattern. Dark stripe continues in front from eye. Snout dark. Spinous dorsal fin with small dark spot posteriorly. Soft dorsal fin with minute dark spots arranged in 2 horizontal rows; anal fin with traces of dark color; lower half of caudal fin dark. Pectoral and pelvic fins without dark color (Jordan and Snyder, 1901c).

Length to 65 mm (Tomiyama, 1936).

Distribution: In the Sea of Japan known from Peter the Great Bay (rivers) (Taranetz, 1937b: 150); near Wonsan and Pusan (Mori, 1952: 143); Sea of Japan coast of Hokkaido (Ueno, 1971: 88); Sado Island (Honma, 1952: 225); Niigata (Jordan, Tanaka and Snyder, 1913: 352); Toyama Bay (Katayama, 1940: 23); Wakasa Bay (Takegawa and Morino, 1970: 383); San'in region (Mori, 1956a: 25); north coast of Kyushu (Tabeta and Tsukahara, 1967: 299). In the Pacific Ocean from the central part of Honshu southward (Matsubara, 1955: 839). Yellow Sea (Wang and Wang, 1935: 186); Gulf of Chihli (Bohai) (Zhang et al., 1955: 211).

384   ## 7. Genus *Chloea* Jordan and Snyder, 1901

Chloea Jordan and Snyder, Proc. U.S. Nat. Mus., 24, 1901: 78 (type: *Gobius castaneus* O'Shaughnessy) Berg, Ryby Presnykh Vod..., 1949: 1071.

Body covered with moderate-sized or minute scales arranged in 53-69 transverse rows. Sides of head naked. Corners of mouth extend only up to vertical from anterior margin of eye. Tongue notched anteriorly. First dorsal fin with 6-8 unbranched rays, second dorsal with 10-12, and anal fin with 10-12. Swim bladder present. Supraorbital sensory canal consists only of part located above posterior margin of eye; anterior and posterior parts replaced by papillae. Curved series of papillae located under eye;

cheeks without transverse series of papillae, but with several longitudinal rows (Figure 269) (Berg, 1949).

Sea of Japan, Yellow Sea, and southern part of the Sea of Okhotsk. Japan.

1. *Chloea castanea* (O'Shaughnessy, 1875) (Figure 298)

Gobius castaneus O'Shaughnessy, Ann. Mag. Nat. Hist., ser. 4, 15, 1875: 145 (Nagasaki).

Chloea castanea, Jordan and Snyder, Proc. U.S. Nat. Mus., 24, 1901: 79.

Chloea nakamurae Jordan and Richardson, Proc. U.S. Nat. Mus., 33, 1907: 265, fig. 3 (Japan, Echigo Province, Niigata Prefecture).

Chaenogobius macrognathus (non Bleeker) Berg, Ezegodn. Zool. Muzeya Rossiisk Akad. Nauk, 19, 1914: 560 (in part).

Chloea senbae Tanaka, Zool. Mag., 28, 1917: 228 (see: Jordan and Hubbs, 1925: 307).

Chloea laevis (non Steindachner, 1879) Shmidt, Tr. Tikhookeansk. Kom. Akad. Nauk SSSR, 1931: 132, fig. 22.

Chloea castanea, Taranetz, Dokl. Akad. Nauk SSSR, 1933: 2, 1934: 398; Tr. Zool. Inst. Akad. Nauk SSSR, 4, 1936: 517, figs. 9, 10. Berg, Ryby Presnykh Vod..., 1949: 1072, fig. 801.

1629. Hakodate. 1863. Maksimovich. More than 6 specimens.

1630. Hakodate. 1863. Maksimovich. More than 6 specimens.

23174. Aniva Bay, Busse Inlet. August 21, 1961. P.Yu. Shmidt. More than 6 specimens.

D VI-VIII, 10–12; A 10–12; *squ.* 53–69; vertebrae 35–37. Head length 28 to 30% and body depth 18.5 to 20.6% in standard length. Sides and back with minute brown irregular shaped spots; spots on sides tend to form longitudinal row (Taranetz, 1934).

Mode of life described by Dotu (1954: 133).

Length to 50 mm (Taranetz, 1934).

Distribution: In the Sea of Japan known from Peter the Great Bay (Taranetz, 1934: 398), Wonsan, Ulchgln (Mori, 1952: 15); southwest Sakhalin, west coast of Hokkaido (Ueno, 1971: 88); Hakodate, Aomori, Akita, Fukui (Jordan and Hubbs, 1925: 307); Niigata and Tsuruga (Jordan and Snyder, 1901c: 79); Sado Island (Honma, 1952: 225); Toyama Bay (Katayama, 1940: 23); Wakasa Bay (Takegawa and Morino, 1970: 383); San'in region (Yanai, 1950: 22); and Tsushima Islands (Arai and Abe, 1970: 93). In the Yellow Sea reported from Mokp'o (Mori, 1952: 15). In the Sea of Okhotsk found near Notoro on Hokkaido (Hikita, 1952: 15).

8. Genus *Chasmichthys* Jordan, 1901

Chasmichthys Jordan, Amer. Naturalist., 35, 1901: 941 (type:

494

Saccostoma gulosus Guichenot, 1882). Fowler, Synopsis..., 14, 1-2, 1961: 49.

Chasmias Jordan and Snyder, Proc. U.S. Nat. Mus., 24, 1901: 761 (type: *C. misakius* Jordan and Snyder) (preocc.).

386 Body fairly slender, moderately elongate. Broad flat head; eyes wideset. Mouth large, horizontal; lower jaw shorter than upper. Upper jaw extends beyond eyes. Broad tongue slightly notched anteriorly.[14] Teeth on jaws arranged in bands. Barbels absent. Gill openings restricted to body sides. Pectoral girdle without fleshy processes. Body covered with very small cycloid scales arranged in 58 to 90 transverse rows. Dorsal fins short; first dorsal fin with 6 rays, second dorsal with about 11, and anal fin with 10 to 11. Caudal fin rounded. In pectoral fin upper rays with free tips. Pelvic fins fused, short and broad (Fowler, 1961).

Coasts of Japan. Two species. Both species known from the Sea of Japan.

Key to Species of Genus Chasmichthys

1 (2). Transverse rows of scales 70 to 80. Pectoral, dorsal, and caudal fins with pattern of dark stripes. Body without numerous small white spots. Dark spot size of eye occurs on base of caudal fin.......................... 1. **C. dolichognathus** (Hilgendorf).

2 (1). Transverse rows of scales 85 to 90. Only dorsal fins with pattern of dark stripes. Body with numerous very small white spots. Dark spot not present on base of caudal fin.........................
.... 2. **C. gulosus** (Guichenot).

1. **Chasmichthys dolichognathus** (Hilgendorf, 1878) (Figure 299)

Gobius dolichognathus Hilgendorf, Sitzber. Ges. Naturf. Freunde, Berlin, 1878: 108 (Tokyo).

Chasmias dolichognathus, Jordan and Snyder, Proc. U.S. Nat. Mus., 24, 1901: 84, fig. 16.

Chasmichthys dolichognathus dolichognathus, Tomiyama, Japan. J. Zool., 7, 1, 1936: 93. Abe, Enc. Zool., 2, Fishes, 1958: 94, fig. 274.

Chasmichthys dolichognathus, Matsubara, Fish Morphol. and Hierar., 1955: 839, fig. 326. Fowler, Synopsis..., 14, 1-2, 1961: 49, fig. 32.

22819. Misaki. April 11, 1901. P.Yu. Shmidt. More than 6 specimens.

22987. Tsuruga. August 26, 1917. V. Rozhkovskii. 1 specimen.

D VI, 11; A 10; *squ.* 70-80 (Abe, 1958).

Differences from the closely related species *C. gulosus* or even subspecies given in the key.

Length to 70 mm (Abe, 1958).

[14]Fowler indicated rounded, but Jordan and Snyder (1901b: 769) described the tongue as very broad, slightly notched. In our specimens the tongue is slightly notched.

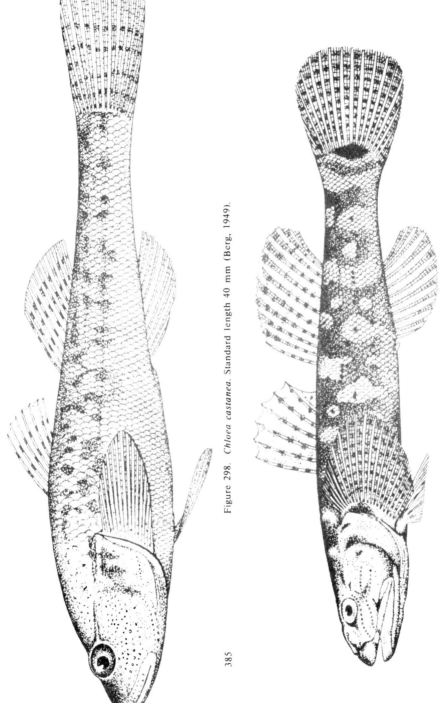

Figure 298. *Chloea castanea.* Standard length 40 mm (Berg, 1949).

385

Figure 299. *Chasmichthys dolichognathus.* Standard length 50 mm (Matsubara, 1955).

385

Distribution: In the Sea of Japan known from Wonsan, Pusan (Mori 1952: 144); Hakodate (Jordan and Snyder, 1901c: 84); Sado Island (Honma, 1952: 225); Toyama Bay (Katayama, 1940: 23); Tsuruga (Shmidt and Lindberg, 1930: 1149); San'in region (Mori, 1956a: 25); and Tsushima Islands (Arai and Abe, 1970: 93). Along the Pacific coast of Japan from Hokkaido to Nagasaki, to Tanega Island (Matsubara, 1955: 839).

2. *Chasmichthys gulosus* (Guichenot, 1882) (Figure 300)

Saccostoma gulosum Guichenot. In: Sauvage, Bull. Soc. Philomath., Paris, 7, 6, 1882: 171 (Japan).

Chasmias misakius Jordan and Snyder, Proc. U.S. Nat. Mus., 23, 1901: 761, pl. 36 (Misaki).

Chasmichthys dolichognathus gulosus, Tomiyama, Japan. J. Zool., 7, 1, 1936: 93. Abe, Enc. Zool., 2, Fishes, 1958: 93, fig. 273.

Chasmichthys gulosus, Jordan and Starks, Proc. U.S. Nat. Mus., 28, 1905: 208. Wang and Wang, Contrib. Biol. Lab. Sci. Soc. China, 11, 6, 387 1935: 189, fig. 18. Matsubara, Fish. Morphol. and Hierar., 1955: 839. Fowler, Synopsis..., 14, 1-2, 1961: 81, fig. 33.

22819. Misaki. April 11, 1901. P.Yu. Shmidt. 6 specimens.

D VI, 11; A 10; *squ.* 85-90 (Abe, 1958).

Characters given in key to species.

Length to 130 mm (Abe, 1958).

Distribution: In the Sea of Japan reported from Wonsan, Pusan (Mori, 1952: 144); Hakodate (Jordan and Snyder, 1901b); Sado Island (Honma, 1952: 225); Toyama Bay (Katayama, 1940: 23); San'in region (Mori, 1956a: 25; Yamaguti Prefecture (Tomiyama, 1936: 93); Kyushu coast 388 (Tabeta and Tsukahara, 1967: 299); and Tsushima Islands (Arai and Abe, 1970: 93). In the Yellow Sea—Gulf of Chihli (Bohai) (Zhang et al., 1955: 208); Cheju-do Island (Uchida and Yabe, 1939: 13); and Tsingtao (Wang and Wang, 1935: 189). Along the Pacific coast of Japan from Tohoku region everywhere southward (Matsubara, 1955: 839).

9. Genus *Parachaeturichthys* Bleeker, 1875

Parachaeturichthys Bleeker, Arch. Néerl. Sci. Nat., 9, 1874: 325 (type: *Chaeturichthys polynema* Bleeker). Fowler, Synopsis..., 14, 3-4, 1961: 223.

Body moderately elongate, slightly compressed laterally. Profile of head round, not flat. Snout longer than eye.[15] Eyes close-set, located in anterior part of head. Mouth slightly oblique, moderate in size; lower jaw protrudes slightly. Tongue blunt. Teeth simple, arranged in bands

---

[15]In the figure given by Zhu (see Figure 301) the snout length is equal to the eye length and the pores on the head are not close-set.

on both jaws; enlarged in outer series but true canines not present. One open pore on each side of snout, one pore in middle of interorbital space, one on each side behind eyes, and several along upper and posterior margin of preopercle. Many small barbels present on chin along lower jaw. Width of interorbital space equal to half diameter of eye. Nostrils without tubules. Gill openings do not continue forward ventrally. Branchiostegal rays 4. Isthmus broad. Body covered with rather large ctenoid scales arranged in 29 to 32 transverse rows. Scales do not extend onto interorbital space anterior to posterior part of snout and on cheeks[16]; scales on head and on anterior part of body cycloid, posteriorly ctenoid. Dorsal fins separate; 6 rays in first dorsal, 11 to 13 in second dorsal, and 9 to 11 in anal fin. Pointed caudal fin longer than head; black spot in upper part of fin base. Free rays not present in upper part of pectoral fins (Fowler, 1961).

One species in the Indian Ocean and western Pacific Ocean; also known from the Sea of Japan.

1. **Parachaeturichthys polynema** (Bleeker, 1853) (Figure 301)

Chaeturichthys polynema Bleeker, Verh. Batavia Genoots., 25, 1853: 44, fig. 4 (Nagasaki).

Parachaeturichthys polynemus, Jordan and Snyder, Proc. U.S. Nat. Mus., 24, 1901: 103.

Parachaeturichthys polynema Herre, Gobies..., 1927. Tomiyama, Japan. J. Zool., 7, 1, 1936: 94. Koumans, Fish Indo-Austr. Arch., 10, 1953: 37, fig. 8. Matsubara, Fish Morphol. and Hierar., 1955: 840. Abe, Enc. Zool., 2, Fishes, 1958: 92, fig. 272. Zhu et al., Ryby Vostochno-Kitaiskogo Morya, 1963: 426, fig. 322.

D VI, 10-11; A 10; *squ.* 27-30 (Abe, 1958).

Characters given in description of genus.

Length to 100 mm (Abe, 1958).

Distribution: In the Sea of Japan reported from Tsuruga (Jordan and Snyder, 1901c: 104). Not reported from the Yellow Sea. From the coast of southern Japan south to Indonesia and further west to East Africa (Matsubara, 1955: 840).

10. [Genus *Lophiogobius* Günther, 1873]

Lophiogobius Günther, Ann. Mag. Nat. Hist., ser. 4, 12, 1873: 241 (type: *L. ocellicauda* Günther). Fowler, Synopsis..., 14, 3-4, 1961: 226.

Ranulina Jordan and Starks, Proc. U.S. Nat. Mus., 31, 1906: 522 (type: *R. fimbriidens* Jordan and Starks).

Body rather elongate, almost cylindrical, with a very slender caudal

---

[16]Scales shown on cheeks in Figure 301.

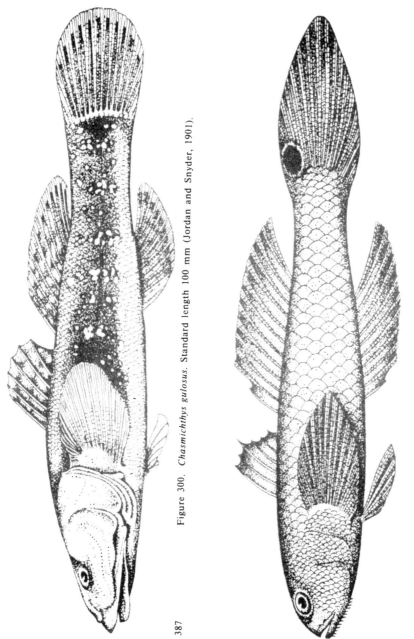

Figure 300. *Chasmichthys gulosus*. Standard length 100 mm (Jordan and Snyder, 1901).

387

Figure 301. *Parachaeturichthys polynema*. Standard length 95 mm (Zhu et al., 1962).

387

peduncle. Head large, flat. Snout length equal to 3 times diameter of eye; eyes set in anterior half of head. Mouth broad, oblique, extends beyond eye; lower jaw protrudes slightly. Teeth in outer series rather large, slanted, rarely deep-set, and jut out entirely from under lips; inner second row comprises smaller pointed teeth with cusps directed inward. Large broad tongue with almost truncate tip. Palatines smooth. Sides of head covered with papillae and lower side with very large number of barbels from chin to posterior margin of preopercle. Width of interorbital space 3 times greater than diameter of eye. Gill openings broad. Isthmus narrow. Branchiostegal rays 5. Transverse rows of scales 36 to 40. Head behind eyes, preopercle, and opercle covered with ctenoid scales. Inner margin of pectoral girdle without fleshy processes. Dorsal fins well separated; first dorsal with 7 and second dorsal with 15 to 17 rays. Caudal fin moderate in size, slightly pointed. Large pectoral fins almost equal to head length, their bases without scales, and upper rays without free tips. Pelvic fins rather large, not attached to belly posteriorly (Fowler, 1961).

East China and yellow seas. One species.

1. [*Lophiogobius ocellicauda* Günther, 1873] (Figure 302)

Lophiogobius ocellicauda Günther, Ann. Mag. Nat. Hist., (4), 12, 1873: 241 (Shanghai). Herre, Gobies..., 1927: 272, pl. 22, fig. 1. Tomiyama, Japan. J. Zool., 7, 1, 1936: 94. Matsubara. Fish Morphol. and Hierar., 1955: 840. Fowler, Synopsis..., 14, 3-4, 1961: 227, fig. 62. Zhu et al., Ryby Vostochno-Kitaiskogo Morya, 1963: 427, fig. 323.

Runulina fimbriidens Jordan and Starks, Proc. U.S. Nat. Mus., 31, 1906; 523, fig. 3 (Port Arthur). Jordan and Metz, Mem Carnegie Mus., 6, 1, 1913: 58, fig. 57.

D VII, 16-17; A 17-18; *squ.* 38-40 (Fowler, 1961).

Characters of species given in description of genus.

Length to 115 mm (Zhu et al., 1963).

Distribution: Absent in the Sea of Japan. In the Yellow Sea known from the Gulf of Chihli (Bohai) (Zhang et al., 1955: 212); Port Arthur (Fowler, 1961: 228); and Namp'o (Mori, 1952: 148). In the East China Sea off the coasts of Tszyansi Province (Zhu et al., 1963: 427).

11. Genus *Synechogobius* Gill, 1862

Synechogobius Gill, Ann. Lyceum Nat. Hist., New York, 7, 1862: 146 (type: *Gobius hasta* Temminck and Schlegel). Fowler, Synopsis..., 14, 3-4, 1961: 214.

Actinogobius Bleeker, Arch. Néerl. Sci. Nat., 9, 1874: 319 (type: *Gobius ommaturus* Richardson).

Body moderately elongate, compressed laterally. Head cylindrical.

Snout twice diameter of eye. Eyes set in anterior half of head. Mouth almost horizontal; jaws equal in size. One very short barbel[17] on each side of symphysis of lower jaw. Teeth on both jaws in 2-3 rows; canines absent. Tongue rounded. Mucous canal continues along lower jaw up to posterior margin of preopercle. Nostrils without tubules. Gill openings slightly broader than base of pectoral fin. Isthmus broad. Inner margin of pectoral girdle without fleshy processes. Body anteriorly covered with cycloid and posteriorly with ctenoid scales. Transverse rows of scales 70 to 90. Thorax and belly covered with scales. Scales also present on head behind eyes. Dorsal fins separate; first dorsal with 8 to 9 rays, second dorsal with 18 to 20, and anal fin with 15 to 17. Caudal fin long, pointed.

390 Pectoral fins without free rays on upper side, bases covered with scales. Pelvic fins fused, rather long (Fowler, 1961).

East China and South China seas and waters of Japan. One species also reported from the Sea of Japan.

1. *Synechogobius hasta* (Temminck and Schlegel, 1845) (Figure 303)

Gobius hasta Temminck and Schlegel, Fauna Japonica, Poiss., 1845: 144, pl. 75, fig. 1 (Nagasaki).

Synechogobius hasta, Jordan and Snyder, Proc. U.S. Nat. Mus., 24, 1901c: 102. Matsubara, Fish Morphol. and Hierar., 1955: 836. Fowler, Synopsis..., 14, 3-4, 1961: 215 (synonyms and description).

391 *Acanthogobius hasta*, Tomiyama, Japan. J. Zool., 7, 1 1936: 85. Abe, Enc. Zool., 2, Fishes, 1958: 98, fig. 286.

Gobius ommaturus Richardson, Voy. Sulphur, Fishes, 1845: 146, pl. 55, figs. 1-4 (mouth of Yangtze Changjiang River).

Acanthogobius ommaturus, Herre, Gobies..., 1927: 266.

Actinogobius ommaturus, Koumans, Fish Indo-Austr. Arch., 10, 1953: 27.

35576. Yellow Sea, Sinjung. May, 1956. Academy of Sciences, China. 1 specimen.

36404. Yellow Sea, Changjiang. June 12, 1957. E.F. Gur'yanova. 1 specimen.

36405. Yellow Sea, Tsingtao. June 24, 1957. E.F. Gur'yanova. 3 specimens.

D VIII-IX, 19-21; A 16-18; *squ.* 80-90 (Abe, 1958).

Characters given in description of genus.

Length to 242 mm (Herre, 1827).

Distribution: In the Sea of Japan known from Pusan. Yellow Sea—Gulf of Chihli (Bohai) (Zhang et al., 1955: 207); Namp'o (Mori, 1952: 145); Inch'on (Mori and Uchida, 1934: 20); Chefoo (Wang and Wang, 1935: 193); East China and South China seas (Zhu et al., 1963: 431-432).

[17]Barbels lobate.

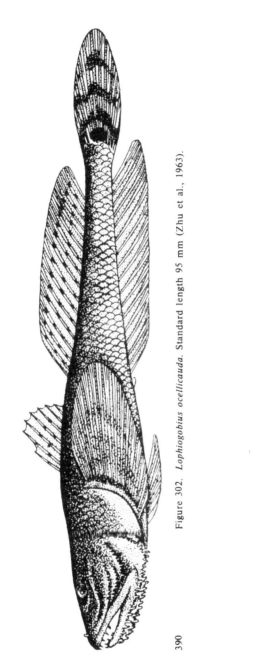

Figure 302. *Lophiogobius ocellicauda.* Standard length 95 mm (Zhu et al., 1963).

390

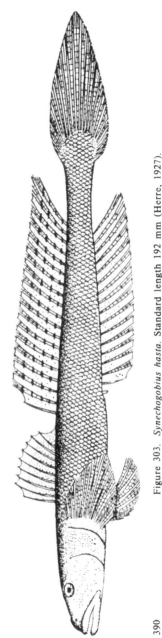

Figure 303. *Synechogobius hasta.* Standard length 192 mm (Herre, 1927).

390

502

12. Genus *Chaeturichthys* Richardson, 1844

Chaeturichthys Richardson, Voy. Sulphur, Fishes, 1844: 54 (type: *C. stigmatias* Richardson). Fowler, Synopsis..., 14, 3-4, 1961: 219.

Body moderately long. Head broad and rounded in profile. Eyes close-set. Mouth moderate in size, slightly oblique. Teeth sharp, arranged in 2 rows on each jaw, larger in outer series; teeth close-set, immovable, curved, and directed obliquely inward. Tongue blunt. Gill openings continue forward ventrally. Isthmus narrow. Body covered with mo-derate-sized, cycloid, easily shed scales. Cheeks covered with scales. Lower jaw with 3 small barbels on each side. Dorsal fins long; first dorsal with 8 rays, second dorsal with 14 to 25, and anal fin with 18 to 21. Caudal fin more or less pointed,[18] its upper and lower rays highly reduced. Pectoral fins without free rays (Fowler, 1961).

Several species in the seas of China and Japan. Three species reported from the Sea of Japan.

Key to Species of Genus *Chaeturichthys*[19]

1 (4). Second dorsal fin with 14 to 17 rays, anal fin with 13 to 14. Transverse rows of scales 35 to 40. Fleshy papillae absent along inner margin of pectoral girdle.

2 (3). Transverse rows of scales about 35. Rounded caudal fin with four to five dark stripes concentrically arranged. Marine .... 1. **C. sciistius** Jordan and Snyder.

3 (2). Transverse rows of scales usually slightly more than 40. Caudal fin elongate, without stripes of uniform dark or gray color. Marine and brackish waters .... 2. **C. hexanema** Bleeker.

392 4 (1). Second dorsal fin with 21 to 25 rays, anal fin with 18 to 20. Transverse rows of scales 47 to 50. Inner margin of pectoral girdle with 3 fleshy papillae (Figure 304) .... 3. **C. stigmatias** Richardson.

1. *Chaeturichthys sciistius* Jordan and Snyder, 1901 (Figure 305)

Chaeturichthys sciistius Jordan and Snyder, Proc. U.S. Nat. Mus., 24, 1901: 107, fig. 22 (Hakodate). Tomiyama, Japan. J. Zool., 7, 1, 1936: 94. Matsubara, Fish Morphol. and Hierar., 1955: 840. Abe, Enc. Zool., 2, Fishes, 1958: 93, fig. 271. Fowler, Synopsis..., 14, 3-4, 1961: 220.

Suruga fundicula Jordan and Snyder, Proc. U.S. Nat. Mus., 24, 1901: 96, fig. 20 (Sagami Bay). Ouchi and Ogata, Rep. Japan. Sea Reg. Fish. Res. Lab., 6, 1960: 183.

---

[18]Caudal fin of *Ch. sciistius* rounded.
[19]From Matsubara, 1955: 840.

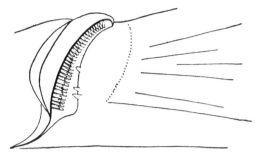

391   Figure 304. Fleshy papillae on inner margin of pectoral girdle of
Chaeturichthys stigmatias.

22829. Misaki, April 2, 1901. P.Yu. Shmidt. 1 specimen.

D VIII, 14-15; A 12-13; P 20-23; *squ.* 31-36 (Jordan and Snyder, 1901c).

Abe has reported 40 to 45 transverse rows of scales (*squ.*) for this species; therefore, the main differences from *C. hexanema* are: presence of 4-5 concentrically arranged stripes on caudal fin and dark spot with diameter more than eye in posterior part of first dorsal fin.

Mode of life described by Japanese researchers (Okada and Suzuki, 1955: 112).

Length to 75 mm (Abe, 1958).

Distribution: In the Sea of Japan known from Pohang (Mori, 1952: 146); Hakodate, Aomori, Tsuruga (Jordan and Snyder, 1901c: 108); Sado Island (Honma, 1963: 22); Toyama Bay (Katayama, 1940: 23). Along the Pacific coast of Japan up to Kyushu (Matsubara, 1955: 840).

2. *Chaeturichthys hexanema* Bleeker, 1853 (Figure 306)

Chaeturichthys hexanema Bleeker, Verh. Batavia Genootsch, 25, 1953: 43, fig. 5 (Nagasaki). Tomiyama, Japan. J. Zool., 7, 1, 1936: 94. Matsubara, Fish Morphol. and Hierar., 1955: 840. Abe, Enc. Zool., 2, Fishes, 1958: 92, fig. 270. Fowler, Synopsis..., 14, 3-4, 1961: 221, fig. 60. Zhu et al., Ryby Vostochno-Kitaiskogo Morya, 1963: 129, fig. 325.

Chaeturichthys hexanemus Jordan and Snyder, Proc. U.S. Nat. Mus., 24, 1901: 106.

32986. Tsuruga. September 3-9, 1917. V. Rozhkovskii. 3 specimens.

36421. Yellow Sea. June 7, 1957. AN SSSR. 4 specimens.

36422. Yellow Sea. June 18, 1957. AN SSSR. 1 specimen.

D VIII, 17; A 13-14; *squ.* 40-46 (Abe, 1958).

Differs from *Ch. sciistius* in more pointed and elongate caudal fin of dark color, absence of transverse concentric stripes on caudal fin, and light-colored first dorsal fin, without dark spot on base and with darkened upper margin.

504

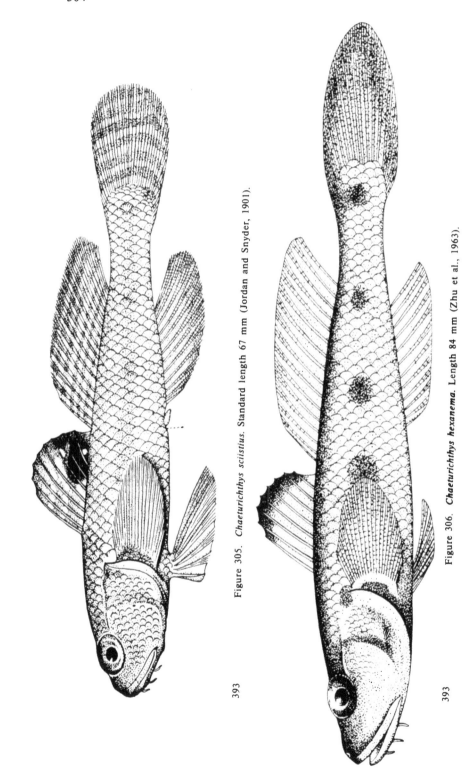

Figure 305. *Chaeturichthys sciistius*. Standard length 67 mm (Jordan and Snyder, 1901).

393

Figure 306. *Chaeturichthys hexanema*. Length 84 mm (Zhu et al., 1963).

393

Mode of life described by Japanese researchers (Dotu, Mito and Ueno, 1955: 359).

Length to 200 mm (Jordan and Snyder, 1901c).

Distribution: In the Sea of Japan known from Pusan (Mori, 1952: 146); Hakodate, Aomori (Jordan and Snyder, 1901c: 107); Sado Island (Honma, 1952: 225); Toyama Bay (Katayama, 1940: 23); Tsuruga (Shmidt and Lindberg, 1930: 1149). In the Yellow Sea—Gulf of Chihli (Bohai) (Zhang et al., 1955: 215). Everywhere in waters of Japan (Matsubara, 1955: 840). China (Zhu et al., 1963: 429).

3. *Chaeturichthys stigmatias* Richardson, 1844 (Figure 307)

Chaeturichthys stigmatias Richardson, Voy. Sulphur, Fishes, 1844: 55 (without indication of locality). Jordan and Snyder, Proc. U.S. Nat. Mus., 394 24, 1901: 105. Tomiyama, Japan. J. Zool., 7, 1, 1936: 95. Matsubara, Fish Morphol. and Hierar., 1955: 841. Fowler, Synopsis..., 14, 3-4, 1961: 222, fig. 61. Zhu et al., Ryby Vostochno-Kitaiskogo Morya, 1963: 428, fig. 324

36423. Yellow Sea, Tsingtao. June 8, 1957. Exped. ZIN AN SSR. 1 specimen.

36424. Yellow Sea, Tsingtao. June 28, 1957. Exped. ZIN AN SSR. 1 specimen.

36425. Yellow Sea, Tsingtao. May 25, 1957. Exped. ZIN AN SSR. 4 specimens.

D VIII, 21-22; A 19-20; P 21-23; *squ.* 48-50 (Zhu et al., 1963).

Differs from other species of this genus in larger number of rays in dorsal and anal fins, and presence of 3 fleshy papillae along inner margin of pectoral girdle (Figure 304).

Length to 210 mm (Zhu et al., 1963).

Distribution: In the Sea of Japan reported from Tsushima Islands (Tamodo, 1970: 199). In the Yellow Sea—Gulf of Chihli (Bohai) (Zhang et al., 1955: 214); Ionamp'o (Mori, 1952: 146); and Chefoo (Wang and Wang, 1935: 194).

13. Genus *Sagamia* Jordan and Snyder, 1901

Sagamia Jordan and Snyder, Proc. U.S. Nat. Mus., 24, 1901: 100 (type: *S. russula* Jordan and Snyder). Matsubara, Fish Morphol. and Hierar., 1955: 84.

Ainosus Jordan and Snyder, Proc. U.S. Nat. Mus., 24, 1901: 109 (type: *A. geneionema* Hilgendorf).

This genus is close to *Chaeturichthys,* differing from it in larger number of very small barbels (about 10 pairs on lower jaw versus 5 pairs), free rays in upper part of pectoral fin, blunt caudal fin, and narrower isthmus.

One species near coasts of Japan. Also known from the Sea of Japan.

1. *Sagamia geneionema* (Hilgendorf, 1879) (Figure 308)

Gobius geneionema Hilgendorf, Sitzber. Ges. Naturf. Freunde, Berlin, 1879: 108 (Tokyo).

Ainosus geneionemus, Jordan and Snyder, Proc. U.S. Nat. Mus., 24, 1901: 109.

Sagamia russula, Jordan and Snyder, Proc. U.S. Nat. Mus., 24, 1901: 100, fig. 21 (Misaki).

Sagamia geneionema, Tomiyama, Japan. J. Zool., 7, 1, 1936: 95. Matsubara, Fish Morphol. and Hierar., 1955: 841. Abe, Enc. Zool., 2, Fishes, 1958: 92, fig. 269.

Ainosus geneionema, Fowler, Synopsis..., 14, 3-4, 1961: 225.

6458. Tokyo. 1882. Snyder. Two specimens.

22826. Misaki. April 9, 1901. P.Yu. Shmidt. 6 specimens.

22830. Misaki. April 11, 1901. P.Yu. Shmidt. 3 specimens.

22831. Nagasaki. June 14, 1901. P.Yu. Shmidt. 1 specimen.

23110. Nagasaki. November 26, 1897. A. Bunge. 1 specimen.

D VIII, 16; A 14; P 20; *squ.* 62. Head length 3.5 and depth 4 times in standard length. Depth of caudal peduncle 3, snout 3.2, and upper jaw 3 times in head length. Diameter of eye equal to snout length. Mouth oblique, jaws ' equal; upper jaw does not protrude and extends to vertical from anterior margin of orbit. Teeth simple, arranged in 2 distinct rows on each jaw; teeth of outer row enlarged; posterior teeth on lower jaw resemble canines. Gill openings continue slightly forward. Gill rakers on first arch 2 + 9. Lower jaw and anterior part of throat with 24 thin barbels; longest barbels sometimes shorter than diameter of pupil. Four rows of minute pores located on cheek under eye. Head naked except for occiput and upper margin of operculum, which are covered with minute scales.

Fins large. Dorsal fins separate. Caudal fin truncate. Pectoral fins pointed, their upper rays with silk-like tips.

396   Middle of sides of body with 6-7 large dark spots; series of similar spots located below this row. Upper part of head and body with small elongate spots with blurred outline. Suborbital space and cheeks with 4 oblique dark spots. Dorsal fin with 3-4 dark stripes. Membrane between last two rays of first dorsal fin with black spot. Caudal fin with transverse stripes (Jordan and Snyder, 1901c: 109).

The barbels in our specimen No. 22826 were very small and difficult to discern; upper rays of pectoral fin with free tips. These rays are not shown in the figure given by Abe. In other respects our specimen fully conforms to the description of *Ainosus geneionemus* given by Jordan and Snyder (1901c: 109).

Length to 90 mm (Abe, 1958).

Distribution: In the Sea of Japan reported from Sado Island (Honma, 1963: 22); Toyama Bay (Katayama, 1940: 23); and Tsushima Islands (Arai

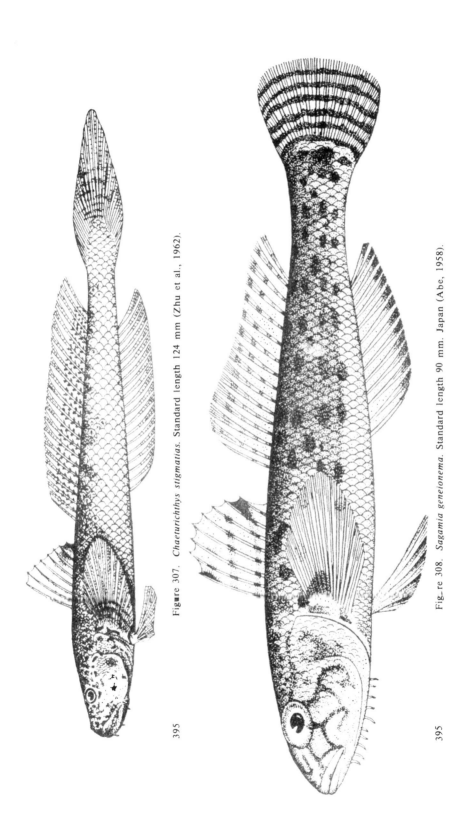

Figure 307. *Chaeturichthys stigmatias*. Standard length 124 mm (Zhu et al., 1962).

Figure 308. *Sagamia geneionema*. Standard length 90 mm. Japan (Abe, 1958).

395

395

508

and Abe, 1970: 93). In Japan from the central part of Honshu southward (Matsubara, 1955: 841). Cheju-do Island (Mori, 1952: 147).

3. Subfamily Luciogobiinae[20]

First dorsal fin absent or reduced; rays less than 5 (usually 3). Teeth on lower jaw arranged in several rows. Body naked or covered with minute scales embedded in skin. Head flat; cheeks inflated. Soft dorsal and anal fins generally moderate in length. Pelvic fins exceptionally absent in the genus *Expedio*.

Seven genera, mostly found along the shores of China and Japan. All seven reported from the Sea of Japan.

Key to Genera of Subfamily Luciogobiinae

1 (6). First dorsal fin present, with 3 unbranched rays.
2 (3). Body relatively deep, about 5 times in its length depth more or less equal to length of postorbital space. Origin of first dorsal fin at vertical with middle of pectoral fin. Scales well developed in posterior part of body, extending up to base of pectoral fin in form of narrow stripe......... 1. **Astrabe** Jordan and Snyder.
3 (2). Body low, its depth more than 9 times in its length; depth less than length of postorbital space. Origin of first dorsal fin distinctly behind pectoral fin. Body almost naked or scales present on caudal peduncle and above lateral line.
4 (5). Body depth about 10 times in length. Origin of second dorsal fin almost at vertical with origin of anal fin; bases of fins almost equal. Mouth large, slightly more than diameter of eye. ....
.... 2. **Clariger** Jordan and Snyder.
5 (4). Body depth about 14 times in length. Origin of second dorsal fin notably anterior to origin of anal fin; base of anal fin much shorter than base of second dorsal fin. Mouth very short, almost equal to diameter of eye.....................................
.... 3. **Eutaeniichthys** Jordan and Snyder.
6 (1). First dorsal fin absent.
397 7 (8). Isthmus narrow; gill openings continue forward ventrally. Posterior margin of maxilla not concealed in skin...........
.... 4. **Leucopsarion** Hilgendorf.
8 (7). Isthmus broad; gill openings do not continue forward ventrally. Posterior margin of maxilla concealed in skin.
9 (12). Pelvic fins present, moderate in size.
10 (11). Posterior part of body as well as anterior part without scales.
.... 5. **Luciogobius** Gill.

[20]Fowler, 1961: 228 (with additions).

11 (10). Posterior part of body covered with scales..... 6. **Inu** Snyder.
12 (9). Pelvic fins absent........................ 7. **Expedio** Snyder.

1. Genus *Astrabe* Jordan and Snyder, 1901

Astrabe Jordan and Snyder, Proc. U.S. Nat. Mus., 24, 1901: 119 (type: *A. lactisella* Jordan and Snyder).

Body robust, caudal peduncle deep. Teeth on both jaws simple; canines absent; teeth absent on vomer. Gill openings do not continue forward ventrally; isthmus broad. Two low fleshy lobes on inner margin of pectoral girdle. Head naked; skin distinctly wrinkled. Scales small, embedded in skin, cover posterior part of body and continue up to base of pectoral fin in a narrow strip. Dorsal fins separate; first dorsal fin with 3 rays, second dorsal with 11, and anal fin with 10; all three fins covered with thick skin. Upper rays of pectoral fins simple, with free tips. Pelvic fins I 5, all connected; fins not attached to belly posteriorly. Color dark, with distinct white spots (Jordan and Snyder, 1901c).

One species found off the coasts of Japan; also known from the Sea of Japan.

1. ***Astrabe lactisella*** Jordan and Snyder, 1901 (Figure 309)
Astrabe lactisella Jordan and Snyder, Proc. U.S. Nat. Mus., 24, 1901: 119, fig. 26 (Misaki).
D III, 11; A 10; P 24 (Jordan and Snyder, 1901c).
Characters given in description of genus and in key to genera.
Length to 55 mm (Tomiyama, 1936: 53).
Distribution: In the Sea of Japan known off Sado Island (Honma, 1963: 22) and in Toyama Bay (Katayama, 1940: 22). Pacific coast of Japan (Matsubara, 1955: 842).

2. Genus *Clariger* Jordan and Snyder, 1901

Clariger Jordan and Snyder, Proc. U.S. Nat Mus., 24, 1901: 120 (type: *C. cosmurus* Jordan and Snyder).

Externally similar to *Luciogobius*, but differs in presence of short spinous dorsal fin consisting of three thin unbranched rays. Body elongate, head broad and flat. Cycloid scales present in small number only on caudal peduncle (sometimes absent). Several small barbels present (sometimes) under eyes (Jordan and Snyder, 1901c).

Three species found off the coasts of Japan. One species reported from the waters of the Sea of Japan.

1. ***Clariger cosmurus*** Jordan and Snyder, 1901 (Figure 310)
Clariger cosmurus Jordan and Snyder, Proc. U.S. Nat. Mus., 24, 1901:

Figure 309. *Astrabe lactisella*. Standard length 28 mm (Jordan and Snyder, 1901).

398

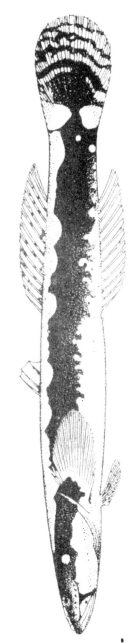

Figure 310. *Clariger cosmurus*. Standard length 37 mm (Jordan and Snyder, 1901).

398

121, fig. 27 (Misaki). Matsubara, Fish Morphol. and Hierar., 1955: 843, fig. 366 A, B.

398 Differs from other species in presence of traces of minute cycloid scales along axis of caudal peduncle which continue forward in form of single row up to origin of first dorsal fin. Typical color: back pale, sides with broad longitudinal stripe with wavy margins. Five small barbels under eye.

Length to 40 mm (Tomiyama, 1936: 54).

Distribution: In the Sea of Japan known from Hakodate (Snyder, 1912: 444)[21]; Sado Island (Honma, 1970: 72); Toyama Bay (Katayama, 1940: 22); and near Tsushima Islands (Arai and Abe, 1970: 93).

399 ### 3. Genus *Eutaeniichthys* Jordan and Snyder, 1901

Eutaeniichthys Jordan and Snyder, Proc. U.S. Nat. Mus., 24, 1901: 122 (type: *E. gilli* Jordan and Snyder).

Body elongate, compressed laterally, covered with rudimentary scales embedded in skin. Head short. Mouth small, slightly oblique; chin does not protrude forward. Teeth simple. Isthmus broad. Barbels absent. Dorsal fins wide-set; first dorsal with 3 rays, second dorsal with 17, and anal fin with 12. Base of second dorsal fin much longer than that of anal fin and originates much anterior to origin of anal fin. Pelvic fins well developed; caudal fin slightly pointed (Jordan and Snyder, 1901c).

One species found off the coasts of Japan; also known from the Sea of Japan.

1. *Eutaeniichthys gilli* Jordan and Snyder, 1901 (Figure 311)

Eutaeniichthys gilli Jordan and Snyder, Proc. U.S. Nat. Mus., 24, 1901: 122, fig. 28 (Tokyo). Tomiyama, Fig. and Descr., Fishes Japan, 58, 1958: 1206, pl. 234, fig. 591 (Sakata, Yamagata Prefecture, Sea of Japan).

D III, 18; A 11.

Body very long and thin; depth 11 times in body length; head 6.5 times in body length. Snout blunt, almost equal to diameter of eye; jaws equal. Eyes directed slightly upward. Interorbital space narrow. Mouth oblique, oral opening continues to vertical from anterior margin of pupil. Teeth simple, curved, arranged in 2-3 rows on each jaw; canines absent. Gill openings do not continue forward ventrally. Fleshy lobes absent on inner margin of pectoral girdle. Barbels absent on head. Head naked (Jordan and Snyder, 1901c).

Mode of life described by Dotu (1955a: 338).

Length to 50 mm (Tomiyama, 1936).

Distribution: In the Sea of Japan reported from brackish waters of

---

[21]Honma et al. (1972: 53) have reported this species for Yamaguti Prefecture in the Sea of Japan.

512

Yamagata, Toyama, and Yamaguti prefectures (Tomiyama, 1958c: 1206); Pusan (Mori, 1952: 145); and Tsushima Islands (Tomoda, 1970: 199). In the Yellow Sea—Masan (Mori, 1972: 145); Gulf of Chihli (Bohai) (Zhang et al., 1955: 222). Pacific coasts of Japan (Matsubara, 1955: 842).

4. Genus *Leucopsarion* Hilgendorf, 1880

Leucopsarion Hilgendorf, Monatsber. Akad. Wiss., Berlin, 1880: 340 (type: *L. petersi* Hilgendorf). Jordan and Snyder, Proc. U.S. Nat. Mus., 24, 1901: 125.

Body elongate, compressed laterally, without scales, transparent. Head short, depressed; cheeks slightly dilated. Eyes slightly convex. Mouth rather large, conical, oblique. Teeth simple. Barbels absent. Tongue without notch. Isthmus narrow. Gill openings continue forward ventrally. Spinous dorsal fin absent. Soft dorsal fin moderate in size, well separated from short caudal fin. Anal fin longer than dorsal fin and originates well ahead of it. Pectoral fins rather long. Pelvic fins very small, completely connected, and form scaly disk in which rays are difficult to discern. Shape of disk as in *Gobius,* but fins not so well developed (Jordan and Snyder, 1901c).

Small transparent fishes found in estuaries of Japan. One species. Also known from the Sea of Japan.

401 ### 1. *Leucopsarion petersi* Hilgendorf, 1880 (Figure 312)

Leucopsarion petersi Hilgendorf, Monatsber. Akad. Wiss., Berlin, 1880: 340 (southern Japan). Jordan and Snyder, Proc. U.S. Nat. Mus., 24, 1901: 125, fig. 31.

25798. Okama, Wakasa Bay. March, 1935. Kobayashi. 1 specimen.
D 13; A 17; P 15 (Jordan and Snyder, 1901c).
Description given in characters of genus and in key to genera.
Length to 55 mm (Tomiyama, 1936: 54).

Distribution: In the Sea of Japan known off Pusan (Mori, 1952: 147); Sado Island (Honma, 1963: 22); Toyama Bay (Mori, 1956a: 25); San'in region (Katoh et al., 1956: 323); north coast of Kyushu Island (Tabeta and Tsukahara, 1967: 299); and Tsushima Islands (Shibata, 1968: 27; Tomoda, 1979: 199). Near the Pacific coast of Japan from Aomori to Kagoshima (Matsubara, 1955: 843).

5. Genus *Luciogobius* Gill, 1859

Luciogobius Gill, Proc. Acad. Nat. Sci. Philad., 11, 1859: 146 (type: *L. guttatus* Gill). Fowler, Synopsis..., 14, 3-4, 1961: 230.

Body elongate, moderately compressed laterally, not covered with scales. Head long, low, flat; cheeks dilated. Eyes set in upper half of head.

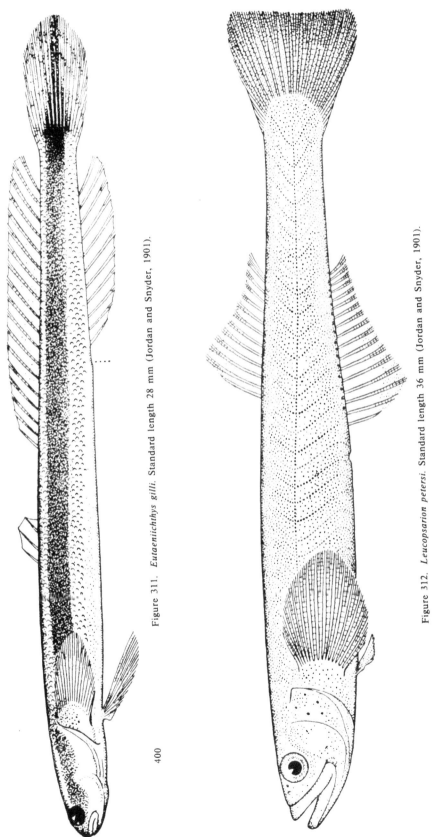

Figure 311. *Eutaeniichthys gilli.* Standard length 28 mm (Jordan and Snyder, 1901).

Figure 312. *Leucopsarion petersi.* Standard length 36 mm (Jordan and Snyder, 1901).

400

Mouth fairly large, conical; mouth slightly oblique. Outer row of teeth on lower jaw continues only up to half its length; teeth usually arranged in several rows. Teeth of outer row on upper jaw enlarged; canines absent. Barbels not present. Tongue with notch in front. Width of interorbital space and length of snout almost equal to diameter of eye. Gill openings do not continue forward and separated by broad isthmus. Inner margin of pectoral girdle without fleshy lobes. Spinous dorsal fin absent. Soft dorsal fin with 10 to 15 rays; base often rather short and located opposite anal fin; latter with 11 to 12 rays. Caudal fin short, rounded, not confluent with anal and dorsal fins. Pectoral fins fairly large, upper rays without filamentous tips. Pelvic fins very short, small, their rays difficult to discern; fused, fins form round disk. Color dark (Fowler, 1961).

The taxonomy of this genus has been poorly analyzed. We therefore provide a brief description of all the species known in Japan, using the latest information given by the Japanese ichthyologist Arai (1970).

Eight species. Small gobies from the silted coast of the Sea of Japan, the Yellow Sea, and the East China Sea, as well as from estuaries of rivers, caves, and wells. Six species known in Japan, two of which live in caves and wells. Four species known from the Sea of Japan.

Key to Species of Genus Luciogobius[22]

1 (8). Eyes normal and readily visible.
2 (5). Dorsal fin with 11 to 17 rays, anal with 12 to 17 rays. Pectoral fins with free rays.
3 (4). D 11–13; A 12–15; vertebrae 34–38. Pectoral fin with one free ray. . . . . 1. **L. guttatus** Gill.
4 (3). D 15–17; A 15–17; vertebrae 39–40. Pectoral fin with three to seven free rays. . . . . 2. **L. grandis** Arai.
5 (2). Dorsal fin with 7 to 10 rays, anal with 8 to 10. Pectoral fin without free rays.
6 (7). Vertebrae 42; D 7–9; A 8–10. Pelvic fins small, lobate. Body depth 10–12 times in standard length . . . .
. . . . 3. **L. elongatus** Regan.
7 (6). Vertebrae 32; D 8–10; A 9–10. Pelvic fins normal, rounded. Body depth 8.0 to 9.5 times in standard length. . . . .
. . . . 4. **L. saikaiensis** Dotu.
8 (1). Eyes reduced. Dorsal fin high, with 10 to 11 soft rays. Body without minute dark spots. Benthic or cavernicolous fishes.
9 (10). Eyes small, covered with skin. Body length 13 times its depth and 5 times length of head. Body pallid with distinct pigment as well as pale color. . . . . [**L. pallidus** Regan].

[22]From Dotu, 1957a: 70; Arai, 1970: 203 (with additions).

10 (9). Eyes rudimentary. Body length 9 times its depth and 3.5 times length of head. Pigment absent on body ... [**L. albus** Regan].

1. *Luciogobius guttatus* Gill, 1859 (Figure 313)

Luciogobius guttatus Gill, Proc. Acad. Nat. Sci. Philad., 11, 1859: 146 (Simoda, Idzu Province). Jordan and Snyder, Proc. U.S. Nat. Mus., 24, 1901: 123, fig. 29. Tomiyama, Japan. J. Zool., 7, 1, 1936: 51, fig. 10A. Soldatov and Lindberg, Obzor..., 1930: 431. Berg, Ryby Presnykh Vod..., 3, 1949: 1124. Matsubara, Fish Morphol. and Hierar., 1955: 844, fig. 308. Fowler, Synopsis..., 1961: 230, fig. 65. Arai, Bull. Nat. Sci. Mus., Tokyo, 13, 2, 1970: 203.

22835. Misaki. April 11, 1901. P.Yu. Schmidt. 6 specimens.

36410. Yellow Sea, Tsingtao. May 23, 1957. E.F. Gur'yanova. 1 specimen.

36411. Yellow Sea, Tiang-tsin. June 13, 1957. O.A. Skarlato. 1 specimen.

36412. Yellow Sea, Tsingtao. June 23, 1957. Exped. ZIN AN SSSR. 2 specimens.

39724. Kunashir Island, Golovnino. June 28, 1969. A.N. Golikov. 11 specimens.

40065. Peter the Great Bay, Sobol Inlet. August 24, 1928. N.S. Khranilov. 9 specimens.

40949. Shikotan Island. August 2, 1949. Poletika. 1 specimen.

D 11-13; A 12-15 (Arai, 1970).

Differs from other species of the genus, in addition to well-developed eyes, in presence of 1 free ray in pectoral fin. Specimens in our collection fully conform to these characters. Length to 95 mm (Tomiyama, 1936). Mode of life described by Dotu (1957b: 93).

Distribution: In the Sea of Japan known from Peter the Great Bay (Soldatov and Lindberg, 1930: 431); Pusan (Mori, 1952: 142); Otaru (Tomiyama, 1936: 51); Hakodate (Jordan and Snyder, 1901c: 124); Sado Island (Honma, 1963: 22): Toyama Bay (Mori, 1956a: 42); and Tsushima Islands (Arai and Abe, 1970: 94). In the Yellow Sea—from Masan and Cheju-do Island (Mori, 1952: 142). Along the Pacific coast of Japan from Hakodate to Nagasaki (Shmidt, 1931b: 137).

2. *Luciogobius grandis* Arai, 1970 (Figure 314)

Luciogobius grandis Arai, Bull. Nat. Sci. Mus., Tokyo, 13, 2, 1970: 199, figs. 1, 2 (Tsushima Islands).

D 17 (15-17); A 16 (15-17); P (15-17); V I, 5; vertebrae (without hypurals) 40 (39-40) (Arai, 1970).

Differs from other species of the genus in: 1) larger number of vertebrae (39–40); 2) presence of 3 to 7 rays with free tips in pectoral fins; 3) larger number of rays in dorsal fin (15–17) and in anal fin (15–17); 4) smaller size of orbit (2.3 to 3.6 times in interorbital space); 5) nature

404 of arrangement of pores on head (Figure 315); and 6) larger body dimensions.

Length to 93.6 mm (Arai, 1970).

Distribution: In the Sea of Japan reported from Fukui (Fukui Prefecture), Ullyudo Island (Dajelet) in the southern part of the sea, and near Tsushima Islands (Arai). Along the Pacific coast of Japan from Izu Province.

3. *Luciogobius elongatus* Regan, 1905

Luciogobius elongatus Regan, Ann. Mag. Nat. Hist., (7) 15, 1905: 23 (Inland Sea of Japan). Jordan and Hubbs, Mem. Carnegie Mus., 10, 2, 1925: 309 (Noo). Arai, Bull. Nat. Sci. Mus., Tokyo, 13, 2, 1970: 203.

443 *Addendum*: Under the synonymy of this species *L. guttatus guttatus* Tomiyama should also be included (*Japan. J. Zool.*, 7, 1, 1936: 52, fig. 10, B).

404 D 7–9; A 8–10; vertebrae 42 (Arai, 1970).

Differs from *L. guttatus* in lower body, depth 10 to 12 times or even more in body length (versus 6.5 to 9.0 times); smaller head length, 7 to 8 times in body length (versus 4 to 5); and smaller number of rays in the dorsal fin (7 to 9 versus 11 to 13) and in the anal fin (8 to 10 versus 12 to 15).

Standard length to 42 mm (Arai, 1970).

Distribution: In the Sea of Japan found in Primor'e (Soviet Gavan) (Popov, 1933a: 148). In Japan known from the Inland Sea (Regan, 1905: 23).

4. *Luciogobius saikaiensis* Dotu, 1957 (Figure 316)

Luciogobius saikaiensis Dotu, J. Fac. Agric. Kyushu Univ., 11, 1, 1957: 69, fig. 1 (Amakusa Islands off Kyushu). Arai, Bull. Nat. Sci. Mus., Tokyo, 13, 2, 1970: 203.

D 8–10; A 9–10; P 18; vertebrae 32 (Arai, 1970).

Body cylindrical anteriorly and posteriorly moderately compressed laterally. Head broader than body, flat, and muscles on sides and top bulge so much that depression formed behind occiput. Interorbital space broad, slightly concave, with narrow transverse fleshy crest. Upper jaw continues posteriorly to vertical with posterior margin of eye. Mouth almost vertical anteriorly, but horizontal posteriorly. Teeth very small, arranged in narrow bands on both jaws. Five fleshy barbels under eye, and one larger barbel in front of eye; pair of barbels on snout. Head and body entirely naked. Membranes of dorsal and anal fins fleshy. Color in alcohol

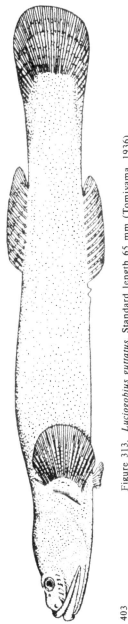

Figure 313. *Luciogobius guttatus*. Standard length 65 mm (Tomiyama, 1936).

403

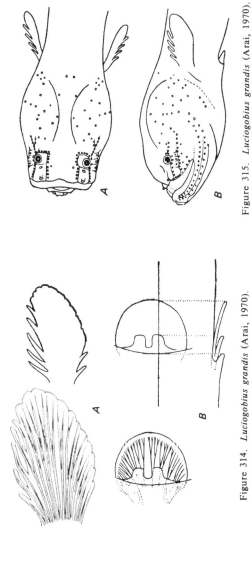

Figure 315. *Luciogobius grandis* (Arai, 1970). A—dorsal view; B—lateral view.

Figure 314. *Luciogobius grandis* (Arai, 1970). A—shape of pectoral fin; B—shape of pelvic fin.

blackish, with white spots; fins light-colored with dark stripes (Dotu, 1957a).

Mode of life described by Dotu and Mito (1958: 419).

Length to 41 mm (Dotu, 1957a).

Distribution: In the Sea of Japan reported from Tsushima Islands (Dotu and Mito, 1958: 424). Described from the coast of Amakusa Islands and reported from Nagasaki region (Dotu, 1957a).

5. *Luciogobius pallidus* Regan, 1940 (Figure 317)

Luciogobius pallidus Regan, Ann. Mag. Nat. Hist., (11) 5, 1940: 462–465 (Shimane Prefecture, Japan). Arai, Bull. Nat. Sci. Mus., Tokyo, 13, 2, 1970: 203.

Luciogobius guttatus guttatus (non Gill), Tomiyama, Japan. J. Zool., 7, 1, 1936: 51, fig. 10, D.

D 10–11; A 11; vertebrae 33–34 (Arai, 1970).

This species and the next, *L. albus,* are found in caves or at the bottom of wells; as such, they are almost devoid of eyes and more or less colorless. Dorsal fin higher compared to that in *L. guttatus* and *L. elongatus.* *L. pallidus* differs from *L. albus* in presence of transparent pigment on body and lesser body depth, which is about 13 times in body length, and head 5 times in body length versus 3.5 times in *L. albus.*

Standard length to 34 mm (Arai, 1970).

Distribution: In the basin of the Sea of Japan known from Shimane Prefecture, and from the Pacific side of Japan in Wakayama and Koti prefectures (Matsubara, 1955: 844).

6. *Luciogobius albus* Regan, 1940 (Figure 318)

Luciogobius albus Regan, Ann. Mag. Nat. Hist. (11), 5, 1940: 462–465 (Shimane Prefecture, Japan). Tomiyama, Japan. J. Zool., 7, 1, 1936: 51, fig. 10, C. Arai, Bull. Nat. Sci. Mus., Tokyo, 13, 2, 1970: 203.

D 10; A 10; vertebrae 30 (Arai, 1970).

Differences between this species and *L. pallidus* given in description of latter.

Mode of life described by Dotu (1963: 1).

Standard length to 30 mm (Arai, 1970).

Distribution: In the basin of Sea of Japan known from Shimane Prefecture, where it dwells in caves.

6. Genus *Inu* Snyder, 1909

Inu Snyder, Proc. U.S. Nat. Mus., 36, 1909: 607 (type: *I. koma* Snyder, 1909).

Similar to the genus *Luciogobius* Gill, but differs in the presence of minute cycloid scales in the posterior part of the body.

Two species. One known from the limits of the Sea of Japan.

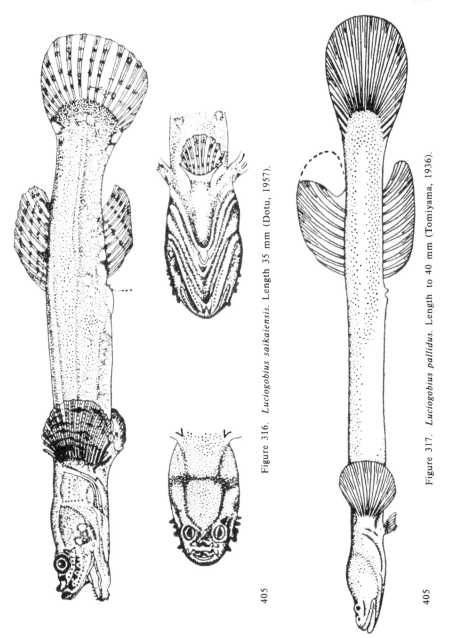

Figure 316. *Luciogobius saikaiensis.* Length 35 mm (Dotu, 1957).

405

Figure 317. *Luciogobius pallidus.* Length to 40 mm (Tomiyama, 1936).

405

1. *Inu koma* Snyder, 1909 (Figure 319)

Inu koma Snyder, Proc. U.S. Nat. Mus., 36, 1909: 607 (Misaki); 42, 1912: 445, pl. 60, fig. 2.

Luciogobius koma, Tomiyama, Japan. J. Zool., 7, 1, 1936: 52.

D 11; A 12 (Snyder, 1909).

Differs from the closely related species *I. ama* (Snyder, 1909) in distinct but small crests or folds on head and back anterior to fin; folds located along sides of shallow depression at median line of back. Folds also present in *I. ama* but far less distinct.

Length to 40 mm (Tomiyama, 1936: 53).

Distribution: In the Sea of Japan reported from Fukui Prefecture (Matsubara, 1955: 843); San'in region (Katoh et al., 1956: 323); and Pusan (Matsubara, 1955: 843). Pacific coast of Japan (Tomiyama, 1936: 53).

7. Genus *Expedio* Snyder, 1909

Expedio Snyder, Proc. U.S. Nat. Mus., 36, 1909: 606 (type: *Expedio parvulus* Snyder, 1909).

Differs from the genera *Luciogobius* and *Inu* in absence of pelvic fins.

One species, known from the Sea of Japan.

1. *Expedio parvulus* Snyder, 1909 (Figure 320)

Expedio parvulus Snyder, Proc. U.S. Nat. Mus., 36, 1909: 606 (Misaki); 42, 1912: 445, pl. 61, fig. 1. Matsubara, Fish Morphol, and Hierar., 1955: 845. Dotu, J. Fac. Agric. Kyushu Univ., 11, 1, 1957: 75.

Luciogobius parvulus, Tomiyama, Japan. J. Zool., 7, 1, 1936: 51. Arai, Bull. Nat. Sci. Mus., Tokyo, 13, 2, 1970: 203.

D 10; A 10 (Snyder, 1909).

This goby resembles the longer specimens of *Luciogobius guttatus,* but differs from them in absence of pelvic fins.

Length, 37 mm (Snyder, 1909).

Distribution: In the Sea of Japan reported from Fukui Prefecture (Matsubara, 1955: 845). Described from Misaki.

4. Subfamily Apocrypteinae

Body distinctly elongate. Head compressed laterally, covered for the most part with scales on top and sides. Teeth on upper jaw arranged in 1 row; teeth of lower jaw not vertical, but slightly horizontal. Each 408 side of symphysis of lower jaw with one upwardly pointed canine. Scales cycloid. Base of second dorsal fin elongate. Pelvic fins completely fused, attached only at base to belly. Gill openings narrow and moderate in size (Koumans, 1953: 246).

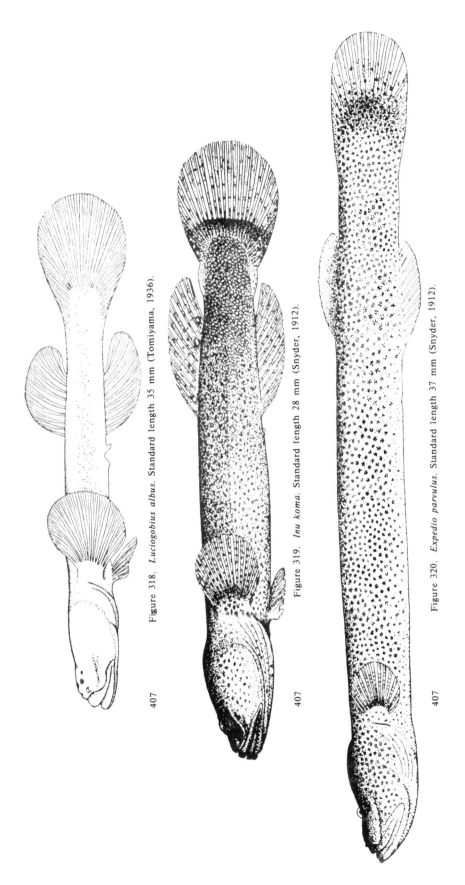

Figure 318. *Luciogobius albus*. Standard length 35 mm (Tomiyama, 1936).

Figure 319. *Inu koma*. Standard length 28 mm (Snyder, 1912).

Figure 320. *Expedio parvulus*. Standard length 37 mm (Snyder, 1912).

407

407

407

Indian Ocean and western part of the Pacific Ocean. Eight genera. Two recorded from Yellow Sea; one species acclimatized in the Sea of Japan.

Key to Genera of Subfamily Apocrypteinae

1 (2). Lower free eyelid absent. Height of first dorsal fin equal to or less than length of base........ 1. [**Apocryptodon** Bleeker].
2 (1). Lower free eyelid present. Height of first dorsal fin distinctly more than length of base. .... 2. **Boleophthalmus** Valenciennes.

1. [Genus *Apocryptodon* Bleeker, 1874]

Apocryptodon Bleeker, Arch. Néerl. Sci. Nat., 9, 1874: 327 (type: *A. madurensis* Bleeker). Koumans, Fish. Indo-Austr., 10, 1953: 253.

Body distinctly elongate, not compressed anteriorly, but laterally compressed posteriorly; covered with easily shed cycloid scales (*squ.* 40-78). Head slightly flat anteriorly. Scales on head behind eyes large, as also on body; scales present along sides of head, under eyes, on cheeks and on gill covers. Interorbital space narrow, less than eye diameter, and snout length slightly more. Nostrils not tubular. Mouth almost horizontal, jaws equal. Teeth on both jaws arranged in 1 row, canine-shaped on upper jaw and truncate or bicuspid on lower jaw and arranged horizontally. Pair of canines located behind symphysis of lower jaw. Upper jaw with notch in which anterior end of lower jaw fits. Tongue rounded, almost completely fused with lower surface of oral cavity. Gill openings almost same size as width of pectoral fin base; isthmus broad. Inner margin of pectoral girdle without fleshy lobes. Dorsal fins close-set; first dorsal with 6 unbranched rays, second dorsal with 1 unbranched and 22 to 27 branched rays; anal fin with 1 unbranched and 21 to 27 branched rays. Pelvic fins fused, rather long. Rays of pectoral fins without free filamentous tips; bases of pectoral fins covered with scales. Caudal fin slightly pointed (Koumans, 1953: 253).

From India to Japan; brackish waters, enters rivers. Several species. Absent in the Sea of Japan, but one species found in the Yellow Sea.

1. [*Apocryptodon madurensis* (Bleeker, 1849)] (Figure 321)

Apocryptes madurensis Bleeker, Verh. Batavia Genoots., 22, 1849: 35 (Madura Island, Indonesia).

Apocryptodon bleekeri Day, Fishes India, 1878: 300, pl. 64, fig. 3 (India). Zhang et al., Ryby Zaliva Bokhai..., 1955: 221, fig. 141.

Apocryptodon madurensis, Koumans, Fish. Indo-Austr. Arch., 10, 1953: 254, fig. 63 (description, synonyms).

D VI, 23; A 23; P 22; *l. l.* 50-55 (Koumans, 1953).

Characters given in description of genus.

Mode of life described by Japanese researchers (Uchida, 1932: 109; Dotu, 1961b: 133).

Length to 77 mm (Koumans, 1953).

Distribution: Absent in the Sea of Japan. In the Yellow Sea found near the coast of Shantung Province (Zhang et al.). South Japan, Indonesia, India (Matsubara, 1955: 845).

2. Genus *Boleophthalmus* Valenciennes, 1837

Boleophthalmus Valenciennes, Hist. Nat. poiss., 12, 1837: 198 (type: *Gobius boddaerti* Pallas). Koumans, Fish. Indo-Austr. Arch., 10, 1953: 257.

Body distinctly elongate, compressed laterally, covered with cycloid scales becoming larger posteriorly (60 to 100 and more). Head slightly 409 flat; skin on head warty, entirely covered with scales or scales rudimentary. Eyes very close-set, movable, moving over dorsal profile of head; lower eyelid well developed. Snout blunt, its length equal to diameter of eye. Mouth slightly oblique. Jaws almost equal. Teeth on both jaws arranged in 1 row; teeth of upper jaw conical, some anterior teeth canine-shaped; teeth on lower jaw arranged slightly horizontally, thickened, and apically curved. Each side of symphysis with one curved canine. Tongue with straight cut, almost entirely fused with lower surface of oral cavity. 410 Barbels absent on head. Gill openings located obliquely, their size equal to width of pectoral fin base. Isthmus broad. Inner margin of pectoral girdle without fleshy lobes. Dorsal fins separate or contiguous at bases. First dorsal with 5 unbranched rays, second dorsal with 1 unbranched and 22 to 27 branched rays. Anal fin with 1 unbranched and 23 to 26 branched rays. Pelvic fins fused, rather long. Pectoral fins without filamentous rays; base muscular, covered with scales. Caudal fin asymmetrical, upper half slightly longer than lower half (Koumans, 1953).

Differs from the genus *Scartelaos* Swainson, 1839 in absence of sharp teeth and distinctly developed, albeit small and rudimentary scales, as well as absence of barbels on lower side of head.

Indian Ocean and western part of Pacific Ocean. Several species. Acclimatized in the Sea of Japan. One species known from the limits of the Yellow Sea.

1. *Boleophthalmus pectinirostris* (Linné, 1758) (Figure 322)

Apocryptes chinensis Osbeck, Amoen Acad., 1754: 20, fig. 23 (Canton). Prelinnaean.

Gobius pectinirostris Linné, Syst. Nat., ed. 10, 1, 1758: 264 (Canton).

Boleophthalmus chinensis, Zhang et al., Ryby Zaliva Bokhai..., 1955: 229, fig. 147.

524

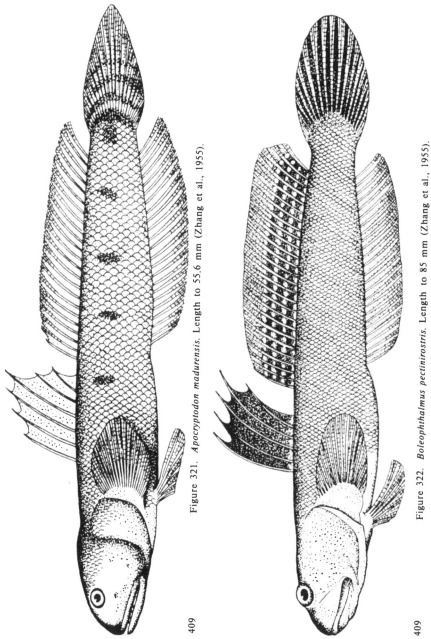

Figure 321. *Apocryptodon madurensis*. Length to 55.6 mm (Zhang et al., 1955).

409

Figure 322. *Boleophthalmus pectinirostris*. Length to 85 mm (Zhang et al., 1955).

409

Boleophthalmus pectinirostris, Koumans, Fish. Indo-Austr. Arch., 10, 1953: 261 (synonyms, description). Zhu et al., Ryby Vostochno-Kitaiskogo Morya, 1963: 435, fig. 331.

1229. Japan. 1862. Schlegel. 1 specimen.

D V, 24-27; A 24-27; P 17-20; *l. l.* ca 100 (Koumans, 1953).

Differs from other species known in the western part of the Pacific Ocean in presence of bluish spots on dorsal fins and absence of oblique stripes on sides of body. In our specimens third ray of first dorsal fin almost reaches end of base of second dorsal fin.

Mode of life described by Enami and Dotu (1961).

Lenght to 176 mm (Enami and Dotu, 1961).

Distribution: In the Sea of Japan not found up to 1950, but later acclimatized in the estuary of the Tatar River near Fukuoka, where it has adapted quite well (Enami and Dotu, 1961: 141).

5. Subfamily Sicydiaphinae

Body elongate and covered with scales or naked. Gill openings not very broad, except in *Aphia*; isthmus broad. Two dorsal fins; base of second fin almost not elongate. Pelvic fins connected, sometimes only at base, and sometimes completely fused with belly. Sometimes teeth present on lower lip. Teeth on lower jaw simple and arranged in single row. Upper jaw also with one row of teeth, behind which sometimes several rows concealed in gums. Head elongate, snout and cheeks naked (Koumans, 1953: 219). Examined under a lens, teeth of upper jaw found to be tri- or bicuspid or bifurcate (Fowler, 1961: 241).

Several genera, widely distributed. One genus recorded from Japan, members of which are found in fresh waters and in estuaries.

1. Genus *Sicyopterus* Gill, 1860

Sicyopterus Gill, Proc. Acad. Nat. Sci. Philad., 1860: 101 (type: *S. stimpsoni* Gill) Koumans, Fish, Indo-Austr. Arch., 10, 1953: 220).

Differs from other genera of this subfamily in pelvic fins fused and forming suctorial disk attached to belly. Body covered with scales. At 411 least 1 canine located behind symphysis on each side of lower jaw (Koumans, 1953).

Fresh waters and mouths of rivers in the Indian Ocean and western part of the Pacific Ocean. Several species. One species found in estuaries of the Sea of Japan.

1. *Sicyopterus japonicus* (Tanaka, 1909) (Figure 323)

Sicydium japonica Tanaka, J. Coll. Sci. Imp. Univ., Tokyo, 27, 8, 1909; 22 (Tosa, Shikoku Island, Japan); Fig. and Descr..., XII, 1913: 203, pl. 56, figs. 209-211; pl. 58, fig. 215.

526

Sicyopterus japonicus, Fowler, Synopsis..., 14, 3-4, 1961: 242, fig. 69 (synonyms, description).

412 D VI, 11; A 11; P 19; *squ.* 59 (Fowler, 1961).

Characters of this only species found in the Sea of Japan and near Japan given in description of family and genus.

Length to 165 mm (Tomiyama, 1936: 99)

Distribution: In the Sea of Japan found in mouths of rivers. In Japan from central Honshu south to Taiwan (China) (Matsubara, 1955: 847). Cheju-do Island (Mori, 1952: 146).

CLXXIII. Family TRYPAUCHENIDAE

Body elongate, covered with quite large cycloid scales. Head naked. Pelvic fins fused and forming disk, sometimes deeply incised, almost up to base. Dorsal fin single; unbranched rays distinguishable. Eyes small. This family differs from Gobiidae mainly in the presence of pouchlike cavity at upper margin of operculum, which is separate from the gill cavity (Figure 324).

Five genera. Indian Ocean and western part of the Pacific Ocean. In the Sea of Japan one genus, in the Yellow Sea another genus.

Key to Genera of Family Trypauchenidae

1 (2). Pelvic fins completely fused, form distinct disk. Canines present.
.... 1. [**Trypauchen** Valenciennes].
2 (1). Pelvic fins incompletely fused, with deep notch on posterior margin of disk. Canines absent...............................
.... 2. **Ctenotrypauchen** Steindachner.

1. [Genus *Trypauchen* Valenciennes, 1837]

Trypauchen Valenciennes, Hist. Nat. Poiss., 12, 1837: 152 (type: *Gobius vagina* Schneider). Koumans, Fish. Indo-Austr. Arch., 10, 1953: 277.

Body distinctly elongate, compressed laterally, covered with cycloid scales. Head compressed laterally, naked, with median crest on occiput. Eyes small. Mouth very oblique; lower jaw protrudes anteriorly. Teeth on both jaws arranged in 2-3 rows; teeth of outer row highly enlarged and canine-shaped. Tongue rounded. Gill openings not very broad; isthmus broad. Inner margin of pectoral girdle without fleshy lobes. Pouchlike cavity present on upper margin of operculum, which is separate from gill cavity. Barbels not present on head. Dorsal fin single, with 6 simple and 41 to 49 branched rays. Anal fin with 40 to 46 rays. Both fins confluent with caudal fin. Pelvic fins small, each with 1 simple and 5 branched rays; fins fused, forming small disk. Pectoral fins also small. Caudal fin slightly pointed (Koumans, 1953).

About 3 species. One species reported from the Yellow Sea.

1. [*Trypauchen vagina* (Bloch and Schneider, 1801)] (Figure 324)

Gobius vagina Bloch and Schneider, Syst. Ichthyol., 1801: 73 (Tranqueber, India).

Trypauchen vagina, Koumans, Fish. Indo-Austr. Arch., 10, 1953: 277 (synonyms, description). Zhu et al., Ryby Yuzhno-Kitaiskogo Morya, 1962: 825, fig. 670. Fowler, Synopsis..., 15, 1-2, 1962: 26 (bibliography).

40032. Tonkinsk Bay. July 9, 1961. E.F. Gur'yanova. 1 specimen.

D VI, 40-49; A I, 39-46; P 15-18; *l. l.* 80-115; *l. tr.* 21 (Koumans, 1953).

Differs from other species in smaller scales (*squ.* 80-115 versus 45-65). Length to 220 mm (Koumans, 1953).

Distribution: Absent in the Sea of Japan. In the Yellow Sea reported from Chefoo (Wang and Wang, 1935: 203).

413   **2. Genus *Ctenotrypauchen* Steindachner, 1867**

Ctenotrypauchen Steindachner, Sitzber. Akad. Wiss. Wien, 55, 1867: 530 (type: *C. chinensis* Steindachner). Koumans, Fish. Indo-Austr. Arch., 10, 1953: 281.

Body greatly elongate, compressed laterally, covered with cycloid scales (about 65). Head compressed laterally, with median crest on occiput. Mouth very oblique; lower jaw protrudes. Teeth in outer series enlarged, but no canines present. Pouchlike cavity on upper margin of operculum separate from gill cavity. Head without barbels. Dorsal fin single, with 6 simple and about 50 branched rays; anal fin with 44 to 49 rays. Both fins confluent with caudal fin. Pelvic fins connected, but deeply incised; membrane present at base. Pectoral fins small. Caudal fin rounded or pointed (Koumans, 1953: 281).

Three species in the Indian Ocean and western part of the Pacific Ocean. One species recorded from the Sea of Japan, and another from the Yellow Sea.

Key to Species of Genus Ctenotrypauchen

1 (2). Head relatively short; its length almost equal to body depth at vertical with origin of anal fin. Pectoral fins fan-shaped......
.... 1. **C. microcephalus** (Bleeker).
2 (1). Head relatively long; its length much greater than body depth at vertical with origin of anal fin. Pectoral fins slightly falcate; upper rays much longer than lower ones. ....
.... 2. [**C. chinensis** Steindachner].

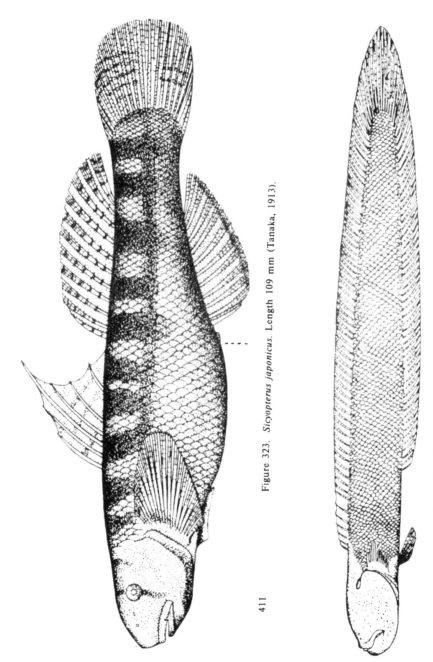

Figure 323. *Sicyopterus japonicus*. Length 109 mm (Tanaka, 1913).

Figure 324. *Trypauchen vagina*. Standard length 117 mm (Zhu et al., 1962).

411

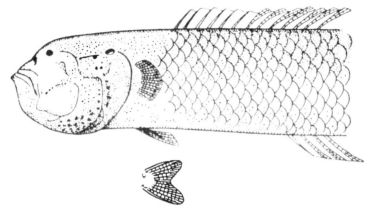

Figure 325. *Ctenotrypauchen microcephalus.* Anterior part of body (Koumans, 1953).

1. *Ctenotrypauchen microcephalus* (Bleeker, 1860) (Figure 325)

Trypauchen microcephalus Bleeker, Acta Soc. Indo-Néerl., 8, 1860: 62 (Sungaidare, Indonesia).

Ctenotrypauchen microcephalus, Koumans, Fish. Indo-Austr. Arch., 10, 1953: 282.

Trypauchen wakae Jordan and Snyder, Proc. U.S. Nat. Mus., 24, 1901: 127, fig. 32 (Wakanoura).

D VI, 50; A I, 44–49; P 17; *l. l.* ca 65 (Koumans, 1953).

Body elongate, but relatively deeper compared to *C. chinensis*; depth 8.75 times in length. Head relatively short, length almost equal to body depth at vertical with anal fin. Eyes very small, covered with skin. Snout length 4 times in head length. Mouth very oblique; lower jaw protrudes forward. Length of maxilla 3 times in head length. Teeth of outer row enlarged; canines absent. Head, occiput, thorax, and belly naked. Scales on body cycloid. Dorsal fin single, spiny rays distinct. Pectoral fin 1/3 head length. Pelvic fins short, 4.5 times in head length, connected, but with deep notch; membrane present at base of fin. Caudal fin rounded, 7.25 times in body length. Body color red; fins transparent with reddish tinge (Koumans, 1953).

Mode of life described by Dotu (1958b: 371).

Length to 180 mm (Koumans, 1953).

Distribution: In the Sea of Japan reported from Toyama Bay (Katayama, 1940: 32) and Sado Island (Honma, 1963: 22). In the Yellow Sea known from Kunsan and Namp'o (Mori, 1952: 149) and Chefoo (Wang and Wang, 1935: 202). Found from southern Japan to Australia and South Africa (Matsubara, 1955: 849).

2. [*Ctenotrypauchen chinensis* Steindachner, 1867] (Figure 326)

Ctenotrypauchen chinensis Steindachner, Sitzber. Akad. Wiss. Wien, 55,

1867: 530, pl. 6, fig. 3-4 (China). Zhang et al., Ryby Zaliva Bokhai...,
1955: 226, fig. 145. Zhu et al., Ryby Vostochno-Kitaiskogo Morya, 1963:
442, fig. 338.

36441. Yellow Sea, Tsingtao. June 3, 1957. Exped. ZIN AN SSSR. 6
specimens.

36442. Yellow Sea, Tsingtao. June 6, 1957. E.F. Gur'yanova. 2
specimens.

36443. Yellow Sea, Tsingtao. June 24, 1957. E.F. Gur'yanova. 1
specimen.

36444. Yellow Sea, Chefoo. June 28, 1957. Exped. ZIN AN SSSR. 6
specimens.

D VI, 50-58; A 42-50; P 14-15; V I, 4; *l. l.* 67-74 (Zhu et al., 1963).

Differences from *C. microcephalus* given in key. Fowler (1962: 25)
included this species in the synonymy of *C. microcephalus*; however,
Chinese ichthyologists (Zhu et al., 1963) consider it an independent
species, although in his publication of 1962 Zhu included the specimen of
Zhang et al. (1955) as a synonym of *M. microcephalus*

Length to 95 mm (Zhu et al., 1963).

Distribution: Absent in the Sea of Japan. In the Yellow Sea reported
from Lyonin, Hebei, and Shangtung provinces (Zhang et al., 1955: 227).
South to Indonesia and the Indian Ocean (Zhu et al., 1963: 442).

CLXXIV. Family GOBIOIDIDAE (TAENIOIDIDAE)—Eel-like Gobies

Body greatly elongate, naked or covered with very small cycloid scales.
Pelvic fins fused and form disk. Dorsal fin single, very long, but with
distinct anterior unbranched rays. Eyes small or indistinguishable.
Isthmus broad. Differs from the family Trypauchenidae mainly in absence
of pouchlike cavity on upper margin of operculum, which is separate from
gill cavity (Taenioninae—Koumans, 1963: 265).

Eight genera. Indian and Pacific oceans. One genus reported from the
Sea of Japan, and another from the Yellow Sea.

Key to Genera of Family Gobioididae

1 (2). Pectoral fins much shorter than pelvic fins, about 3 times
in head length. Anal fin with more than 45 rays. Canines not
present behind symphysis of lower jaw.........................
.... 1. [**Taenioides** Lacépède].

2 (1). Pectoral fins same length as pelvic fins, about 1.5 times in
head length. Anal fin with fewer than 45 rays. Pair of large canines
located behind symphysis of lower jaw........................
.... 2. **Odontamblyopus** Bleeker.

416   **1. [Genus *Taenioides* Lacépède, 1798]**

Taenioides Lacépède, Hist. Nat. Poiss., 2, 1798: 580, 4, 1800: 339 (type: *T. hermannianus* Lac.). Koumans, Fish. Ando-Austr. Arch., 10, 1953: 269.

Body greatly elongate, compressed laterally, covered with very rudimentary scales or naked. Head almost cylindrical in cross section. Eyes reduced, covered. Mouth almost vertical; lower jaw protrudes forward. Lips fimbriate. Teeth on both jaws with truncate cusps or bluntly pointed, arranged in several rows in a band; teeth of outer row enlarged, more wide-set, and resemble canines. Tongue rounded. Gill openings not very large. Isthmus same width as base of pelvic fins. Inner margin of pectoral girdle without fleshy lobes. Head with several row of ridges, radiating from eyes onto cheeks, gill covers, and lower jaw. Several barbels present on lower side of head. Dorsal fins fused, with 5 to 6 unbranched and 38 to 50 branched rays. Anal fin with 45 to 49 rays. Dorsal and anal fins more or less confluent with fairly long caudal fin. Pelvic fins fused, large, elongate. Pectoral fins small (Koumans, 1953).

Several species. Not reported from the Sea of Japan. One species known from the Yellow Sea, off the west coast of the Korean Peninsula—Mokp'o and Kunsan (Mori, 1952: 148).

1. [*Taenioides cirratus* (Blyth, 1860)] (Figure 327)

Amblyopus cirratus Blyth, J. Asiat. Soc., Bengal, 29 1860: 147 (Calcutta).

Taenioides lacepedei (non Temminck and Schlegel) Jordan and Snyder, Proc. U.S. Nat. Mus., 24, 1901: 128, fig. 33 (Wakanoura).

Taenioides snyderi Jordan and Hubbs, Mem. Carnegie Mus., 10, 2, 1925: 310 (Wakanoura).

Taenioides cirratus, Koumans, Fish. Indo-Austr. Arch., 10, 1953: 270, fig. 67 (synonyms, description).

D VI, 43–49; A I, 42–47; P 13 (Koumans, 1953).

Differs from other species of this genus in smaller length of head, which is shorter than distance from base of pelvic fin to vent, and in small number of teeth on jaws; each side of upper jaw with 5 and not 7 teeth.

Mode of life described by Dotu (1958b: 371).

Length to 300 mm (Koumans, 1953).

Distribution: Absent from the Sea of Japan. In the Yellow Sea known from Mokp'o and Kunsan (Mori, 1952: 148). From southern Japan south to the Indian Ocean (Matsubara, 1955: 849).

2. Genus *Odontamblyopus* Bleeker, 1874

Odontamblyopus Bleeker, Arch. Néerl. Sci. Nat., 9, 1874: 330 (type:

532

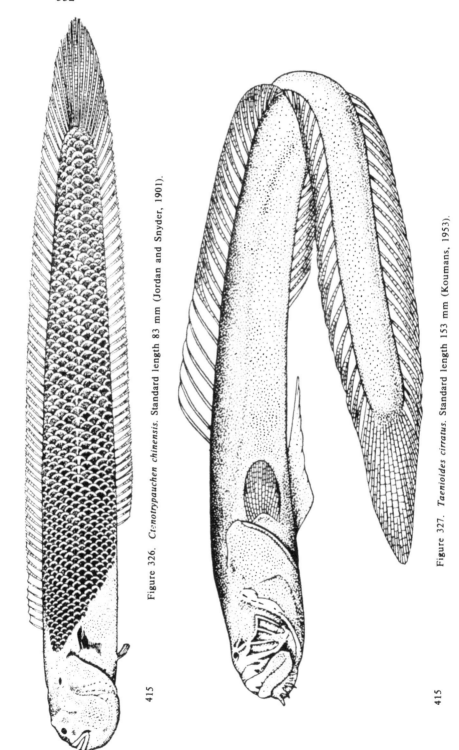

Figure 326. *Ctenotrypauchen chinensis*. Standard length 83 mm (Jordan and Snyder, 1901).

415

Figure 327. *Taenioides cirratus*. Standard length 153 mm (Koumans, 1953).

415

Gobioides rubicundus Hamilton). Koumans, Fish. Indo-Austr. Arch., 10, 1953: 274.

Body greatly elongate, compressed laterally, covered with large number of minute, more or less rudimentary scales. Head compressed laterally, almost entirely covered with similar scales. Eyes very small, almost set at top of head. Interorbital space and snout about twice diameter of eye. Mouth oblique; lower jaw protrudes forward. Teeth on upper jaw arranged in 2 rows, on lower jaw in front in 3 rows, and on sides in 2 rows. Teeth in outer rows of both jaws highly enlarged and wide-set, forming 4 curved canines on each side. Each side of symphysis of lower jaw with one canine. Tongue truncate at tip. Gill openings broad; isthmus same width as base of pelvic fin. Inner margin of pectoral girdle without fleshy lobes. Head with 2 rows of barbels. Dorsal fins fused, with 6 simple and 34 to 40 branched rays; anal fin with 33 to 38 rays. Dorsal and anal fins confluent with caudal fin. Pelvic fins fused and elongate. Pectoral fins same length as pelvic fins. Caudal fin pointed and elongate (Koumans, 1953).

Found in the Indian Ocean to Japan. Two or more species. One species in the Sea of Japan.

417 1. ***Odontamblyopus rubicundus*** (Hamilton, 1822)—Eel-like Goby (Figure 328)

Gobioides rubicundus Hamilton, Gangetic Fishes, 1822: 37, 365, pl. 5, fig. 9 (estuary of Ganges River).

Amblyopus lacepedei Temminck and Schlegel, Fauna Japonica, Poiss., 1845: 146, pl. 75, fig. 2 (Japan). Jordan and Hubbs, Mem. Carnegie Mus., 10, 2, 1925: 310.

Taenioides petschilensis Rendahl. Arkiv für Zool., Stockholm, 16, 2, 1924: 31 (Gulf of Chihli, Yellow Sea)

Taenioides rubicundus, Tomiyama, Japan. J. Zool., 7, 1, 1936: 102, fig. 43 (synonyms).

Odontamblyopus rubicundus, Koumans, Fish. Indo-Austr. Arch., 1953: 275, fig. 68 (synonyms and description).

35575. Yellow Sea. May 27, 1956. Academy of Sciences, China. 2 specimens.

35722. Yellow Sea, Tangu. June 13, 1957. P.V. Ushakov. 5 specimens.

35723. Yellow Sea, Tangu. E.F. Gur'yanova. 3 specimens.

40033. Tonkinsk Bay. July 28, 1961. E.F. Gur'yanova. 3 specimens.

D VI, 35-40; A I, 32-39; P 30 (Koumans, 1953).

Each side of mouth with 4 canines in outer series of upper jaw and 4-6 canines on lower jaw. Pair of canines present behind symphysis of lower jaw. Lower surface of head with 3 barbels on each side, sometimes poorly distinguishable. Pectoral fins of same length as pelvic fins (Koumans, 1953).

Mode of life of adults and larvae described by Japanese researchers (Dotu, 1957c: 101; Dotu and Takita, 1967: 135).

Length to 330 mm (Tomiyama, 1936).

Distribution: In the Sea of Japan found near Fukuoka (Jordan and Hubbs, 1925: 310). In the Yellow Sea near Ionamp'o (Mori, 1952: 148) and Gulf of Chihli (Bohai) (Zhang et al., 1955: 225). South of Japan (Ariake Bay), China, India (Matsubara, 1955: 848).

CLXXV. Family PERIOPHTHALMIDAE—Mudskippers

Body distinctly elongate. Anterior part of head highly truncate. Body and head covered with cycloid or weakly ctenoid scales. Eyes convex, protruding above profile of head; lower eyelid well developed. Teeth on upper jaw arranged in 1-2 rows, on lower jaw in 1 row. Base of second dorsal fin only slightly longer than base of first fin. Pectoral fins with muscular base; fins capable of bending at line of attachment of rays and used for locomotion outside water. Pelvic fins highly variable in shape; sometimes completely fused and form disk, sometimes only connected at bases, and sometimes completely separate without connecting membrane at base. Gill openings not very broad (Koumans, 1953—Periophthalminae).

Found from the west coast of Africa to the Pacific Ocean. Tropical and subtropical waters. Two genera. One known off the coasts of Japan and found in the Sea of Japan.

1. Genus *Periophthalmus* Bloch and Schneider, 1801

Periophthalmus Bloch and Schneider, Syst. Ichthyol., 1901: 63 (type: *P. papilio* Bloch and Schneider). Koumans, Fish. Indo-Austr. Arch., 10, 1953: 200.

Body elongate, slightly compressed laterally and covered with minute (60 to 100) cycloid or weakly ctenoid scales. Head compressed laterally, completely or partly covered with scales behind eyes, on cheeks and on opercula. Eyes close-set, convex, and protrude above dorsal profile; lower eyelid well developed. Snout blunt, approximately as long as eyes. 418 Anterior nostril in form of tubule located on triangular lobe above upper lip. Mouth horizontal; jaws almost equal. Teeth arranged in single row on both jaws, differ in size; anterior teeth more or less canine-shaped. Tongue rounded, almost completely fused with lower surface of oral cavity. Gill openings narrow, continue upward not more than 3/4 base of pectoral fin. Isthmus broad. Dorsal fins not contiguous. First dorsal fin with 8 to 17 rays, second fin with 1 unbranched and 10 to 14 branched rays. Pelvic fins either connected only at bases and their marginal rays completely separate, not forming disk, or rays connected over half their

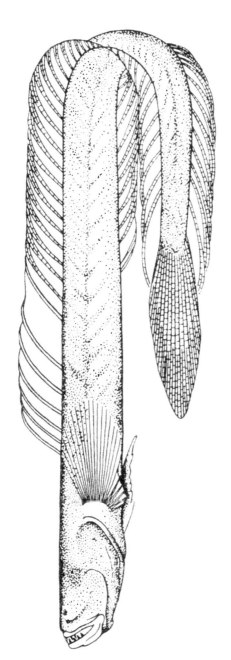

Figure 328. *Ozontamblyopus rubicundus*—eellike goby. Standard length 220 mm (Koumans, 1953).

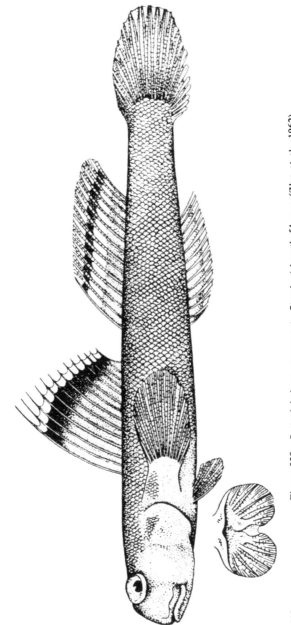

Figure 329. *Periophthalmus cantonensis.* Standard length 51 mm (Zhu et al., 1962).

418

length by short membrane, or rays connected throughout their length,
419 forming disk. Base of pectoral fin with highly developed muscles
(Koumans, 1953).

Differs from the genus *Periophthalmodon* Bleeker, 1874 in teeth on
upper jaw arranged only in one row and not in two.

About 10 species. One species reported from the Sea of Japan.

1. *Periophthalmus cantonensis* (Osbeck, 1757) (Figure 329)

Apocryptes cantonensis Osbeck, Reise nach China, 1757: 171 (Canton).
Tomiyama, Japan. J. Zool., 7, 1, 1936: 99. Koumans, Fish. Indo-Austr.
Arch., 10, 1953: 204 (remarks). Zhu et al., Ryby Vostochno-Kitaiskogo
Morya, 1963: 434, fig. 330.

36433. Yellow Sea. Tiatsin. June 13, 1957. Exped. ZIN AN SSSR. 10
specimens.

D X–XIV, 12–14; A 12–14; P 14; V I, 5; *l. l.* 85–91 (Zhu et al., 1963).

Species of this genus vary notably and some authors (Tomiyama) have
grouped many species into one. Koumans followed Eggert (*Zool.
Jahrbücher, Abt. Systematic,* 67, 1/2, 1935: 29–116, pls. 1–9) by and large,
and reported the occurrence of about 10 species in the waters of Indonesia
and Australia. Since only one species is known from the waters of Japan
and China, we have not listed its differences from the more southern
species.

It should be noted that the name *cantonensis* Osbeck, 1757 is
pre-Linnaean and, therefore, most probably this species should be named
P. koelreuteri Pallas [*Spicilegia Zoologica, Pisces* 1769 (1770): 8, pl. 2, figs.
1–3].

Mode of life described by Uchida (1932: 109).

Length to 105 mm (Tomiyama, 1936).

Distribution: In the Sea of Japan reported from coast of Yamaguti
Prefecture (Yoshida and Ito, 1957: 268) and Pusan (Mori, 1952: 149). In
the Yellow Sea found near the west coast of the Korean Peninsula (Mori,
1952: 149), Gulf of Chihli (Bohai) (Zhang et al., 1955: 228). Known from
the coasts of China (Zhu et al., 1962: 829; 1963. 434).

Bibliography *

420 Abe T. 1953. New, rare, or uncommon fishes from Japanese waters. II. Records of rare fishes of the families Diretmidae, Luvaridae, and Tetragonuridae, with an appendix (Description of a new species, *Tetragonurus pacificus,* from the Solomon Islands), *Japan. J. Ichthyol.,* 3, 1, 39-47, 7 figs.

Abe, T. 1954. New, rare, and uncommon fishes from Japanese waters. V. Notes on the rare fishes of the suborders Stromateoidei and Tetragonuroidei (Berg.), *Japan. J. Ichthyol.,* 3, 3-5, 178; 3, 6, 222, 246.

Abe, T. 1955a. New, rare, or uncommon fishes from Japanese waters. V. Notes on the rare fishes of the suborders Stromateoidei and Tetragonuroidei, *Japan. J. Ichthyol.,* 4, 1-3, 113-118, figs. 1-3.

Abe, T. 1955b. Notes on the adult of *Cubiceps gracilis* from the Western Pacific, *J. Oceanogr. Soc., Japan,* 11, 2, 75-79.

Abe, T. 1959. On the presence of at least two species of *Cubiceps* (Nomeidae, Pisces) in the path of the "Kuro-Shiwo," *Rec. Oceanogr. Works, Japan,* spec. numb., vol. 3, pp. 225-229.

Abe, T. 1963. Unusual occurrence of several species of boreal, amphi-Pacific and bathypelagic fishes in Sagami Bay and adjoining waters during the first half of 1963, a cold-water season in southern Japan, *Bull. Tokai Reg. Fish. Res. Lab.,* vol. 37, pp. 27-35.

Abe, T. and V. Takashima. 1958. Differences in the number and position of two kinds of fin supports of the spinous dorsal in the Japanese mackerels of the genus *Pneumatophorus, Japan. J. Ichthyol.,* 7, 1, 1-11.

Abe, T. and T. Kosakai. 1964. Notes on an economically important but scientifically little-known silver pomfret, *Pampus echinogaster* (Pampidae, Teleostei), *Japan. J. Ichthyol.,* 12, 1/2, 29-31.

Abe, T., S. Kojima and T. Kosakai. 1963. Description of a new nomeid fish from Japan, *Japan. J. Ichthyol.,* 9, 1/2, 31-35.

Akazaki, M. 1957. Biological studies on a dragonet, *Synchiropus altivelis* (T. and S.), *Japan. J. Ichthyol.,* 5, 3/6, 146-152.

Akihito, Prince. 1963. On the scapula of gobiid fishes, *Japan. J. Ichthyol.,* 11, 1/2, 1-26.

Akihito, Prince. 1966. On the scientific name of a gobioid fish named "urohaze," *Japan, J. Ichthyol.,* 13, 4/6, 73-101, 27 figs.

*Some entries incomplete in the Russian text—General Editor.

Akihito, Prince. 1967a. On four species of gobioid fishes of the genus *Eleotris* found in Japan, *Japan. J. Ichthyol.*, 14, 4/6, 135-166, 31 figs.

Akihito, Prince. 1967b. Additional research on the scapula of gobioid fishes, *Japan. J. Ichthyol.*, 14, 4/6, 167-182, 4 figs.

Akihito, Prince. 1969. A systematic examination of gobioid fishes based on the mesopterygoid, postcleithra, branchiostegals, pelvic fins, scapula, and suborbital, *Japan. J. Ichthyol.*, 16, 3, 93-114, 8 figs.

Akihito, Prince. 1971. On the supratemporals of gobioid fishes, *Japan. J. Ichthyol.*, 18, 2, 57-64, 2 figs.

Alverson, D.L. 1961. Ocean temperatures and their relation to albacore tuna (*Thunnus germo*) distribution in waters off the coast of Oregon, Washington, and British Columbia, *J. Fish. Res. Board, Canada*, 18, 6, 1145-1152, ill.

Alverson, F.G. 1963. The food of yellowfin ad skipjack tuna in the eastern Pacific Ocean, *Bull. Inter-Amer. Trop. Tuna Comm.*, vol. 7, pp. 295-396.

Andriyashev, A.P. 1935. Novye dannye o glubokovodnykh rybakh Beringova morya (New data on deep-sea fishes of the Bering Sea). *Dokl. Akad. Nauk SSSR*, 4, 1/2, 105-108, fig.

Andriyashev, A.P. 1937. K poznaniyu ikhtisfauny Beringova i Chukotskogo morei (On the ichthyofauna of the Bering and Chukchi seas). *Issled. Morei SSSR*, vol. 25, pp. 292-355, figs. 1-27, 1 pl.

Andriyashev, A.P. 1938. Obzor roda kruzenshterniella iz sem. bel'dyugovykh [*Krusensterniella* Schmidt (Pisces, Zoarcidae)] s opisaniem novogo vida iz Yaponskogo morya [Review of the genus *Krusensterniella* Schmidt (Pisces, Zoarcidae) with a description of a new species from the Sea of Japan). *Vestn. Dal'nevost. Fil. Akad. Nauk SSSR*, 32, 5, 117-121.

Andriyashev, A.P. 1939a. Ob amfipatsificheskom (yaponooregonskom) rasprostranenii morskoi fauny v severnoi chasti Tikhogo okeana [The amphi-Pacific (Japanese-Oregon) distribution of marine fauna in the northern Pacific Ocean]. *Zool. Zhurn.*, 18, 2, 181-195, figs. 1-4.

Andriyashev, A.P. 1939b. Ocherk zoogeografii i proiskhozhdeniya fauny ryb Beringova morya i sopredel'nykh vod (Zoogeography and Origin of Fish Fauna of the Bering Sea and Adjacent Waters). Leningrad, 187 pp.

Andriyashev, A.P. 1954. Ryby severnykh morei SSSR (Fishes of the Northern Seas of the Soviet Union). Moscow-Leningrad, 556 pp., 300 figs.

Andriyashev, A.P. 1955a. Obzor ugrevidnykh likodov, *Lycenchelys* Gill. (Pisces, Zoarcidae) i blizkie formy morei SSSR i sopredel'nykh vod [Review of eelpouts *Lycenchelys* Gill. (Pisces, Zoarcidae) and related forms in the seas of the Soviet Union and adjacent waters]. *Tr. Zool. Inst. AN SSSR,* vol. 18, pp. 349-384.

Andriyashev, A.P. 1955b. Novye i redkie vidy ryb semeistva bel'-dyugovykh (Pisces, Zoarcidae) s yugo-vostochnogo poberezh'ya Kamchatki (New and rare fish species of the family Zoarcidae from the southeast coast of Kamchatka). *Tr. Zool. Inst. AN SSSR,* vol. 21, pp. 393-400, figs. 1-6.

Andriyashev, A.P. and V.M. Makushok. 1955. *Argyropterus corallinus* (Pisces, Blennioidei)—novaya ryba bez parmykh plavnikov [*Argyropterus corallinus* (Pisces, Blennioidei)—a new fish without paired fins]. *Vopr. Ikhtiologii,* vol. 3, pp. 50-53, figs. 1-2.

Aoyagi, H. 1955. Studies on the coral fishes of the Riu-Kiu Islands. Blenniidae, *Zool. Mag.,* 64, 3, 76-83, ill.

Arai, R. 1964. Sex characters of Japanese gobioid fishes. (I), *Bull . Nat. Sci. Mus. Tokyo,* 7, 3, 295-306.

Arai, R. 1970. *Luciogobius grandis,* a new goby from Japan and Korea, *Bull. Nat. Sci. Mus. Tokyo,* 13, 2, 199-205, pl. I, figs. 1-4.

Arai, R. 1971. Record of the dragonet, *Draculo mirabilis* Snyder, from Hokkaido, Japan, *Japan. J. Ichthyol.,* 18, 1, 33-35, 5 figs.

Arai, R. and T. Abe. 1970. The sea fishes of Tsushima Island, *Japan. Mem. Nat. Sci. Mus. Tokyo.* 3. *Natural History of the Islands of Tsushima,* no. 2, pp. 83-100, pls. 17-18.

Arnold, D.C. 1956. A systematic revision of the fishes of the teleost family Carapidae (Percomorphi, Blennioidea), with descriptions of two new species, *Bull. Brit. Mus. (Nat. Hist.) Zool.,* 4, 6, 15-307, 20 figs.

Ayres, W.O. 1854-1857. Description of new species of Californian fishes, *Proc. Calif. Acad. Sci.,* vol. 1, pp. 23-77.

Barsukov, V.V. 1954. O smene zubov u zubatok (sem. Anarhichadidae) [Teeth modification in wolf-fishes (Anarhichadidae)]. *Dokl. Akad. Nauk SSSR,* 95, 4, 897-899, ill.

Barsukov, V.V. 1959. Sem. zubatok (Anarhichadidae) [Wolf-fishes (Anarhichadidae)]. *Fauna Rossii, Ryby,* 5, 5, 1-171, figs. 1-42, pls. I-XXII.

Barsukov, V.V. 1960. Skorost' dvizheniya ryb (Speed of fish movement). *Priroda,* vol. 3, pp. 103-104.

Basilewsky, S. 1855. Ichthyographia Chinae borealis, *Nouv. Mém. Soc. Imper. Natur. Moscow,* vol. 10, pp. 215-263, 9 pls.

Bean, T.H. 1894 (1893). Description of a new blennioid fish from California, *Proc. U.S. Nat. Mus.,* vol. 16, pp. 699-701, 1 fig.

Beaufort, L.F. and W.M. Chapman. 1951. Percomorphi (concluded), Blennoidea. In: *Fishes of the Indo-Australian Archipelago,* edited by Weber and Beaufort, 1951, vol. IX, 484 pp., 89 figs.

Bell, R.R. 1962. Age determination of the Pacific albacore of the Californian coast, *Calif. Fish and Game,* 48, 3, 39–48, figs. 1–6.

Bell, R.R. 1963. Preliminary age determination of bluefin tuna, *Thunnus thynnus, Calif. Fish and Game,* 49, 4, 307.

Bell, R.R. 1964. Weight-length relationship for bluefin tuna in the California fishery, 1963, *Calif. Fish and Game,* 50, 3, 216–218.

Belloc, G. 1937. Note sur un poisson comestible nouveau de la côte occidentale d'Afrique (*Paracubiceps ledenoisi* nov. gen., nov. sp.), *Rev. Trav. l'Office Pêches Marit.,* 10, 3, 353–356, 4 figs.

Berenbeim, D.Ya. 1968. Vliyanie temperatury vody na sroki neresta atlanticheskoi i tikhookeanskoi skumbrii (Effect of water temperature on the spawning period of Atlantic and Pacific mackerels). *Tr. Kaliningradsk. Tekhn. Inst. Rybn. Prom.,* vol. 20, pp. 53–69.

Berg, L.S. 1940. Sistema ryboobraznykh i ryb, nyne zhivushchikh i iskopaemykh (Classification of present and fossil ichthyoids and fishes). *Ezhegodn. Zool. Muzeya Akad. Nauk SSSR,* 5, 2, 87–517, figs. 1–190.

Berg, L.S. 1948-1949. Ryby presnykh vod SSSR i sopredel'nykh stran. Izd. 4-e (Fresh-water Fishes of the Soviet Union and Adjoining Countries. 4th ed.). *Tr. Zool. Inst. Akad. Nauk SSSR,* pt. 1: 466 pp, 281 figs.; pt. 2: pp. 469–925, figs. 288–674; pt. 3: pp. 929–1382, figs. 675–946; map.

Blackburn, M. 1965. Oceanography and the ecology of tunas. *Oceanogr. and Marine Biol.* London, vol. 3, pp. 299–322.

Bleeker, P. 1854. Fauna ichthyologicae japonicae species novae, *Nat. Tijdschr. Ned. Ind.,* vol. 6, pp. 395–426.

Bleeker, P. 1854-1857. Nieuwe nalezingen op de ichthyologie van Japan, *Verh. Bat. Gen.,* vol. 26, pp. 1–132.

Bloch, M.E. 1795. *Naturgeschichte der ausländischen Fische,* vol. 9, pp. 1–192; *Oder Allgemeine Naturgeschichte der Fische,* vol. 12, pp. 1–192. Berlin.

Boeseman, M. 1947. Revision of the fishes collected by Burger and von Siebold in Japan, *Zool. Meded. (Leiden),* vol. 28, pp. 1–242, pls. 1–5.

Bonaparte, C.L. 1837. *Iconographia della fauna italica, per le quattro classi degli animali vertebrati,* 3, 2, *Pesce.* Roma.

Briggs, J.C. and F.H. Berry. 1959. The Draconettidae—a review of the family with a description of a new species, *Copeia,* 2, pp. 123–133.

Brock, V.E. 1954. Some aspects of the biology of the aku, *Katsuwonus pelamis,* in the Hawaiian Islands, *Pacif. Sci.,* 8, 1, 94–104.

Brock, V.E. 1965. A review of the effects of the environment on tuna, *Spec. Publ. Internat. Commiss. North-West Atlant. Fish.,* vol. 6, pp. 75–92.

Bullis, H.R. and F.J. Mather. 1956. Tunas of the genus *Thunnus* of the northern Caribbean, *Amer. Mus. Novitates,* vol. 1765, pp. 1–12.

542

Calkins, T.P. and W.L. Klawe, 1963. Synopsis of biological data on black skipjack, *Euthynnus lineatus* Rishinouye, 1920, *FAO. Fish. Repts.*, 2, 6, 130-146.

Cavaliere, A. 1963. Studi sulla biologia a pesca di *Xiphias gladius* L., *Boll. Pesca, Piscicolt. e Idrobiol.*, 18, 2, 143-170.

Chapman, W.M. and L.D. Townsend. 1938. The osteology of *Zaprora silenus* Jordan, *Ann. Mag. Nat. Hist.*, 11, 2, 89-117, figs. 1-10.

Cheeseman, T.F. 1876. Notes on the swordfish (*Xiphias gladius*), *Trans. New Zeal. Inst.*, vol. 8, pp. 219-220.

Chigirinskii, A.I. 1970. Raspredelenie ikry i lichinok stavridy i skumbrii v Vostochno-Kitaiskom more (Distribution of the eggs and larvae of horse mackerel in the East China Sea). Sb. *Issledovaniya po Biologii i Promyshlennoi Okeanografii.* Vladivostok, pp. 67-77.

Chyung, M.K. 1961. *Illustrated Encyclopedia: The Fauna of Korea.* 2. *Fishes.* Ministry of Education, Korea (Seoul), vol. IV, 861 pp., 311 pls. (72 col.), text-figs. (in Korean).

Clemens, H.B. 1961. The migration, age, and growth of Pacific albacore (*Thunnus germo*), 1951-1958, *Calif. Fish and Game, Fish Bull.*, vol. 115, pp. 1-128, 56 figs.

Clemens, H.B. 1962. A model of albacore migration in the North Pacific Ocean. *FAO. World Sci. Meet. Biol. Tunas, La Jolla, USA*, sec. 4a, 5, exp. paper 31, pp. 1-12.

Clemens, H.B. 1966. Tagging experiments on albacore and bluefin tuna in the North Pacific. *Proc. the Eleventh Pac. Sci. Congress, Tokyo, 7.*

Clemens, W.A. and G.V. Wilby. 1961. Fishes of the Pacific coast of Canada, *Bull. Fish. Res. Bd., Canada*, pp. 1-443, figs. 1-281, pl. 1.

Clemens, H.B. and R.A. Iselin. 1962. Food of Pacific albacore in the California Fishery. *FAO. World Sci. Meet. Biol. Tunas, La Jolla, USA*, sec. 5. exp. paper 30, pp. 1-13.

Clemens, H.B. and W.L. Craig. 1965. An analysis of Californian albacore fishery, *Calif. Fish and Game, Fish Bull.*, vol. 128, pp. 1-301, 176 figs.

Clemens, H.B. and G.A. Flittner. 1969. Bluefin tuna migrate across the Pacific Ocean, *Calif. Fish and Game*, 55, 2, 132-135, 2 figs.

Clothier, Ch.R. 1950. A key to some southern California fishes based on vertebral characters, *Calif. Fish and Game, Fish Bull.*, vol. 79, pp. 1-83, pl. 23, text-figs, 1-21.

Collette, B.B. 1962. A preliminary review of the tunas of the genus *Thunnus.* In: Pacific Tuna Biology Conference August 14-17, 1961, Honolulu, Hawaii, *U.S. Fish and Wildl. Surv., Spec. Sci. Rep., Fish.*, vol. 415, no. 24.

Collette, B.B. 1966. The genera of scombrid fishes. *Proc. the Eleventh Pac. Sci. Congress, Tokyo, 7.*

Collette, B.B. and R.H. Gibbs. 1963. Preliminary field guide to the mackerel and tuna-like fishes of the Indian Ocean (Scombridae). Edit. U.S. Nat. Mus., 5, Rotoprint, pp. 1–48, ill.

Copley, H. 1936. The swordfish of the Indian Ocean, *Ceylon Sea Anglers Club Quart.*, vol. 1, pp. 19–23.

Cuvier, G. and A. Valenciennes. 1833. *Histoire naturelle des poissons.* Paris, vol. 9, 512 pp.

Cuvier, G. and A. Valenciennes. 1837. *Histoire naturelle des poissons.* Paris, vol. 12, 508 pp.

D'Aubenton, F. and M. Blanc. 1965. Étude systématique et biologique de *Scomberomorus sinensis* (Lacépède, 1802), *Bull. Mus. Nat. Hist. Nat., Paris,* 2nd ser., 37, 1, 233, figs. 1–5.

Davidoff, E.B. 1963. Size and year class composition of catch, age and growth of yellowfin tuna in the eastern tropical Pacific Ocean, 1951–1961, *Bull. Inter-Amer. Trop. Tuna Comm.*, 250, 8, 201.

Dekhnik, T.V. 1959. Razmnozhenie i razvitie yaponskoi skumbrii *Pneumatophorus japonicus* (Houttuyn) u beregov yuzhnogo Sakhalina [Reproduction and development of chub mackerel *Pneumatophorus japonicus* (Houttuyn)]. *Issled. Dal'nevost. Morei SSSR,* 6, 2, 97–108, figs. 1–3.

Deraniyagala, P.E. 1933. The Istiophoridae and Xiphiidae of Ceylon, *Spolia Zeylon,* vol. 36, pp. 137–142.

Dotsu [Dotu], Y. 1954. On the life history of a goby, *Chaenogobius castanea* O'Shaugnessy, *Japan. J. Ichth.*, 3 (3/4), 5, 133–138, 4 figs.

Dotsu [Dotu], Y. and T. Takita. 1967. Induced spawning by hormone operation, egg development, and larvae of blind gobioid fish, *Odontamlyopus rubicundus, Bull. Fac. Fisher. Nagasaki Univ.*, vol. 25, pp. 135–144, 3 figs.

Dotsu [Dotu], Y., S. Arima and S. Mito. 1965. The biology of eleotrid fishes *Eviota abax* and *Eviota zonura, Bull. Fac. Fisher. Nagasaki Univ.,* vol. 18, pp. 41–49, 3 figs., 1 pl.

Dotu, Y. 1955a. On the life history of a gobioid fish, *Eutaeniichthys gilli* Jordan and Snyder, *Bull. Biogeogr. Soc. Japan*, vol. 16–19, pp. 338–344, 6 figs.

Dotu, Y. 1955b. The life history of a goby, *Chaenogobius urotaenia* (Hilg.), *Sci. Rep. Fac. Agric. Kyushu Univ.,* 15, 3, 367–374.

Dotu, Y. 1957a. A new species of goby with a synopsis of the species of the genus *Luciogobius* Gill and its allied genera, *J. Fac. Agric. Kyushu Univ.*, 11, 1, 69–76, 1 pl.

Dotu, Y. 1957b. The life history of the goby *Luciogobius guttatus* Gill, *Sci. Bull. Fac. Agric. Kyushu Univ.*, 16, 1, 93–100, 6 figs.

Dotu, Y. 1957c. On the bionomics and life history of the eellike goby,

Odontamblyopus rubicundus (Hamilton), *Sci. Bull. Fac. Agric. Kyushu Univ.*, 16, 1, 101-110, 9 figs.

Dotu, Y. 1957d. The bionomics and life history of the goby *Triaenopogon barbatus* (Günther) in the innermost part of Ariake, *Sci. Bull. Fac. Agric. Kyushu Univ.*, 16, 2, 261-274.

Dotu, Y. 1958a. The bionomics and life history of two gobioid fishes, *Tridentiger undicervicus* and *Tridentiger trigonocephalus* (Gill) in the innermost part of Ariake Sound, *Sci. Bull. Fac. Agric. Kyushu Univ.*, 16, 3, 343-358, 7 figs.

Dotu, Y. 1958b. The bionomics and larvae of the two gobioid fishes, *Ctenotrypauchen microcephalus* (Bleeker) and *Taenioides cirratus* (Blyth), *Sci. Bull. Fac. Agric. Kyushu Univ.*, 16, 3, 371-380, 4 figs.

Dotu, Y. 1959. The life history and bionomics of the gobiid fish *Aboma lactipes* (Hilgendorf), *Bull. Fac. Fisher. Nagasaki Univ.*, vol. 8, pp. 196-201, 3 figs., pl. 19.

Dotu, Y. 1961a. The bionomics and life history of the gobioid fish *Rhinogobius giurinus* (Rutter), *Bull. Fac. Fisher. Nagasaki Univ.*, vol. 10, pp. 120-125, 2 figs. pl. 16.

Dotu, Y. 1961b. The bionomics and life history of the gobioid fish *Apocryptodon bleekeri* (Day), *Bull. Fac. Fisher. Nagasaki Univ.*, vol. 10, pp. 133-139, 2 figs. pl. 28.

Dotu, Y. 1963. On the blind gobioid fish, *Luciogobius albus* Regan, *Zool. Mag., Tokyo*, vol. 72, pp. 1-5, 3 figs.

Dotu, Y. and S. Mito. 1955. On the breeding habits, larvae, and young of a goby, *Acanthogobius flavimanus* (Temm. et Schl.), *Japan. J. Ichth.*, 4, 4 (5, 6), 153-161, 5 figs.

Dotu, Y. and S. Mito. 1958. The bionomics and life history of the gobioid fish *Luciogobius saikaiensis* Dotu, *Sci. Bull. Fac. Agric. Kyushu univ.*, 16, 3, 419-426, fig. 1, pl. 20.

Dotu, Y. and S. Fujita. 1959. The bionomics and life history of the eleotrid fish *Eleotris oxycephala* Temm. et Schl., *Bull. Fac. Fisher. Nagasaki Univ.*, vol. 8, pp. 191-196, pl. 18.

Dotu, Y. and T. Tsutsumi. 1959. The reproductive behavior in the gobiid fish *Pterogobius elapoides* (Günther), *Bull. Fac. Fisher. Nagasaki Univ.*, vol. 8, pp. 186-190, fig. 1, pl. 16.

Dotu, Y. and S. Mito. 1963. The bionomics and life history of the eleotrid fish *Asteropteryx semipunctatus* Rüppell, *Bull. Fac. Fisher. Nagasaki Univ.*, vol. 15, pp. 10-16, 3 figs., 1 pl.

Dotu, Y., S. Mito and M. Ueno. 1955. The life history of a goby, *Chaeturichthys hexanema* Bleeker, *Sci. Bull. Fac. Agric. Kyushu Univ.*, 15, 3, 359-365, 33 figs.

Duncker, G. and E. Mohr. 1939. Revision der Ammodytidae, *Mitt. Zool. Mus. Berlin*, 24, 1, 8-31, 4 figs.

Ehrenbaum, E. 1924. Scombriformes, *Rep. Danish Oceanogr. Exped.*, *1908-1910*, 2, *Biology*, A, 11, 1-42, figs. 1-10.

Einarsson, H. 1951. The postlarval stages of sand eels (Ammodytidae) in Faeroe, Iceland, and W. Greenland waters, *Acta Natur. Islandica, Reykjavik*, 1, 7, 1-75, 54 figs., 2 pls.

Enami, A. and Y. Dotu. 1961. Transplantation of the gobioid fish *Boleophthalmus chinensis* (Osbeck) from the inner part of Ariake Sound to Fukuoka City, *Bull. Fac. Fisher. Nagasaki Univ.*, vol. 10, pp. 141-147, pls. XIX-XXI.

Eschmeyer, W.N. 1963. A deep-water-trawl capture of two swordfish (*Xiphias gladius*) in the Gulf of Mexico, *Copeia*, 3, p. 590.

Esipov, B.V. 1928. O tuntsakh (The tunas). *Ukrain'skii Mislivets'ta Ribalka*, vol. 1, pp. 67-68.

Euphrasen, B.A. 1788. Beskrifning På trenne fiskar, *Vetensk. Akad. Nya Handl. Stockholm*, vol. 9, pp. 51-55, pl. 9.

Fedorova, L.P. 1968. Nekotorye dannye o morfologicheskikh osobenno-styakh i raspredelenii ikry i lichinok vostochnoi skumbrii (Some observations on the morphological peculiarities and distribution of the eggs and larvae of the Japanese mackerel). *Rybnoe Khozyaistovo*, vol. 4, pp. 14-15.

Fink, B.D. 1966. Tuna tagging in the eastern tropical Pacific Ocean. *Proc. the Eleventh Pac. Sci. Congress, Tokyo*, 7.

Fitch, J.E. 1960. Swordfish, *Xiphias gladius, Calif. Fish and Game*, pp. 63-64, 2 figs.

Fitch, J.E. and P.M. Roedel. 1963. A review of the frigate mackerels (genus *Auxis*) of the world, *FAO, Fish Repts.*, 3, 6, 1329-1342.

Flittner, G.A. 1966. Bluefin tuna in the North Pacific Ocean. *Proc. the Eleventh Pac. Sci. Congress, Tokyo*, 7.

Fowler, H.W. 1933. Description of a new long-finned tuna (*Semathunnus guildi*) from Tahiti, *Proc. Acad. Nat. Sci., Philadelphia*, vol. 85, pp. 163-164, pl. 12.

Fowler, H.W. 1936. The marine fishes of West Africa, *Bull. Amer. Mus. Nat. Hist.*, 70, 2, 607-1493.

Fowler, H.W. 1943. Description and figures of new fishes obtained in Philippine seas and adjacent waters by the U.S. Bur. Fish. Steam. *Albatross*. Contrib. to the biology of the Philipp. arch. and adjc. waters, *Bull. U.S. Nat. Mus.*, 100 (14, 1), 53-91, figs. 4-25.

Fowler, H.W. 1958-1962. A synopsis of the fishes of China, *Quart. J. Taiwan Mus.* Part VIII. Blennioid and related fishes, 1958, 11, 3-4, 147-339, figs. 1-47; 1959, 12, 1-2, 67-97, 9 figs.; Part IX. Gobioid fishes, 1960, 13, 3-4, 91-161, figs. 1-31; 1961, 14, 1-2, 49-87, figs. 32-50; 3-4, 203-250, figs. 51-71; 1962, 15, 1-2, 1-77, figs. 72-92.

Fowler, H.W. and B.A. Bean. 1929. The fishes of the series Capriformes, Ephippiformes, and Squamipennes, collected by the U.S. Bur. Fish. Steam. *Albatross* chiefly in Philipp. seas and adjc. waters, *Bull. U.S. Nat. Mus.*, 100, 8, 1-352, figs. 1-25.

Franz, V. 1910. Die japanischen Knochenfische der Sammlungen Haberer und Doflein, *Abhandi. Math.-Phys. Klasse. Akad. Wiss.*, 4 supple., vol. 1, pp. 1-135, pls. 1-11.

Fraser-Brunner, A. 1949. On the fishes of the genus *Euthynnus, Ann. Mag. Nat. Hist.*, (12) 2, 622-627, 2 text-figs.

Fraser-Brunner, A. 1950. The fishes of the family Scombridae, *Ann. Mag. Nat. Hist.*, (12) 3, 131-163, figs.

Frey, H.W. (ed.). 1971. California's living marine resources and their utilization, *California Fish and Game,* pp. 1-148 (82-92).

Fujino, K. 1966. Instructions for collecting blood and serum samples from tuna fish, *FAO, Fish. Circ.*, no. 26, pp. 1-5.

Fujino, K. 1967. Review of subpopulation studies on skipjack tuna, *Proc. West. Assoc. Game Fish Comm.*, vol. 47, pp. 349-371.

Fujino, K. 1969a. Atlantic skipjack tuna genetically distinct from Pacific specimens, *Copeia,* 3, pp. 626-629, 2 figs.

Fujino, K. 1969b. Skipjack subpopulation identified by genetic characteristics in the western Pacific. *Proc. Coop. Study Kuroshio and Adjac. Reg. Symp. Honolulu. Hawaii,* 29, 1968.

Fujino, K. and T. Kang. 1968a. Serum esterase groups of Pacific and Atlantic tunas, *Copeia,* 1, 56-63.

Fujino, K. and T. Kang. 1968b. Transferring groups of tuna, *Genetics,* vol. 59, pp. 79-91.

Fujino, K. and T. Kazama. 1968. The "Y"-system of skipjack tuna blood groups, *Vox Sanguinis,* 14, 5, 383-395.

Geft, V.N. 1970. Dannye po biologicheskoi Kharakteristike i raspro-deleniyu zheltoperogo tuntsa v priekvatorial'noi chasti Tikhogo okeana (On the biological characteristics and distribution of yellowfin tuna in the equatorial Pacific). In: Sb. *Issledovanie Biologii i Promyshlennoi Okeanografii.* Vladivostok, pp. 49-57.

Gehringer, J.W. 1956. Observations on the development of the Atlantic sailfish *Istiophorus americanus* (Cuvier) with notes on an unidentified species of istiophorid, *Fish. Bull. Fish and Wildlife Serv.*, 57, *Fishery Bull.*, vol. 110, pp. 139-171.

Genovese, S. 1962. Sul regime alimentare di *Thunnus thynnus* (L.), *Boll. Pesca, Piscicolt. e Idrobiol.,* vol. 15, p. 2.

Gibbs, R.H. and B.B. Collette. 1966a. Comparative anatomy and systematics of tunas, genus *Thunnus, Fish. Bull. U.S. Fish and Wildlife Serv.*, 66, 1, 65-130, 35 figs.

Gibbs, R.H. and B.B. Collette. 1966b. The species of tunas. Genus *Thunnus. Proc. the Eleventh Pac. Sci. Congress, Tokyo,* 7.

Gilbert, C.H. 1904. Notes on fishes from the Pacific coast of North America, *Proc. Calif. Acad. Sci.*, ser. 3, *Zool.*, 3, 9, 255-271, pls. 25-29.

Gilbert, C.H. 1905. Fishes collected by the U.S. Fish. Steam. *Albatross* in southern California in 1904, *Proc. U.S. Nat. Mus.*, 48, 2075, 305-380, pls. 14-22, 24 figs.

Godsil, H.C. 1945. The Pacific tunas, *Calif. Fish and Game,* 31, 4, 185-194, figs. 56-61.

Godsil, H.C. 1954. A descriptive study of certain tunalike fishes, *Calif. Fish and Game, Fish Bull.*, vol. 97, pp. 411-413.

Godsil, H.C. 1955. A description of two species of bonito, *Sarda orientalis* and *S. chiliensis* and consideration of relationships within the genus, *Calif. Fish and Game, Fish Bull.*, vol. 99, p. 42.

Godsil, H.C. and R.D. Byers. 1944. A systematic study of the Pacific tunas, *Calif. Fish and Game, Fish Bull.*, vol. 60, pp. 1-131, figs. 1-71.

Godsil, H.C. and E.K. Holmberg. 1950. A comparison of the bluefin tunas, genus *Thunnus,* from New England, Australia, and California, *Calif. Fish and Game, Fish Bull.*, vol. 77, pp. 1-55, 15 figs.

Godsil, H.C. and E.C. Greenhood. 1951. A comparison of the populations of yellowfin tuna, *Neothunnus macropterus,* from the eastern and central Pacific, *Calif. Fish and Game, Fish Bull.*, vol. 82, pp. 1-32.

Golenchenko, A.P. 1960. Mech-ryba (Swordfish). *Priroda,* vol. 4, p. 115.

Gorbunova, N.N. 1965a. O nereste skumkrievidnykh ryb (Pisces, Scombroidei) v Tonkinskom zalive Yuzno-Kitaiskogo morya [Spawning of scombroids (Pisces, Scombroidei) in the Gulf of Tonkin of the South China Sea]. *Tr. Inst. Okeanol. Akad. Nauk SSSR,* vol. 80, pp. 165-176.

Gorbunova, N.N. 1965b. O nakhozhdenii lichinok skumbriovidnykh ryb (Pisces, Scombroidei) v vostochnoi chasti Indiiskogo okeana [Distribution of the larvae of scombroids (Pisces, Scombroidei) in the eastern part of the Indian Ocean]. *Tr. Inst. Okeanol. Akad. Nauk SSSR,* vol. 80, pp. 32-35, 2 figs.

Gorbunova, N.N. 1965c. Sroki i usloviya razmnozheniya skumkriovidnykh ryb (Pisces, Scombroidei) [Periods and conditions of reproduction of scombroids (Pisces, Scombroidei)]. *Tr. Inst. Okeanol. Akad. Nauk SSSR,* vol. 80, pp. 36-61.

Gorbunova, N.N. 1969a. Raiony razmnozheniya i pitanie lichinok mech-ryby [*Xiphias gladius* L. (Pisces, Xiphiidae)] [Areas of reproduction and feeding of the larvae of swordfish, *Xiphias gladius* L. (Pisces, Xiphiidae)]. *Vopr. Ikhtiologii,* 9, 3, 474-488.

Gorbunova, N.N. 1969b. O dvukh tipakh lichinok makerelevidnogo tuntsa roda *Auxis* (Pisces, Scombroidei)[Two types of larvae of the frigate mackerel, genus *Auxis* (Pisces, Scombroidei)]. *Vopr. Ikhtiologii,* 9, 6, 1036-1046.

Gosline, W.A. 1955. The osteology and relationships of certain gobioid fishes, with particular reference to the genera *Kraemeria* and *Microdesmus, Pacif. Sci.,* 9, 2, 158-170.

Gosline, W.A. 1960. Hawaiian lava-flow fishes, pt. 4. *Snyderidia canina* Gilbert, with notes on the osteology of ophidioid families, *Pacif. Sci.,* 14, 4, 373-381.

Gosline, W.A. 1963. Notes on the osteology and systematic position of *Hypoptychus dybowskii* Steindachner and other elongated Perciformes fishes, *Pacific Sci.,* 17, 1, 90-101, 8 figs.

Gosline, W.A. 1968. The suborders of perciform fishes, *Proc. U.S. Nat. Mus.,* 124, 3647, 1-78 (17-28, 65).

Graham, J. and J. McGary. 1961. Investigation of the potential albacore resource of the central North Pacific, *Comm. Fish. Rev.,* 23, 11, 1-7, ill.

Gratsianov, V.I. 1907. Opyt obzora ryb Rossiskoi imperii v sistema-ticheskom i geograficheskom otnoshenii (Attempt at a taxonomic and geographic review of fishes from the Russian empire). *Tr. Otdela Ikhtiologii Russkogo Obshchestava Akklimatizatsii Zhivotnykh i Rastenii,* vol. 4, pp. i-xxx+1-567.

Gregory, W.R. and G.M. Conrad. 1937. The comparative osteology of swordfish (*Xiphias*) and sailfish (*Istiophorus*), *Amer. Mus. Novitates,* vol. 952, pp. 1-25, 12 figs.

Grey, M. 1953. Fishes of the family Gempylidae, with records of *Nesiarchus* and *Epinnula* from the western Atlantic and descriptions of two new subspecies of *Epinnula orientalis, Copeia,* 3, pp. 135-141.

Grey, M. 1955. The fishes of the genus *Tetragonurus* Risso, *Dana Rep.,* vol. 41, pp. 1-75, 16 figs.

Grey, M. 1960. Description of a western Atlantic specimen of *Scombrolabrax heterolepis* Roule, and notes on fishes of the family Gempylidae, *Copeia,* pp. 210-215, 8 figs.

Grinols, R.B. 1969. Atlantic skipjack tuna, genetically distinct Pacific specimens, *Copeia,* 3, pp. 626-629.

Gudger, E.W. 1938. Tales of attacks by the ocean gladiator: How the swordfish *Xiphius gladius* wrecks occasional vengeance by spearing the dories of the fishermen who persecute him, *Proc. U.S. Nat. Hist.,* vol. 41, pp. 128-137.

Guitart, N.D. 1964. Biologia pesquera dei Emperador o Pez de Espada, *Xiphias gladius* Linnaeus (Telostomi: Xiphiidae en las Aguas de

Cuba), *Acad. Ciencias Republica Cuba. Poeyana Instituto de Biologia,* ser. 2, 1, 37, figs. 1-16, pls.

Haedrich, R.L. 1967. The stromateoid fishes: Systematics and a classification, *Bull. Mus. Comp. Zool. Harvard Univ., Cambridge,* 135, 2, 31-139, 56 figs.

Haedrich, R.L. 1969. A new family of aberrant stromateoid fishes from the equatorial Indo-Pacific, *Dana Rep.,* vol. 76, pp. 1-14, 10 figs.

Hamre, J. 1963. Size and composition of tuna stocks, *FAO. Fish. Rep.,* 3, 6, 1023-1039.

Hennemuth, R. 1961. Year class, abundance, mortality, and yield per recruit of yellowfin tuna in the Eastern Pacific Ocean, 1954-1959, *Bull. Inter-Amer. Trop. Tuna Comm.,* 6, 1, 3-32, ill.

Herre, A.W. 1927. Gobies of the Philippines and the China Sea, *Bureau of Science. Manila. Monograph.,* 23, 1, 1-352, 6 figs., pls. 1-30.

Herzenstein, S. 1890. Ichthyologische Bemerkungen aus dem Zoologischen Museum der Kaiserlichen Akademie der Wissenschaften, *Mélanges Biologiques Tirés du Bulletin de l'Académie Impériale des Sciences de St. Pétersbourg,* vol. XIII, 1, 113-126; 2, 1890, 127-141; 3, 1892, 219-235.

Higgins, B.E. 1967. The distribution of juveniles of four species of tunas in the Pacific Ocean. *FAO. Proc. Indo-Pac-Fish. Council., Bangkok, 12 Sess.,* sect. 2, pp. 79-99, figs. 1-2.

Hikita, T. 1950. Notes on the fish fauna of Volcano Bay in Hokkaido. I, *Sci. Rep. Hokkaido Fish. Hatchery,* 5, 2, 1-13.

Hikita, T. 1952. Notes on the fishes and aquatic animals found in Lake Notoro in Hokkaido, *Sci. Rep. Hokkaido Fish. Hatchery,* 7, 1/2, 1-18, 3 pls.

Hikita, T. and T. Hikita. 1950. On a new wrymouth fish found in Japan, *Japan. J. Ichth.,* 1, 2, 140-142, fig. 1.

Hobbs, K.L. 1929. A new species of *Centrolophus* from Monterey Bay, California, *J. Wash. Acad. Sci.,* 19, 20, 460-461.

Honma, Y. 1952. A list of the fishes collected in the province of Echigo, including Sado Island, *Japan. J. Ichth.,* 2, 3, 138-145; 2, 4-5, 220-229.

Honma, Y. 1954-1957. On the rare bottom fishes found in the vicinity of Echigo Province and Sado Island in the Sea of Japan. I, *J. Fac. Sci. Niigata Univ.,* 1954, 2, 1, 1-5; II-1. c., 1955, 2, 2, 45-48; III-1. c., 1957, 2, 4, 103-109, 6 figs.

Honma, Y. 1955-1957. A list of the fishes found in the vicinity of Sado Marine Biological Station. I, *J. Fac. Sci. Niigata Univ.* (2), 1955, 2, 2, 49-60; II-1. c., 1956, 2, 3, 79-87; III-1. c., 1957, 2, 4, 111-116.

Honma, Y. 1956. On the thyroid gland of the sailfish, *Histiophorus orientalis* (Temminck and Schlegel), *Bull. Japan. Soc. Sci. Fish.,* 21, 9, 1016-1018, 2 figs.

Honma, Y. 1957. On the pituitary gland of a northern Japanese blenny, *Stichaeus grigorjewi* Herzenstein, *Japan. J. Ichth.*, 5, 3-6, 93-98, figs. 1-2.

Honma, Y. 1963. Fish fauna (Agnatha, Chondrichthyes, Osteichthyes) of Sado Island, Sea of Japan, *Publ. Sado Mus.*, vol. 5, pp. 12-32.

Honma, Y. and Ch. Sugihara. 1963. A revised list of the blennioid and ophidioid fishes of the suborder Blenniina obtained from the waters of Sado Island, including the area of Yamagata Prefecture, Sea of Japan, *Bull. Sado Mus.*, vol. 11, pp. 5-9.

Honma, Y. and T. Kitami. 1967. A list of the fishes found in the vicinity of Sado Marine Biological Station. IV, *Sci. Rep. Niigata Univ.*, Ser. D (*Biol.*). vol. 4, pp. 59-74, figs. 1-10.

Honma, Y. and T. Kitami. 1970. List of fishes in the vicinity of Sado Mar. Biol. Stat., *Sci. Rep. Niigata Univ.*, ser. D (*Biol.*), vol 7, pp. 63-86.

Honma, Y.R., R. Mizusawa and M. Okiyama. 1972. Further additions to "A list of the fishes in the province of Echigo, including Sado Island." IX. *Bull. Biogeograph. Soc. Japan,* 28, 4, 47-57.

Honma, Y. et al. 1955-1972. Further additions to "A list of the fishes collected in the province of Echigo, including Sado Island." I, *Japan. J. Ichth.,* 1955, 4, 4-6, 212-217, 2 figs.; II−1. c., 4, 4-6, 218-222; III−1. c., 4, 4-6, 223-228; IV−1. c., 1956, 1-2, 59-60; V−1. c., 1957, 6, 4-6, 109-112; VI−1. c., 1959, 7, 5-6, 139-144; VII−1. c., 1962, 9, 1-6, 127-134; VIII−1. c., 1966, 14, 1-3, 53-61; IX. *Bull. Biogeograph. Soc. Japan,* 1972, 28, 4, 47-57.

Hotta, H. 1955. Seasonal distribution and growth of the frigate mackerel *Auxis tapeinosoma* Bleeker along the Pacific coast of Japan, *Bull. Tohoku Reg. Fish. Res. Lab.,* vol. 4, pp. 120-126.

Hotta, H., T. Abe and Y. Takashima. 1958. Notes on the head bones of Japanese mackerels of the genus *Pneumatophorus, Bull. Tohoku Reg. Fish. Res. Lab.,* vol. 12, pp. 101-105, 2 figs.

Hubbs, C.L. 1927. Notes on the blennioid fishes of western North America. Papers Michigan Acad. Sci., Arts and Letters, 7 (1926), 351-394.

Hubbs, Cl. 1944. Species of the circumtropical fish genus *Brotula, Copeia,* 3, pp. 162-173.

Hubbs, Cl. 1952. A contribution to the classification of blennioid fishes of the family Clinidae, with a partial revision of the eastern Pacific forms, *Bull. Stanford Univ., Ichth.,* 4, 2, 42-165, figs. 1-64.

Hubbs, Cl. 1953. Revision and systematic position of blennioid fishes of the genus *Neoclinus, Copeia,* 1, pp. 11-23, figs. 1-16.

Hubbs, C.L. and R.L. Wisner, 1953. Food of marlin in 1951 off San Diego, California, *Calif. Fish and Game,* 39, 1, 127-133, fig. 1, pl. 1.

Il'in, B.S. 1927. Opredelitel' bychkov (fam. Gobiidae) Azovskogo i ch꞉꞉nogo morei [Identification of gobies (family Gobiidae) from the Azₒv and Black seas]. *Tr. Azovsko-Chernom. Nauchno-Prom. Eksp.*, vol. 2, pp. 126-143, 1 pl. fig.

Imamura, T. and S. Hashitani. 1957. On the food habits of four fishes in Marsh Hinuma, *Bull. Fac. Lib. Arts. Ibaraki Univ. Nat. Sci.*, vol. 7, pp. 45-56.

Inoue, Motoo, Amano Ryohei, Iwasaki Yukinoba and Yamauti Minoru. 1968. Studies on environments alluring skipjack and other tunas. II. On the driftwoods accompanied by skipjack and tunas, *Bull. Japan. Soc. Sci. Fish.*, 34, 4, 283-287.

Ishigaki, T. and Y. Kaga. 1957. Fishery biological studies of sandlance (*Ammodytes personatus* Girard) in waters around Hokkaido. I. Especially on the structure of the population, *Bull. Hokkaido Reg. Fish. Res. Lab.*, vol. 16, pp. 13-38.

Ishikawa, A.C. 1904. Notes on some new or little-known fishes of Japan. I, *Proc. Dept. Nat. Hist., Tokyo Imp. Mus.*, 6, 1, 1-17, 7 pls.

Ito, Shoichi. 1970. Marine fauna of Teradomari coast, Niigata Prefecture, *Bull. Niigata Pref. Biol. Soc. Educ.*, vol. 6, pp. 21-36.

Iversen, E.C. 1956. Size variation of central and eastern Pacific yellowfin tuna, *U.S. Fish and Wildl. Serv., Spec. Sci. Rep. Fish.*, 174:00 [*sic*].

Iversen, R.T.B. 1962. Food of albacore tuna, *Thunnus germo* (Lacépède) in the central and northeastern Pacific, *U.S. Fish and Wildl. Serv. Fish. Bull.*, 62, 214, 459-481.

Iwai, T. 1963. Sensory capulae found in newly hatched larvae of *Blennius yatabei* Jordan and Snyder, *Bull. Japan. Soc. Sci. Fish.*, vol. 29, pp. 503-506, 2 figs.

Iwai, T. and I. Nakamura. 1964a. Branchial skeleton of the bluefin tuna, with special reference to the gill rays, *Bull. Nisaki Mar. Biol. Inst., Kyoto Univ.*, vol. 6, pp. 21-25, 1 fig.

Iwai, T. and I. Nakamura. 1964b. Olfactory organs of tunas with special reference to their systematic significance, *Bull. Misaki Mar. Biol. Inst., Kyoto Univ.*, vol. 7, pp. 1-8, 3 figs.

Iwai, T., J. Nakamura and K. Matsubara. 1965. Taxonomic study of tunas, *Misaki Mar. Biol. Inst., Kyoto Univ., Spec. Rep.*, vol. 2, pp. 1-51, figs. 1-24.

Jensen, A.S. 1952. Recent finds of Lycodinae in Greenland waters, *Medd. Greenland,* 142, 7, 1-28, 2 pls.

Johnson, C.R. 1971. Revision of the callionymid fishes referable to the genus *Callionymus* from Australian waters, *Mem. Queensland Mus.*, 16, 1, 103-140, 26 figs.

Jones, S. 1958. Notes on eggs, larvae, and juveniles of fishes from Indian waters. I. *Xiphius gladius* Linnaeus, *Indian J. Fish.*, 5, 2, 357-361.

Jones, S. 1961. Notes on eggs, larvae, and juveniles of fishes from Indian waters, *Indian J. Fish.*, 8, 1, 107-120, figs. 1-15.

Jones, S. 1962. Distribution of larval billfishes (Xiphiidae and Istiophoridae) in the Indo-Pacific with special reference to the collections made by the Danish Dana Expedition. *Proc. Symp. Scombroid Fishes, Mar. Biol. Ass. India*, pt. I, pp. 483-498.

Jones, S. 1963. Synopsis of biological data on the northern bluefin tuna *Kishinoella tonggol* (Bleeker) 1851 (Indian Ocean), *FAO. Fish. Rep.*, vol. 2, pp. 862-876.

Jones, S. and E.G. Silas. 1960. Indian tunas. A preliminary review with a key for their identification, *Indian J. Fish.*, 7, 2, 369-393, 15 figs.

Jones, S. and E.G. Silas. 1962. On fishes of the subfamily Scomberomorinae (family Scombridae) from Indian waters, *Indian J. Fish.*, 8, 1, 189-206, 9 figs.

Jones, S. and M. Kumaran. 1963. Distribution of larval tuna collected by the Carlsberg Foundation's Dana Expedition (1928-1930) from the Indian Ocean, *FAO. Fish. Rept.*, 3, 6, 1753-1774.

Jones, S. and E.G. Silas. 1963. Synopsis of biological data on skipjack *Katsuwonus pelamis* (L.) 1758 (Indian Ocean), *FAO. Fish. Rep.*, 2, 6, 663-694.

Jordan, D.S. (1902) 1903. Supplementary note on *Bleekeria mitsukurii* and on certain Japanese fishes, *Proc. U.S. Nat. Mus.*, vol. 26, pp. 693-696, 3 figs. 1 pl.

Jordan, D.S. and B.W. Evermann. 1896-1900. Fishes of North and Middle America, *Bull. U.S. Nat. Mus.*, 47, 1-4, ccxiv, 1-3313, pls. 1-392.

Jordan, D.S. and B.W. Evermann. 1926. A review of the giant mackerellike fishes, tunnies, spearfishes, and swordfishes. *Occasion. Papers. Calif. Acad. Sci.*, vol. 12, pp. 1-113, figs.

Jordan, D.S., B.W. Evermann and H.W. Clark. 1930 (1928). Check list of fishes and fishlike vertebrates of North and Middle America north of the northern boundary of Venezuela and Colombia, *Rep. U.S. Comm. Fisher.*, 2, 1, 1-670.

Jordan, D.S. and H.W. Fowler. 1902a. A review of the Chaetodontidae and related families of fishes found in the waters of Japan, *Proc. U.S. Nat. Mus.*, vol. 25, p. 560.

Jordan, D.S. and H.W. Fowler. 1902b. A review of the ophidioid fishes of Japan, *Proc. U.S. Nat. Mus.*, vol. 25, pp. 743-766, figs. 1-6.

Jordan, D.S. and H.W. Fowler. 1903. A review of dragonets (Callionymidae) and related fishes of Japan, *Proc. U.S. Nat. Mus.*, vol. 26, pp. 939-959.

Jordan, D.S. and C.L. Hubbs. 1925. Records of fishes obtained by D.S. Jordan in Japan, 1922, *Mem. Carnegie Mus.*, 10, 2, 93–346, 7 pls.

Jordan, D.S. and C.W. Metz. 1913. A catalog of the fishes known from the waters of Korea, *Mem. Carnegie Mus.*, 6, 1, 1–65, 65 figs. pls. 1–10.

Jordan, D.S. and A. Seale. 1906. Descriptions of six new species of fishes from Japan, *Proc. U.S. Nat. Mus.*, vol. 30, pp. 143–148, 6 figs.

Jordan, D.S. and J.O. Snyder. 1900. A list of fishes collected in Japan by Keinosuke Otaki and the United States Fish Commission Steamer *Albatross*, with descriptions of fourteen new species, *Proc. U.S. Nat. Mus.*, vol. 23, pp. 335–380, IX–XX.

Jordan, D.S. and J.O. Snyder. 1901a. Descriptions of nine new species of fishes contained in museums of Japan, *J. Coll. Sci. Univ., Tokyo*, 15, 2, 301–311, pls. XV–XVII.

Jordan, D.S. and J.O. Snyder, 1901b. List of fishes collected in 1883 and 1885 by Pierre Louis Jouy and preserved in the United States National Museum, with descriptions of six new species, *Proc. U.S. Nat. Mus.*, vol. 23, pp. 739–769, figs. 31–38.

Jordan, D.S. and J.O. Snyder. 1901c. A review of the gobioid fishes of Japan, with descriptions of twenty-one new species, *Proc. U.S. Nat. Mus.*, vol. 24, pp. 33–132, 33 figs.

Jordan, D.S. and J.O. Snyder. 1902a. A review of the blennioid fishes of Japan, *Proc. U.S. Nat. Mus.*, vol. 25, pp. 441–504, figs. 1–28.

Jordan, D.S. and J.O. Snyder. 1902b. On certain species of fishes confused with *Bryostemma polyactocephalum*, *Proc. U.S. Nat. Mus.*, vol. 25, pp. 613–618, figs. 1–3.

Jordan, D.S. and J.O. Snyder. 1904. Notes on the collections of fishes from Oahu Island and Laysan Island, Hawaii, with descriptions of four new species, *Proc. U.S. Nat. Mus.*, vol. 27, pp. 939–948.

Jordan, D.S. and E.Ch. Starks. 1895 (1896). Fishes of the Puget Sound, *Proc. Calif. Acad. Sci.*, (2), 5, 2, 785–855, pls. 57–63.

Jordan, D.S. and E.C. Starks. 1905. On a collection of fishes made in Korea by Pierre Louis Jouy, with descriptions of new species, *Proc. U.S. Nat. Mus.*, vol. 28, pp. 193–212.

Jordan, D.S. and S. Tanaka. 1927. Notes on new and rare fishes of the fauna of Japan, *Ann. Carnegie Mus.*, 17, 3/4, 385–394, pl. 34.

Jordan, D.S., S. Tanaka and J.O. Snyder. 1913. A catalogue of the fishes of Japan, *J. Coll. Sci. Univ., Tokyo*, 33, 1, 1–431, 396 figs.

Jordan, D.S. and W.F. Thompson. 1914. Record of the fishes obtained in Japan in 1911, *Mem. Carnegie Mus.*, 6, 4, 205–313, pls. 24–42, 87 figs.

Joseph, J. 1963. Fecundity of yellowfin tuna (*Thunnus albacares*) and skipjack (*Katsuwonus pelamis*) from the eastern Pacific Ocean, *Bull. Inter-Amer. Trop. Tuna Comm.*, 7, 4, 257–292, figs. tbls.

554

Joseph, J. 1966. Distribution and migration of yellowfin tuna. *Proc. the Eleventh Pac. Sci. Congress, Tokyo,* 7.

Joseph, J.F., G. Alverson, B. Fink and E.B. Davidoff. 1964. A review of the population structure of yellowfin tuna, *Thunnus albacares,* in the eastern Pacific Ocean, *Bull. Inter-Amer. Trop. Tuna Comm.,* vol. 9, pp. 55-112.

Kadzawara and Ito. 1953. Obraz zhizni Scombridae (Mode of life of Scombridae). *Fisher. Sci. Japan.* ser. 7, pp. 1-131, 17 figs. (in Japanese).

Kaganovskii, A.G. 1951. Migratsii skumbrii (*Pneumatophorus japonicus*) v Yaponskow more [Migrations of chub mackerel (*Pneumatophorus japonicus*) in the Sea of Japan]. *Izv. Tikhookeansk Nauchno-Issled. Inst. Rybn. Khoz. i Okeanografii.* Vladivostok, vol. 35, pp. 61-79, figs. 1-4.

Kaganovskii, A.G., P.A. Starovoitov and I.V. Kizevetter. 1947. Skumbriya (The Mackerel). Moscow.

Kamohara, T. 1938. Gemplyidae of Japan, *Annot. Zool. Japan.,* 17, 1, 45-50, pl. III, figs. 1-4.

Kamohara, T. 1940. Scombroidei (exclusive of Carangiformes), *Fauna Nipponica,* 152, 5, 1-225, figs. 1-102.

Kamohara, T. 1954. A review of the family Brotulidae found in the waters of prov. Tosa, Japan, *Rep. USA Mar. Biol. Stat.,* 1, 2, 1-14.

Kamohara, T. and T. Yamakawa. 1965. Fishes from Amami-Oshima and adjacent regions, *Rep. USA Mar. Biol. Stat.,* 12, 2, 1-27.

Kask, J.L. 1966. Future problems in tagging experiments on tunas and billfishes. *Proc. the Eleventh Pac. Sci. Congress, Tokyo,* 7.

Katayama, M. 1940. A catalogue of the fishes of Toyama Bay, *Toyama Haka Butsugaku-Koishi,* vol. 3, pp. 1-28.

Katayama, M. 1941. A new blennioid fish from Toyama Bay, *Zool. Mag.,* 53, 12, 591-593, 1 fig.

Katayama, M. 1943. On two new ophidioid fishes from the Sea of Japan, *Annot. Zool. Japan.,* 22, 2, 101-104, 2 figs.

Katayama, M. 1952a. Record of the fishes of northern Japan obtained off Tajima, *Bull. Educ. Yamaguchi Univ.,* 2, 1, 1-7.

Katayama, M. 1952b. A record of *Ariomma lurida* Jordan and Snyder from Japan, with notes on its systematic position *Japan. J. Ichth.,* 2, 1, 31-34, 2 figs.

Katoh, G., I. Yamanaka, A. Ouchi and T. Ogata. 1956. Progress report of cooperative research on trawl fishery resources in the Sea of Japan, *Bull. Japan Sea Reg. Fish. Res. Lab.,* vol. 4, pp. 1-330, figs. 1-148.

Khikita, T. and M. Khirosi [Hikita, T. and M. Hirosi]. 1952. Fishes of the northwestern part of the Sea of Japan. *Kom. po Issl. Ryby Promyshl. Sev. Chasti Yaponskogo Morya.* Otaru, pp. 1-70, 15 figs. (in Japanese).

Kikawa, S. 1962. Studies on the spawning activity of the Pacific tunas *Parathunnus mebachi* and *Neothunnus macropterus*, by the gonad index examination, *Occ. Rep. Nankai Reg. Fish. Res. Lab.*, 1.

Kikawa, S. 1963a. Synopsis of biological data on bonito *Sarda orientalis* Temminck and Schlegel, 1842, *FAO. Fish. Rep.*, 2, 6, 147–156.

Kikawa, S. 1963b. Synopsis on the biology of the little tuna *Euthynnus yaito* Kishinouye, 1923, *FAO. Fish. Rep.*, 2, 6, 218–240.

Kikawa, S. 1966. Spawning potential of bigeye and yellowfin tuna. *Proc. the Eleventh Pac. Sci. Congress, Tokyo*, 7.

Kikawa, S. and M.G. Ferraro. 1967. Maturation and spawning of tunas in the Indian Ocean. *FAO. Proc. Indo-Pacific Fisher. Congress, 12 Sess. Honolulu*, vol. 2, pp. 65–78, figs. 1–6.

Kimuro. 1953. Migration of chub mackerel *Pneumatophorus japonicus* (Houttuyn) on the basis of results of tagging in Japanese waters, *Bull. Japan. Soc. Sci. Fish.*, 19, 4, 415–423 (in Japanese).

King, Joseph E. and I.I. Ikehara. 1956. Comparative study of food of bigeye and yellowfin tuna in the central Pacific, *U.S. Fish. and Wildl. Serv. Fish. Bull.*, 57, 108, 61–85.

Kishinouye, K. 1915. The tunnies, *Proc. Sci. Fisher., Assoc. Tokyo*, 1, 1, 1–24, pls.

Kishinouye, K. 1923. Contributions to the comparative study of the so-called scombroid fishes, *J. Coll. Agric. Univ., Tokyo*, 8, 3, 293–475, pls. 13–34, 26 text-figs.

Kitakata, M. 1957. Fishery biological studies of sandlances (*Ammodytes personatus* Girard) in waters around Hokkaido. 2. On the age and growth, *Bull. Hokk. Reg. Fish. Res. Lab.*, vol. 16, pp. 39–48, 5 figs., pl.

Klawe, W.L. 1961. Notes on larvae, juveniles, and spawning of bonito (*Sarda*) from the eastern Pacific Ocean, *Pacific Sci.*, 15, 4, 487–493.

Klawe, W.L. and M.P. Miyake. 1967. An annotated bibliography on the biology and fishery of the skipjack tuna *Katsuwonus pelamis* of the Pacific Ocean, *Bull. Inter American Tropical Tuna Commission, La Jolla, California*, 12, 4, 139–363.

Kner, R. 1868. Folge neuer Fische aus dem Museum der Herren Joh. Cäs. Godeffroy und Sohn in Hamburg, *Sitzb. Akad. Wiss.*, 58, 1, 293–356, figs. 1–9.

Kobayashi, K. 1961a. Primary record of *Psenes maculatus* Lütken from the Notrth Pacific, *Bull. Fac. Fish., Hokkaido Univ.*, 11, 4, 191–194.

Kobayashi, K. 1961b. Young of the wolf-fish *Anarhichas orientalis* Pallas, *Bull. Fac. Fish. Hokkaido Univ.*, 12, 1, 1–4, ill.

Kobayashi, K. 1961c. Larvae and young of the sandlance *Ammodytes hexapterus* Pallas from the North Pacific, *Bull. Fac. Fisher. Hokkaido Univ.*, 12, 2, 111–120, 2 figs.

Kobayashi, K. 1962. Ichthyofauna of Oshoro Bay and adjacent waters, *Bull. Fac. Fisher. Hokkaido Univ.*, 12, 4, 253-264, 1 fig.

Kobayashi, K. and T. Ueno. 1956. Fishes from the northern Pacific and from Bristol Bay, *Bull. Fac. Fish. Hokkaido Univ.*, 6, 4, 239-265, figs. 1-9

Kobayashi, K. and K. Abe. 1963. Studies of the larvae and young of fishes from the boundary zones off the southeastern coast of Hokkaido, Japan, *Bull. Fac. Fish. Hokkaido Univ.*, [?], pp 165-179, 11 figs.

Kolesnikov, V.G., Yu.A. Torin and N.Z. Khlystov. 1961. O vliyanii okeanologicheskikh uslovii no respredelenii zheltoperogo tuntsa (Effect of the oceanological conditions on the distribution of yellowfin tuna). *Balt. Nauchno-Issled. Inst. Morsk. Rybn. Khoz. i Okeanografii (BaltNIRO)*, vol. 7, pp. 31-34.

Koumans, F.P. 1953. Gobioidea. In: *Fishes of the Indo Australian Archipelago*, edited by Weber and Beaufort, vol. 10, pp. 1-423, 95 figs.

Kramer, D. 1960. Development of eggs and larvae of Pacific mackerel and distribution and abundance of larvae 1952-1956, *Fish. Bull.*, 60, 174, 393-439, ill.

Kume, S. 1966. Distribution and migration of bigeye tuna in the Pacific Ocean. *Proc. the Eleventh Pac. Sci. Congress, Tokyo*, 7.

Kun, M.S. 1951. Pitanie skumbrii v Yaponskom more (Food of mackerel in the Sea of Japan). *Izv. Tikhookeansk. Nauchno-Issled. Inst. Morsk. Rybn. Khoz. i Okeanografii*. Vladivostok, vol. 34, pp. 67-79, figs. 1-3.

Kun, M.S. 1954. Osobennosti pitaniya segoletok i vzrosloi skumbrii (Peculiarities in feeding of fingerlings and adult mackerel). *Izv. Tikhookeansk, Nauchno-Issled. Inst. Rybn. Khoz. i Okeanografii*. Vladivostok, vol. 42, pp. 95-108, figs. 1-5.

Kundlius, M. 1964. Raspredelenie, povedenie i sposoby lova skumbrii v severo-zapadnoi chasti Tikhogo okeana (Distribution, behavior, and methods of mackerel fishing in the northwestern part of the Pacific Ocean). *Rybnoe Khozyaistvo*, vol. 12, pp. 9-13, figs. 1-4.

Kurogane, K. 1959. Morphometric comparison of the albacore from the Indian and Pacific oceans, *Rec. Oceanogr. Works Japan*, 5, 1, 68-84, ill.

La Monte, F.R. 1955. A review and revision of marlins of the genus *Makaira, Bull. Amer. Mus. Nat. Hist.*, 107, 3, 319-358, 9 pls.

Latysh, L.V. and A.S. Sokolovskii. 1972. Materialy o pitanii lichinok, val'kov i molodi skumbrii (*Scomber japonicus* Houttuyn) v zone techniya Kuro-Sivo [On the food of larvae, fingerlings, and fry of chub mackerel (*Scomber japonicus* Houttuyn) in the zone of the Kuru-Shima Current]. Sb. *Issledovanie Biologii Ryb i Promyehlennoi Okeanografii*. Vladivostok, pp. 114-120.

Lindberg, G.U. 1927. Promyslovye ryby Dal'nego Vostoka i ikh ispol'zovanie (Commercial fishes of the Far East and their utilization). *Tr., 1-i Konf. po Izuch. Proizvod. Sil Dal'nego Vostoka.* Khabarovsk-Vladivostok, vol. 4, pp. 19-59.

Lindberg, G.U. 1936. Materialy po rybam Primor'ya (Data on fishes of Primor'e). *Tr. Zool. Inst. Akad. Nauk SSSR,* vol. 3, pp. 393-407, figs. 1-10.

Lindberg, G.U. 1937. O sistematike i resprostranenii peschanok roda *Ammodytes* (Pisces) [Systematics and distribution of sand eels of the genus *Ammodytes* (pisces)]. *Vestn. Dal'nevost. Fil. Akad. Nauk SSSR.* Vladivostok, vol. 27, pp. 85-93.

Lindberg, G.U. 1938. O novykh rodakh i vidakh ryb sem. Blenniidae (Pisces), blizkikh k rodu *Anoplarchus* [New genera and species of fishes of the family Blenniidae (Pisces), close to the genus *Anoplarchus*). *Tr. Gidrobiol. Eksp. Zool. Inst. Akad. Nauk SSSR na Yaponskom More.* Moscow-Leningrad, vol. 1, pp. 499-514, figs. 1-6.

Lindberg, G.U. 1947. Predveritel'nyi spisok ryb Yaponskogo morya (Preliminary list of fishes of the Sea of Japan). *Izv. Tikhookeansk. Nauchno-Issled. Inst. Rybn. Khoz. i Okeanografii.* Vladivostok. vol. 25, pp. 125-206.

Lindberg, G.U. 1955. O nakhozdenii rybki-drakochika *Draculo mirabilis* Snyder (Pisces, Callionymidae) v zalive Pos'et u Vladivostoka [Occurrence of *Draculo mirabilis* Snyder (Pisces, Callionymidae) in Pos'et Bay off Vladivostok]. *Tr. Zool. Inst. Akad. Nauk SSSR,* vol. 18, pp. 385-388.

Lindberg, G.U. 1971. Opredelitel' i kharakteristika semeistv ryb mirovoi fauny (Keys and Characteristics of the Fish Families in World Fauna). Leningrad, 470 pp., 986 figs.

Lindberg, G.U. and G.D. Dul'keit. 1929. Materialy po rybam Shantarskogo morya (Data on fishes of Shantar Islet). *Izv. Tikhookeansk. Nauchno-Prom. St.* Vladivostok, 3, 1, 1-140, ill., map.

Lindberg, G.U. and A.P. Andriyashev, 1938. Obzor geograficheskikh form dal'nevostochnogo bychka *Icelus spiniger* Gilb. (Review of the geographic forms of *Icelus spiniger* Gilb.). *Tr. Gidrobiol. Eksp. Zool. Inst. Akad. Nauk SSSR na Yaponskom More.* Moscow-Leningrad, vol. 1, pp. 515-525, figs. 1-4.

Lindberg, G.U. and M.I. Legeza. 1959. Ryby yaponskogo morya i sopredel'nykh chastei Okhotskogo i Zheltogo morei. I: Amphioxi, Petromyzones, Myxini, Elasmobranchii, Holocephali (Fishes of the Sea of Japan and Adjacent Parts of the Sea of Okhotsk and the Yellow Sea. I: Amphioxi, Petromyzones, Myxini, Elasmobranchii, Holocephali). Moscow-Leningrad, vol. 1, 207 pp., 108 figs.

Lindberg, G.U. and M.I. Legeza. 1965. Ryby yaponskogo morya i sopredel'nykh chastei Okhotskogo i Tsheltogo morei. II. Teleostomi. XII. Acipenseriformes—XXVIII. Polynemiformes (Fishes of the Sea of Japan and Adjacent Parts of the Sea of Okhotsk and the Yellow Sea. II: Teleostomi. XII. Acipenseriformes—XXVIII. Polynemiformes). Moscow-Leningrad, vol. 2, 391 pp., 324 figs.

Lindberg, G.U. and Z.V. Krasyukova. 1969. Ryby Yaponskogo morya i sopredel'nykh chastei Okhotskogo i Zheltogo morei. III. Teleostomi. XXIX. Perciformes. 1. Percoidei (XC. Sem. Serranidae—CXLIV. Sem. Champsodontidae) [Fishes of the Sea of Japan and Adjacent Parts of the Sea of Okhotsk and the Yellow Sea. III: Teleostomi. XXIX: Perciformes. 1 Percoidei (XC. Family Serranidae—CXLIV. Family Champsodontidae)]. Leningrad, vol. 3, 480 pp., 431 figs.

Lindberg, G.U. and Z.V. Krasyukowa [Krasyukova]. *Fishes of the Sea of Japan and the Adjacent Areas of the Sea of Okhotsk and the Yellow Sea.* Part III. *Teleostomi. XXIX. Perciformes. 1. Percoidei (XC. Fam. Serranidae—CXLIV. Fam. Champsodontidae).* Jerusalem, 498 pp. 431 figs.

Lindberg, G.U. et al. 1959. Spisok fauny morskikh vod Yuzhnogo Sakhalina in Yushnykh Kuril'skikh ostrovov (Fauna of the marine waters of southern Sakhalin and the southern Kuril Islands). *Issled. Dal'nevost Morei SSSR,* vol. 6, pp. 173-257 (Ryby, pp. 247-256).

Lowe, R.T. 1843. Notices of fishes newly observed or discovered in Madeira during the years 1840, 1841, and 1842, *Proc. Zool. Soc., London,* vol. 11, pp. 81-95.

Macleay, W. 1885. A remarkable fish from Lorde Howe Island, *Proc. Linn. Soc. New South Wales,* vol. 10. pp. 718-720.

Magnuson, J.J. 1966. a comparative study of the function of continuous swimming. *Proc. the Eleventh Pac. Sci. Congress, Tokyo,* 7.

Magnuson, J.J. and J.H. Prescott. 1966. Courtship, locomotion, feeding, and miscellaneous behaviour of Pacific bonito (*Sarda chiliensis), Animal Behaviour,* 14, 1, 54-67, ill.

Maksimov, V.P. 1969. Pitani bol'sheglazogo tunstsa (*Thunnus obesus* Lowe) i mech-ryby (*Xiphias gladius* L.) vostochnoi chasti tropicheskoi Atlantiki [Food of bigeye tuna (*Thunnus obesus* Lowe) and swordfish (*Xiphias gladius* L.) from the eastern part of the tropical Atlantic]. *Tr. Atlant. Nauchno-Issled. Inst. Morsk. Rybn. Khoz. i Okeanografii,* vol. 25, pp. 87-99.

Makushok, V.M. 1958. Morfologicheskie osnovy sistemy stikheevykh i blizkikh k nim semeistv ryb (Stichaeoidae, Blennioidei, Pisces) [Morphological basis for the classification of pricklebacks and closely related fish families (Stichaeoidae, Blennioidei, Pisces)]. *Tr.. Zool. Inst. Akad. Nauk SSSR,* vol. 25, pp. 1-129, figs. 1-83.

Makushok, V.M. 1961a. Dopolnitel'nye dannye po morfologii i Sistema-tike krivorotov (Cryptacanthodidae, Blennioidei, Pisces) [Additional data on the morphology and classification of wrymouths (Cryptacan-thodidae, Blennioidei, Pisces)]. *Tr. Inst. Okeanol. Akad. Nauk SSSR,* vol. 43, pp. 184-197, figs. 1-4.

Makushok, V.M. 1961b. Nekotorye osobennosti stroeniya seismosen-sornoi sistemy severnykh blenniid (Stichaeoidea, Blennioidei, Pisces) [Some structural peculiarities of the seismosensory system of northern blenniid fishes (Stichaeoidea, Blennioidei, Pisces)]. *Tr. Inst. Okeanol. Akad. Nauk SSSR,* vol. 43, pp. 225-269, figs. 1-9.

Makushok, V.M. 1961c. Gruppa Neozoarcinae i ee mesto v sisteme (Zoarcidae, Blennioidei, Pisces) [The Neozoarcinae group and its place in classification (Zoarcidae, Blennoidei, Pisces)]. *Tr. Inst. Okeanol. Akad. Nauk SSSR,* vol. 43, pp. 198-224, figs. 1-6.

Manacop, P.R. 1958. A preliminary systematic study of the Philippine chub mackerels, family Scombridae, genera *Pneumatophorus* and *Rastrelliger, Philipp. J. Fish.,* 4, 2, 79-101.

Markina, A.D. 1959. Nekotorye dannye po biologii stikheye Grigor'eva (Some notes on the biology of Grigorev's prickleback). *Izv. Tikhookeansk. Nauchno-Issled. Inst. Rybn. Khoz. i. Okeanografii.* Vladivostok, vol. 47, p. 188.

Marschal, E. 1963. Description des stades postlarvaires et juveniles de *Neothunnus albacora* (Lowe) de l'Atlantique tropico-oriental, *FAO. Fish. Rep.,* 3, 6, 1797-1811.

Martinsen, G.V., I.G. Smyslov and V.E. Tishin. 1965. Tuntsy, ikh biologiya, promysel i obrabotka. Obzor (Tunas, Their Biology, Fishery, and Processing. A Review). VNIRO, Moscow, 122 pp., 50 figs.

Mather, F.J. 1963a. Tunas (genus *Thunnus*) of the western North Atlantic. Part II. Description, comparison and identification of species of *Thunnus* based on external characters, *FAO. Fish. Rep.,* 3, 6, 1155-1157,

Mather, F. 1963b. Tunas (genus *Thunnus*) of the western North Atlantic. Part III. Distribution and behavior of *Thunnus* species, *FAO. Fish. Rep.,* 3, 6, 1159-1161.

Mather, J. and R.H. Gibbs. 1958. Distribution of the Atlantic bigeye tuna, *Thunnus obesus,* in the Caribbean Sea, *Copeia,* vol. 3, p. 237.

Mather, F.J. and H.A. Schuck. 1960. Growth of bluefin tuna of the western North Atlantic, *U.S. Fish and Wildl. Serv., Fish. Bull.,* 61, 179, 39-52, 17 figs.

Mather, F.J. and M.R. Bartlett. 1966. Results of tagging experiments on tunas and billfishes conducted by the Woods Hole Oceanographic Institution. *Proc. the Eleventh Pac. Sci. Congress, Tokyo,* 7.

Matsubara, K. 1932. A new blennioid fish from Tyosen, *Bull. Japan. Soc. Sci. Fish.*, 1, 2, 1-3, 1 fig.

Matsubara, K. 1936. A new and a rare ophidioid fish from Japan, *J. Imper. Fish. Inst.*, 31, 2, 115-118, figs. 1-2.

Matsubara, R. 1943. Ichthyological annotations from the depth of the Sea of Japan. I. On a new blennioid fish, *Leptoclinus triocellatus*, *J. Sigenkagaku Kenkyusyo*, 1, 1, 37-40, fig. 1, pl. 1.

Matsubara, K. 1950. Identity of the wrymouth fish *Lyconectes ezoensis* with *Cryptacanthoides bergi, Japan. J. Ichthyol.*, 1, 3, 207.

Matsubara, K. 1953. On a new pearlfish, *Carapus owasianus,* with notes on the genus *Jordanicus* Gilbert, *Japan. J. Ichthyol.*, 3, 1, 29-32, ill.

Matsubara, K. 1955. *Fish Morphology and Hierarchy.* Tokyo, 1605 pp., 267 text-figs., 135 pls., 461 figs. (in Japanese).

Matsubara, K. and T. Iwai. 1951a. On an ophidioid fish, *Petroschmidtia toyamensis* Katayama, with some remarks on the genus *Petroschmidtia, Bull. Japan. Soc. Sci. Fish., Tokyo*, 16, 12, 104-111, 5 figs.

Matsubara, K. and T. Iwai. 1951b. *Lycodes japonicus,* a new ophidioid fish from Toyama Bay, *Japan. J. Ichthyol.*, 1, 6, 368-375, ill.

Matsubara, K. and T. Iwai, 1952. Studies on some Japanese fishes of the family Gempylidae, *Pacific Science,* 6, 3, 193-212, 12 figs.

Matsubara, K. and A. Ochiai. 1952. Two new blennioid fishes from Japan, *Japan. J. Ichthyol.*, 2, 4-5, 206-213, figs. 1-2.

Matsubara, K. and T. Iwai. 1958. Anatomy and relationships of the Japanese fishes of the family Gempylidae, Mem. Coll. *Agric. Kyoto Univ. Fisher.* ser., Spec. no., pp. 23-54, 14 figs.

Matsubara, K. and T. Iwai. 1959. Description of a new sandfish, *Kraemeria sexradiata,* from Japan, with special reference to its osteology, *J. Wash. Acad. Sci.,* vol. 49, pp. 27-32, 3 figs.

Matsumoto, W.M. 1959. Descriptions of *Euthynnus* and *Auxis* larvae from the Pacific and Atlantic oceans and adjacent seas, *Dana Rept.,* 9, 50, 1-34.

Matsumoto, W.M. 1960. Notes on the Hawaiian frigate mackerel of the genus *Auxis, Pacific Sci. Univ. Hawaii,* 14, 2, 173-177, 3 figs.

Matsumoto, W.M. 1961. Collection and descriptions of juvenile tunas from the central Pacific. [*sic*]

Matsumoto, W.M. 1966. Identification of tuna larvae. *Proc. the Eleventh Pac. Sci. Congress, Tokyo,* 7.

Matsumoto, W.M. et al. 1969. Pacific bonito (*Sarda chiliensis)* and skipjack tuna (*Katsuwonus pelamis*) without stripes, *Copeia,* 2, pp. 397-398.

Matsumoto, W.M. et al. 1972. On the clarification of larval tuna identification, particularly in the genus *Thunnus, Inter-Amer. Trop. Tuna Comm., Fish. Bull.,* 70, 1, 1-17, figs. 1-6, 5 tabs.

McAllister, D.E. 1968. Evolution of branchiostegals and classification of teleostome fishes, *Bull. Nat. Mus. Canada,* vol. 221, pp. 1-239, 21 pls.

McAllister, D.E. and R.J. Krejsa. 1961. Placement of prowfishes, Zaproridae, in the superfamily Stichaeoidae, *Nat. Hist. Mus., Canada,* vol. 11, pp. 1-4.

McHugh, J.L. 1952. The food of albacore (*Germo alalunga*) off California and Baja, California, *Bull. Scripps Inst. Oceanogr., Univ. Calif.,* 6, 4, 161-172, figs. 1-4.

Merrett, N.R. 1970. Gonad development in billfish (Istiophoridae) from the Indian Ocean, *J. Zool. London,* vol. 160, pp. 355-370.

Merrett, N.R. and C.H. Thorp. 1965. A revised key to scombroid fishes of East Africa, with new observations on their biology, *Ann. Mag. Natur. Hist.,* vol. 8, pp. 367-384.

Metelkin, L.I. 1957. Promysel tuntsov (Tuna Fishery). Vladivostok, 63 pp., 22 figs.

Miller, P.J. 1973. The osteology and adaptive features of *Rhyacichthys aspro* (Teleostei, Gobioidei) and the classification of gobioid fishes, *J. Zool. London,* vol. 171. pp. 397-434.

Mimura, K. 1964. Synopsis of biological data on yellowfin tuna *Neothunnus macropterus* Temminck and Schlegel, 1842 (Indian Ocean), *FAO. Fish. Rep.,* 2, 6, 319-349.

Mimura, K. and Staff. 1962. Synopsis of biological data on yellowfin tuna *Neothunnus macropterus* (Indian Ocean). *World Sci. Meet. Biol. Tunas, Calif., Spec. Synopsis,* no. 10, pp. 2-14.

Miyazaki, J. 1940. Studies on the Japanese common goby, *Acanthogobius flavimanus* (T. and S.), *Bull. Japan. Soc. Sci. Fisher.,* 9, 4, 159-180 (in Japanese)

Moiseev, P.A. 1957. O biologicheskikh osnovakh rybmego khozyaistva v zapadnvi chasti Tikhogo okeana (Biological basis of fisheries in the western part of the Pacific Ocean). *Dokl. II Plenuma Kom. Rybokhoz, Issled. Zapadnoi Chasti Tikhogo Okeana,* pp. 5-24.

Moore, H.L. 1951. Estimation of age and growth of yellowfin tuna (*Neothunnus macropterus*) in Hawaiian waters by size frequencies, *U.S. Fish and Wildl. Serv., Fish. Bull.,* 52, 65, 133-149.

Mori, T. 1952. Check list of the fishes of Korea, *Mem. Hyogo Univ. Agric.,* 1, 3, 1-228.

Mori, T. 1956a. Fishes of San-in District, including Oki Islands and its adjacent waters (southern Sea of Japan), *Mem. Hyogo Univ. Agric.,* 2, 3, 1-62.

Mori, T. 1956b. On the bottom fishes of Yamato Bank in the central Sea of Japan, with descriptions of two new species, *Sci. Rep. Hyogo Univ. Agric.*, 2, 2, *Nat. Sci.*, 29-32.

Mori, T. and K. Uchida. 1934. A revised catalogue of fishes of Korea, *J. Chosen Nat. Hist. Soc.*, vol. 19, pp. 1-23 (in Japanese).

Morice, J. 1953a. Essai systématique sur les familles des Cybiidae, Thunnidae et Katsuwonidae, poissons Scombroides, *Rev. Trav. Off. Pêches Marit.*, 18, 1, 35-63, 10 figs.

Morice, J. 1953b. Un caractére systématique pouvant servir à séparer les espéces de Thunnidae atlantiques, *Rev. Trav. Off. Pêches Marit.*, 18, 1, 65-74, figs. 1-6.

Morrow, J. E. and S.J. Harbo. 1969. A revision of the sailfish genus *Istiophorus, Copeia,* vol. 1, pp. 34-44, figs. 1-15.

Munro, I.S.R. (1948) 1950. The rare gempylid fish *Lepidocybium flavobrunneum* (Smith), *Proc. Roy. Soc. Queensland,* 60, 3, 31-41, 1 pl., 3 text-figs.

Myaksha, A.F. 1964. Tuntsy in mech-ryba kak promyshlennoe syr'e (Tunas and swordfishes as raw material for fish processing). *Izv. Tikhookeansk. Nauchno-Issled. Inst. Rybn. Knoz. i Okeanografii.* Vladivostok, vol. 55, p. 197.

Nakamura, E.L. 1965. Food and feeding habits of skipjack tuna (*Katsuwonus pelamis*) from the Marquesas and Tuamotu Islands, *Trans. Amer. Fish. Soc.*, 94, 3, 236-242.

Nakamura, E.L. 1969. A review of field observations on tuna behavior, *FAO. Fish. Rep.*, 2, 62, 59-68.

Nakamura, E.L. and W.M. Matsumoto. 1966. Distribution of larval tunas in Marquesan waters, *U.S. Fish and Wildl. Serv., Fish. Bull.*, 66, 1, 1-12, figs 1-5.

Nakamura, H. 1950. The food habits of yellowfin tuna (*Neothunnus*), *U.S. Fish. and Wildl. Serv., Spec. Sci. Rep., Fish.*, 23.

Nakamura, H. 1955. Report of investigation of spearfishes of Formosan waters, *U.S. Fish. and Wildl. Serv., Spec. Rep., Fish.*, vol. 153, pp. 1-46.

Nakamura, H. 1966. Biological studies of tunas and sharks in the Pacific Ocean. *Proc. the Eleventh Pac. Sci. Congress, Tokyo,* 7.

Nakamura, H. 1969. Tuna distribution and migration, *London Fish. News,* 76 pp., ill.

Nakamura, H. and H. Yamanaka. 1959. Relation between the distribution of tuna and the ocean structure, *J. Oceanogr. Soc. Japan.*, 15, 3, 1-2.

Nakamura, H. et al. 1951. Notes on the life-history of the swordfish *Xiphias gladius, Japan. J. Ichthyol.*, 1, 4, 264-271, figs. 1-4.

Nakamura, I, 1965. Relationships of fishes referable to the subfamily

Thunnidae on the basis of the axial skeleton, *Bull. Misaki Mar. Biol. Inst. Kyoto Univ.,* vol. 8, pp. 7-38, ill.

Nakamura, I. 1968. Juveniles of the striped marlin *Tetrapturus audax* (Phillip), *Mem. Coll. Agric. Kyoto Univ.,* vol. 94, pp. 17-29, figs. 1-8.

Nakamura, I. 1969. Big catches of longtail tuna in Wakasa Bay, Sea of Japan, *Japan. J. Ichthyol.,* 16, 4, 1960-161. 1 fig.

Nakamura, I. and Y. Warashina. 1965. Occurrence of bluefin tuna, *Thunnus thynnus* L., in the eastern Indian Ocean and eastern South Pacific Ocean, *Rep. Nankai Reg. Fish. Res. Lab.,* vol. 22, pp. 1-20, ill., bibl.

Nakamura, I., T. Iwai and K. Matsubara. 1968. A review of the sailfish, spearfish, marlin, and swordfish of the world, *Misaki Mar. Biol. Inst. Kyoto, Univ., Spec. Rep.,* vol. 4, pp. 1-95, 26 figs.

Nichols, J.T. and F.R. La Monte. 1937. Notes on swordfish at Cape Breton, Nova Scotia, *Amer. Mus. Novitates,* vol. 901, pp. 1-7.

Nielsen, J.G. 1969. Systematics and biology of the Aphyonidae (Pisces, Ophidioidea). Afhandl..., Kobenhavn Univ. Galathea, Rep. 10, pp. 1-90. Reprint.

Norman, J.R. 1957. *A Draft Synopsis of the Orders, Families, and Genera of Recent Fishes and Fishlike Vertebrates.* Edit. British Mus. (Nat. Hist.), 649 pp.

Novikov, Yu.V. 1957. Sluchai poimki *Xesurus scalprum* v vodakh Primor'ya (Cases of catching *Xesurus scalprum* in the waters of Primor'e). *Izv. Tikhookeansk. Nauchno-Issled. Inst. Rybn. Khoz. i Okeanografii.* Vladivostok, vol. 44, pp. 245-246.

Ochiai, A., Ch. Arago and M. Nakajima. 1955. A revision of the dragonets referable of the genus *Callionymus* found in the waters of Japan, *Publ. Seto Mar. Biol. Lab.,* V. 1, 96-132, 19 figs.

Ochiai, A. and K. Mori. 1965. Studies on the Japanese butterfish referable to the genus *Psenopsis, Bull. Misaki Mar. Biol. Inst. Kyoto Univ.,* vol. 8, pp. 1-6, pl. 1, figs. 1-2.

Okada, Y. 1955. *Fishes of Japan.* Tokyo, 1-434 + 28 pp., 391 figs.

Okada, Y. and H. Ikeda. 1938. Notes on the fresh-water fishes of Tohoku District in the collection of the Saito Ho-on Kai Museum, *Res. Bull. Saito Ho-on Kai Mus.,* 15, 5, 85-139, figs. 1-16, pls. 4-7.

Okada, Y. and K. Matsubara. 1938. *Keys to the Fishes and Fishlike Animals of Japan.* Tokyo and Osaka, 1-584 pp., 1-113 pls.

Okada, Y. and K. Suzuki. 1954. A new blennioid fish from Japan, *Rep. Fac. Fisher., Pref. Univ. Mie,* 1, 3, 227-228.

Okada, Y. and K. Suzuki. 1955. Biometrical studies of the gobioid fish *Chaeturichthys sciistius* Jordan and Snyder in the breeding season, *Rep. Fac. Fish. Univ. Mic.,* vol. 2, pp. 112-123.

Okhryamkin, D.I. 1931. Nekotorye vyvody po izucheniyu pitaniya skumbrii (Some conclusions from a study of the food of mackerals). *Rybnoe Khozyaistvo Dal'nego Vostoka,* nos. 1-2, pp. 53-55.

Orang, C. 1961. Spawning of yellowfin tuna and skipjack in the Far East tropical Pacific, as learned from studies of gonad development, *Bull. Inter-Amer. Trop. Tuna Comm.,* vol. 5, p. 6.

Orang, C.J. and B.D. Fink. 1963. Migration of a tagged bluefin tuna across the Pacific Ocean, *Calif. Fish and Game,* vol. 49, pp. 307-309.

Orel, P.Kh. 1926. Tuntsovyi promysel (Tuna Fishery). *Sov. Primor'ye,* vol. 5, pp. 59-69.

Oshoro Maru Cruise, 1969. *Data Rec. Oceanogr. Obs. Expl. Fish.,* vol. 13, pp. 361-394.

Osipov, V.G. 1960. O rasprostranenii, biologii i promysel tikhookeanskikh tuntsov (Distribution, biology, and fisheries of Pacific tunas). *Tr. Soveshch. Ikhtiol. Kom. Akad. Nauk SSSR,* vol. 10, pp. 188-194.

Osipov, V.G. 1965a. Osobennosti biologii melkikh tuntsov i perspektivy i promyslov v vodakh tikhogo ikh vostochnoi chasti Indiiskogo okeanov (Biological Peculiarities of Little Tunas and Prospects of Their Fisheries in the Waters of the Pacific Ocean and Eastern Part of the Indian Ocean). Moscow.

Osipov, V.G. 1965b. Raspredelenie i zapasy krupnykh tuntsov, mechoobraznykh i shel'fovykh ryb v vostochnoi chasti Indiiskogo okeana (Distribution and reserves of large tunas, swordfishes, and shelf fishes in the eastern part of the Indain Ocean). Sb. *Nauchno-Tekhn. Inform. VNIRO,* vol. 5, pp. 8-14.

Osipov, V.G. 1966. Rasprostranenie i usloviya obitaniya tuntsov v severozapadnoi chasti Tikhogo okeana (Distribution and living conditions of tunas in the northwestern part of the Pacific Ocean). Sb. *Nauchno-Tekhn. Inform. VNIRO,* vol. 5, pp. 3-8.

Osipov, V.G. 1967. Nekotorye osobennosti rasprostraneniya tuntsov i drugikh krupnykh pelagicheskikh ryb v Tikhom i Indiiskom okeanakh (Some peculiarities of distribution of tunas and other large pelagic fishes in the Pacific and Indian oceans). Avtoref. Kand. Diss., Vladivostok.

Osipov, V.G. 1968a. Biologiya i promysel tuntsov i drugikh pelagicheskikh ryb severo-vostochnoi chasti Indiiskogo okeana (Biology and fisheries of tunas and other pelagic fishes from the northeastern part of the Indian Ocean). *Tr. Vsesoyuzn. Nauchno-Issled. Inst. Morsk. Rybn. Khoz. i Okeanografii (VNIRO),* vol. 44, pp. 300-322, figs. 1-5.

Osipov, V.G. 1968b. O vertikal'nom raspredolenii zheltopergo (*Neothunnus albacora*) i bol'sheglazogo (*Parathunnus obesus*) tuntsov [Vertical distribution of yellowfin tuna (*Neothunnus albacora*) and bigeye tuna (*Thunnus obesus*)]. *Zool. Zhurn.,* 47, 8, 1192-1197.

Osipov, V.G. 1968c. Okeanskie pelagicheskie ryby (Pelagic Ocean Fishes). Vladivostok, 63 pp., 41 figs.

Osipov, V.G. 1968d. Nekotorye osobennosti raspredeleniya tuntsov i drugikh pelagicheskikh ryb v severo-zapadnoi chasti Indiiskogo okeana (Some peculiarities of distribution of tuna and other pelagic fishes in the northwestern part of the Indian Ocean). *Vopr. Ikhtiologii,* 8, 1 (48), 31–38, figs. 1–6.

Osipov, V.G. 1970. Nekotorye osobennosti biologii i promysla tuntsov i drugikh pelagicheskikh ryb v severo-zapadnoi chasti Indiiskogo okeana (Some peculiarities of the biology and fisheries of tuna and other pelagic fishes in the northwestern part of the Indian Ocean). *Izv. Tikhookeansk Nauchno-Issled. Inst. Rybn. Khoz. i Okeanografii.* Vladivostok, vol. 69, p. 331, figs. 1–11.

Osipov, V.G. and A.F. Myaksha. 1961. Biologicheskaya i tekhnologicheskaya kharakteristiki nekotorykh tikhookean-skikh tuntsov i mechenobraznykh (Biological and technological characteristics of some Pacific tunas and swordfishes). *Rybnoe Khozyaistovo,* vol. 1, pp. 53–58.

Osipov, V.G., V.S. Dolbish and I.V. Kizevetter. 1963. Tuntsy (Tunas). Vladivostok, pp. 1–68.

Osipov, V.G., I.V. Kizevetter and A.V. Zhuravlev. 1964. Tuntsy i mecheobraznye Tikhogo i Indiiskogo okeanov (Tunas and Swordfishes of the Pacific and Indian oceans). Moscow, 74 pp., 16 figs.

Otsu, T. 1960. Albacore migration and growth in the North Pacific Ocean as estimated from tag recoveries, *Pac. Sci.,* 14, 3, 257–266.

Otsu, T. and K.N. Uchida. 1959a. Study of age determination by hard parts of albacore from central North Pacific and Hawaiian waters, *U.S. Fish. and Wildl. Serv., Fish. Bull.,* 59, 150, 353–363.

Otsu, T. and K.N. Uchida. 1959b. Sexual maturity and spawning of albacore in the Pacific Ocean, *U.S. Fish. and Wildl. Serv., Fish. Bull.,* 59, 148, 287–305.

Otsu, T. and R.J. Hansen. 1962. Sexual maturity and spawning of albacore in the central South Pacific Ocean, *U.S. Fish. and Wildl. Serv., Fish. Bull.,* 62, 204, 151–161.

Otsu, T. and K.N. Uchida. 1963. Model of the migration of albacore in the North Pacific Ocean, *U.S. Fish. Wildl. Serv., Fish. Bull.,* 63, 1, 33–44.

Otsu, T. and H.O. Yoshida. 1967. Distribution and migration of albacore (*Thunnus alalunga*) in the Pacific Ocean. *FAO. Proc. Indo-Pac. Fisher. Council 12th Session, Honolulu, Hawaii, USA.* Section 2. *Technical Papers, Bangkok,* pp. 49–64, figs. 1–9.

Ouchi, A. 1963. The bottom fish fauna on Hyotan and Mukoze Banks in the northern Sea of Japan, *Bull. Japan Sea. Reg. Fish. Res. Lab.*, vol. 11, pp. 129-132.

Ouchi, A. and T. Ogata. 1960. Studies on the animal distribution in areas closed to trawl fishing in the northern Sea of Japan, *Rep. Japan Sea. Reg. Fish. Res. Lab.*, vol. 6, pp. 183-189.

Ovchinnikov, V.V. 1963. Parusnik (Sailfish). *Rybnoe Khozyaistvo*, vol. 11, pp. 7-9.

Ovchinnikov, V.V. 1964. Pitanie parusnika u zapadnogo poberezh'ya Afriki (Food of sailfish off the African coast). *Tr. Atlant. Nauchno-Issled. Inst. Morsk. Rybn. Khoz. i Okeanografii,* vol. 11, pp. 36-44.

Ovchinnikov, V.V. 1969. Migratsii mech-ryby, parusnikov i marlinov (Migrations of swordfish, sailfish, and marlin). *Tr. Atlant. Nauchno-Issled. Inst. Morsk. Rybn. Khoz. i Okeanografii,* vol. 25, pp. 210-212.

Ovchinnikov, V.V. 1970. Mech-ryba i parusnikovya (Sword-fish and Sailfish). Kaliningrad, 105 pp. 40 figs.

Parin, N.V. 1958. Redkie pelagicheskie ryby severo-zapadnoi chasti Tikhogo okeana (*Taractes steindachneri, Palinurichthys japonicus* i *Centrolophus lockingtoni*) [Rare pelagic fishes from the northwestern part of the Pacific Ocean (*Taractes steindachneri, Palinurichthys japonicus,* and *Centrolophus lockingtoni*)]. *Vopr. Ikhtiologii,* 12, 11, 162-170.

Parin, N.V. 1967. Skumbrievidnye ryby otkrytogo okeana (Scombroids of the open sea). In: *Tikhii Okean. Biologiya.* Kn. 3. *Ryby Otkrytykh Vod.* Moscow, pp. 88-128, figs. 16-29.

Parin, N.V. 1968. Ikhtiofauna okeanskoi epipelagiali (Ichthyofauna of the Oceanic Epipelagic Zone). Moscow, 185 pp., 56 figs.

Pavlenko, M.N. 1910. Ryby zaliva Petr Velikii (Fishes of Peter the Great Bay). Kazan', 95 pp., 13 figs., map.

Pavlenko, M.N. 1919 (1920). Rybolovstvo v zal. Petra velikogo (Fishing in Peter the Great Bay). *Mater. Izuch. Rybolovstva na Dal'nem Vostoke,* vol. 1.

Phelan, J.E. 1966. Bluefin age and growth estimates. Proc. 17th Pacific Tuna Conf. (Mimeographed).

Philipps, W.J. 1941. New rare fishes from New Zealand, *Trans. Proc. Roy. Soc. N.Z.,* 71, 3, 231-246 (see Abe, 1953: 42).

Pinchuk, V.I. [n.d]. Ob *Alectridium aurantiacum* Gilbert et Burke Komandorskikh i *Pseudoalectrias tarasovi* (Popov) Kuril'skikh ostrovov, a takzhe o neobychnom ekzemplyare *Stichaeopsis epallax* (Jordan et Snyder) [*Alectridium aurantiacum* Gilbert and Burke of the Komandor Islands and *Pseudoalectrias tarasovi* (Popov) of the Kuril Islands, as well as an unusual specimen of *Stichaeopsis epallax* (Jordan and Snyder)]. *Vopr. Ikhtiologii,* 14, 6 (89), 948.

Popov, A.M. 1931a. K poznaniya fauny ryb Okhotskogo morya (On the fish fauna of the Sea of Okhotsk). *Issled. Morei SSSR,* vol. 14, pp. 121-154, pls. I-II.

Popov, A.M. 1931b. Tikhookeanskaya zubatka *Anarhichas orientalis* Pall. (Pisces), ee sistematicheskoe polozhenie i rasprostranenie, s zamechaniyami o zubatkakh SSSR [The Bering wolffish *Anarhichas orientalis* Pall. (Pisces), its taxonomic position and distribution, with notes on the wolffishes of the Soviet Union]. *Dokl. Akad. Nauk SSSR,* vol. 14, pp. 380-386, 1 ill.

Popov, A.M. 1931c. O novom rode ryb *Davidojordania* (Zoarcidae, Pisces) v Tikhom okeane [A new genus of fishes, *Davidojordania* (Zoarcidae, Pisces) in the Pacific Ocean]. *Dokl. Akad. Nauk SSSR.* Moscow, pp. 210-215.

Popov, A.M. 1933a. K ikhtiofaune Yaponskogo morya (On the ichthyofauna of the Sea of Japan). *Issled. Morei SSSR.* vol. 19, pp. 139-155, figs. 1-5, pls.

Popov, A.M. 1933b. Fishes of Avatcha Bay on the southern coast of Kamchatka, *Copeia,* vol. 2, pp. 59-67.

Popov, A.M. 1935. Novyi rod i vid *Lycozoarces hubbsi* n. sp. (Pisces, Zoarcidae) Okhotskogo morya [A new genus and species, *Lycozoarces hubbsi* n. sp. (Pisces, Zoarcidae) from the Sea of Okhotsk]. *Dokl. Akad. Nauk SSSR,* 4 (9), 6/7, 285-286, 1 fig.

Popov, A.M. 1936. O rodakh *Davidojordania* Popov i *Bilabria* (Pisces, Zoarcidae) [The genera *Davidojordania* Popov and *Bilabria* (Pisces, Zoarcidae)]. *Dokl. Akad. Nauk SSSR,* 1, 2, 95.

Postel, E. 1969. Presentation des Thons, *Pêches Maritime,* 48, 1095, 397-415.

Probatov, A.N. 1951. O proniknovenii teplolyubivykh ryb v vody Sakhalina (Penetration of warm-water fishes into Sakhalin waters). *Dokl. Akad. Nauk SSSR,* 77, 1, 145-147.

Proceedings of the world scientific meeting on the biology of tunas and related species. 1964. *FAO. Fish. Rep.,* 4, 6, 1853-2200.

Pushkareva, N.F. 1960. Materialy po plodovitosti i razvitiya polovykh produktov skumbrii (Data on the fecundity and development of gonads in mackerel). *Izv. Tikhookeansk. Nauchno-Issled. Inst. Rybn. Khoz. i Okeanografii.* Vladivostok, vol. 46, pp. 79-84.

Pushkov, P.A. 1913. Ryby promysly Dal'nego Vostoka v 1912 g. (Fishes in the commercial catches of the Far East in 1912). *Mater. Izuch. Priamursk. Kraya,* no. 14.

Radovich, J. 1962. Effects of water temperature on the distribution of some scombrid fishes along the Pacific coast of North America. *FAO World Sci. Meet. Biol. Tunas, La Jolla, USA,* Sec. 4, *Exp. Papers,* vol. 27, pp. 1-19, figs. 1-12.

Rass, T.S. 1936. Sistematisch-morphologische studien über zwei naheverwandte arten: *Lumpenus fabricii* (C.V.). und *Lumpenus medius* Reinh. (Pisces, Blenniidae), *Acta Zool.*, vol. 17, pp. 395-463, figs. 1-16.

Rass, T.S. 1948. Mirovoi promysel vodnykh zhivotnykh (World Fisheries of Aquatic Animals). Moscow, 64 pp. 19 figs.

Rass, T.S. 1960. Promyslovo-geograficheskie kompleksy. Atlanticheskogo i Tikhogo okeanov i ikh sophotavlenie (Comparison of the fishery-geographic complexes of the Atlantic and Pacific Oceans). *Tr. Inst. Okeanol. Akad. Nauk SSSR*, vol. 31, pp. 3-17.

Rass, T.S. 1965a. Promyslovaya ikhtiofauna i rybnye resursy Indiiskogo okeana (Commercial ichthyofauna and fish resources of the Indian Ocean). *Tr. Inst. Okeanol. Akad. Nauk SSSR*, vol. 80, pp. 3-31, figs. 1-4 [*sic*].

Rass, T.S. 1965b. Ryby Indiiskogo okeana i Yugo-vostochnoi Azii (Fishes of the Indian Ocean and seas of Southeast Asia). *Tr. Inst. Okeanol. Akad. Nauk SSSR*, vol. 80, pp. 1-31, figs. 1-4 [*sic*].

Regan, C.T. 1912. The classification of blennioid fishes, *Ann. Mag. Natur. History*, 10, 57, 265-280, figs. 1-4.

Reid, E.D. 1936. Revision of fishes of the family Microdesmidae, with description of a new species, *Proc. U.S. Nat. Mus.*, vol. 84, pp. 55-72, 12 figs.

Reintjes, I.W. and I.E. King. 1953. Food of yellowfin tuna in the central Pacific, *U.S. Fish. and Wildl. Serv., Fish. Bull.*, 54, 81, 91-110.

Richards, S.W., A. Perlmutter and D.C. McAneny. 1963. A taxonomic study of the genus *Ammodytes* from the east coast of North America (Teleostei: *Ammodytes*), *Copeia*, vol. 2, pp. 358-377.

Richardson, J. 1844. Fishes in Zool., *Voy. Sulphur*, vol. 1, pp. 51-150, pls. 35-63.

Richardson, J. (1845) 1846. Report on the ichthyology of the seas of China and Japan, *Rep. British Assoc. Adv. Sci.*, vol. 15, pp. 187-320.

Risso, A. 1810. *Ichthyologie de Nice*. F. Schoell, Paris, pp. i-xxxvi+1-388, 51 pls.

Rivas, L.R. 1961. A review of tuna fishes of the subgenera *Parathunnus* and *Neothunnus* (genus *Thunnus*)., *Ann. Mus. Storia Nat.*, Genova, vol. 72, pp. 126-148.

Rivero, L.H. and M.J. Fernande. 1954. Estados larvales y juveniles del bonito (*Katsuwonus pelamis*), *Torreia*, vol. 22, pp. 1-14, figs. 1-11.

Robins, J.P. 1963. Synopsis of biological data on bluefin tuna [*Thunnus thynnus maccoyii* (Castelnau), 1872], *FAO. Fish. Rep.*, 6, 2, 562-585.

Robins, J.P. 1966. Distribution and migration of southern bluefin tuna, *Thunnus thynnus maccoyii* (Castelnau) in the Australasian region. *Proc. the Eleventh Pac. Sci. Congress, Tokyo*, 7.

569

Roedel, Ph.M. 1953. Common ocean fishes of the California coast, *Calif. Fish. and Game, Fish. Bull.*, vol. 91, pp. 1-184, figs.

Ronquillo, Inocencio A. 1963. A contribution to the biology of Philippine tunas, *FAO. Fish. Rep.*, 3, 6, 1752.

Rothschild, B.J. 1967. Estimates of the growth of skipjack tuna (*Katsuwonus pelamis*) in the Hawaiian Islands, *Proc. Indo-Pac. Fish. Coun.*, vol. 12, pp. 100-111.

Rothschild, B.J. and M.Y.Y. Yang. 1970. Apparent abundance, distribution, and migrations of albacore, *Thunnus alalunga*, on the north Pacific longline grounds, *U.S. Fish. and Wildl. Serv., Spec. Rep., Fish.*, 623, 6, 1-37.

Roux, Ch. 1961. Resumé des connaissance actuelles sur *Katsuwonus pelamis* (L.), *Bull. Inst. Pêches Marit., Maroc.*, vol. 7, pp. 33-53.

Royce, W.F. 1957. Observations of spearfishes of the central Pacific, *U.S. Fish. and Wildl. Serv., Fish. Bull.*, 57, 124, 497-554, figs. 1-27.

Royce, W.F. 1964. A morphometric study of yellowfin tuna, *Thunnus albacares* (Bonnaterre), *U.S. Fish. and Wildl. Serv., Fish. Bull.*, 63, 2, 395-443.

Rumyantsev, A.I. 1950. Krupnye eksemplyary vostochnogo tuntsa v zalive Petra Velikogo (Large specimens of Oriental bluefin tuna in Peter the Great Bay). *Izv. Tikhookeansk. Nauchno-Issled. Inst. Rvbn. Khoz. i Okeanografii.* Vladivostok, vol. 32, pp. 160.

Rumyantsev, A.I. and I.V. Kizevetter. 1949. Tuntsy (Kratkie svedeniya po bidogii, promyslu i obrabotke tuntsov Tikhogo okeana) (Tunas. Some Information on the Biology, Fishery, and Classification of Tunas of the Pacific Ocean). Vladivostok, 63 pp. 24 figs.

Sakamoto, K. (Matsubara) 1930. Two new species of fishes from the Sea of Japan, *J. Imp. Fish. Inst.*, 26, 1, 15-19.

Sato, S. and K. Kobayashi. 1956. Bottom fishes of Volcano Bay, Hokkaido. 1. A taxonomic study, *Bull. Hokkaido Reg. Fish. Res. Lab.*, vol. 13, pp. 1-19.

Scaccini, A. 1966. Studio dei caratteri differenziali dei primi stadi in alcune specie di tunnidi, *Arch. Zool. Ital.*, 51, 2, 1053-1061.

Schaefer, M.B., G.C. Broadhead and C.J. Orang. 1963. Synopsis of the biology of yellowfin tuna [*Thunnus* (*Neothunnus*) *albacares* (Bonnaterre), 1788] (Pacific Ocean), *FAO. Fish. Rep.*, 2, 6, 538-561.

Schaefer, M.B. and J.C. Marr. 1948. Contributions to the biology of Pacific tunas, *U.S. Fish. and Wildl. Serv., Fish. Bull.*, 51, 44, 187-196.

Scheer, D. 1961. Neues vom Schwertfisch, *Aquarien und Terrarien*, vol. 12, pp. 370-372, 2 figs.

Schmidt, P.J. [Shmidt, P.Yu]. 1930. Fishes of the Riu-Kiu Islands, *Tr. Tikhookeansk. Kom. Akad. Nauk. SSSR*, vol. 1, pp. 19-156, 8 figs. 6 pls.

Schmidt, P.J. [Shmidt, P.Yu.]. 1931a. A list of fishes collected in Japan and China by Dr. A. Bunge and N. Grebnitzky. *Izv. Akad. Nauk SSSR, Otd. Mater. i Estestv. Nauk,* pp. 101-123, 5 figs.

Schmidt, P.J. [Shmidt, P.Yu.]. 1931b. Fishes of Japan collected in 1901. *Tr. Tikhookeansk. Kom. Akad. Nauk SSSR,* vol. 2, pp. 1-176, figs. 1-30.

Schmidt, P.J. [Shmidt, P.Yu.]. 1936b. On the genera *Davidojordania* Popov and *Bilabria* n. (Pisces, Zoarcidae). *C.R. Acad. Sci. USSR.* Moscow, pp. 97-100, 1 text-fig.

Schmidt, P.J. [Shmidt, P.Yu.] and G.U. Lindberg. 1930. A list of fishes collected in Tsuruga (Japan) by W. Roszkowski. Izv. Akad. Nauk SSSR, vol. 10, pp. 1135-1150, fig. 1.

Schmidt, P.J. and A.P. Andriashev [Shmidt, P.Yu. and A.P. Andriyashev]. 1935. A Greenland fish in the Okhotsk Sea, *Copeia,* 2, pp. 57-60, figs. 1-2.

Schultz, L.P. and L.P. Woods. 1948. A new name for *Synchiropus altivelis* Regan, with a key to the genera of the fish family Callionymidae, *J. Washington Acad. Sci.,* 38, 12, 419-420.

Schultz, L.P., W.M. Chapman, E.A. Lachner and L.P. Woods. 1960. Fishes of the Marshall and Marianas Islands, 2, *Bull. U.S. Nat. Mus.,* vol. 202, pp. 1-438, pls. 75-123, figs. 91-132.

Scott, W.B. and S.N. Tibbo. 1968. Food and feeding habits of swordfish (*Xiphias gladius*) in the western North Atlantic, *J. Fish. Res. Bd. Canada,* 25, 5, 903-919, figs. 1-3.

Serventy, D.L. 1956. The southern bluefin tuna, *Thunnus thynnus maccoyii* (Castelnau), in Australian waters, *Austr. J. Mar. and Freshwater Res.,* vol. 7, pp. 1-43.

Shabotinets, E.I. 1968. Opredelenie vozrasta tuntsov Indiiskogo okeana (Determination of age of tunas from the Indian Ocean). *Tr. Vsesoyuzn Nauchno-Issled. Inst Morsk. Rybn. Khoz. i Okeanografii,* vol. 64, pp. 374-376.

Shaw, G. and F.P. Nodder. 1792. *The Naturalist's Miscellany or Colored Figures of Natural Objects, Drawn and Described ... from Nature.* London, 24 vols., 1790-1813.

Shepers, H. 1936. *Japan's Seefischerei.* Breslau, 228 pp.

Shibata, Y. 1968. A list of the fresh- and brackish-water fishes of Tsushima, preserved in the Osaka Museum of Natural History, *Bull. Osaka Mus. Nat. Hist.,* vol. 21, pp. 19-29, 2 figs.

Shimada, B.M. 1951a. Contribution to the biology of tunas from the western equatorial Pacific, *U.S. Fish. and Wildl. Serv., Fish. Bull.,* 52, 62, 112-119.

Shimada, B.M. 1951b. An annotated bibliography on the biology of Pacific tunas, *U.S. Fish. and Wildl. Serv., Fish. Bull.,* 52, 58, 1-58.

Shingu, C. 1966. Distribution and migration of the southern bluefin tuna. *Proc. the Eleventh Pac. Sci. Congress, Tokyo, 7*.

Shiogaki, M. and Y. Dotsu. 1973a. The spawning behavior of *Tripterygion etheostoma, Japan. J. Ichthyol.*, 20, 1, 36–41, figs. 1–4.

Shiogaki, M. and Y. Dotsu. 1973b. The egg development and larva rearing of the tripterygiid blenny, *Tripterygion etheostoma, Japan. J. Ichthyol.*, 20, 1, 42–46, figs. 1–3.

Shmidt, P.Yu. 1904. Ryby vostochnykh morei Rossiskoi imperii (Fishes in the Eastern Seas of the Russian Empire). Izd. Russk Geograf. Obshch., 466 pp., 6 pls.

Shmidt, P.Yu. 1936a. O rodakh *Dravidojordania* Popov i *Bilabria* n. (Pisces, Zoarcidae) [Genera *Davidojordania* Popov and *Bilabria* n. (Pisces, Zoarcidae)]. *Dokl. Akad. Nauk SSSR*, 1 (10), 2 (79), 93–96.

Shmidt, P.Yu. 1950. Ryby Okhotskogo mory (Fishes of the Sea of Okhotsk). *Tr. Tikhookeansk. Kom. Akad. Nauk SSSR*, vol. 6 pp. 1–379, 51 figs., 20 pls.

Shmidt, P.Yu and A.Ya.Taranetz. 1934. O novykh yuzhnykh elementakh v faune ryb severnoi chasti yaponskogo morya (New southern elements in the fish fauna of the northern part of the Sea of Japan). *Dokl. Akad. Nauk SSSR*, 2, 9, 591–595.

Shomura, R.S. and B.A. Keala. 1962. Growth and sexual dimorphism in growth of bigeye tuna (*Thunnus obesus*). *World Sci. Meet. Biol. Tunas, La Jolla, California, USA*, 2, *Exp. Paper*, 24.

Silas, E.G. 1962. The taxonomy and biology of the oriental bonito, *Sarda orientalis* (Temminck and Schlegel). *Symposium on Scombroid Fishes*, pt. 1 pp. 1–26, figs. 7 pl. II.

Silas, E.G. 1963. Synopsis of biological data on oriental bonito, *Sarda orientalis* (Temminck and Schlegel), 1842 (Indian Ocean), *FAO. Fish. Rep.*, 2, 6, 834–861.

Siro, Issi, 1947. Ryby yuzhnogo Sakhalina (Fishes of southern Sakhalin). *Sakhalinsk. Otd. Timhookeansk. Nauchno-Issled. Inst. Rybn. Khoz. i Okeanografii.* Vladivostok (translated into Russian by Pak).

Sivasubramaniam, K. 1966. Predators and competitors of tunas. *Proc. the Eleventh Pac. Sci. Congress, Tokyo, 7*.

Smith, H.M. 1902. Description of a new species of blenny from Japan, *Bull. U.S. Fish. Comm.*, pp. 93–94.

Smith, J.L.B. 1948. New clinid fishes from the southwestern Cape (South Africa), with notes on fishes, *Ann. Mag. Natur. History*, (11) 14 732–736, 2 text-figs.

Smith, J.L.B. 1955a. Fishes of the family Carapidae in the western Indian Ocean, *Ann. Mag. Natur. History*, (12) 8, 401–416, 8 figs.

Smith, J.L.B. 1955b. The genus *Pyramodon* Smith and Radcliffe, *Ann. Mag. Natur. History*, (12) 8, 545–550, figs. 1–2.

572

Snyder, J.O. 1909. Descriptions on new genera and species of fishes from Japan and Riu-Kiu Island, *Proc. U. S. Nat. Mus.*, vol. 36, pp. 597-610.

Snyder, J.O. 1911. Description of new genera and species of fishes from Japan and Riu-Kiu Island, *Proc. U. S. Nat. Mus.*, vol. 40, pp. 529-549.

Snyder, J.O. 1912. Japanese shore fishes collected by the United States Bureau of Fisheries steamer *Albatross* Expedition of 1906, *Proc. U. S. Nat. Mus.*, vol. 42, pp. 399-450, pls. 51-61.

Sokolovskii, A.S. 1970. Nekotorye dannye o vozraste i roste yaponskoi skumbrii (*Scomber japonicus* Houttuyn) severozapadnoi chasti Tikhogo okeana [Some data on the age and growth of chub mackerel (*Scomber japonicus* Houttuyn) in the northwest part of the Pacific Ocean]. Sb. *Issledovanie po Biologii Ryb i Promyshlennoi Okeanografii.* Vladivostok, pp. 58-66.

Sokolovskii, A.S. 1971. Nekotorye cherty biologii skumbrii (*Scomber japonicus* Houttuyn) severo-zapadnoi chasti Tikhogo okeana [Some biological peculiarities of chub mackerel (*Scomber japonicus* Houttuyn) from the northwest part of the Pacific Ocean]. *Izv. Tikhookeansk. Nauchno-Issled. Inst. Rybn. Khoz. i Okeanografii.* Vladivostok, vol. 79, pp. 58-67, figs. 1-5, pls. 2.

Sokolovskii, A.S. 1972. Dinamika chislennosti skumbrii (*Scomber japonicus* Houttuyn) v severo-zapadnoi chasti Tikhogo okeana [Population dynamics of chub mackerel (*Scomber japonicus* Houttuyn) in the northwest part of the Pacific Ocean]. Sb. *Issledovanie Biologii Ryb i Promyshlennoi Okeanografii.* Vladivostok, pp. 161-183.

Soldatov, V.K. 1915. A new genus of Blenniidae from Peter the Great Bay. *Ezhegodn. Zool. Muzeya Rossiisk. Akad. Nauk,* vol. 20, pp. 635-637, 1 fig.

Soidatov, V.K. 1917a. Notes on two new species of *Lycodes* from the Okhotsk Sea. *Ezhegodn. Zool. Muzeya Rossiisk. Akad. Nauk,* vol. 22, pp. 112-117, 1 fig.

Soldatov, V.K. 1917b. Description of a new species of *Krusensterniella* Schmidt. *Ezhegodn. Zool. Muzeya Rossiisk. Akad. Nauk.* vol. 23, pp. 157-159, 1 fig.

Soldatov, V.K. 1917c. On a new genus and three new species of Zoarcidae. *Ezhegodn. Zool. Muzeya Rossiisk. Akad. Nauk.* vol. 23, pp. 160-163.

Soldatov, V.K. 1927. Note on two little-known genera and species from Shantar Islands (Okhotsk Sea). Sb. *v Chest' Knipovicha.* Moscow, pp. 399-404, figs.

Soldatov, V.K. and M. Pavlenko. 1915. A new genus of the family Blenniidae—*Kasatkia* g. n. *Ezhegodn. Zool. Muzeya Rossiisk. Akad. Nauk,* vol. 20, pp. 638-640, 1 fig.

Soldatov, V.K. and G.U. Lindberg. 1930. Obzor ryb dal'nevostochnykh morei (Review of fishes from seas of the Far East). *Izv. Tikhookeansk. Nauchn. Inst. Rybn. Khoz.,* Vladivostok, vol. 5, pp. i-xlvii + 1-576, figs. 1-76.

Springer, V.G. 1964. Review of "A Revised Classification of Blennioid Fishes of the Family Chaenopsidae" by J.S. Stephens, *Copeia, 3,* pp. 591-593.

Springer, V.G. 1967. Revision of the circumtropical shore-fish genus *Entomacrodus* (Blenniidae: Salariinae), *Proc. U.S. Nat. Mus.,* vol. 122, pp. 1-150, pls. 1-30.

Springer, V.G. 1968. Osteology and classification of fishes of the family Blenniidae, *Bull. U.S. Nat. Mus.,* vol. 284, pp. 1-85, 16 figs., 11 pls.

Springer, V.G. and W.F. Smith-Vaniz. 1972. Mimetic relationships involving fishes of the family Blenniidae, *Smiths. Contr. Zool. Wash.,* vol. 112, pp. 1-36, pls. 1-7.

Starks, E.C. 1910. The osteology and mutual relationships of fishes of the family Scombridae, *J. Morph., Philad.,* vol. 21, pp. 77-99, figs. 1-2, 3 pls.

Starks, E.C. 1911. Osteology of certain scombroid fishes, *Stanford Univ. (California),* vol. 5, pp. 1-41, ill.

Steindachner, F. 1876. Ueber einige neue oder seltene Fischarten aus dem atlantischen, indischen und stillen Ocean. Ichthyologische Beitrage 5, *Sitzb. Akad. Wiss., Wien,* 74, 1, abth., 49-240, 15 pls.

Steindachner, F. 1879 (1880). Ichthyologische Beitrage 8, *Sitzb. Akad. Wiss., Wien,* 80, (1), 6, 119-191, figs. 1-3.

Steindachner, F. 1880 (1881). Über einige Fischarten aus dem nördlichen Japan, gesammelt vom Professor Dybowski. (III). Ichth. Beitr. 9, *Stizb. Akad. Wiss., Wien,* 82, 1, 1-29, figs. 1-5.

Strasburg, D. 1960. Estimates of larval tuna abundance in the central Pacific, *U.S. Fish. and Wildl. Serv., Fish. Bull.,* 167, 60, 231-250.

Strasburg, D.W. 1969. Billfishes of the central Pacific Ocean, *U.S. Fish. and Wildl. Serv. Biol. Lab. Honolulu, Hawaii,* pp. 1-11, figs. 7.

Sun' Tszi-Zhen'. 1960. Lichinki i mal'ki tuntsov, parusnikov i mech-ryby (Thunnidae, Istiophoridae, Xiphiidae) tsentral'noi i zapadnoi chastei Tikhogo okeana [Larvae and juveniles of tuna, sailfish, and swordfish (Thunnidae, Istiophoridae, Xiphiidae) from the central and western parts of the Pacific Ocean]. *Tr. Inst. Okeanol. Akad. Nauk SSSR,* vol. 41, pp. 175-191, figs. 1-7.

Suvorov, E.K. 1935. Novy rod i dvo novykh vida ryb semeistva Zoarcidae iz Okhotskogo morya (A new genus and two new species of fish of the family Zoarcidae from the Sea of Okhotsk). *Izv. Akad. Nauk SSSR.* Moscow, pp. 435-440, ill.

Svetovidov, A.N. 1964. Ryby Chernogo morya (Fishes of the Black Sea). Moscow-Leningrad, 550 pp. 191 figs.

Swainson, W. 1839. *Natural History of Fishes, Amphibians, and Reptiles,* vol. 2, pp. 1-452, figs. 1-135.

Sylva, D.P. 1963. Billfish round-up, *Sea Frontiers,* 9, 2, 85-91.

Tabeta, O. and H. Tsukahara. 1967. Ecological studies of fishes stranded on the beach along the coast of the Tsushima Current. I: Fishes and other animals recorded during the first half of 1965 in northern Kyushu, *Bull. Japan. Soc. Sci. Fish.,* 33, 4, 295-302.

Takagi, K. 1952. A critical note on the classification of *Chaenogobius urotaenia* and its two allies, *Japan. J. Ichthyol.,* 2, 1, 14-22, 2 figs.

Takagi, K. 1957. Descriptions of some new gobioid fishes of Japan with a proposition on the sensory line system as a taxonomic character, *J. Tokyo Univ. Fish.,* 43, 1, 97-126 (in Japanese).

Takagi, K. 1963. *Studies of Gobioid Fishes in Japanese Waters: Comparative Morphology, Phylogeny, Taxonomy, Distribution, and Bionomics.* Tokyo, 1-3+1-273 pp., 1-47 figs. (in Japanese).

Takagi, K. 1966a. Taxonomic and nomenclatural status in chaos for the gobioid fish, *Chaenogobius annularis* Gill, 1858. I: Review of the original description, with special reference to estimation of the upper jaw relative length as a taxonomic character, *J. Tokyo Univ. Fish.,* 52, 1, 17-27, fig. 1 (in Japanese).

Takagi, K. 1966b. Taxonomic and nomenclatural status in chaos for the gobioid fish, *Chaenogobius annularis* Gill 1858. II: Specific heterogeneity of *C. annularis* Gill senus Towiyama, with description of the genus *Rhodoniichtys* gen. nov., *J. Tokyo Univ. Fish.,* 52, 2, 29-46, 5 figs. (in Japanese).

Takagi, K. 1966c. Distribution and ecology of gobioid fishes in Japanese waters, *J. Tokyo Univ. Fish.,* 52, 2, 83-127, figs. 1-3 (in Japanese).

Takegawa, Y. and H. Morino. 1970. Fishes from Wakasa Bay, Sea of Japan, *Publ. Seto Mar. Biol. Lab.,* 17, 6, 373-392.

Talbot, F.H. and M.J. Penrith. 1963. Synopsis of biological data on species of the genus *Thunnus* (sensu lato) in South Africa, *FAO. Fish. Rep.,* 2, 6, 608-646.

Tanaka, S. 1908. Notes on a collection of fishes made by Prof. Ijima in the southern parts of Sakhalin, *Ann. Zool. Japan.* 6, 4, 235-254, pls.

Tanaka, S. 1911-1930. *Figures and Descriptions of Fishes of Japan, Including Riu-Kiu Islands, Bonin Island, Formosa, Kuril Islands. Korea, and Southern Sakhalin.* Tokyo, i-xlviii+1-960 pp., 190 pls.

Tanaka, S. 1931. On the distribution of fishes in Japanese waters, *J. Fac. Sci. Imp. Univ.,* 4, 3 (1), 1-90.

Tanoue, T. 1961. Studies on the relationships between the drifting distributions of mackerel larvae (*Pneumatophorus tapeinocephalus* Bleeker) and the environmental factors. II: On the larvae and the sea conditions in the surface and middle layers around Osumi Islands, *Bull. Japan. Soc. Sci. Fish.,* 27, 12, 1041-1046, ill.

Tanoue, T. and Tamari. 1960. Studies on the relationships between the drifting distributions of mackerel larvae (*Pneumatophorus tapeinocephalus* Bleeker) and the environmental factors. I: On the larvae collected and the sea conditions around Osumi Islands, *Bull. Japan. Soc. Sci. Fish.*, 26, 9, 882–886, figs. 1–4.

Tanoue, T., K. Yoji and T. Yoichiro. 1960. On the spawning season of the mackerel *Pneumatophorus tapeinocephalus* Bleeker in three different regions—East China Sea, Satsunan, and Izu, *Bull. Japan, Soc. Sci. Fish.*, 26, 3, 277–283, ill.

Taranetz, A.Ya. 1934. Kratkii obzor ryb roda *Gymnogobius* s opisaniem odnogo novogo vida i zametkami o nekotorykh blizkikh rodakh (Brief review of fishes of the genus *Gymnogobius*, with a description of a new species and notes on some closely related genera). *Dokl. Akad. Nauk SSSR*, pp. 397–400.

Taranetz, A.Ya. 1935. Nekotorye izmeneniya v sistematike ryb Sovetskogo Dal'nego Vostoka s zametkami ob ikh rasprostraneni (Some changes in the classification of fishes of the Soviet Far East, with notes on their distribution). *Vestn. Dal'nevost. Fil. Akad. Nauk SSSR*, vol. 13, pp. 89–101.

Taranetz, A.Ya. 1936. Kratkii obzor rodov sem. Blenniidae, rodstvennykh *Stichaeus*, iz Beringova, Okhotskogo i Yaponskogo morei (Brief review of genera of the family Blenniidae related to *Stichaeus* from the Bering Sea, Sea of Okhotsk, and Sea of Japan). *Dokl. Akad. Nauk SSSR*, 1, 3 (80), 141–144.

Taranetz, A.Ya. 1937a. K poznaniyu ikhtiofauny Sovetskogo Sakhalina (Ichthyofauna of Soviet Sakhalin). *Izv. Tikhookeansk. Nauchno-Issled. Inst. Rybn. Khoz. i Okeanografii*. Vladivostok, vol. 12, pp. 1–50, figs. 1–7.

Taranetz, A.Ya. 1937b. Kratkii opredelitel' ryb Sovetskogo Dal'nego Vostoka i prilezhashchikh vod (Abridged Keys to Fishes of the Soviet Far East and Adjacent Waters). *Izv. Tikhookeansk. Nauchno-Issled. Inst. Rybn. Khoz. i Okeanografii.* Vladivostok, vol. 11, pp. 1–200, 103 figs., map.

Taranetz, A.Ya. 1938a. O novykh nakhodkakh yuzhnykh elementov v ikhtiofaune severo-zapandnoi chasti Yaponskogo morya (New finds of southern elements in the ichthyofauna of the northwestern part of the Sea of Japan). *Vestn. Dal'nevost. Fil. Akad. Nauk SSSR*. Vladivostok, 28, 1, 113–130, figs. 1–6.

Taranetz, A.Ya. 1938b. Morskie i presnovodnye promyslovye bogatstva DVK (Marine and fresh-water fishery resources of Far East). *Vestn. Dal'nevost. Fil. Akad. Nauk SSSR*, 30, 3, 143–188.

Taranetz, A.Ya. 1958. Opisanie *Soldatovia polyactocephala* (Pallas), soderzhashcheesya v neopublikovannoi rukopisi A.Ya. Tarantsa "O rybakh baseina severo-zapadnoi chasti Tikhogo okeana, opisannykh Pallasom" [Description of *Soldatovia polyactocephala* (Pallas) in the unpublished manuscript of A.Ya. Taranets entitled "Fishes from the Basin of the Northwestern Part of the Pacific Ocean Described by Pallas"). Cited by Makushok in *Tr. Zool. Inst. Akad. Nauk SSSR,* vol. 25, pp. 118-119, fig. 83.

Taranetz, A.Ya. and A.P. Andriyashev. 1934. O novom roda i vide *Petroschmidtia albonotata* (Zoarcidae, Pisces) iz Okhotskogo morya [New genus and species, *Petroschmidtia albonotata* (Zoarcidae, Pisces) from the Sea of Okhotsk]. *Dokl. Akad. Nauk SSSR,* 11, 2 (8), 506-512, 2 figs.

Taranetz, A.J. and A.P. Andriashew [Taranets, A.Ya. and A.P. Andriyashev]. 1935. Vier neue fischarten der Gattung *Lycodes* Reinh. aus dem Ochotskischen Meer, *Zool. Anz.,* 112, 9-10, 242-253, figs. 1-7.

Temminck, C.J. and H. Schlegel. 1842-1850. Pisces. In *Fauna Japonica, Poiss.* edited by P.F. Siebold. Leiden, pp. 1-323, pls. 1-160.

Thomas, P.T. and M. Kumaran. 1963. Food of Indian tunas; *FAO. Fish. Rep.,* 3, 6, 1659-1667.

Thompson, W.F. 1917. Temperature and the albacore, *Calif. Fish. and Game,* 3, 4, 153-159.

Tiang Yir Hang. 1957. O skumbrii Koreiskogo zaliva (Mackerels of the Korean Gulf). *Nauchno-Issled. Vodn. Prom. Vostochnykh Morei.* Moscow.

Tokarev, A.K. 1948. Skumbriya yaponskogo morya (Mackerel from the Sea of Japan). *Rybnoe Khozyaistvo,* vol. 6, pp. 43-47, figs. 1-2.

Tomiyama, I. 1934. Four new species of gobies of Japan, *J. Fac. Sci. Imp. Univ., Tokyo,* 3, 3, 325-334, 4 figs.

Tomiyama, I. 1936. Gobiidae of Japan, *Japan. J. Zool.,* 7, 1, 37-112, 44 figs.

Tomiyama, I. 1950a. On a Japanese blennioid fish, *Dasson elegans* (Steindachner), *Zool. Mag., Tokyo,* 59, 9, 220, 2 figs.

Tomiyama, I. 1950b. On a Japanese blennioid fish, *Blennius yatabei* (Jordan and Snyder), *Żool. Mag., Tokyo,* 59, 9, 221-222, 3 figs.

Tomiyama, I. 1951. On a Japanese blennioid fish, *Dasson trossulus* (Jordan and Snyder), *Zool. Mag., Tokyo,* 60, 8, 159-161, 7 figs.

Tomiyama, I. 1952a. Additional notes on *Dasson elegans* (Steindachner) and *Blennius yatabei* Jordan and Snyder, *Zool. Mag., Tokyo,* vol. 61, pp. 9-10, 2 figs.

Tomiyama, I. 1952b. On a Japanese blennioid fish, *Dasson japonicus* (Bleeker), *Zool. Mag., Tokyo,* vol. 61, pp. 10-12, 2 figs.

577

Tomiyama, I. 1954a. *Icticus pellucidus* (Lütken) (Nomeidae). In: *Fig. and Descr. Fishes Japan,* vol. 50, pp. 1002–1007, pls. 199, 200, figs. 539–542.

Tomiyama, I. 1954b. *Psenes arafurensis* Günther (Nomeidae). In: *Fig. and Descr. Fishes Japan,* vol. 50, pp. 1008–1011, pl. 201, fig. 543.

Tomiyama, I. 1958a. *Parioglossus dotui* (new species). In: *Fig. and Descr. Fishes Japan,* vol. 57, p. 1179, pl. 230, fig. 582.

Tomiyama, I. 1958b. *Vireosa hanae* Jordan and Snyder. In: *Fig. and Descr. Fishes Japan,* vol. 58, pp. 1200–1205, pl. 233, figs. 589, 590.

Tomiyama, I. 1958c. *Eutaeniichthys gilli* Jordan and Snyder. In: *Fig. and Descr. Fishes Japan,* vol. 58, pp. 1206–1209, pl. 234, fig. 591.

Tomiyama, I. 1958d. *Lumpenus macropus* Matsubara and Ochiai (Pholidae). In: *Fig. and Descr. Fishes Japan,* vol. 59, pp. 1236–1239, pl. 237, fig. 597.

Tomiyama, I. 1959. Secondary sexual characters and supplementary notes on a Japanese blennioid fish, *Istiblennius enosimae* (Jordan and Snyder), *Zool. Soc. Japan, Zool. Inst., Tokyo Univ.,* 32, 4, 225–228, 3 figs.

Tomiyama, I. 1972. List of the fishes preserved in the Aitsu Marine Biological Station, Kumamoto University, with notes on some interesting spacies and descriptions of two new species, *Publ. Amakusa Mar. Biol. Lab.,* 3, 1, 1–21, 9 figs.

Tomiyama, I. and T. Abe. 1958. *Encyclopaedia Zoologica, Illustrated in Color,* vol. 2, *Fishes.* Tokyo, 306 pp., 912 figs. (in Japanese).

Tomodo, Y. 1970. A preliminary study of the fresh-water fish fauna in Iki-Tsushima Islands, *Mem. Nat. Sci. Mus., Tokyo,* vol. 3, pp. 199–210, 1 pl.

Torin, Yu.A. 1969. Vertikal'noe raspredelenie i temperaturnye usloviya obitaniya bol'sheglazogo tuntsa (*Thunnus obesus*) v yugo-vostochnoi Atlantike [Vertical distribution of, and temperature conditions for, bigeye tunas (*Thunnus obesus*) in the southeastern Atlantic]. *Tr. Atlant. Nauchno-Issled. Inst. Morsk. Rybn. Khoz. i Okeanografii,* vol. 25, pp. 115–119.

Tsutsumi, T. and Y. Dotu. 1961. The reproductive behavior in gobioid fish (*Pterogobius zonoleucus* Jordan and Snyder), *Bull. Fac. Fisher. Nagasaki Univ.,* vol. 10, pp. 149–154, fig. 1, pl. 22.

Tucker, D.W. 1956. Studies on trichiuroid fishes. 3: A preliminary revision of the family Trichiuridae, *Bull. Brit. Mus. Nat. Hist. Zool.,* 4, 3, 73–130, text-figs. 1–23, pl. 10.

Tyler, J.C. 1970. Osteological aspects of interrelationships of surgeon fish genera (Acanthuridae), *Proc. Acad. Nat. Sci., Philad.,* 122, 2, 87–124, 23 figs.

578

Uchida, K. 1932. Life histories of *Boleophthalmus pectinirostris* and *Periophthalmus cantonensis, Ann. Rep. Japan. Assoc. Adv. Sci.*, 7, 2, 109-117 (in Japanese).

Uchida, K. and H. Yabe. 1939. The fish fauna of Saisyu-to (Quelpart Island) and its adjacent waters, *J. Chosen Nat. Hist. Soc.*, vol. 25, pp. 1-16.

Uchida, R.N. 1963. Synopsis of biological data on frigate mackerel [*Auxis thazard* (Lacépède) 1802] (Pacific Ocean), *FAO. Fish. Rep.*, 2, 6, 241-273.

Ueno, T. 1954a. Studies on the deep-water fishes from off Hokkaido and adjacent regions. I: On a rare fish, *Zaprora silenus* Jordan, found off Kushiro, Hokkaido, *Japan. J. Ichthyol.*, 3, 2, 79-82.

Ueno, T. 1954b. First record of a strange bathypelagic species, referable to the genus *Centrolophus* (Centrolophidae, Stromateiformes) from Japanese waters, with remarks on the specific differentiation, *Bull. Fac. Fish. Hokkaido Univ.*, 5, 3, 240-247, figs. 1-4.

Ueno, T. 1965a. On two rare pelagic fishes, *Luvarus imperialis* and *Rachycentron canadum*, recently captured at Yoichi, Hokkaido, *Japan. J. Ichthyol.*, vol. 12, pp. 99-103, 3 figs.

Ueno, T. 1965b. The stromateid fishes (suborder Stromateoidei) captured from waters of Hokkaido, *Sci. Rep. Hokkaido Fish. Exp. Sta.*, vol. 4, pp. 1-22, 6 figs. (in Japanese).

Ueno, T. 1965-1966. Ryby v vodakh Khokkaido i prilezhatsikh vod (Fishes in the Waters of Hokkaido and Adjacent Waters), vol. 1, pp. 1-14, fig.

Ueno, T. 1971. List of the marine fishes from waters of Hokkaido and its adjacent regions, *Sci. Rep. Hokkaido Fish. Exper. Station*, vol. 13, pp. 61-102.

Ueno, T. and K. Abe. 1964. Studies on deep-water fishes from off Hokkaido and adjacent regions, 3-7, *Bull. Hokkaido Reg. Fish. Res. Lab.*, vol. 28, pp. 1-22, figs. 1-14.

Ueno, T. and K. Abe. 1968. On rare newly found fishes from waters of Hokkaido (III), *Japan. J. Ichthyol.*, 15, 1, 36-37.

Ueyanagi, S. 1963. A study of the relationships of the Indo-Pacific istiophorids, *Rep. Nankai Reg. Fish. Res. Lab.*, vol. 17, pp. 151-165, figs. 7, pls. 2.

Ueyanagi, S. 1966a. Feeding habits of tunas. *Proc. the Eleventh Pac. Sci. Congress, Tokyo*, 7.

Ueyanagi, S. 1966b. The distribution and migration of the skipjack. *Proc. the Eleventh Pac. Sci. Congress, Tokyo*, 7.

Vedenskii, A.P. 1951. Materialy po biologii skumbrii yaponskogo morya (Data on the biology of mackerel of the Sea of Japan). *Izv. Tikhookeansk. Nauchno-Issled. Inst. Rybn. Khoz. i Okeanografii.* Vladivostok, vol. 34, pp. 47-66.

Vedenskii, A.P. 1953. Biologiya skumbriya yaponskogo morya (Biology of mackerel of the Sea of Japan). Avtoref. Kand. Diss., Vladivostok, 24 pp.

Vedenskii, A.P. 1954a. Raspredelenie i povedenie dal'nɛvostochnoi skumbrii v yaponskom more (Distribution and behavior of the Far East mackerel in the Sea of Japan). *Tret'ya Ekol. Konf. Tez. Dokl.* Kiev, vol. 4, pp. 63–64.

Vedenskii, A.P. 1954b. Biologiya dal'nevostochnoi skumbrii v yaponskom more (Biology of the Far East mackerel in the Sea of Japan). *Izv. Tikhookeansk. Nauchno-Issled. Inst. Rybn. Khoz. i Okeanografii.* Vladivostok, vol. 42, pp. 1–94, figs. 1–12.

Vedenskii, A.P. 1962. Sostoyanie zapasov skumbrii i perspektivy ee promysla (State of the mackerel stocks and prospects of fishery). Sb. *Dokl. II Plenuma Kom. Rybokhoz Issled. Zap. Chasti Tikhogo Okeana.* Moscow, pp. 103–108.

Vyskrebentsev, B.V. 1969. Dannye po biologii skumbrii *Scomber japonicus colias* Gmelin zapadnogo poberezh'ya Afriki (On the biology of chub mackerel *Scomber japonicus colias* Gmelin from the west African coast). *Tr. Azovo-Chernomorsk. Inst. Morsk. Rybn. Khoz. i Okeanografii,* vol. 29, pp. 144–167.

Waldron, K.D. 1963. Synopsis of biological data on skipjack, *Katsuwonus pelamis* (L.), 1758 (Pacific Ocean), *FAO. Fish. Rep.,* 2, 6, 695–748.

Waldron, K. and J. King. 1963. Food of skipjack in the central Pacific, *FAO. Fish. Rep.,* 3, 6, 1431–1457.

Walford, L.A. 1937. *Marine Game Fishes of the Pacific Coast from Alaska to the Equator.* Univ. Calif. Press, Berkeley, California, 205 pp., pls.

Walters, V. and H. Fierstine. 1964. Measurements of swimming speeds of yellowfin tuna and wahoo, *Nature,* 202, 4928, 208–209.

Wang, K.F. 1935. Study of the teleost fishes of the coastal region of Shangtung. II, *Contrib. Biol. Lab. Sci. Soc. China, Zool.* ser., 10, 9, 393–481, 51 figs.

Wang, K.F. and S.C. Wang. 1935. Study of the teleost fishes of the coastal region of Shangtung. III, *Contrib. Biol. Lab. Sci. Soc. China, Zool.* ser., 11, 6, 165–237, 52 figs.

Watanabe, H. 1958. On the difference in the stomach contents of yellowfin and bigeye tunas from the equatorial Pacific, *Rep. Nankai Reg. Fish. Res. Lab.,* 7.

Watanabe, H. and S. Ueyanagi. 1963. Young of the shortbill spearfish, *Tetrapturus angustirostris* Tanaka, *Rep. Nankai Reg. Fish. Res. Lab.,* vol. 17, pp. 133–136, fig. 1, pls. 2.

Watson, M.E. 1963. Tunas (genus *Thunnus*) of the western North Atlantic, pt. I, *FAO. Fish. Rep.,* 3, 6, 1153–1154.

580

Whitley, G.P. 1940. The second occurrence of a rare fish (*Luvarus*) in Australia, *Rec. Austral. Mus.*, 20, 5, 325-326.

Whitley, G.P. 1943. Ichthyologial notes and illustrations (pt. 2), *Austral. Zool.*, 10, 2, 167-187, text-figs. 1-10.

Wilimovsky, N.J. 1956. A new name, *Lumpenus sagitta,* to replace *Lumpenus gracilis* (Ayres) for a northern blennioid fish (family Stichaeidae), *Stanf. Ichth. Bull.*, 7, 2, 23-24.

Williams, F. 1963a. Synopsis of biological data on little tuna *Euthynnus affinis* (Cantor), 1850 (Indian Ocean), *FAO. Fish. Rep.*, 2, 6, 167-179.

Williams, F. 1963b. Synopsis of biological data on the frigate mackerel *Auxis thazard* (Lacépède), 1802 (Indian Ocean), *FAO. Fish. Rep.*, 2, 6, 157-166.

Wu, H.W. 1930. On *Zoarces tangwangi,* a new eelpout from the Chinese coast, *Contrib. Biol. Lab. Sci. Soc. China, Zool.* ser., 6, 6, 59-63, fig.

Wu, H.W. and K.F. Wang. 1931. Four new fishes from Chefoo, *Contrib. Biol. Lab. Sci. Soc. China, Zool.* ser., 8, 1, 1-7, 4 figs.

Yabe, H. 1951. Larva of the swordfish, *Xiphias gladius, Japan. J. Ichthyol.*, 1, 4, 260-263, fig. 1.

Yabe, H. and S. Ueyanagi. 1962. Contributions of the study of the early life history of tunas, *Occ. Rep. Nankai Reg. Fish. Res. Lab.*, vol. 1, pp. 57-72.

Yabe, H.S. Ueyanagi and H. Watanabe. 1966. Studies on the early life of the bluefin tuna, *Thunnus thynnus,* and on the larva of the southern bluefin tuna, *T. maccoyii, Occ. Rep. Nankai Reg. Fish. Res. Lab.*, vol. 23, pp. 95-129.

Yabe, H., S. Ueyanagi, S. Kikawa and H. Watanabe. 1959. Study of the life history of the swordfish *Xiphias gladius* Linnaeus, *Occ. Rep. Nankai Reg. Fish. Res. Lab.*, vol. 10, pp. 107-171.

Yabuta, Y. and M. Yukinawa. 1958. The growth and age of yellowfin tuna. *Indo-Pacif. Fish. Proc. 7th Session, Bandung, Indonesia.*

Yamanaka, H. 1966a. Abiotic environment relating to the ecology of tunas. *Proc. the Eleventh Pac. Sci. Congress, Tokyo,* 7.

Yamanaka, H. 1966b. Tagging experiments of tunas by Japanese scientists. *Proc. the Eleventh Pac. Sci. Congress, Tokyo,* 7.

Yanai, T. 1950. Fishes of San'in District, *Zool. Mag., Tokyo,* 59, 1, 17-22 (in Japanese).

Yao, M. 1966. The distribution and migration of the skipjack. *Proc. the Eleventh Pac. Sci. Congress, Tokyo,* 7.

Yoshida, H.O. 1965. New Pacific records of juvenile albacore [*Thunnus alalunga* (Bonnaterre)] from stomach contents, *Pacif. Sci.*, 19, 4, 442-450.

Yoshida, H.O. 1966. Skipjack tuna spawning in the Marquesas Islands and Tuamotu Airchipelago, *U.S. Fish. and Wildl. Serv., Fish. Bull.,* 65, 2, 497–488.

Yoshida, H.O. and T. Ito. 1957. Fish fauna of the Sea of Japan, *J. Shimonoseki Coll. Fish.,* vol. 6, pp. 261–270.

Yoshida, H.O. and T. Otsu. 1963. Synopsis of biological data on the albacore *Thunnus germo* (Lacépède) (Pacific and Indian oceans). *FAO. Fish. Rep.,* 6, 2, 274–318.

Yoshida, H.O. and E.L. Nakamura. 1965. Notes on schooling behavior, spawning, and morphology of Hawaiian frigate mackerels, *Auxis thazard* and *Auxis rochei, Copeia,* 1, pp. 111–114.

Yuen, H.S.H. 1967. Yellowfin tuna spawning in the central equatorial Pacific, *U.S. Fish. and Wildl. Serv., Fish. Bull.,* 57, 112, 251–264.

Yuen, H.S.H. 1970. Behavior of skipjack tuna, *Katsuwonus pelamis,* as determined by tracking with ultrasonic devices, *J. Fish. Res. Board Canada,* 27, 11, 2071–2079.

Yukinawa, M. and Y. Yabuta. 1967. Age and growth of the bluefin tuna, *Thunnus thynnus,* in the North Pacific Ocean, *Rep. Nankai Reg. Fish. Res. Lab.,* 25.

Zhang et al. 1955. Ryby zaliva Bokhai, Zheltoe more (Fishes of the Gulf of Bohai, Yellow Sea). Peking, 353 pp., 260 figs. (in Chinese).

Zharov, V.L. 1965. O temperature tela tuntsov (Thunnidae) i nekotorykh drugikh ryb otryada okuneobraznykh Perciformes tropicheskoi Atlantiki [Body temperature of tunas (Thunnidae) and some other fishes of the order Perciformes from the tropical Atlantic]. *Vopr. Ikhtiologii,* 5, 1, 157–163.

Zharov, V.L. 1966. Zavisimost' raspredeleniya skoplenii tuntsov ot okeanologicheskoi struktury vod v nekotorykh raionakh tropicheskoi chasti Atlanticheskogo okeana (Dependence of the distribution of tuna concentrations on the oceanological structure of waters in some regions of the tropical Atlantic). *Tr. Vsesoyuzn. Nauchno-Issled. Inst. Morsk. Rybn. Khoz. i Okeanografii,* vol. 60, pp. 135–142.

Zharov, V.L. 1967. Sistema skombroidnykh ryb (podotryad Scombroidei, otr. Perciformes) [Classification of scombroid fishes (suborder Scombroidei, order Perciformes)]. *Vopr. Ikhtiologii,* 7, 2, 209–224.

Zharov, V.L. 1970a. Zheltoperyi tunets (*Thunnus albacares* Bonnaterre) Atlanticheskogo okeana [Yellowfin Tuna (*Thunnus albacares* Bonnaterre) of the Atlantic Ocean]. Kaliningrad, 120 pp., 24 figs.

Zharov, V.L. 1970b. Razmery, vozrast i rost zheltoperogo tuntsa (*Thunnus albacares* Bonnaterre) Atlanticheskogo okeana [Size, age, and growth of yellowfin tuna (*Thunnus albacares* Bonnaterre) of the Atlantic Ocean]. *Tr. Atlant. Nauchno-Issled. Inst. Morsk. Rybn. Khoz. i Okeanografii,* vol. 25, pp. 19–40, figs. 1–10.

582

Zharov, V.L. 1970c. Razmnozhenie zheltoperogo tuntsa (*Thunnus albacares* Bonnaterre) Atlanticheskogo okeana [Breeding of yellowfin tuna (*Thunnus albacares* Bonnaterre) of the Atlantic Ocean]. *Tr. Atlant. Nauchno-Issled. Inst. Morsk. Rybn. Khoz. i Okeanogr.*, vol. 25, pp. 41–62, figs. 1–6.

Zharov, V.L. 1970d. Pitanie zheltoperogo tuntsa (*Thunnus albacares* Bonnat.) Atlanticheskogo okeana [Food of yellowfin tuna (*Thunnus albacares* Bonnaterre) of the Atlantic Ocean]. *Tr. Atlant. Nauchno-Issled. Inst. Morsk. Rybn. Khoz. i Okeanografii*, vol. 25, pp. 62–86.

Zharov, V.L., Yu.L. Karpechenko and G.V. Martinsen. 1961. Tuntsy i drugie ob"ekty tuntsovogo promysla (Tunas and Other Objects of the Tuna Fisheries). Moscow, 114 pp., 43 figs. 5 pls.

Zharov, V.L., Yu. Zherebenkov, Yu. Kadil'nikov and V. Kuznetsov. 1964. Tuntsy i ikh promysel v Atlanticheskom okeana (Tunas and Their Fishery in the Atlantic Ocean). Kaliningrad, 180 pp., 104 figs.

Zhu et al. 1962. Ryby Yuzhno-Kitaiskogo morya (Fishes of the South China Sea). Peking, 1184 pp., 860 figs. (in Chinese).

Zhu et al. 1963. Ryby Vostochno-Kitaiskogo morya (Fishes of the East China Sea). Peking, 642 pp., 442 figs. (in Chinese).

Zhudova, A.M. 1969. Lichinki skombroidnykh ryb (Scombroidei, Perciformes) tsentral'noi chasti Atlanticheskogo okeana [Larvae of scombroid fishes (Scombroidei, Perciformes) from the central part of the Atlantic Ocean]. *Tr. Atlant. Nauchno-Issled. Inst. Morsk. Rybn. Khoz. i Okeanografii,* vol. 25, pp. 101–108, figs. 1–6.

Zvyagina, O.A. 1961. Raspredelenie ikry skumbrii ...i pelengasa ...v zalive Petra Velikogo (Distribution of the eggs of mackerel ... and pelengas ... in Peter the Great Bay). *Tr. Inst. Okeanol. Akad. Nauk SSSR,* vol. 43, pp. 328–336, pls. and figs.

Index of Latin Names of
Genera and Species [1]*

---

[1]Page numbers in bold print indicate description of the given form; page numbers marked with an asterisk indicate figure.

*Original Russian page numbers have been given in the left margin of the English translation—General Editor.

586

588

600

Milton Keynes UK
Ingram Content Group UK Ltd.
UKHW030901141024
449569UK00025B/1283

9 789061 914150